Computer Simulation of Thermal Plant Operations

Peter O'Kelly

Computer Simulation of Thermal Plant Operations

Springer

Peter O'Kelly
Costec Systems Pty Ltd.
Unit 66 42-46, Wattle Road
Brookvale (Sydney)
Australia

ISBN 978-1-4614-4255-4 ISBN 978-1-4614-4256-1 (eBook)
DOI 10.1007/978-1-4614-4256-1
Springer New York Heidelberg Dordrecht London

Library of Congress Control Number: 2012950340

Printed on acid-free paper

Springer is part of Springer Science+Business Media (www.springer.com)

Contents

Chapter 1
Introduction

Unlike many other books on a technical subject with a fairly substantial mathematical orientation, this book has not evolved from several years of teaching the subject material to undergraduate and graduate courses but rather from many years of development, design, delivery and performance testing of a wide variety of power station and utility plant training simulators.

Mathematical simulation covers a very large number and diversity of applications. At a fundamental level, the broad field of mathematical simulation divides into two major areas. The first of these, and the subject of this book, is concerned with *continuous* processes, that is, processes whose states and outputs continuously follow their changing inputs and environmental circumstances. They are generally described by some form of differential equations expressed in terms of continuous time. They include most chemical and power generation processes and other diverse fields such as meteorology and weather forecasting. Possibly the best known example is that of flight simulators for pilot training and certification. These incorporate mathematical modelling of aircraft flight and structural dynamics and other aircraft systems coupled to a realistic or even facsimile reproduction of the aircraft's cockpit equipment and layout. On the other hand, *discrete event simulation* is concerned with processes whose behaviour can best be described as a sequence of discrete events, such as the arrival of cars at an intersection, the arrival of ships at a port, the passage of widgets along a production line or the arrival and processing of telephone calls at a call centre. These processes are characterised by a strong stochastic aspect and are described mathematically in the form of transition events with defined probabilities, durations and outcomes. They can be applied to economies, health and medicine, manufacturing scheduling, transport scheduling and coordination and such like.

Continuous mathematical process modelling, both static and dynamic, is well established in many chemical process fields, particularly the paper and petrochemical industries, and within the power generation industry. As a tool for the analysis of *thermal* plant dynamics and control and for the implementation of high fidelity simulators for operator training, simulation techniques have, broadly speaking,

P. O'Kelly, *Computer Simulation of Thermal Plant Operations*,
DOI 10.1007/978-1-4614-4256-1_1,

been restricted to large central power stations, with the largest scope of simulated plant and the highest performance standards demanded from those using nuclear power generation. There is now an increasing interest in the application of these methodologies and devices to other process plant thermal, steam and power utilities. These tend to use smaller steam generators with a wide variety of fuel types—coal, oil and gas—also biofuels and a variety of combustible by-products.

As the training simulator business matured, it adopted higher levels of system organisation and repeatability. The early use of assembler coding and fixed-point arithmetic, even for large simulators, produced teams of engineer-programmers, low productivity rates—measured by lines of code—and time-consuming and difficult debugging. The use of assembler was dictated by the need to achieve maximum execution speeds as these simulators were expected to run in "real time", and computing power was limited. Fixed-point arithmetic gave way to a universal use of floating point as appropriate hardware and better compilers became available. The use of modern structured languages such as C++ and Java and graphical configuration tools has led to the development of reusable mathematical modelling libraries which have brought with them a substantial reduction in engineering hours spent on simulator development and debugging. The combination of cheaper but more powerful computing systems and reduced engineering hours has seen a marked reduction in the base price of every category of power and process plant simulator. The price reduction can also be attributed in no small part to the move from hard-wired control panels to screen-based operator interfaces in the plant control room as plant control moved to screen-based DCS systems.

Automated configuration of simulation modelling systems has become the norm in recent years. Complex simulations can be assembled from standard suites of models with little or no new code. These have predefined links to other modules and to their associated data. Data required by each module is also predefined and is usually readily available or can be easily derived from plant design information. Combined with proven libraries, tools reduce engineering time and configuration errors.

All of which brings us to this book. It is the author's attempt to present an overview of the objectives and methods of practical thermal plant mathematical simulation to engineers who either want to know or need to know. This includes engineers involved in the design, specification, procurement, performance testing and use of power and process utility plant training simulators, or who wish to use simulation as a tool for plant dynamic behaviour analysis, or who wish to build realistic plant simulations for the development and test of plant automation systems. The increasing use of software tools and their pre-configured models is moving the user away from the underlying mathematical principles and physical bases of the models. This book is an attempt to reacquaint the tool user with some of these principles.

A primary objective has been to demonstrate that the computational methods and mathematical assumptions commonly employed derive from and are consistent with the underlying fundamental physical source equations and to derive all of the modelling equations in terms of actual physical data while avoiding, as far as possible, artificial constants of no clear physical meaning, sometimes unkindly called fudge factors. While these have their place by substituting for substance when substance is unachievable, because of either process or computational complexity, they should be used sparingly and then only when their significance is clear and predictable.

This book concerns itself with broad issues of plant *operation* as opposed to a minute examination of some aspect of plant component *design*. An example of the latter might be the use of computational fluid dynamics to predict the flight paths of coal particles through a furnace in order to map the distribution of particle density, heat flux and flow velocities throughout the combustion chamber for different firing configurations and load rates. These issues lie outside the scope of this book, not being of an *operational* nature. The information they yield, however important it may be to the optimum design of the component, is not of immediate concern to operating staff and management and does not of itself influence the design and conduct of operating procedures. To distinguish these approaches, we might speak of *macro simulation* to describe an operational emphasis and *micro simulation* to describe the other.

Simulation as an engineering analysis tool is widely used. Mathematical modelling is routinely used as an aid in the design of critical mechanical components such as steam turbines, gas turbines and compressors. Detailed pressure and flow modelling is used in the design and optimisation of hydraulic piping networks. Complex modelling of heat fluxes and their distribution is an essential element of boiler and superheater design. For each of these and many other fields, standard commercial software packages are available. In many cases, equipment manufacturers have developed their own proprietary analysis tools, specifically tailored to their own products and requirements. As a general rule, these packages provide very detailed information on steady-state or static conditions but rarely do they handle unsteady transient conditions.

The simulation methods described in this book are not intended for design analysis, in the sense of the previous paragraph. They are well suited, and have been applied successfully, to the implementation of operator training simulators, of any scope, and to the analysis of a wide variety of plant operations, both normal and abnormal, over the full range of operating phases, from cold shut-down, through start-up, to full load and back down again, with all manner of variations and transients in between.

The book starts with a general review of simulation concepts. It describes a simulation-oriented classification of thermal plant systems and suggests a typical structure of a simulation run-time environment. Chapter 3 presents a review of the numerical methods used for implementation and solution of the modelling equations derived later in the book. Chapter 4 brings together the fundamental thermodynamic concepts which serve as the starting point for many of the later model derivations. Chapter 5 discusses the statement, reduction and solution methods of the three basic conservation equations of mass, momentum and energy.

Chapters 6–16 discuss simulation modelling of specific plant components and cover static components (pipes, ducts, valves and dampers), turbomachines (fans, pumps and compressors), heat exchangers, furnaces, boilers and steam generators, steam turbines, condensers—water and air-cooled—deaerators and feedwater heaters. Chapter 17 is concerned with the simulation of pressure/flow networks, both incompressible and compressible. Other chapters discuss specific matters such as heat transfer processes (Chap. 8) and the calculation of heat conduction through thick-walled components (Chap. 10). Chapter 13 surveys empirical methods for the treatment of two-phase steam flow.

While having an obvious mathematical orientation, the book seeks to relate these theoretical aspects to the practical design, construction and operating principles of the equipment. To that end, each mathematical development is accompanied by a discussion and description of the relevant physical plant.

The book covers a wide range of equipment types and modelling methods but gives no more than a brief snapshot of any one. For each field touched by this coverage, there exists a large body of specialised knowledge and available material to which the reader is referred should more detail of any of these areas be sought. No attempt has been made to provide a comprehensive review of the literature of any of these subject areas. The literature references provided with the book are limited to those which have contributed directly to the development of the book's own material.

Chapter 2
Plant Simulation Modules and Functional Groups

Most industrial process plant, including power generation and process utility plant, can be conveniently divided into:

1. Simple or complex components
2. Interconnecting piping and duct networks

This facilitates the definition of a modular organisation of plant components for which libraries of common standard modelling modules can be developed, together with their individual data sets and computational methods. Modularity is the key to an effective systematic organisation of components and their interconnection into sub-systems. Given a modular system of building blocks, we can approach any configuration of plant as a set of discrete units or components linked together by a network of fluid conduits containing control, heat, and momentum transfer elements.

A small process utility steam system is shown by Fig. 2.1. Four gas-fired boilers supply high-pressure steam to turbine drives and through controlled letdown valves to a number of low-pressure consumers. One turbine is configured to exhaust to the low-pressure system, the others exhausting to condensers. This is typical of the steam utility plant which might be found in an oil refinery, steelworks or chemical process plant.

Branches containing pipework, valves, and steam turbines connect to nodes and to peripheral devices such as boilers, condensers, tanks, and a deaerator.

2.1 Components

A plant component may take one of the following forms:

1. A simple mechanical unit of plant (pump, valve, header, pipe element, pressure loss element, gearbox, etc)
2. A simple electrical unit of plant (line, RLC element, circuit breaker, etc)

P. O'Kelly, *Computer Simulation of Thermal Plant Operations*,
DOI 10.1007/978-1-4614-4256-1_2,

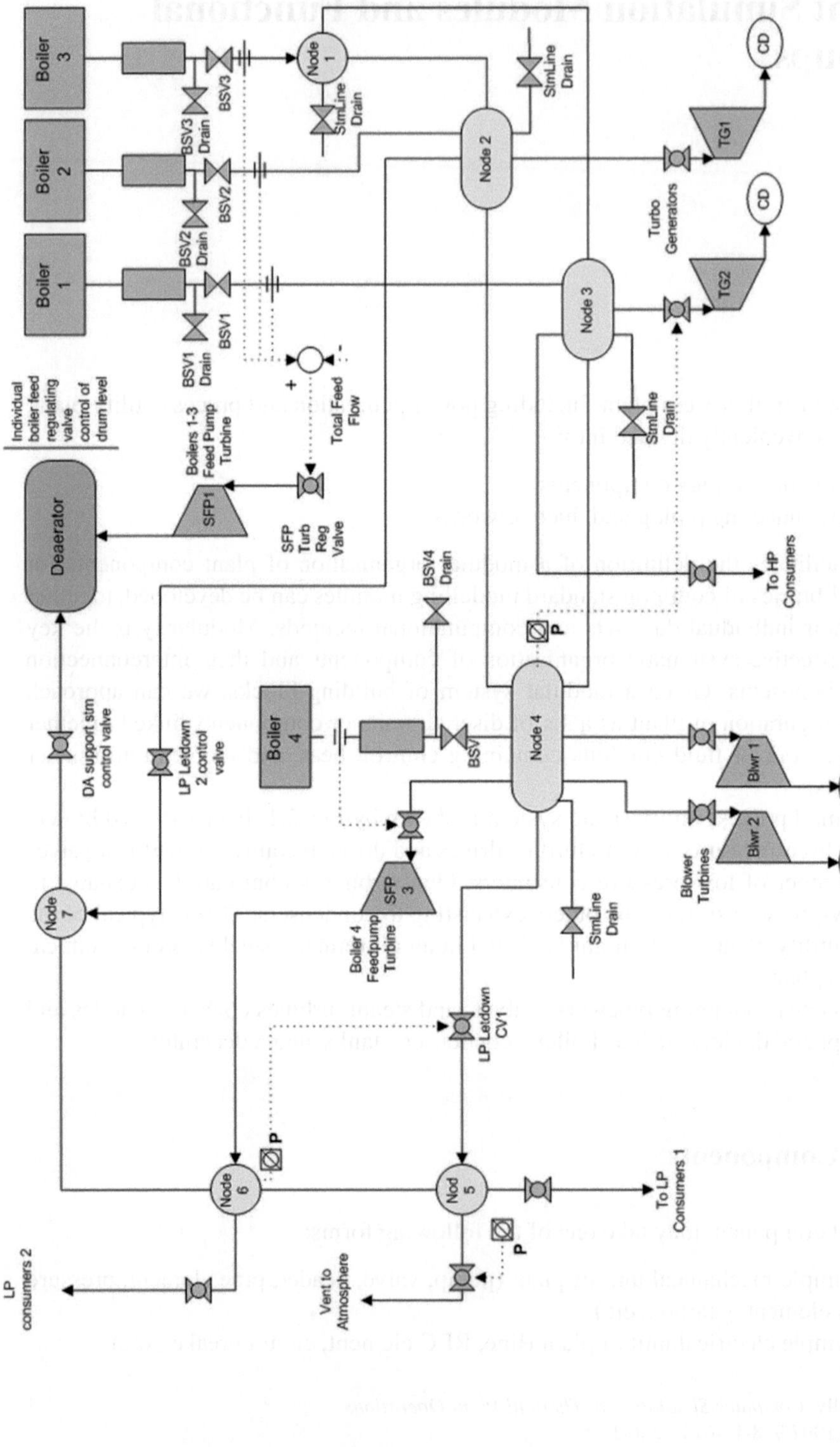

Fig. 2.1 A typical process plant steam system

3. A complex unit of mechanical plant (single-cylinder steam turbine, furnace, evaporator, drum, deaerator, condenser, feedwater heater, flash tank/standpipe, cooling tower, precipitator, gas turbine, bearing, etc)
4. A complex unit of electrical plant (motor, generator, transformer)
5. A compound unit of plant, being a complex assembly of components (e.g. multi-cylinder turbine, multi-effect evaporator, gas turbine, multi-element shaft, heat recovery boiler)
6. An automation interfacing unit (transducer, switchgear, drive/actuator positioner, tap changer)

Broadly speaking, plant equipment can be grouped into systems identified by the principle working medium, allowing us to identify, for example:

(a) Hydraulic systems
(b) Gas systems
(c) Steam systems
(d) Solids handling equipment

Using modern object-oriented coding methods, the implementation code of a component simulation module can be organised as a self-contained object and a specific instance of each created as required by the application scope. Each instance is configured to match the specific application by the setting of a (sometimes large) number of model parameters and configuration options. Component objects can be organised into libraries.

2.2 Networks

A network consists of:

- Branches, a branch being a contiguous arrangement of series-connected components through which the working fluid flows while exchanging heat and performing useful work
- Nodes, being points of connection of branches with each other
- Links, being points of connection to external interfacing components or atmosphere

Nodes and links can be thought of as two flavours of the same species and referred to individually as *internal* or *external* nodes.

Networks are ephemeral and come into existence only as required by a specific application. There are no libraries of networks. Networks are built up as the interconnection of simple components into branches. Individual networks are joined together by connection to discrete components at their external nodes or interfacing points. Discrete components will usually contain significant mass or energy storage and can serve as stable sources of boundary conditions for the more fragile networks which typically feature small volumes, coupled non-linearities and rapid transients, the right ingredients for numerical stability problems. Most industrial plant "systems" can be considered as networks for simulation purposes. Using the working fluid as base for classification, we can identify the following typical discrete networks:

- *Hydraulic systems*
  - Feedwater
  - Condensate
  - Lubrication oil
  - Power oil
  - Fuel oil
  - Condenser cooling water
  - Miscellaneous cooling water
- *Air and gas systems*
  - Fuel gas
  - Flue gas
  - Combustion air (primary, secondary and tertiary)
  - Air extraction
  - Compressed air
- *Steam systems*
  - Main
  - Auxiliary
  - Extraction
  - Gland sealing
  - Sootblowing
  - Tank farm heating
  - Steam line tracing

From this classification of components and networks, we can conclude that:

- Major components are static, fixed geometry objects which may be modelled in a generic fashion and configured to suit a specific implementation instance.
- Networks are dynamic, have no fixed geometry and must be built up from appropriate objects to suit each application.
- Networks are spatially distributed, sometimes over considerable distances, and their dynamic models must be capable of representing this spatial extension.
- Components are generally not spatially distributed although, in some cases, internal spatial effects need to be considered explicitly if their dynamic behaviour is to be accurately reproduced.
- Differing methods of simulation equation development and solution will be used for components and networks.

Library models have configuration options giving the user access to a variety of physical modelling features, such as the orientation of a tank or feedwater heater, vertical or horizontal. A less obvious example is the choice of the heat transfer correlation to apply to a specific instance of a heat exchanger. The tank orientation selection requires the user to know no more than the physical plant arrangement. The other requires the user to have some knowledge of the thermodynamics of the process and the significance of the choice to the simulation behaviour.

2.3 Simulation Model Organisation

Throughout this book, we will seek to represent the physical behaviour of process systems mathematically by some differential/difference/algebraic equation system. The describing equations will establish the causal relationships which exist physically among two sets of quantities, which we will term "inputs" and "outputs", via a third set, the internal "states" of the process. The physical data of the plant and its environment will be encapsulated within a fourth set of quantities which we will call simply "data". These relationships are illustrated by the following diagram:

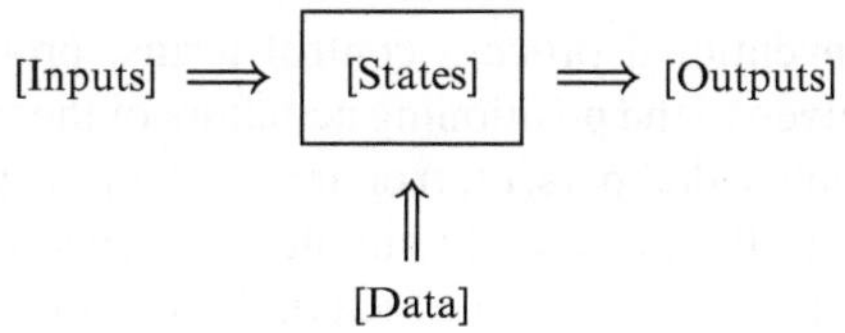

States These are quantities whose values completely define the state of the process at any time. Conversely, and sometimes perversely (think of weather forecasting), full knowledge of the state of the process at any time presupposes knowledge of all individual states. States are grouped together into state vectors or arrays. The order of the process is equal to the number of its internal states which is the dimension of its state vector.

The time evolution of a process is described by the time evolution of its states. In the usual mathematical modelling approach, each state is associated with a single differential or difference equation through which its rate of change with respect to time and/or spatial variables is related to itself, to other states and to the process inputs. "Solution" of the model, which means execution of the computer code which implements the model, generates the time and spatial evolution of the states moving in response to the changing inputs. If the inputs are unchanging, the states autonomously move towards some terminal state. If this terminal state is finite and stationary, the process is said to be stable; otherwise, it is unstable. "Finite" here means asymptotic approach to numerical values which lie within the range of real numbers which can be handled by the computing system.

Most physical systems are stable and quasi-deterministic. "Deterministic" means that the number and selection of states is sufficient to permit prediction of the future evolution of the plant, given full knowledge of the present state and of all inputs, including disturbances. Since most disturbances occurring in physical plant are at least partially stochastic in nature, the time and spatial evolution of real plant states is never totally predictable. As all physical systems are essentially of infinite order, any representation of them by a finite number of states will introduce approximations and deviations by the model from the true physical behaviour. It is the art and science of simulation to find the best compromise between state dimensionality, which bears on computational load, and fidelity of model representation. The distinction between fidelity and accuracy should be well noted.

Outputs In process control terms, outputs are those quantities which can be measured or estimated. They are usually produced by the plant instrumentation and are used for control, operator information and general data processing purposes. In simulation terms, in which only models exist, outputs are quantities derived from states and inputs. They are used to link models by providing the inputs needed by other models, downstream in the execution sequence. They may be sent to peripheral processing systems such as the plant data processing computer. Outputs are related to their source variables via algebraic equations or decision variables. They do not have internal states of their own.

Inputs Again, in conventional process control terms, process inputs are either actuation commands given to the positioning actuators of the various control devices of the plant (valves, pumps, dampers, etc.) or are the defining parameters of material and other flows entering the process. In simulation terms, the outputs and inputs define the set of matching quantities via which individual models communicate with each other or which define the variables passing across model boundaries. In most cases, but not always, one model's output array provides elements of the input arrays of one or more other models.

Storage of States and Initialization States must be stored with their models and must be given initial values prior to the execution of the model. States must be consistent, that is, all states of a given model must have been calculated at the same time instant and must be consistent with the model's inputs at that time instant. If

$$x^{n+1} = x^n + \Delta x$$

then all elements of Δx must have the same time reference (step n or $n + 1$). In practice, this consistency requirement may be relaxed to some extent without serious consequences for the computation provided that, for the ith element of Δx, $|\Delta x_i| \ll |x_i|$. This tacitly ignores any other conditions which may be imposed by considerations of numerical stability.

Data This is the set of numerical values of all coefficients of the equations the model will use. It can be either:

1. Static data, which, once calculated, does not change as the process evolves.
2. Dynamic data, which is recalculated and updated either every computation cycle or periodically as conditions elsewhere within the simulation change.

Static data is derived from the physical design parameters of the plant (dimensions, geometries, fixed characteristics, environmental parameters, etc.). Dynamic data is calculated from changing process conditions such as the physical properties of materials and of the working media, composition of the working media, heat exchange coefficients and friction factors which depend on thermodynamic and flow states, tank levels which depend on tank geometries and pump head/flow relationships which depend on speed. These dynamically varying properties and parameters are the usual sources of process non-linearities and can undergo large and rapid variations.

The physical data required by the simulation modules should be meaningful and available from conventional sources. Not infrequently, some parameters requiring user setting are obscure to say the least and the user can be left wondering what they mean and what value might be appropriate. An incorrectly or inappropriately set parameter can adversely affect the quality of the analysis results or the performance of the simulation. It's probably fair to say that the use of the more complex or more advanced modelling packages requires an equally advanced level of understanding from the user, both of the mathematical methods themselves and of their relevance to the modelled systems.

2.4 A Simulation Runtime Environment

To obtain useful information from simulation models, they must be installed in a user-friendly runtime environment. For best results, this will be an interactive environment in which the user can select and load a project, adjust parameter values, save complete snapshots of all model states for later recall and reuse, start and stop execution of the simulation models and other functions at will, monitor and record selected calculated variables from within the simulation while it is still executing ("running") and vary inputs to the simulation, such as control system set points, in much the same way as a plant operator.

These features can be provided by an environment which includes at least the following components:

- A configurable modelling system
- A configurable control and automation system
- A configurable user interface, including measured value display, trending and possibly alarms
- A simulation management system
- A communication system which links these four

Figure 2.2 shows the logical links or data flows between software processes, each implementing one of these system tasks. Communication between tasks is handled by an underlying TCP/IP network with data traffic managed via some form of structured data exchange protocol such as OPC ("OLE[1] for Process Control").

[1]OLE is a trademark of Microsoft Corp.

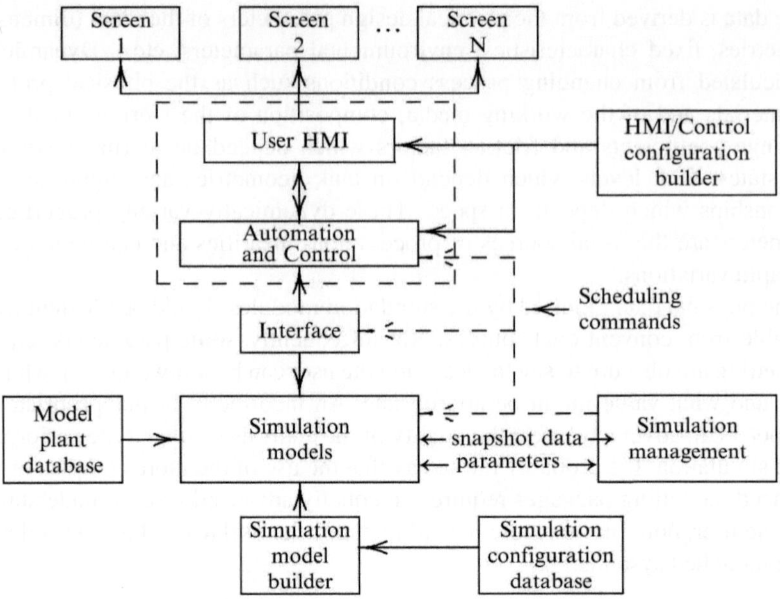

Fig. 2.2 A possible simulation runtime architecture

Chapter 3
Numerical Methods

The literature is replete with books and other publications dealing with the numerical solution of ordinary (ODE) and partial (PDE) differential equations. Some specialised fields have developed their own communities and have evolved into self-contained disciplines such as computational fluid dynamics (CFD) and finite element analysis (FEA). In addition to describing the methods themselves, the literature concerns itself with issues of stability, error propagation, local linearisation, adaptive step sizes and so on. Inherent to much of this literature are the tacit assumptions that (a) the solution is being sought to a specific equation or set of equations, with given boundary conditions, (b) the time step size is completely under the control of the analyst and (c) computation effort is not a limiting consideration. None of these assumptions is completely valid for the applications covered by this book, for which complex simulation systems are built up, possibly using automatic configuration tools and generating time evolutions of a large number of process variables into an indefinite future. If these are to be generated in real time, the time step is fixed at some small value—generally less than 100 ms—consistent with the process dynamics and actuator positioning accuracy. For most process plant simulation purposes, the underlying physical solution is stabilised by the inherent process feedbacks—physical systems, on the whole, are stable, but not always!—or by the actions of control systems, and significant numerical errors do not build up despite the open-ended nature of the procedure.

That having been said, differential equations of many sorts must be solved, either singly or as organised sets. This chapter considers some of the more useful numerical methods available for this purpose.

The material of this chapter will provide the reader with all of the mathematical equation solving techniques and background needed for the later material of this book. If the reader wishes to delve more deeply into these techniques and the issues related to them, as is to be recommended, the references *Numerical Recipes* by Press et al. [1], *Matrix Computations* by Golub and Van Loan [2] and *Numerical Mathematics* by Quarteroni et al. [4] provide extensive coverage.

P. O'Kelly, *Computer Simulation of Thermal Plant Operations*,
DOI 10.1007/978-1-4614-4256-1_3,

3.1 The Derivation of Simulation Model Equations

The mathematical simulation models described in this book are all derived in terms of three types of mathematical relationships:

- PDE and ODEs, which relate the rate of change of a variable, with respect to time and/or one or more spatial coordinates, to itself and other variables
- Algebraic equations, which define static relationships among sets of variables
- Functional relationships, which are reserved for the expression of material properties and plant component characteristics; they are usually available as one- or two-dimensional tabulations or as appropriate fitting functions

Calculation of the time and spatial evolution of the simulation state and other variables requires the solution of differential equations in space and time while preserving the algebraic and functional relationships.

The selection of a numerical procedure is guided by considerations of stability, steady-state accuracy, dynamic accuracy and the treatment of boundary conditions. Of these, the most important is stability as an unstable solution method can never satisfy any accuracy criteria.

A numerical procedure generates a sequence of numerical values of the state and other variables which describe the dynamic behaviour of the simulated plant. The procedure advances the solution from a given initial condition to a new value defined one discrete-time step later. For the scalar differential equation

$$\frac{\mathrm{d}y}{\mathrm{d}x} = f(y, u)$$

starting from a known value y^0, the numerical procedure produces a series of contiguous solutions $y^1, y^2, \ldots, y^n, y^{n+1} \ldots$ defined at discrete points $x = x_0 + \Delta x, x_0 + 2\Delta x, \ldots, x_0 + n\Delta x, x_0 + (n+1)\Delta x \ldots$, in response to the external input u defined at the same x values. The ordinate x can be a spatial dimension, for example, z, or elapsed time t. The series y^n generates the evolution of the process variable y as a sequence of discrete values spaced at intervals Δx.

Many model derivations are directed at the formulation of a set of time derivatives $\mathrm{d}y_i/\mathrm{d}t$ in terms of the set of process variables (y_i, u_j) where the y_i are state variables and u_j are forcing inputs. The function $f(y_i, u_j)$ is usually a non-linear function of its arguments and is evaluated at each time step using values of y_i and u_j known from the preceding step. Two fundamentally different cases arise, defined by the prevailing boundary conditions:

- Initial-value problems for which only initial boundary conditions are specified at $t = 0, z = 0$
- Two-point boundary value problems, for which the solution must satisfy two boundary conditions, usually at two spatial points, known at the nth step

The first of these is frequently found in engineering digital computer simulation. All of the information needed to advance the solution from the current nth step to the next $(n + 1)$th step is available at the nth step. The solution is "self-starting" once the complete state is defined at the first step. The equation can be solved using a direct integration method or can be solved using an integration procedure.

Two-point boundary value problems arise naturally in spatially distributed systems described by PDEs for which boundary conditions are specified at two spatially separated points. The numerical solution of these problems is not self-evident and is generally computationally intensive. They can be reduced to a set of ODEs by an appropriate spatial discretisation scheme or reduced to an algebraic system by time and space discretisation. Applied to the very important case of networks—piping and duct systems of considerable spatial extent and complexity—the resulting algebraic scheme can grow to very large dimensions.

3.2 Classification of Differential Equations

Differential equations are classified as being of *order* "n" or *degree* "n" and either *linear* or *non-linear*. The *order* of the differential equation is the highest order of the differential. The *degree* of the differential equation is the highest power to which the differential is raised. Thus, the equation

$$\frac{d^4 y}{dx^4} + 3\,y = f(x)$$

is of fourth order, while

$$\left(\frac{dy}{dx}\right)^3 + 3\,y = f(x)$$

is of third degree.

A linear equation is one in which the dependent variable and its derivatives appear only to the first degree, while a non-linear equation allows these to appear to higher degrees or in functional form. Thus, the first of the equations above is linear and the second non-linear. Also,

$$\frac{d^4 y}{dx^4} + 3\,y^3 = f(x)$$

and

$$y\frac{d^4 y}{dx^4} + \ln(y) = f(x)$$

are non-linear.

Linear equations are extensively studied in the literature because they feature two very desirable properties. First, the rule of superposition applies, that is, if $y_1(x)$

and $y_2(x)$ are solutions of a linear equation, then so also is $y_1(x) + y_2(x)$. Second, closed form analytical solutions are available for both scalar and vector equations. Non-linear equations rarely have analytical solutions and must be solved by purely numerical schemes. These almost always involve some form of local linearisation which frequently means fitting a truncated Taylor series expansion to the non-linear terms.

3.3 Solution Methods for Initial-Value Problems

3.3.1 *Analytical Solutions*

The simplest form of a scalar linear differential equation is

$$\tau \frac{\mathrm{d}x}{\mathrm{d}t} + x = b\,u. \tag{3.1}$$

The analytical solution of Eq. 3.1 is obtained by use of the integrating factor $\mathrm{e}^{-t/\tau}$.

$$x(t) = c_0 \mathrm{e}^{-t/\tau} + \mathrm{e}^{-t/\tau} \frac{1}{\tau} \int^{t} \mathrm{e}^{t'/\tau} b\,u(t')\mathrm{d}t' \quad \text{for} \quad t \geq 0. \tag{3.2}$$

3.3.2 *Numerical Procedures of Solution*

Consider the following differential equation which relates the time derivative of x to a continuous differentiable function of x and an external input or forcing function $u(t)$, a function of the independent variable t.

$$\frac{\mathrm{d}x}{\mathrm{d}t} = f(x, u(t)).$$

Replace the continuous derivative by its first-order discrete-time equivalent.

$$\frac{\mathrm{d}x}{\mathrm{d}t} \approx \frac{x^{n+1} - x^n}{\Delta t} = f(x, u(t)) + \mathcal{O}(\Delta t)^2$$

$\mathcal{O}(\Delta t)^2$ denotes terms of order $(\Delta t)^2$ and higher.

If the function $f(x, u(t))$ is denominated at the nth step, the resulting procedure is called *explicit* integration. The simplest form of explicit integration is the well-known Euler integration rule.

$$x^{n+1} = x^n + \Delta t\, f(x^n, u^n).$$

Despite its simplicity, this form is only conditionally stable and should be considered for practical computation only when the conditions for stability have been established. These disadvantages are overcome by an *implicit* formulation in which x in $f(x, u(t))$ is denominated at the $(n+1)$th step. Then,

$$x^{n+1} = x^n + \Delta t\, f(x^{n+1}, u^n).$$

In the case of a multi-variable system, $\mathbf{x}$, $\mathbf{u}$ and $\mathbf{f}(\mathbf{x}, \mathbf{u})$ are vectors, and we have

$$\mathbf{x}^{n+1} = \mathbf{x}^n + \Delta t\, \mathbf{f}(\mathbf{x}^{n+1}, \mathbf{u}^n).$$

Since $\mathbf{f}(\mathbf{x}, \mathbf{u})$ is usually non-linear in $\mathbf{x}$, direct solution of this equation for $\mathbf{x}^{n+1}$ is rarely possible. Local linearisation of $\mathbf{f}(\mathbf{x}, \mathbf{u})$ allows us to write

$$\mathbf{x}^{n+1} = \mathbf{x}^n + \Delta t \left(\mathbf{f}(\mathbf{x}^n, \mathbf{u}^n) + \left.\frac{\partial \mathbf{f}}{\partial \mathbf{x}}\right|_{\mathbf{x}=\mathbf{x}^n} (\mathbf{x}^{n+1} - \mathbf{x}^n) \right) + \mathcal{O}(\Delta t)^2.$$

Here, $\partial \mathbf{f}/\partial \mathbf{x}$ is the Jacobian matrix of the partial derivatives of $\mathbf{f}(\mathbf{x}, \mathbf{u})$ evaluated for $\mathbf{x} = \mathbf{x}^n$.

If $\partial \mathbf{f}/\partial \mathbf{x}$ is known and neglecting the higher-order terms, the solution may be formulated as

$$\mathbf{x}^{n+1} \left[\mathbf{I} - \Delta t\, \frac{\partial \mathbf{f}}{\partial \mathbf{x}} \right] = \mathbf{x}^n + \Delta t \left(\mathbf{f}(\mathbf{x}^n, \mathbf{u}^n) - \frac{\partial \mathbf{f}}{\partial \mathbf{x}} \mathbf{x}^n \right),$$

where $\mathbf{I}$ is the unit matrix. The linearised implicit equation to be solved for $\mathbf{x}^{n+1}$ may be reformulated as

$$\mathbf{x}^{n+1} = \left[\mathbf{x}^n + \Delta t \left(\mathbf{f}(\mathbf{x}^n, \mathbf{u}^n) - \frac{\partial \mathbf{f}}{\partial \mathbf{x}} \mathbf{x}^n \right) \right] \left[\mathbf{I} - \Delta t \frac{\partial \mathbf{f}}{\partial \mathbf{x}} \right]^{-1}. \tag{3.3}$$

All of the terms on the right-hand side of this equation are known at the nth time step, and $\mathbf{x}^{n+1}$ is readily found, if $[\mathbf{I} - \Delta t\, \partial \mathbf{f}/\partial \mathbf{x}]^{-1}$ exists and is non-zero.

For a linear matrix equation

$$\frac{d\mathbf{x}}{dt} = \mathbf{A}\,\mathbf{x} + \mathbf{B}\,\mathbf{u}(t) = \mathbf{f}(\mathbf{x}, \mathbf{u})$$

the Jacobian $\partial \mathbf{f}/\partial \mathbf{x} = \mathbf{A}$, and Eq. 3.3 becomes

$$[\mathbf{I} - \Delta t \mathbf{A}]\, \mathbf{x}^{n+1} = \mathbf{x}^n + \Delta t \mathbf{B}\, \mathbf{u}^n. \tag{3.4}$$

The right-hand side of this equation is completely known at the nth time step, and the equation may be solved for x^{n+1}. Note that *solving* for x^{n+1} avoids the need to invert the coefficient matrix. Equation 3.4 provides a stable, efficient and compact method for the solution of the general linear matrix differential equation and, for most cases of practical interest to process plant simulation, is adequately accurate.

Returning now to Eq. 3.1, this may be converted to difference equation form by the replacement of the derivative by a first-order difference approximation. Explicit and implicit algorithms are derived as

Explicit Euler

$$\frac{\tau}{\Delta t}(x^{n+1} - x^n) = -x^n + b\,u^n$$

$$x^{n+1} = x^n\left(1 - \frac{\Delta t}{\tau}\right) + \frac{\Delta t}{\tau} b\,u^n.$$

Implicit Euler

$$\frac{\tau}{\Delta t}(x^{n+1} - x^n) = -x^{n+1} + b\,u^n$$

$$x^{n+1}\left(1 + \frac{\tau}{\Delta t}\right) = \frac{\tau}{\Delta t}x^n + b\,u^n$$

$$x^{n+1} = x^n\left(\frac{1}{1 + \Delta t/\tau}\right) + \left(\frac{\Delta t/\tau}{1 + \Delta t/\tau}\right) b\,u^n.$$

Stability of the solution requires that x^n remain bounded as $n \to \infty$. This will be the case if the magnitude of the coefficient of x^n be always ≤ 1. For the explicit Euler case, this imposes the constraint

$$\left|1 - \frac{\Delta t}{\tau}\right| \leq 1$$

or

$$0 \leq \frac{\Delta t}{\tau} \leq 2.$$

For the implicit Euler case, there is no constraint as $1/(1+\Delta t/\tau)$ is always ≤ 1. The implicit formulation is therefore always stable. However, under some circumstances, such as a rapid and large transient change, it can generate large dynamic errors as it is only first-order accurate in time.

A third procedure may be followed. This is similar to the pole/zero matching procedure used by control engineers to secure certain desirable features of the closed-loop response of a process control loop. Without going through all of the manipulations required, we can say that the transient and steady-state solutions of the original Eq. 3.1 can be matched by an assumed form of solution constructed to have the same roots (poles) as the original equation. The result is a difference equation which yields the best fit to the dynamics of the original equation and gives the best dynamic accuracy of the three alternatives.

$$x^{n+1} = \mathrm{e}^{-\Delta t/\tau} x^n + (1 - \mathrm{e}^{-\Delta t/\tau}) b\,u^n \tag{3.5}$$

which fits the step response of the original equation exactly. It is also *absolutely* stable for any choice of Δt for any $\tau > 0$.

This classical first-order lag process is a useful tool in process simulation. Later, we shall encounter reference to it under the function name `fnlag` which implements Eq. 3.5.

To finish this tale, note that the polynomial series expansion of $e^{-\Delta t/\tau}$ is

$$e^{-\Delta t/\tau} = 1 - (\Delta t/\tau) + \frac{1}{2}(\Delta t/\tau)^2 - \cdots$$

Assuming that $(\Delta t/\tau)$ is small ($\ll 1$), we may neglect all terms of higher powers of $(\Delta t/\tau)$. Then, Eq. 3.5 may be written as

$$x^{n+1} = (1 - \Delta t/\tau)x^n + (\Delta t/\tau)B\,u^n$$

which is identical to the explicit Euler form. For this reason, we can say that the Euler differencing approximation is *first order* as it represents only terms to the first order of $(\Delta t/\tau)$ in the series expansion of $e^{-\Delta t/\tau}$.

3.3.3 The Matrix Exponential

Any nth-order linear (or appropriately linearised) ODE can be reorganised as a single vector equation of dimension n, expressed as

$$\frac{d\mathbf{x}}{dt} = \mathbf{A}\,\mathbf{x}(t) + \mathbf{B}\,\mathbf{u}(t), \tag{3.6}$$

where **A**(n x n) and **B**(n x m) are coefficient matrices. With $\mathbf{\Phi} = e^{\Delta t\mathbf{A}}$ and **I** the unit matrix, the vector equivalent of Eq. 3.5 is

$$\mathbf{x}^{n+1} = (\mathbf{I} + \mathbf{A\Phi})\mathbf{x}^n + \Delta t\,\mathbf{\Phi}\mathbf{B}\,\mathbf{u}^n. \tag{3.7}$$

This requires the calculation of the matrix exponential $e^{\Delta t\mathbf{A}}$. It is beyond the scope of this book to discuss the various methods available (e.g. using eigenvalue analysis), but, from a computational point of view, the recommended method is that based on a Padé approximant, as described by Golub and Van Loan ([2], algorithm 11.3.1).

3.4 Semi-implicit Methods

The explicit and implicit formulations above are the two extreme ends of the general formulation of semi-implicit methods. Consider Eq. 3.6 and assume that the rate of change over the interval $(t \to t + \Delta t)$ can be represented as the mean value $\theta\,\dot{\mathbf{x}}^n + (1-\theta)\,\dot{\mathbf{x}}^{n+1}$ for $0 \le \theta \le 1$ and $\dot{\mathbf{x}} = \mathrm{d}\mathbf{x}/\mathrm{d}t$. Explicit integration (forward differencing) corresponds to $\theta = 1$ and is only conditionally stable. Full implicit integration (backward differencing) corresponds to $\theta = 0$ and while always stable can suffer from large dynamic errors. Choosing a value of θ between these two extremes can achieve a trade-off between stability and dynamic accuracy. For example, choosing θ to be 0.5 gives a centre-differencing scheme which, for first-order differencing, is most accurate dynamically but marginally stable.

Replacing the time derivative in Eq. 3.6 by its usual first-order approximation gives

$$\mathbf{x}^{n+1} - \mathbf{x}^n = \mathbf{A}\left[\theta\Delta t\,\mathbf{x}^n + (1-\theta)\Delta t\,\mathbf{x}^{n+1}\right] + \Delta t\mathbf{B}\,\mathbf{u}^n$$

from which we can write

$$\begin{aligned}[\mathbf{I} - \theta\Delta t\mathbf{A}]\,\mathbf{x}^{n+1} &= [\mathbf{I} + \mathbf{A}(1-\theta)\Delta t]\,\mathbf{x}^n + \Delta t\mathbf{B}\,\mathbf{u}^n \\ &= \hat{\mathbf{x}}^n. \end{aligned} \tag{3.8}$$

Noting from this equation that

$$(\mathbf{x}^{n+1} - \hat{\mathbf{x}}^n) = \theta\Delta t\,\mathbf{A}\,\mathbf{x}^{n+1}$$

it follows that

$$\hat{\mathbf{x}}^{n+1} = \frac{1}{\theta}\mathbf{x}^{n+1} + \frac{\theta - 1}{\theta}\hat{\mathbf{x}}^n + \Delta t\mathbf{B}\,\mathbf{u}^{n+1}. \tag{3.9}$$

Solution using a semi-implicit scheme is a two-step procedure requiring the creation and initialisation of an auxiliary variable $\hat{\mathbf{x}}$. At step 1, the updated state vector is calculated as the solution of Eq. 3.8, using the stored value of the auxiliary variable $\hat{\mathbf{x}}^n$. At step 2, $\hat{\mathbf{x}}^n$ is replaced by the update calculated from Eq. 3.9.

3.5 Numerical Integration Procedures

Other solution methods are available for initial-value problems in which the right-hand side of the differential equation is approximated by a truncated Taylor series expansion valid for the time interval $(t \to t + \Delta t)$. These may be classified as:

1. Those which calculate the derivative terms of the Taylor expansion, using explicit functions
2. Those which estimate those same terms, using some form of finite differencing
3. Those which combine a set of low-order formulae to obtain a best fit to the solution

In each case the method attempts to improve on a low-order approximation by adding higher-order terms which cancel out error propagation terms left by the lower-order formulae.

Methods of the first type are of little interest unless the system is linear or weakly non-linear, the Taylor series converges rapidly, and explicit functions are available for the computation of the higher derivatives. This is generally not the case for an arbitrary physical system.

Methods of the second type are not much used in process simulation but are widely used in general scientific computing. The most successful of these are *predictor–corrector* methods such as Adams–Bashforth. These are related to the formulae used for numerical integration (such as Simpson's Rule) and, like those, require a good deal of user judgment for good results, based on a priori knowledge of the functional form of the equations.

Which leaves only the third type which, happily, can deliver good results with little prior knowledge of the precise form of the equations. These methods are generally classified as second, third or higher order, depending on the number of Taylor series terms implicitly included.

3.5.1 The Method of Heun

This averages the slope of the solution at the beginning and end points of the time interval. For the differential equation $\mathrm{d}y/\mathrm{d}x = F(x, y)$ and a discrete increment Δ of x, a Taylor expansion of $F(x, y)$ yields

$$\begin{aligned} y^{n+1} &= y^n + \Delta F^n + \frac{\Delta^2}{2}\left[\frac{\partial F}{\partial x} + \frac{\partial F}{\partial y}F^n\right] + \mathcal{O}(\Delta^3) \\ &= y^n + \frac{\Delta}{2}F^n + \frac{\Delta}{2}\left[F^n + \Delta\frac{\partial F}{\partial x} + \Delta\frac{\partial F}{\partial y}F^n\right] + \mathcal{O}(\Delta^3). \end{aligned}$$

Noting that

$$F(x + \Delta, y + \Delta F^n) = F^n + \Delta\frac{\partial F}{\partial x} + \Delta\frac{\partial F}{\partial y}F^n + \mathcal{O}(\Delta^2)$$

we have

$$y^{n+1} = y^n + \frac{\Delta}{2}F + \frac{\Delta}{2}F(x + \Delta, y + \Delta F(x, y)) + \mathcal{O}(\Delta^3).$$

With

$$k_1 = \Delta\, F(x, y) \quad \text{and} \quad k_2 = \Delta F(x + \Delta, y + k_1)$$

Heun's method advances the solution y^{n+1} from the previous value y^n using the scheme

$$y^{n+1} = y^n + (k_1 + k_2)/2. \tag{3.10}$$

The method of Heun is sometimes called a second-order Runge–Kutta scheme, for obvious reasons.

It is interesting to note that this method delivers the same result as the better known *mid-point* method, which uses the gradient of F at the mid-point of the interval. Both of these methods match the expansion of the original equation up to the second term.

This method is quite well suited to the treatment of the physical components discussed in this book. The function $F(x, y)$ and its derivative are known explicitly (albeit sometimes locally linearised) and need be calculated only once each time step. Where step sizes are small in comparison to the dominant time constants, dynamic accuracy is more than adequate.

3.5.2 Runge–Kutta

One of the most popular numerical ODE solution procedures is the standard fourth-order Runge–Kutta procedure. This is well documented in standard texts and will not be described in detail here. Its application requires the following steps:

1. Compute $k_1 = \Delta F(y^n, x^n)$.
2. Compute $k_2 = \Delta F((y^n + \frac{k_1}{2}), (x^n + \frac{\Delta}{2}))$.
3. Compute $k_3 = \Delta F((y^n + \frac{k_2}{2}), (x^n + \frac{\Delta}{2}))$.
4. Compute $k_4 = \Delta F((y^n + k_3), (x^n + \Delta))$.
5. Update the original y^n to $y^{n+1} = y^n + (k_1 + 2k_2 + 2k_3 + k_4)/6$.

This procedure requires four computations of the derivatives at each time step. Providing that the requisite computing power is available, the fourth-order Runge–Kutta method provides a robust method which is fourth-order accurate in time.

The Runge–Kutta method is not unconditionally stable, and most textbooks dealing with serious computing (e.g. [3]) advise that Runge–Kutta not be used without adaptive step sizing. Unfortunately, real-time simulation does not allow this, and Runge–Kutta should be used in such cases with some caution.

3.5.3 Illustrative Examples

Using two different examples of differential equations we will illustrate and compare the performance of the following three direct methods of integration of first-order differential equations:

- First-order explicit Euler
- Heun's method
- Fourth-order Runge–Kutta

These are not equations which might be expected to arise from the modelling of physical systems, but they are sufficiently messy to serve as a good test and illustration of the methods. They also have the advantage of having analytical solutions against which to compare the numerical methods.

The differential equation

$$\frac{dy}{dx} = -x\,(y+1) = F(x,y)$$

has the solution for all $x \geq 0$

$$y = 1 + c\,e^{-x^2/2},$$

where c is an arbitrary constant selected to match a boundary condition.

The evolution of the solution of this equation with an arbitrary choice of $c = 9$ for each of the three methods is set out in the following table for a step increment Δ in x of 0.20. Both Heun and fourth-order Runge–Kutta (RK4) show excellent agreement with the true solution over the whole range. The explicit Euler shows the development of instability for $x > 0.25$ which grows unbounded for larger x. For $\Delta < 0.2$ no oscillations in the Euler solution are apparent, but accuracy is poor for $x > 1.5$. For $\Delta < 0.15$ Heun agrees with the true solution to the third decimal place over the whole range and Runge–Kutta to the fifth decimal place.

Had this equation been written with time t in place of the anonymous x, it could be described as a linear equation with time-varying coefficients.

$$\frac{dy}{dt} = -t\,(y+1)$$

and for $t \geq 0$

$$y = 1 + c\,e^{-t^2/2}.$$

The solution y remains bounded and tends to a constant value 1 as $t \to \infty$.

x	True solution	Euler	Heun	RK4
0.00	10.000	10.000	10.000	10.000
0.50	8.9425	9.6481	8.9394	8.9425
1.00	6.4588	8.0219	6.4654	6.4588
1.50	3.9219	4.7471	3.9607	3.9221
2.00	2.2180	1.6224	2.2817	2.2186
2.50	1.3954	0.4241	1.4546	1.3962
3.00	1.1000	0.9329	1.1369	1.1006
3.50	1.0197	1.2768	1.0367	1.0201
4.00	1.0030	0.8506	1.0092	1.0032
4.50	1.0004	0.9130	1.0023	1.0004
5.00	1.0000	1.2620	1.0006	1.0000

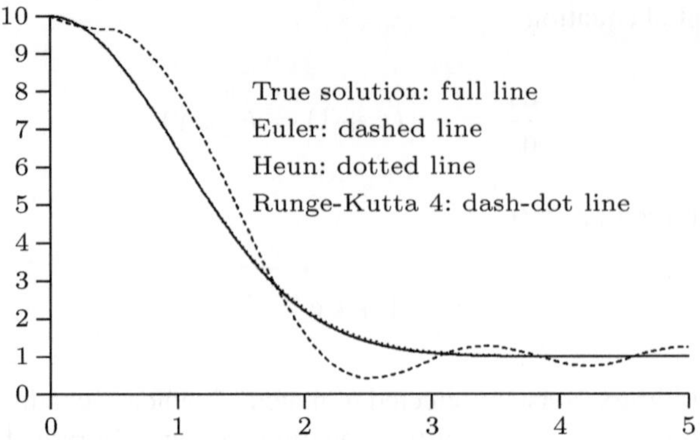

The non-linear differential equation

$$\frac{dy}{dx} = x\,y\left(\frac{1+y}{1-x^2}\right) = y(1+y)\left(\frac{x}{1-x^2}\right) = F(x,y)$$

has the solution

$$y^{-1} = -1 + c\sqrt{1-x^2}$$

and c is again an arbitrary constant. This equation has a singularity at $|x| = x_{sing} = \frac{1}{c}\sqrt{c^2-1}$ and $y \to -1$ as $x > x_{sing} \to 1$. The gradient $F(x, y)$ undergoes rapid and extreme fluctuations, including changing sign, in the vicinity of the singularity.

The evolution of the solution with $c = 4$ ($x_{sing} = 0.968$) for each of the three methods is set out in the following table for a step increment in x of 0.05. Both Heun and RK4 show agreement with the true solution to the second decimal place for $x < 0.6$. The explicit Euler is hard pressed to maintain even this accuracy for any significant range of x and becomes unusable for $x > 0.85$. While both Heun and Runge–Kutta begin to exhibit significant errors for $x > 0.85$, neither shows any advantage for higher values of x. Both would benefit by a reduction in step size as the gradient increases for higher values of x, but all methods fail for x in the near vicinity of the singularity.

x	True solution	$F(x,y)$	Euler	Heun	RK4
0.05	0.3339	0.0223	0.3339	0.3339	0.3339
0.15	0.3384	0.0695	0.3396	0.3384	0.3384
0.25	0.3481	0.1251	0.3507	0.3480	0.3480
0.35	0.3640	0.1981	0.3685	0.3640	0.3639
0.45	0.3888	0.3047	0.3960	0.3887	0.3885
0.55	0.4272	0.4808	0.4391	0.4269	0.4266
0.65	0.4903	0.8223	0.5112	0.4894	0.4885
0.75	0.6076	1.6746	0.6518	0.6045	0.6026
0.85	0.9032	5.2656	1.0511	0.8825	0.8812
0.95	4.0161	196.284	11.6122	2.5066	3.0623

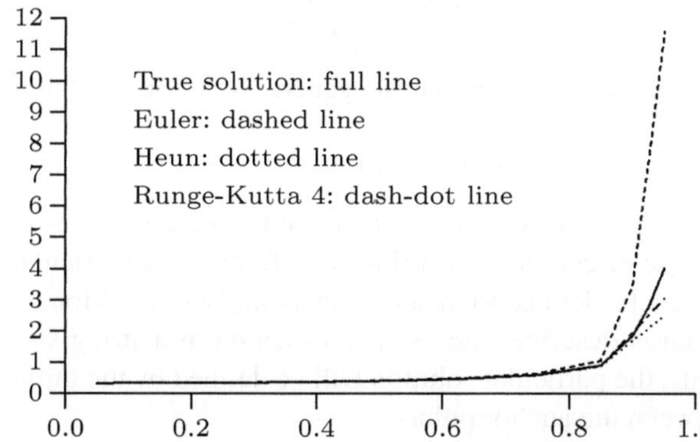

3.6 Systems Described by Partial Differential Equations

While ODEs describe the gradient of a single scalar or vector variable in the direction of a single coordinate, which for dynamic physical systems is most frequently time, a single PDE can describe gradients of several variables in several coordinate directions, including time. The general solution of an ODE is a single function of the independent coordinate and includes unknown constants equal in number to the boundary conditions. The form of the solution is fixed by the original equation. The general solution of a PDE is a family of functions, which can be arbitrary and which together define a multidimensional surface within the coordinate system. A specific solution requires the identification not just of a set of constants but of a set of specific functions, equal in number to the order of the PDE.

PDEs can be solved by the use of analytical or discretisation methods. Analytical solutions can be developed for some boundary condition problems. They can expose structural features of the solution otherwise hidden but only rarely can they provide computationally useful solutions and then frequently as infinite series which require truncation to deliver answers having practical utility. Discretisation methods, applied to both the time and spatial coordinates, are the most commonly used methods for solution of differential equations, both ordinary and partial. However, while useful for the computation of actual solutions, they rarely provide much insight into the structural behaviour of the solutions or the underlying physical principles.

3.6.1 Analytical Solution of Partial Differential Equations

We will give only a very brief introduction to this subject here, which is treated in more depth in Sect. 5.1. Consider the undamped *wave equation* in one spatial dimension

$$\frac{\partial^2 \phi}{\partial x^2} = \frac{1}{c^2}\frac{\partial^2 \phi}{\partial t^2}. \tag{3.11}$$

This is a PDE of second order whose general solution contains two arbitrary functions,

$$\phi = f_1(t - x/c) + f_2(t + x/c).$$

These can be *any* functions of the argument and represent waves travelling in the forward and reverse directions with velocity c. To obtain a particular solution, specific functions must be defined to match given boundary conditions. For example, if the original equation describes the oscillatory action of a string stretched between two anchor points, the particular solution will be defined by the initial displacement of the string between the anchor points.

The wave Eq. 3.11 supports two waves travelling with equal velocity in both directions and can thus support standing waves. The next equation is a general form called an advection equation and describes, for example, the distribution of pressure along the flow path of a working fluid moving with velocity c. The general solution of

$$\frac{\partial p}{\partial t} = -c\frac{\partial p}{\partial z} \tag{3.12}$$

is $p(z,t) = f(z - c\,t)$ and describes a wavefront which propagates the inlet boundary condition forwards at velocity c. It does not support standing waves.

3.6.2 Solution of PDEs by Spatial Discretisation

A numerical procedure for the solution of Eq. 3.12 may be developed by replacing each partial derivative by an appropriate discretisation. We will replace the time derivative by its now familiar first-order discrete-time equivalent and use a second-order approximation for the spatial derivative, what Roache [6] calls the FTCS method (FTCS = "Forward Time Centred Space").

$$\frac{p_j^{n+1} - p_j^n}{\Delta t} = -c\,\frac{p_{j+1} - p_{j-1}}{2\,\Delta z}. \tag{3.13}$$

The physical significance of spatial discretisation is the replacement of the spatial continuum by a contiguous series of discrete elements or cells of length Δz. The working fluid is implied to have homogeneous properties at all points within a cell. The cell index j takes on all values $j \in 1, N$ where N is the number of cells between the flow section boundaries.

p_{j-1}	p_j	p_{j+1}
$\dot{m}_{j-1} \leftrightarrow$	$\dot{m}_j \leftrightarrow$	$\dot{m}_{j+1} \leftrightarrow$

Equation 3.13 can be rearranged as

$$p_j^{n+1} = p_j^n - \left(\frac{c}{2}\frac{\Delta t}{\Delta z}\right)(p_{j+1} - p_{j-1}). \tag{3.14}$$

We can treat Eq. 3.14 as either explicit or implicit by denominating the right-hand bracketed terms at the nth (explicit) or $(n+1)$th (implicit) time step. From the earlier discussion we might expect that the explicit form will be unstable and the implicit form stable. To show that this is indeed the case we use the *von Neumann method* [1] to analyse the stability properties of each algorithm. This assumes that the solution sequence p_j for the jth cell can be represented by a set of *modes* with a frequency dependence on the spatial dimension and an exponential time dependence

$$p_j^k = B_j^n \, \mathrm{e}^{i(jk)},$$

where $\mathrm{e}^{i(jk)} = \cos(jk) + i\sin(jk)$ and $i = \sqrt{-1}$. The stability criterion states that the numerical procedure will be stable if for all k, $|B^{n+1}| \le |B^n|$, that is, all exponential modes will die out.

For the explicit case:

$$B^{n+1}\mathrm{e}^{i(kj)} = B^n \, \mathrm{e}^{i(kj)} - \left(\frac{c}{2}\frac{\Delta t}{\Delta z}\right)\left(B^n \, \mathrm{e}^{ik(j+1)} - B^n \, \mathrm{e}^{ik(j-1)}\right)$$

from which we have

$$\begin{aligned} \frac{B^{n+1}}{B^n} &= 1 - \left(\frac{c}{2}\frac{\Delta t}{\Delta z}\right)(\mathrm{e}^{ik} - \mathrm{e}^{-ik}) \\ &= 1 - i\left(\frac{c}{2}\frac{\Delta t}{\Delta z}\right)\sin(k). \end{aligned} \tag{3.15}$$

Since the absolute value of the right-hand side is always > 1 for any k, this procedure is indeed unstable.

For the implicit case:

$$p_j^{n+1} = p_j^n - \left(\frac{c}{2}\frac{\Delta t}{\Delta z}\right)\left(p_{j+1}^{n+1} - p_{j-1}^{n+1}\right).$$

Rearrange this equation by bringing all terms at time $(n+1)$ to the left-hand side.

$$-\left(\frac{c}{2}\frac{\Delta t}{\Delta z}\right)p_{j-1}^{n+1} + p_j^{n+1} + \left(\frac{c}{2}\frac{\Delta t}{\Delta z}\right)p_{j+1}^{n+1} = p_j^n. \tag{3.16}$$

After modal substitution and some manipulation, this yields

$$\frac{B^n}{B^{n+1}} = 1 + \left(\frac{c}{2}\frac{\Delta t}{\Delta z}\right)(\mathrm{e}^{ik} - \mathrm{e}^{-ik}) \tag{3.17}$$
$$= 1 + i\,c\left(\frac{\Delta t}{\Delta z}\right)\sin(k).$$

The right-hand side of Eq. 3.18 is always > 1, and the absolute magnitude of B^{n+1}/B^n is always < 1. The implicit procedure therefore is always stable. Stability, however, says little about *dynamic accuracy*, and while the implicit method offers almost certain stability, accuracy considerations will dictate limits on the choice of Δt and Δz.

3.6.3 The Courant Stability Condition

We have seen that the explicit method applied to Eq. 3.12 will be unconditionally unstable. The Lax method [8] adds a synthetic diffusion term to the original equation to give

$$\frac{p_j^{n+1} - p_j^n}{\Delta t} = -c\frac{p_{j+1} - p_{j-1}}{2\,\Delta z} + \frac{(\Delta z)^2}{2\Delta t}\frac{\partial^2 p}{\partial z^2}. \tag{3.18}$$

This has the *stable* explicit solution algorithm

$$p_j^{n+1} = (p_{j+1}^n + p_{j-1}^n)/2 - \frac{c\Delta t}{2\Delta z}\left(p_{j+1}^n - p_{j-1}^n\right)$$

if

$$\frac{|c|\Delta t}{\Delta z} \le 1.$$

This is the *Courant condition*, and the left-hand term is the *Courant number*. Applied to a fluid flow of specific volume v, mass flow rate $\dot{m}$, flow cross-sectional area A and cell volume V, the Courant condition becomes

$$\frac{|\dot{m}|v}{A}\frac{\Delta t}{\Delta z} \le 1$$

or

$$\frac{|\dot{m}|v}{V}\Delta t \le 1$$

since $V = A\Delta z$.

The Courant number has a physical significance. Define τ_f to be the cell filling time given by $M/\dot{m}$ where M is the mass of working medium in the cell. Replacing M by V/v, the Courant condition is seen to imply that the computation interval Δt

must be less than the cell filling time. Physically, this means that, for numerical stability, the computational rate of pressure propagation cannot exceed the physical propagation rate. For example, if the maximum $\dot{m}$ expected is 400 kg/s and the specific volume v is 1.2 m^3/kg,[1] then, for a Δt of 0.25 s, the minimum cell volume for stability will be 120 m^3, much larger than the physical volume. This reduces to 24 m^3 for a step time of 50 ms. For $v = 0.018$ m^3/kg[2] and a mass flow of 400 kg/s, the minimum cell volume for a time step of 50 ms is 0.54 m^3.

The Courant stability criterion applies to any conditionally stable explicit time/space differencing scheme. While fourth-order Runge–Kutta is fourth-order accurate in time, if combined with a first-order space differencing scheme, it is subject to the same stability condition.

3.6.4 Solution of Equations in Implicit Form

Expanded for all $j \in (1, N)$, Eq. 3.16 may be written as the matrix equation

$$\mathbf{C}\,\mathbf{p^{n+1}} = \mathbf{p^{n}}. \tag{3.19}$$

The coefficient matrix $\mathbf{C}$ is tridiagonal as it has non-zero coefficients on only the main and its two adjacent diagonals. This has several advantages which are discussed in the next section.

Boundary conditions are included by noting that, for the first and last rows of the array, the index j moves outside its defined range $(1, N)$ by referring to $j = 0$ and $j = N + 1$. Additional information is needed to complete the definition of the solution. Define p_{lbc} at the left-hand boundary ($j = 0$) and p_{rbc} at the right ($j = \mathrm{N} + 1$). Then,

$$j = 1: \quad p_1^{n+1} - \left(\tfrac{K}{2}\tfrac{\Delta t}{\Delta z}\right) p_2^{n+1} = p_1^n - \left(\tfrac{K}{2}\tfrac{\Delta t}{\Delta z}\right) p_{lbc}^{n+1}$$

$$j = N: \quad \left(\tfrac{K}{2}\tfrac{\Delta t}{\Delta z}\right) p_{N-1}^{n+1} + p_N^{n+1} = p_N^n + \left(\tfrac{K}{2}\tfrac{\Delta t}{\Delta z}\right) p_{rbc}^{n+1}.$$

Adding the boundary condition vector to the Eq. 3.19 gives the complete system

$$\mathbf{C}\,\mathbf{p^{n+1}} = \mathbf{p^{n}} + \mathbf{B}\,\mathbf{p_{bcs}}, \tag{3.20}$$

where $\mathbf{B}$ is the $(N \times 2)$ coefficient matrix

---

[1] Air at normal atmospheric pressure and temperature.

[2] Saturated steam at a pressure of 10 MPa.

$$\begin{bmatrix} -\frac{K}{2}\frac{\Delta t}{\Delta z} & 0 \\ 0 & 0 \\ \cdots & \cdots \\ 0 & \frac{K}{2}\frac{\Delta t}{\Delta z} \end{bmatrix}$$

and

$$\mathbf{p_{bcs}} = \begin{bmatrix} p_{lbc} \\ p_{rbc} \end{bmatrix}$$

is the vector of boundary conditions. Solution of Eq. 3.20, given an initial distribution of pressure $\mathbf{p}_j^n$ for all $j \in$ (1,N) and a pair of boundary pressures, yields the updated set of pressures at the next time step for all cells and simultaneously satisfies the two-point boundary conditions.

3.7 Solution of Linear Matrix Equations

We have seen how linear matrix equations arise naturally from the application of discretisation methods to the solution of a wide variety of differential equations. Equation 3.20 is representative of matrix equations to be solved. From a computational viewpoint, the preferred methods are those which are stable and accurate and require the least number of arithmetic operations. In this section, we will examine some of the available methods and the features of the structure of the coefficient matrix **C** which guide the selection of the best method.

3.7.1 *Matrix Inversion*

The formal solution of the equation $\mathbf{Ax} = \mathbf{b}$ is simply $\mathbf{x} = \mathbf{A}^{-1}\mathbf{b}$ where $\mathbf{A}^{-1}$ is the inverse of **A**. It would seem that all that is needed to solve the original equation is to invert the coefficient matrix **A**, presumed square and non-singular. In most cases this is about the most undesirable action to be taken and should be avoided where possible. It is almost always possible. For any but the smallest matrices inversion is time consuming and can be inaccurate. For these reasons numerous methods have been developed for the solution of equations which do not require the inversion of the coefficient matrix. Some of these are described in the following sections.

Specifically, for (2×2) and (3×3) matrices, explicit formulae are available for the calculation of the inverse. Similar calculations can be developed for a (4×4), but as the matrix dimension increases beyond 3, the scale of calculation quickly grows beyond reasonable bounds. These formulae are derived easily from Cramer's Rule which expresses the solution of a matrix equation in terms of the minors, cofactors and determinant of the coefficient matrix.

Inversion of a (2 x 2) Matrix

$$\begin{bmatrix} a_{11} & a_{12} \\ a_{21} & a_{22} \end{bmatrix}^{-1} = \frac{1}{D}\begin{bmatrix} a_{22} & -a_{12} \\ -a_{21} & a_{11} \end{bmatrix},$$

where the determinant $D = a_{11}a_{22} - a_{21}a_{12}$.

Inversion of a (3 x 3) Matrix

$$\begin{bmatrix} a_{11} & a_{12} & a_{13} \\ a_{21} & a_{22} & a_{23} \\ a_{31} & a_{32} & a_{33} \end{bmatrix}^{-1} = \frac{1}{D}\begin{bmatrix} c_{11} & c_{12} & c_{13} \\ c_{21} & c_{22} & c_{23} \\ c_{31} & c_{32} & c_{33} \end{bmatrix},$$

where

$$\begin{aligned} c_{11} &= a_{22}a_{33} - a_{32}a_{23} \quad & c_{12} &= -(a_{12}a_{33} - a_{32}a_{13}) \\ c_{21} &= -(a_{33}a_{21} - a_{31}a_{23}) \quad & c_{22} &= a_{33}a_{11} - a_{31}a_{13} \\ c_{31} &= a_{33}a_{21} - a_{31}a_{22} \quad & c_{32} &= -(a_{32}a_{11} - a_{31}a_{12}) \end{aligned}$$

$$\begin{aligned} c_{13} &= a_{12}a_{23} - a_{22}a_{13} \\ c_{23} &= -(a_{23}a_{11} - a_{21}a_{13}) \\ c_{33} &= a_{22}a_{11} - a_{21}a_{12} \end{aligned}$$

and the determinant D is given from

$$D = a_{11}c_{11} + a_{21}c_{12} + a_{31}c_{13}.$$

3.7.2 *Coefficient Matrix Types*

Some of the more common standard matrix types are defined as follows:

Symmetric A matrix **A** is symmetric if

$$\mathbf{A}^T = \mathbf{A},$$

where T denotes matrix transpose. Expressed in terms of the individual matrix elements, symmetry implies $a_{ij} = a_{ji}$ for all i, j spanning the matrix.

Positive Definite A matrix **A** is positive definite if for any **x**

$$\mathbf{x}^{\mathbf{T}}\,\mathbf{A}\,\mathbf{x} > 0.$$

The following statements are true of a positive-definite matrix:

- Every main diagonal element is positive and non-zero.
- The matrix is diagonal dominant, that is, the largest element of the ith row is the ith diagonal element.
- The matrix is non-singular (invertible) and all of its eigenvalues are positive.
- The inverse matrix is also invertible.
- All principal submatrices are positive definite.

A positive-definite matrix does not need to be symmetric. However, if it is, certain computational advantages accrue, which will be discussed in a later section.

Tridiagonal A matrix **A** is tridiagonal if it has non-zero coefficients on only the main and its immediately adjacent diagonals. Thus,

$$a_{ij} \neq 0 \text{ for only } j = i, \quad j = i - 1, \quad j = i + 1.$$

Block Tridiagonal This is a tridiagonal matrix whose diagonally arranged elements are themselves matrices. This structure offers some computational advantages over the general LU decomposition methods.

Banded A matrix is banded if its non-zero elements are ordered exclusively along rows, columns or diagonals or a mix of these. There is a wide variety of matrix structures which might be called banded, and no general procedure is available for all of them. However, certain banded structures do have efficient procedures available. One such is the so-called periodic structure which is a tridiagonal matrix with non-zero elements in the top right and bottom left corners. These matrices arise in the analysis of systems with periodic boundary conditions, being systems which "loop back" on themselves.

Sparse A sparse matrix is one with relatively few non-zero elements compared to its dimensions. The term could include banded matrices, and the tridiagonal matrix is a special case of a sparse matrix for which a well-developed solution procedure is available. In general, the treatment of sparse matrices is directed at taking advantage of the "sparsity" to reduce coefficient storage requirements and speed up the solution process. They tend to arise in the treatment of very large systems having many branches and relatively few interconnection points. Much of the early work on sparse matrix analysis was undertaken for the analysis of large-scale electricity distribution networks, involving matrices having thousands of rows and columns.

3.7.3 Solution Methods

Tridiagonal Matrices

Solution of $\mathbf{A}\,\mathbf{x} = \mathbf{b}$, where the coefficient matrix **A** is tridiagonal, is very efficient. So much so that it is often worth decomposing a problem into a form which can exploit the tridiagonal structure even if the decomposition introduces a need for iteration to the final solution. Coefficient storage requirements for even large matrices are minimal as only the diagonals need be stored. The solution routine is extremely efficient as computation time is proportional to the matrix dimension N rather than related to N^2 or even N^3 as with many other solution methods. The tridiagonal routine is simple and is described in several sources, such as Golub and Van Loan [2] and Press et al. [1]. Solution is a two-step process involving forward computation of intermediate values followed by reversed-order back substitution. High-precision arithmetic is recommended for large systems.

Cholesky Decomposition

If the coefficient matrix $\mathbf{A}$ is positive definite and symmetric, the equation

$$\mathbf{A}\,\mathbf{x} = \mathbf{b}$$

is particularly easily and efficiently solved by *Cholesky decomposition*. This constructs a lower triangular matrix $\mathbf{L}$ such that

$$\mathbf{A} = \mathbf{L}\mathbf{L}^T.$$

This algorithm is described in detail in [1]. It is efficient, needing on average around half as many operations as the general lower–upper (LU) decomposition, and stable. The usual algorithm includes a test for positive-definiteness of the coefficient matrix and will fail if this condition is not satisfied.

Lower–Upper Decomposition

If the coefficient matrix is not one of these well-defined types it will generally be amenable to solution by one or other standard LU (lower–upper) decomposition methods. These assume that the coefficient matrix $\mathbf{A}$ can be decomposed into a lower and an upper triangular matrix such that

$$\mathbf{A} = \mathbf{L} \cdot \mathbf{U}.$$

We can then write

$$\mathbf{A}\,\mathbf{x} = (\mathbf{L}\,\mathbf{U})\,\mathbf{x} = \mathbf{L}\,(\mathbf{U}\,\mathbf{x}) = \mathbf{b}.$$

After solving for $\mathbf{y} = \mathbf{U}\,\mathbf{x}$ from

$$\mathbf{L}\mathbf{y} = \mathbf{b}$$

we solve for the solution $\mathbf{x}$ from

$$\mathbf{U}\mathbf{x} = \mathbf{y}.$$

This routine is very efficient (though not as efficient as the previous two) and stable. It requires the order of $\frac{1}{3}N^3$ flops for a single-shot solution. In the absence of any special structural features of the coefficient matrix which might suggest some other method, LU decomposition is the method of choice.

3.7.4 Improving the Solution of an Algebraic Equation

A simple form of iteration can be used to improve the accuracy of a one-shot solution. Assume that x is the true solution of the equation

$$\mathbf{A}\,\mathbf{x} = \mathbf{b}. \tag{3.21}$$

Assume that the calculated solution $\hat{\mathbf{x}}$ differs from the true solution by a small amount $\delta\mathbf{x}$. Then, we can assume that $\hat{\mathbf{x}}$ is the true solution of the slightly modified equation

$$\mathbf{A}\,\hat{\mathbf{x}} = \mathbf{b} + \delta\mathbf{b} \tag{3.22}$$

or

$$\mathbf{A}(\mathbf{x} + \delta\mathbf{x}) = \mathbf{b} + \delta\mathbf{b}. \tag{3.23}$$

Subtracting Eq. 3.21 from 3.23 yields

$$\mathbf{A}\,\delta\mathbf{x} = \delta\mathbf{b} \tag{3.24}$$

from which we have

$$\mathbf{A}\,\delta\mathbf{x} = \mathbf{A}\,\hat{\mathbf{x}} - \mathbf{b}. \tag{3.25}$$

Since $\hat{\mathbf{x}}$ is known from the computed solution of Eq. 3.21, the right-hand side of Eq. 3.25 is known, and the equation can be solved explicitly for $\delta\mathbf{x}$. The improved solution $\mathbf{x}$ is then given as $\hat{\mathbf{x}} + \delta\mathbf{x}$. It is important to note that, unless $\hat{\mathbf{x}}$ is well removed from $\mathbf{x}$, the two terms of the right-hand side of Eq. 3.25 will be almost equal and double precision arithmetic is essential for their calculation.

This method can be used to improve the solution of an equation of the form of Eq. 3.21 whose coefficient matrix $\mathbf{A}$ has been formed using some form of linearisation involving elements of $\mathbf{x}$ or other coupled variables.

3.7.5 Solution of Nearly Tridiagonal Equations

A "nearly" tridiagonal system of equations is one whose coefficient matrix is predominantly tridiagonal but with a small number of off-tridiagonal non-zero elements. We will encounter such equations in Sect. 17.2 dealing with complex flow networks. These create systems of equations with large, sparse, diagonally dominant coefficient matrices and are well suited to solution by this method.

The equation to be solved is $\mathbf{A}\,\mathbf{x} = \mathbf{b}$, and the method proceeds as follows. Split the coefficient matrix $\mathbf{A}$ into two parts.

$$\mathbf{A} = \mathbf{A}_{td} + \mathbf{A}_{od},$$

where $\mathbf{A}_{td}$ is a tridiagonal matrix and $\mathbf{A}_{od}$ is the matrix of off-diagonal coefficients. Then

$$(\mathbf{A}_{td} + \mathbf{A}_{od})\ \mathbf{x} = \mathbf{b}$$

or

$$\mathbf{A}_{td}\,\mathbf{x} + \mathbf{A}_{od}\,\mathbf{x} = \mathbf{b}.$$

Denote by $\hat{\mathbf{x}}$ the solution of the tridiagonal equation

$$\mathbf{A}_{td}\,\hat{\mathbf{x}} = \mathbf{b}. \tag{3.26}$$

We assume further that the true solution of the original equation $\mathbf{x}$ consists of the tridiagonal solution $\hat{\mathbf{x}}$ plus a correction $\delta\mathbf{x}$. Thus

$$\mathbf{x} = \hat{\mathbf{x}} + \delta\mathbf{x}. \tag{3.27}$$

The original equation can now be written

$$(\mathbf{A}_{td} + \mathbf{A}_{od})(\hat{\mathbf{x}} + \delta\mathbf{x}) = \mathbf{b} \tag{3.28}$$

which expands to

$$\mathbf{A}_{td}\,\hat{\mathbf{x}} + \mathbf{A}_{td}\,\delta\mathbf{x} + \mathbf{A}_{od}\,\hat{\mathbf{x}} + \mathbf{A}_{od}\,\delta\mathbf{x} = \mathbf{b}. \tag{3.29}$$

Given Eq. 3.26, this can be reduced to

$$(\mathbf{A}_{td} + \mathbf{A}_{od})\,\delta\mathbf{x} = -\mathbf{A}_{od}\,\hat{\mathbf{x}}. \tag{3.30}$$

This is identical to the original equation, with $\mathbf{x}$ replaced by $\delta\mathbf{x}$ and b replaced by $-\mathbf{A}_{od}\,\hat{\mathbf{x}}$. We now assume that $\delta\mathbf{x}$ can be replaced in its turn by

$$\delta\mathbf{x} = \delta\hat{\mathbf{x}} + \delta\hat{\mathbf{x}}_1,$$

where $\delta\hat{\mathbf{x}}$ is the solution of the tridiagonal equation

$$\mathbf{A}_{td}\,\delta\hat{\mathbf{x}} = b1 \tag{3.31}$$

with

$$-\mathbf{A}_{od}\,\hat{\mathbf{x}} = b1. \tag{3.32}$$

This leads to the residual equation

$$(\mathbf{A}_{td} + \mathbf{A}_{od})\,\delta\mathbf{x}_1 = -\mathbf{A}_{od}\,\delta\hat{\mathbf{x}} = b2 \tag{3.33}$$

and so on. This sequence of steps can be used to set up an iterative procedure. Given the solution $\hat{\mathbf{x}}$ to Eq. 3.26, a series of corrections $\delta\hat{\mathbf{x}}$ can be calculated by repeated application of Eqs. 3.31 and 3.32. The final solution is obtained as $\hat{\mathbf{x}}$ plus the sum of the iterated $\delta\hat{\mathbf{x}}$. The iteration is terminated when some terminating condition is satisfied, either some appropriate norm $||\delta\hat{\mathbf{x}}||$ of $\delta\hat{\mathbf{x}}$ is less than a prescribed value or the maximum iteration count has been exceeded.

Computational experience suggests this is an effective scheme for the solution of the type of equations which arise from the spatial analysis of large complex flow networks. The number of iterations depends on the number and magnitude of the off-tridiagonal coefficients but is generally less than 10 or 20. Given that only tridiagonal equations need be solved for each iteration, the method can tolerate a large number of iterations and remain competitive with the otherwise preferred LU decomposition.

This approach can be used expeditiously for a general class of sparse matrix equations, provided that a tridiagonal "core" is available. The number of iterations needed to satisfy a given convergence condition will increase as the density of the off-tridiagonal elements increases.

3.8 Singular Value Decomposition

No discussion of matrix methods would be complete without mention of the method of singular value decomposition (SVD). Although not strictly a method for the solution of equations, it is a powerful method for sorting out the practical problems that can arise in the analysis of physical systems, particularly those dealing with measured data.

All of the solution methods presented in this work assume that the coefficient matrix of the equation to be solved is non-singular, in the sense of the existence of a non-zero determinant. This assumption implies, among other things, that the coefficient matrix is square (equal numbers of rows and columns) and all elements on the main diagonal are non-zero. It is also assumed that the solution defined by the equation system is bounded and that the numerical procedure will produce a stable (not diverging) solution sequence. Simply put, it is assumed that the defining equations are well conditioned. It might be expected that the designer of a mathematical model will ensure that the model satisfies these requirements and no problems will arise. However, should the designer ever wish to compare the model results with actual plant data, he may find that matching the two brings up problems of mismatched dimensions, poor data and ambiguous associations between measured data and model state variables. SVD can help resolve these inconsistencies.

SVD concerns itself with singular and ill-conditioned systems. It provides a means for the identification and quantification of ill-conditioning and for the calculation of solutions which give a best fit to the solution to an ill-conditioned problem.

3.9 Iterative Methods

The preceding sections have treated the solution of linear matrix equations. Each method delivers the solution from a single calculation step, which may involve one or more intermediate steps. The solution method requires no prior knowledge of the solution, other than its existence.

An important class of solution methods treats non-linear equations. These methods require prior knowledge of a starting point (an estimate of the solution) and generate a series of iterates which approach the true solution with increasing accuracy as the iteration proceeds. The process is terminated when a predetermined error criterion is satisfied or a preset iteration count is exceeded.

Iteration-based methods assume:

1. A starting condition from which the iteration process can reach the true solution
2. A stable convergence to the true solution

In some cases, a direct solution will be used to generate a starting solution and the iteration method used to refine it.

For linear equations, the second condition is always satisfied. The first condition will also be satisfied, but if convergence is slow, the method may not deliver the desired solution within an acceptable time or iteration count.

In numerical analysis, iteration methods are commonly used in many areas, such as:

- Finding the roots of non-linear equations or the zero-crossing points of non-linear functions
- Finding the extreme (maxima and minima) values of non-linear functions
- Solving two-point boundary value problems

Finding extreme values of functions arises in the field of process control and optimisation, particularly as it relates to the design of optimising controllers. The method is also used in the process of refining simulation model parameters to match field data. Within the scope of this book, two-point boundary value problems, which occur frequently, are handled either by the development of an analytic solution which embeds the boundary conditions or by an appropriate structuring of the spatial discretisation of the describing PDEs. Both approaches can avoid the need for iteration.

3.9.1 Newton–Raphson Iteration

By far, the most widely used iterative method for solving non-linear equations or finding the roots of non-linear functions is the Newton–Raphson method, also sometimes known as the method of steepest descent.

Consider the scalar equation $f(x) = 0$. The Taylor expansion of $f(x)$ in a small neighbourhood around the point $\hat{x}$ is

$$f(\hat{x} + \epsilon) \approx f(\hat{x}) + \epsilon f'(\hat{x}) + \frac{\epsilon^2}{2} f''(\hat{x}) + \ldots$$

If ϵ is sufficiently small, we can ignore the higher power terms and

$$f(\hat{x} + \epsilon) \approx f(\hat{x}) + \epsilon f'(\hat{x}).$$

Setting $f(\hat{x} + \epsilon) = 0$ projects the tangent to $f(x)$ at $\hat{x}$ onto the x axis. We then have

$$\epsilon = -\frac{f(\hat{x})}{f'(\hat{x})} \quad \text{if } f'(\hat{x}) \neq 0.$$

The iteration procedure, which updates the current solution x^n to produce the next iterate x^{n+1}, is

$$x^{n+1} = x^n - \frac{f(\hat{x})}{f'(\hat{x})}.$$

If the initial value x^0 has been chosen "close enough" to the true solution, and if the function $f(x)$ is well behaved in the sense of convergence behaviour, the iterates x^n will converge quadratically to the true solution. This means that in the vicinity of the true solution, the solution accuracy doubles with each new iteration.

The Newton–Raphson descent method is a powerful and robust method, but it has certain restrictions:

- The function $f(x)$ must be available in a computationally compatible form (analytical or algorithmic).
- It must be differentiable at least once, and the differential must be available in a computationally compatible form (analytical or algorithmic).
- Its derivative cannot approach zero during the iteration process.
- It must be concave in the vicinity of the solution, that is, the second derivative $f''(x)$ must be negative.

The following figures illustrate the influence of the shape of the function whose zero is being sought. The concave shape on the left produces good, non-oscillatory convergence to the solution, as shown by the dotted lines which plot the iterating descent to the solution. The convex shape on the right shows slow oscillatory convergence, if the iteration converges at all and does not either diverge or settle into a stable limit cycle. It is to be noted that this function is well behaved, with no intermediate peaks and monotonic behaviour in the vicinity of the solution.

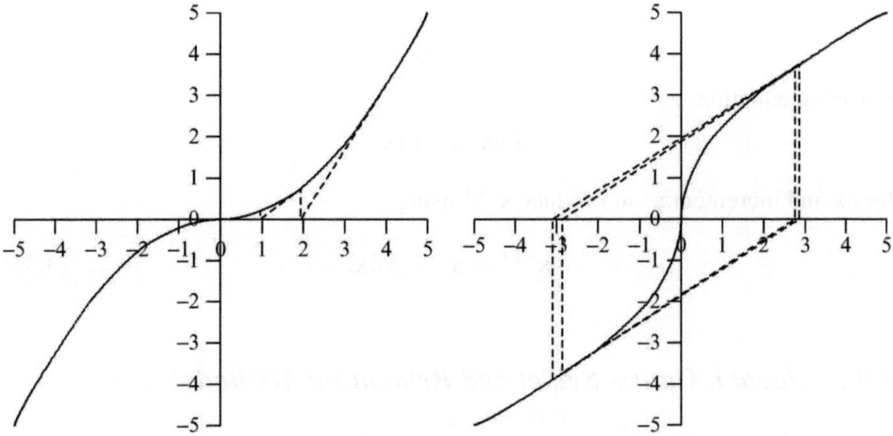

It is generally difficult to guarantee that each of the convergence conditions will be met when applying the Newton–Raphson method generically. However, in the world of process plant simulation, it is fairly easy to ensure that the conditions applying to the specific case are appropriate.

In practice, it is advisable to introduce a "convergence factor" to control convergence behaviour. The actual algorithm used is

$$x^{n+1} = x^n - \beta \frac{f(\hat{x})}{f'(\hat{x})}. \tag{3.34}$$

The factor β can take any value in the range (0,1). β can be set to a fixed value, determined by experiment, or it can be made adaptable and varied automatically to suit the observed convergence behaviour.

All iterative descent methods move the iterates along the *gradient* of $f(x)$. However, the steps do not need to take the maximum step in this direction, and even a small value of β will advance the iterates towards the desired result, but the convergence rate may deteriorate. A small increment in the right direction could move the next iterate to a point at which the gradient produces a better behaved increment.

If there are concerns that divergent or erratic behaviour may be introduced by the second derivative $f''(x)$ changing sign along the way, the gradient $f'(x)$ can be calculated once at a point for which it is known that the second derivative is negative and this value used for each update thereafter.

Extension of Newton–Raphson to Vector Functions

The extension of the scalar results of the preceding section to the vector case is straightforward. Consider the vector equation $\mathbf{F}(\mathbf{x}) = 0$. The Jacobian matrix $\mathbf{J}$ is defined as

$$J_{ij} = \frac{\partial F_i}{\partial x_j}.$$

Solve the equation

$$\mathbf{J}\,\delta\mathbf{x} = -\mathbf{F}(\mathbf{x})$$

for $\delta\mathbf{x}$ and increment $\mathbf{x}^n$ to produce $\mathbf{x}^{n+1}$ using

$$\mathbf{x}^{n+1} = \mathbf{x}^n + \beta\,\delta\mathbf{x}. \tag{3.35}$$

3.9.2 *Jacobi, Gauss–Seidel and Relaxation Methods*

Consider the linear matrix equation of dimension n_x

$$\mathbf{A}\,\mathbf{x} = \mathbf{b}. \tag{3.36}$$

Split the coefficient matrix $\mathbf{A}$ into two submatrices as follows.

$$\mathbf{A} = \mathbf{P} - \mathbf{N}.$$

Then, Eq. 3.36 can be structured to define an iteration sequence

$$\mathbf{P}\,\mathbf{x}^{n+1} = \mathbf{N}\,\mathbf{x}^n + \mathbf{b}$$

or

$$\mathbf{x}^{n+1} = \mathbf{P}^{-1}\mathbf{N}\,\mathbf{x}^n + \mathbf{P}^{-1}\mathbf{b}. \tag{3.37}$$

With a *residual* or error vector defined at the nth iteration as

$$\mathbf{r} = \mathbf{b} - \mathbf{A}\,\mathbf{x}^n$$

Eq. 3.36 can be solved by the iterative sequence

$$\mathbf{x}^{n+1} = \mathbf{x}^n + \mathbf{P}^{-1}\mathbf{r}. \tag{3.38}$$

To be effective, the split of $\mathbf{A}$ into $\mathbf{P}$ and $\mathbf{N}$ must simplify the inversion of $\mathbf{P}$. An obvious and common choice is to choose $\mathbf{P}$ to be the diagonal matrix $\mathbf{D}$ composed of the elements of the main diagonal of $\mathbf{A}$. Then, if a_{ii} is the ith diagonal element of $\mathbf{A}$, the ith element of $\mathbf{D}^{-1}$ is $1/a_{ii}$. The iteration scheme of Eq. 3.37 may then be written as

$$\mathbf{x}^{n+1} = [\mathbf{I} - \mathbf{D}^{-1}\mathbf{A}]\,\mathbf{x}^n + \mathbf{D}^{-1}\,\mathbf{b}. \tag{3.39}$$

Since $\mathbf{D}^{-1}$ is diagonal this method allows us to calculate the elements of $\mathbf{x}^{n+1}$ sequentially, using the following element-by-element procedure.

$$x_i^{n+1} = \frac{1}{a_{ii}}\left[b_i - \sum_{j=1,\,j\neq i}^{n_x} a_{ij}x_j^n\right] \tag{3.40}$$

advancing i from $1 \rightarrow n_x$. The bracketed term is the residual for the ith element of $\mathbf{x}$ at the nth iteration.

There are two possible approaches to this procedure. The first uses only elements of $\mathbf{x}^n$ to calculate each new iterate $\mathbf{x}^{n+1}$ and uses the new $\mathbf{x}^{n+1}$ only when all elements of $\mathbf{x}^n$ have been updated. When executed in this manner the procedure defined by Eq. 3.40 is the Jacobi method. Alternatively we can replace the individual elements of $\mathbf{x}^{\mathbf{n}}$ as they are updated, noticing that the lower indexed elements of $\mathbf{x}$ are updated before the higher. When executed in this manner the procedure is the Gauss–Seidel method and is written on an element-by-element basis as

$$x_i^{n+1} = \frac{1}{a_{ii}}\left[b_i - \sum_{j=1}^{i-1} a_{ij}x_j^{n+1} - \sum_{j=i+1}^{n_x} a_{ij}x_j^n\right] \tag{3.41}$$

advancing i from $1 \rightarrow n_x$.

Either method may be varied by introduction of a *relaxation* parameter β. This can be used to vary the quantum of the correction added at each iteration to the previous iterate to calculate the updated value. For the Jacobi method we then have

$$x_i^{n+1} = \frac{\beta}{a_{ii}}\left[b_i - \sum_{j=1,j\neq i}^{n_x} a_{ij}x_j^n\right] + (1-\beta)x_i^n \tag{3.42}$$

advancing i from $1 \rightarrow n_x$. For the Gauss–Seidel method we have

$$x_i^{n+1} = \frac{\beta}{a_{ii}}\left[b_i - \sum_{j=1}^{i-1} a_{ij}x_j^{n+1} - \sum_{j=i+1}^{n_x} a_{ij}x_j^n\right] + (1-\beta)x_i^n \tag{3.43}$$

advancing i from $1 \rightarrow n_x$. If $\beta < 1$ the method is called *under-relaxation*, while for $\beta > 1$ it is called *over-relaxation*.

Convergence of any of these methods is dependent on certain structural features of the original coefficient matrix $\mathbf{A}$. In general, convergence of linear equations is guaranteed if $\mathbf{A}$ is symmetric and positive definite and if $0 < \beta < m$ where

$m \le 2$. How much less than 2 is dependent on the norm of $\mathbf{D}^{-1}\mathbf{A}$. In the case of non-linear equations or equations whose linearisation depends on elements of $\mathbf{x}$, these general convergence conditions will not always apply, and each case must be individually investigated. It can be said that, in general, under-relaxation produces better stability but slower convergence than either the no relaxation ($\beta = 1$) or over-relaxation methods. It should be noted that both the Jacobi and Gauss–Seidel methods converge slowly, as does even the over-relaxation method for any value of β much removed from the optimal. The reader is referred to [4] for a more detailed discussion of this subject.

3.10 Table Lookup

Tabulated data and functions are widely used in simulation. In engineering simulation probably the most common usage is the tabulation of physical properties of materials as a function of one or more process variables. For example, steam properties—such as temperatures, specific volumes and saturation properties—are tabulated as a function of one or more thermodynamic variables such as pressure and enthalpy. Tables can be used to record the throat area of important valves as a function of stem position where these cannot be fitted by standard characteristics. Turbine control valves often require this treatment. Tabulations of the static characteristics of measurement transducers are used to "linearise" the transducer output. Empirically defined model coefficients can be made functions of other variables to improve the model's ability to follow large process state variations where a purely physical model might be computationally intractable. Non-analytic terms in differential equations can be replaced by tabulations to facilitate solution of the equation. In short, tables are useful, and it is important to have routines for rapidly accessing their contents.

In earlier times, when computer memory was expensive and capacities were limited, tables tended to be small, and accuracy was achieved by sophisticated interpolation routines, often using non-linear functions to model the behaviour of the data between tabulated entries. With computing memory no longer a cost obstacle larger tables can be used to ensure high accuracy, and simpler linear interpolation routines will usually suffice.

3.10.1 One-Dimensional Table Lookup

A table can be regarded as a sparse reference grid overlaid on a continuous variable or function space. A single-dimensioned table consists of a list of values of some reference variable, or possibly only an index, and an associated single output value.

The reference variable list is monotonic and single-valued (no repeats), usually arranged in ascending order or increasing value. The first step to accessing the table is to *locate* the entry in the reference list closest (in the sense of increasing index) to the input variable. The second task is to *interpolate* the desired output variable within the bracketed box.

There are several well-established location routines. The simplest and usually the fastest is bisection. Starting at the low end of the table this jumps to the mid-point and tests to identify the relevant half (upper or lower) of the table. It then bisects this half and repeats the process until convergence.

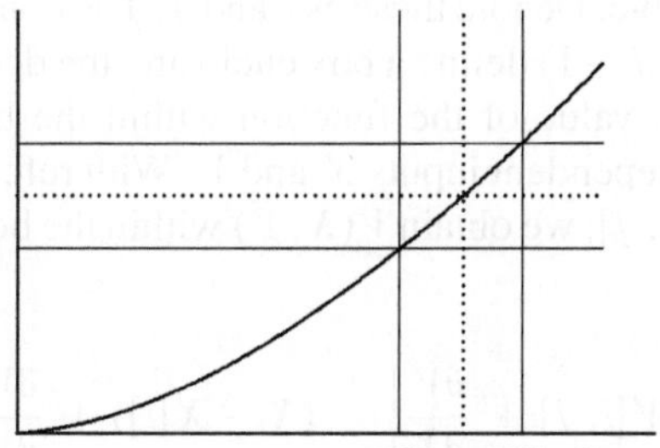

The speed of the search can be improved if the "last time" index is stored and the new search started from there next time. This is not of much help if the "last time" and "next time" searches are not related but can give a major speed boost if they are.

The interpolation task is executed when the bracketing entries have been identified. Assume that the table lists a set of values of the variable *Y(X)*, indexed by *j*, as a set of tabulated values *Y*[*j*]. Denote the corresponding tabulated values of *X* as *X*[*j*]. Then, for a given input *X*, linear interpolation returns,

$$Y(X) = Y[j] + \left(\frac{Y[j+1] - Y[j]}{X[j+1] - X[j]}\right)(X - X[j]).$$

The fraction will be recognised as the gradient of Y with respect to X, $\partial Y/\partial X$, evaluated at $(X[j], Y[j])$, and $(X - X[j])$ is the increment in X.

This works well if the variation of Y within each table entry pair is monotonic but can introduce error if the true gradient of X changes sign between $Y[j]$ and $Y[j+1]$. An interpolation formula can be used if this is not the case. This fits a low-order polynomial to a small set of points surrounding the table location and computes the interior value from the polynomial fit. The polynomial might be fitted, for example, to points at indices $j-1$, j, $j+1$ and $j+2$, assuming all points lie within the range of the table. In general, n points can identify an $(n-1)$th-order polynomial. Linear interpolation (order 1) uses 2 points. A quadratic could be fitted through three points and so on. A cubic interpolation polynomial should be the highest order ever necessary. Polynomial fitting within the table is clearly computationally more intensive and should only be considered if appropriate *a priori* structuring of the table is not possible.

3.10.2 Two-Dimensional Table Lookup

A one-dimensional table describes a single-valued relationship between the output and its input, $y = f(x)$. A two-dimensional table describes a surface as a function of two inputs, $v = f(x, y)$.

Assume that the table search procedure has identified the table index of each input reference variable. Denote these as i and j. The combinations (i, j), $(i, j+1)$, $(i+1, j)$ and $(i+1, j+1)$ define a box enclosing the desired value of the tabulated function. Denote the value of the function within the box as $V(X, Y)$, given the values of the two independent inputs X and Y. With reference to its tabulated value at (i, j), denoted $V[i, j]$, we obtain $V(X, Y)$ within the box at (X,Y) from the Taylor expansion,

$$V(X, Y) = V[i, j] + \left.\frac{\partial V}{\partial X}\right|_{[i,j]} (X - X[i]) + \left.\frac{\partial V}{\partial Y}\right|_{[i,j]} (Y - Y[j]),$$

where

$$\left.\frac{\partial V}{\partial X}\right|_{[i,j]} = \frac{V(X[i+1], Y[j]) - V(X[i], Y[j])}{X[i+1] - X[i]}$$

and

$$\left.\frac{\partial V}{\partial Y}\right|_{[i,j]} = \frac{V(X[i], Y[j+1]) - V(X[i], Y[j])}{Y[j+1] - Y[j]}.$$

Extension to higher-order tabulations follows directly from the two-dimensional procedure. A 3-dimensional tabulation of $V(X, Y, Z)$ is simply

$$V(X, Y, Z) = V[i, j, k] + \left.\frac{\partial V}{\partial X}\right|_{[i,j,k]} (X - X[i]) + \left.\frac{\partial V}{\partial Y}\right|_{[i,j,k]} (Y - Y[j]) + \left.\frac{\partial V}{\partial Z}\right|_{[i,j,k]} (Z - Z[k])$$

with the gradients,

$$\left.\frac{\partial V}{\partial X}\right|_{[i,j,k]} = \frac{V(X[i+1], Y[j], Z[k]) - V(X[i], Y[j], Z[k])}{X[i+1] - X[i]}$$

$$\left.\frac{\partial V}{\partial Y}\right|_{[i,j,k]} = \frac{V(X[i], Y[j+1], Z[k]) - V(X[i], Y[j], Z[k])}{Y[j+1] - Y[j]}$$

$$\left.\frac{\partial V}{\partial Z}\right|_{[i,j,k]} = \frac{V(X[i], Y[j], Z[k+1]) - V(X[i], Y[j], Z[k])}{Z[k+1] - Z[k]}.$$

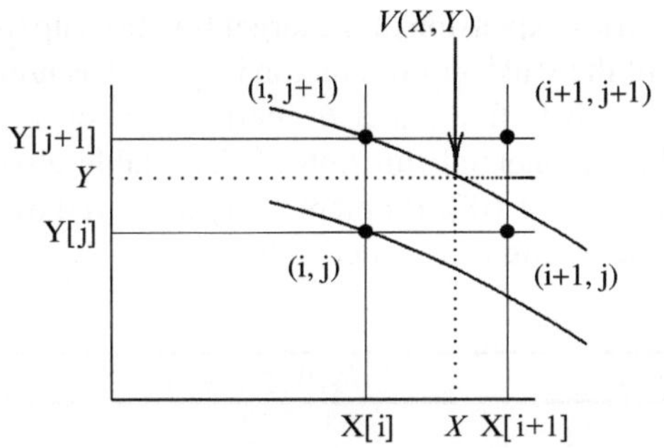

The comment regarding interpolation accuracy of the one-dimensional procedure applies equally to the two- and higher-dimensioned procedures. In this case we are talking about a field of values rather than a single value, and unless the field has been badly defined by the construction of the table, the Taylor series approach will deliver good results. Accuracy can be improved if needed by using higher-order terms in the Taylor expansion, but this should rarely be necessary.

3.11 Treatment of a Pure Time Delay

Transport delays occur in many areas of process plant simulation handling material and fluid flow. They are created by the time taken to physically move a material, and with it its properties and composition, over a distance. The delay is a dead time equal to the distance moved divided by the flow velocity.

An example is the relationship between inlet and outlet temperatures of a heat and mass transport process. The transport delay between these two can be represented by the frequency domain transfer function,

$$T_x = k \frac{\mathrm{e}^{-\tau s}}{1 + \tau_t s} T_i,$$

where k is the fractional temperature change through the process, τ is the flow transport time and τ_t is the temperature response time constant, each a function of the flow velocity ν.

A simple method is available which allows the effective treatment of transport delays of significant duration, with little computational complexity or storage requirements. The solution is provided by the Padé approximation to the exponential function.[3] The underlying principle of the Padé approximation is to match the coefficients of two polynomials, expressed as the ratio

$$\frac{\sum_{j=0}^{n} a_j x^j}{\sum_{k=0}^{m} b_k x^k}$$

[3]We first met the Padé approximation in Sect. 3.3.3 in connection with the matrix exponential.

to the known power series expansion of the target function, up to some given number of terms. The order of the Padé approximation is quoted as n/m.

The following table lists the a and b coefficients for several orders of Padé approximation to the exponential function e^x. The table also quotes the error to the true value of e for $x = 1$ ([8], Problem 12.1) delivered by each. The reduction in fitting error with increasing Padé order is dramatic.

$n = m$	$a_j, 0 \le j \le n$	$b_k, 0 \le k \le m$	Error
1	2 1	2 1	−0.28
2	12 6 1	12 −6 1	0.004
3	120 60 12 1	120 −60 12 −1	−2.8e−5
4	1,680 840 180 20 1	1,680 −840 180 −20 1	1.1e−7
5	30,240 15,120 3,360 420 30 1	30,240 −15,120 3,360 −420 30 −1	−2.8e−10

Selecting, say, $n = m = 2$ allows us to write,

$$e^x \approx \frac{12 + 6x + x^2}{12 - 6x + x^2}.$$

A few spot checks suffice to demonstrate that this yields an excellent approximation to the true value of e^x for $x < 1$ but loses accuracy for larger values of the argument. This is a feature of most power series-based approximation methods, even those which are formally convergent. Specifically, for the exponential function, this need not be a problem as for any factor κ

$$e^x = \left(e^{x/\kappa}\right)^\kappa.$$

The scaling factor κ is selected as the smallest integer such that $x/\kappa < 1$. If κ is also a power of 2 the true result can be recovered by a series of multiplications.

It follows that

$$e^{-x} \approx \frac{12 - 6x + x^2}{12 + 6x + x^2} \tag{3.44}$$

which does not suffer accuracy deterioration with increasing x as the denominator always increases more rapidly than the numerator. Scaling is then not required.

Applying this to the pure time delay, we replace x in Eq. 3.44 by τs to give

$$e^{-\tau s} \approx \frac{12 - 6\tau s + \tau^2 s^2}{12 + 6\tau s + \tau^2 s^2}.$$

If $\bar{u}(s)$ is the input to the time delay whose output is $\bar{y}(s)$,

$$\bar{y}(s) = e^{-\tau s}\,\bar{u}(s)$$

and we may write

$$F(s) = \frac{\bar{y}(s)}{\bar{u}(s)} = \frac{12 - 6\,\tau s + \tau^2 s^2}{12 + 6\,\tau s + \tau^2 s^2} \approx \mathrm{e}^{-\tau s}$$

or

$$\bar{y}(s) = \frac{12 - 6\,\tau s + \tau^2 s^2}{12 + 6\,\tau s + \tau^2 s^2}\,\bar{u}(s).$$

Selecting $n = m = 4$ provides a more accurate and stable Padé approximant to the time delay $\mathrm{e}^{-\tau s}$ for longer delay times τ and faster changing inputs $\bar{u}(s)$. Then,

$$\bar{y}(s) = \frac{1{,}680 - 840\tau\, s + 180\tau^2 s^2 - 20\tau^3 s^3 + \tau^4 s^4}{1{,}680 + 840\tau\, s + 180\tau^2 s^2 + 20\tau^3 s^3 + \tau^4 s^4}\,\bar{u}(s).$$

This can be rearranged as

$$s^4(\bar{y} - \bar{u}) = 1{,}680\tau^{-4}(\bar{y} - \bar{u}) - 840\tau^{-3}s(\bar{y} + \bar{u}) + 180\tau^{-2}s^2(\bar{y} - \bar{u}) \\ - 20\tau^{-3}s^3(\bar{y} + \bar{u}). \tag{3.45}$$

Introduce the new frequency domain variables $\bar{e}_1, \bar{e}_2, \bar{e}_3$, and $\bar{e}_4$. The following relationships between these four variables can be developed from Eq. 3.45.

$$\begin{aligned}
s\bar{e}_1 &= -0.168\,(\bar{y} + \bar{u}) \\
s\bar{e}_2 &= -10\tau^{-1}\bar{e}_1 - 0.84\,(\bar{y} + \bar{u}) \\
s\bar{e}_3 &= -10\tau^{-1}\bar{e}_2 - 1.80\,(\bar{y} - \bar{u}) \\
s\bar{e}_4 &= -10\tau^{-1}\bar{e}_3 - 2.0\,(\bar{y} + \bar{u}) \\
\bar{e}_4 &= \frac{\tau}{10}(\bar{y} - \bar{u})
\end{aligned}$$

The last of these relationships can be used to eliminate $\bar{y}$ from the other expressions to give

$$\begin{aligned}
s\bar{e}_1 &= 1.68\tau^{-1}\bar{e}_4 \\
s\bar{e}_2 &= -10\tau^{-1}\bar{e}_1 - 8.40\tau^{-1}\bar{e}_4 - 1.68\bar{u} \\
s\bar{e}_3 &= -10\tau^{-1}\bar{e}_2 + 18.0\tau^{-1}\bar{e}_4 \\
s\bar{e}_4 &= -10\tau^{-1}\bar{e}_3 - 20.0\tau^{-1}\bar{e}_4 - 4.00\bar{u}.
\end{aligned}$$

The operator s in the frequency domain corresponds to the time derivative in the time domain. This set of equations can then be rewritten in terms of the time domain variables e_1, e_2, e_3 and e_4 as the following differential equation system:

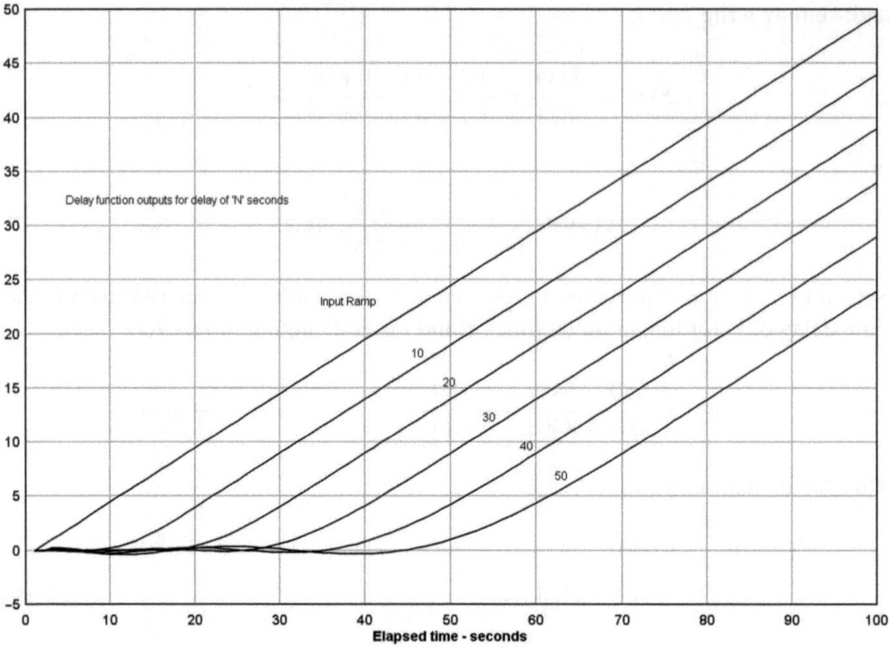

Fig. 3.1 Simulation of a pure time delay using a Padé approximant—constant ramp

$$\frac{d}{dt}\begin{bmatrix} e_1 \\ e_2 \\ e_3 \\ e_4 \end{bmatrix} = \begin{bmatrix} 0 & 0 & 0 & 1.68\tau^{-1} \\ -10\tau^{-1} & 0 & 0 & -8.40\tau^{-1} \\ 0 & -10\tau^{-1} & 0 & 18.00\tau^{-1} \\ 0 & 0 & -10\tau^{-1} & -20.00\tau^{-1} \end{bmatrix}\begin{bmatrix} e_1 \\ e_2 \\ e_3 \\ e_4 \end{bmatrix} + \begin{bmatrix} 0 \\ 1.68 \\ 0 \\ -4.00 \end{bmatrix} u. \tag{3.46}$$

The output y is given from

$$y = 10\tau^{-1}e_4 + u. \tag{3.47}$$

Figure 3.1 illustrates the performance of this system in simulating a range of pure time delays, from 10 to 50 s, for a constant ramp input.

It should be noted that this implementation of a pure transport delay does not respond well to step changes of input. The Padé approximant is structured here as the ratio of two polynomials of equal order. Considering Eq. 3.44, if we divide the numerator into the denominator, we obtain

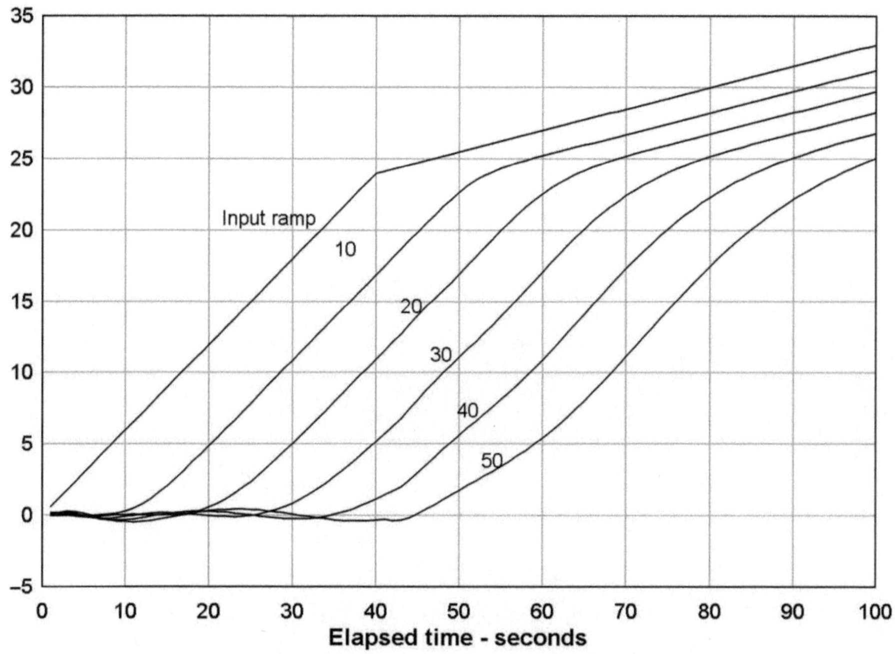

Fig. 3.2 Simulation of a pure time delay using a Padé approximant—variable ramp

$$F(s) = 1 - \frac{12\,\tau s}{12 + 6\,\tau s + \tau^2 s^2} = 1 - 12\tau \frac{s}{(s + 3/\tau)^2 + 3/\tau}.$$

The inverse Laplace transform of $F(s)$ is $f(t)$ where

$$f(t) = \delta(t) - 12\tau\, \mathrm{e}^{-(3/\tau)t} \left[\cos(\sqrt{3}/\tau)t - \sqrt{3}\,\sin(\sqrt{3}/\tau)t\right].$$

The presence of the impulse function $\delta(t)$ in the time function indicates that this rational function, with both polynomials of equal order, performs an implicit differentiation of the input. In practice, this turns out to be less of a problem than might at first appear, but slight smoothing of the input is recommended. As Fig. 3.2 shows, the delay function can handle abrupt changes in the *gradient* of the input without problem though the corner loses its sharpness as the time delay increases.

It should be noted that this approach to dead-time simulation applies only to systems with a fixed time delay τ. Should τ change at any time, the output variable will change since τ is used directly for its calculation via Eq. 3.47. It will regain its correct value but will show a transient error.

Chapter 4
Thermodynamic and Transport Properties of Materials

The principal interest of this book is the simulation of thermal plant. Fundamental to the understanding and description of thermal plant is the physics of thermodynamics. Enough has been written on thermodynamics to fill several libraries, and a reasonable prior knowledge of this subject can be assumed of the reader. However, to place this treatment within the existing framework of thermodynamics and to identify the areas of direct relevance, this chapter will provide a review of some selected areas of an otherwise huge subject.

4.1 Basic Concepts

The thermodynamic state of a substance is defined by the instantaneous values of a set of quantities which completely and uniquely define the state. At the most fundamental level these are the mass M of the substance, which derives from the number of molecules of the substance present, occupying a volume V, and the internal energy U. Internal energy derives from two molecular processes. The first is the level of molecular motion or more properly, agitation, since molecular motion is considered to be random and chaotic. Since the molecules have mass, this motion represents kinetic energy, and its contribution to the internal energy of the substance is the average kinetic energy of all molecules present. The second is a consequence of the arrangement of molecules within the substance, specifically their proximity to each other, and is a form of strain or potential energy.

In practice it is more convenient to analyse thermodynamic processes, not in terms of these fundamental quantities but rather in terms of a set of derived quantities. While mass remains a frequently used quantity, it is often more convenient to use specific mass or density, being the mass per unit volume, or alternatively its inverse, the specific volume, being the volume per unit mass. We introduce the concept of pressure p, being the force exerted by the molecules striking (or being repelled by) the wall containing the process. Thus pressure has dimensions of force per unit area.

P. O'Kelly, *Computer Simulation of Thermal Plant Operations*,
DOI 10.1007/978-1-4614-4256-1_4,

We introduce a new quantity enthalpy H, defined as $H = U + pV$. Since engineering thermodynamics concerns itself with change away from some defined datum this definition is more useful in incremental form as $\Delta H = \Delta U + p\,\Delta V$ or, in terms of specific quantities (per unit mass), $\Delta h = \Delta u + p\,\Delta v$. In differential form we have $\mathrm{d}h = \mathrm{d}u + \mathrm{d}(pv) = \mathrm{d}u + p\mathrm{d}v + v\mathrm{d}p$. Since $\mathrm{d}q = \mathrm{d}u + p\mathrm{d}v$, this may be written as $\mathrm{d}q = \mathrm{d}h - v\mathrm{d}p$ where $\mathrm{d}q$ is the infinitesimal change in total heat crossing the boundaries of the system.

An important thermodynamic quantity is temperature T and is a measure of the ability of the substance to exchange heat. Given that p, v and R are absolute quantities, the ideal gas equation of state or gas law $pv = RT$ serves to define the absolute temperature T,[1] used in degrees Kelvin.

Finally we introduce the concept of entropy. Although not used in any significant way within this book, entropy is a fundamental thermodynamic concept, important to the proper understanding of the relationship between heat and work. Entropy s is defined in incremental form by the Gibbs equation

$$T\,\mathrm{d}s = \mathrm{d}u + p\,\mathrm{d}v,$$

where $\mathrm{d}u$ is an increment in internal energy and $p\,\mathrm{d}v$ is the increment in compression energy. With $\mathrm{d}q = \mathrm{d}u + p\,\mathrm{d}v$ we can write

$$\mathrm{d}s = \frac{\mathrm{d}q}{T}.$$

Replacing $\mathrm{d}q$ by its equivalent $c_m\,\mathrm{d}T$, where c_m is some "specific heat" coefficient and depends on the process, we may write

$$s_2 - s_1 = c_m \int_{T_1}^{T_2} \frac{\mathrm{d}T}{T} = c_m \ln\left(\frac{T_2}{T_1}\right).$$

This defines the *change* in entropy over the complete thermodynamic cycle. The absolute value of entropy is defined by the third law of thermodynamics to be zero at zero absolute temperature (0 K). However, the absolute value of entropy is of little interest and may be defined to be zero at some other temperature, such as 0 °C, from which reference entropy changes may be calculated.

According to Avogadro's law, every gas at the same temperature and pressure contains the same number of molecules. The amount of mass [kg] in one *mole* of any gas is therefore $\overline{M}$ kg where $\overline{M}$ is the molecular weight of the gas. One mole of any gas occupies the same volume $\overline{V}$ at the same temperature and pressure. The ideal equation of state can then be written in terms of molar masses.

$$p\overline{V} = \overline{M}RT$$

[1]Denoted elsewhere in this book as $\mathcal{T}$.

Since for the same p and T, $\overline{V}$ is the same for all gases, the product $\overline{M}R$ will be the same for all gases and is known as the universal gas constant **R**, with the value 8.315 kJ/kmol/K. The gas constant R for any gas with molar mass $\overline{M}$ is therefore $\mathbf{R}/\overline{M}$.

The following examples are of interest to combustion systems:

For dry air the molecular weight is 28.9627 kg/kmol and $R = 0.2869$ kJ/kg/K.
For furnace flue gas of typical composition $R = 0.2810$ kJ/kg/K.
For moist air the gas constant is given by

$$R = 0.2869 + 0.1746 \frac{x_{H_2O}}{1 + x_{H_2O}} \text{ kJ/kg/K},$$

where x_{H_2O} is the mass fraction of moisture in the air.

For steam the molecular weight is 18.016 kg/kmol and $R = 0.4619$ kJ/kg/K.

Throughout this book, we will make consistent use of the SI system of units, with mass M in kg, volume V in m^3 and pressure p in Pascals Pa with units g/(m s^2). In place of the SI standard unit Pascal, we will always use the engineering unit kiloPascal (kPa = 1000 Pa) as this is more widespread in industry and better suited to the pressure ranges with which we will be working. We will occasionally use megaPascal (MPa = Pa $\cdot$ 10^6).[2] Pressure will always be quoted as kPa absolute unless identified as gauge pressure (pressure above ambient, kPag). We will generally use specific volume v m^3/kg in preference to density ρ kg/m^3, though there are exceptions. Specific enthalpy h is expressed as enthalpy per unit mass, with the units kJ/kg.

4.2 Thermodynamic Processes

A variety of relationships among the thermodynamic state variables have been defined, each making differing assumptions about the nature of the process for which the relationship holds true. We will briefly summarise the most important of these.

Isothermal Process An isothermal process is one for which the temperature of the substance at the end of the process is equal to that at the beginning. For an ideal gas,

$$p\,V = \texttt{constant}. \tag{4.1}$$

This is generally known as Boyle's law.

[2]Many formulae require pressure to be expressed in Pascals to be consistent.

Isobaric Process An isobaric process is one during which the pressure of the substance is held constant. For an ideal gas

$$\frac{V}{T} = \texttt{constant}. \tag{4.2}$$

This is generally known as Charles' law for constant pressure.

Isometric Process An isometric process is one during which the volume of the substance is held constant. For an ideal gas

$$pT = \texttt{constant}. \tag{4.3}$$

This is generally known as Charles' law for constant volume.

Adiabatic Process This is a process in which no heat crosses the boundary. It can be reversible or irreversible. A reversible process is one which, having no irreversible losses, arrives back at its starting condition having traversed a complete cycle.

For an adiabatic process,

$$p\,v^{\gamma} = \texttt{constant} \tag{4.4}$$

γ is the adiabatic index and is equal to the ratio of specific heats

$$\gamma = \frac{c_p}{c_v}, \tag{4.5}$$

where c_p and c_v are defined a little later. γ is not constant, being a function of the thermodynamic state, and is specific to each fluid. Since $c_p > c_v$, γ is always > 1.

Isentropic Process A reversible adiabatic process, so called because entropy is not changed (heat neither exchanged nor irreversibly lost). For an isentropic process, the gas law is

$$p\,v^{\gamma} = R\,T, \tag{4.6}$$

where R is the gas constant (specific to each gas and gas mixture).

Polytropic Process All of the processes mentioned above apply to an ideal gas. Although this is satisfactory for most engineering purposes, and the adiabatic assumption gives good results in many cases, real gases are not ideal over large ranges of pressure and/or temperature, and most processes include some degree of irreversible losses. For these, the *polytropic* process assumption can give improved match between theory and practice. The gas law for a reversible process becomes

$$p\,v^{n} = R\,T, \tag{4.7}$$

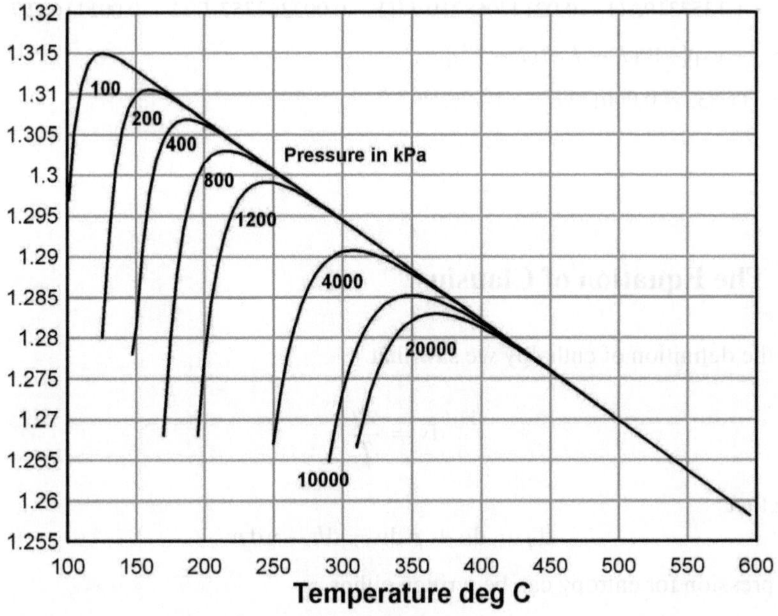

Fig. 4.1 Variation of the adiabatic index for dry steam

where n is the polytropic index. Defining η_p to be the polytropic efficiency (< 1), we have

$$\frac{n}{n-1} = \eta_p \frac{\gamma}{\gamma - 1} \text{ for compression, and} \tag{4.8}$$

$$= \frac{1}{\eta_p} \frac{\gamma}{\gamma - 1} \text{ for expansion.} \tag{4.9}$$

For an irreversible process we can say only that

$$p\, v^n = \texttt{constant}. \tag{4.10}$$

For a polytropic process, a polytropic specific heat capacity c_n can be defined by the heat transfer relationship $\dot{q} = \dot{m}\, c_n \Delta T$ where

$$c_n = c_v \frac{\gamma - n}{1 - n}.$$

The adiabatic index γ for superheated steam is plotted in Fig. 4.1 for pressures in the range 100–20,000 kPa and temperatures 100–600 °C. The values of γ depicted by this diagram have been calculated by the following fitting polynomials [61]. T' is the saturation temperature [°C] at pressure p kPa.

$$x = -1.13533198\,(1 - 0.02147068219\,T'(1 - 0.003367757\,T'(1 - 0.00111662\,T')))$$
$$y = 1 - \exp[-10(T - T')/(2 + T')]$$
$$\gamma = y(1.332 - 0.000123T - x) + x.$$

4.3 The Equation of Clausius

From the definition of enthalpy we saw that

$$\mathrm{d}s = \frac{\mathrm{d}q}{T}.$$

Given that

$$\mathrm{d}q = \mathrm{d}u + p\,\mathrm{d}v = \mathrm{d}h - v\,\mathrm{d}p$$

the expression for entropy can be written either as

$$\mathrm{d}s = \frac{\mathrm{d}h - v\,\mathrm{d}p}{T} \tag{4.11}$$

or

$$\mathrm{d}s = \frac{\mathrm{d}u + p\,\mathrm{d}v}{T}. \tag{4.12}$$

For single-phase (superheated) steam the specific enthalpy h can be expressed as a function of the two independent variables temperature T and pressure p or pressure p and specific volume v. For

$$v = v(T, p)$$

the total differential $\mathrm{d}h$ of enthalpy can be written as

$$\mathrm{d}h = \left.\frac{\partial h}{\partial T}\right|_{p=const} \mathrm{d}T + \left.\frac{\partial h}{\partial p}\right|_{T=const} \mathrm{d}p \tag{4.13}$$

and the entropy expression Eq. 4.11 becomes

$$\begin{aligned}\mathrm{d}s &= \frac{1}{T}\left.\frac{\partial h}{\partial T}\right|_{p} \mathrm{d}T + \frac{1}{T}\left.\frac{\partial h}{\partial p}\right|_{T} \mathrm{d}p - \frac{1}{T} v\,\mathrm{d}p \\ &= \frac{1}{T}\left.\frac{\partial h}{\partial T}\right|_{p} \mathrm{d}T + \frac{1}{T}\left(\left.\frac{\partial h}{\partial p}\right|_{T} - v\right) \mathrm{d}p. \end{aligned} \tag{4.14}$$

Since entropy is completely defined in terms of the two variables T and p we can write the total differential as

$$ds = \left.\frac{\partial s}{\partial T}\right|_p dT + \left.\frac{\partial s}{\partial p}\right|_T dp. \tag{4.15}$$

Equating coefficients of the differentials dT and dP in Eqs. 4.14 and 4.15 yields the following expressions for the partial derivative of s with respect to each of the independent variables.

$$\frac{\partial s}{\partial T} = \frac{1}{T}\left.\frac{\partial h}{\partial T}\right|_p \tag{4.16}$$

and

$$\frac{\partial s}{\partial p} = \frac{1}{T}\left(\left.\frac{\partial h}{\partial p}\right|_T - v\right). \tag{4.17}$$

From Eq. 4.16 we can define the specific heat capacity at constant pressure c_p in terms of the gradient of entropy.

$$c_p = \left.\frac{\partial h}{\partial T}\right|_p = T\left.\frac{\partial s}{\partial T}\right|_p \tag{4.18}$$

Since ds is the total differential of s it follows that

$$\frac{\partial}{\partial p}\left(\frac{\partial s}{\partial T}\right) = \frac{\partial}{\partial T}\left(\frac{\partial s}{\partial p}\right)$$

and therefore, from Eq. 4.16, since $\frac{\partial}{\partial p}\left(\frac{\partial s}{\partial T}\right) = 0$,

$$0 = \frac{1}{T^2}\left(\left.\frac{\partial h}{\partial p}\right|_T - v\right)\frac{1}{T}\frac{\partial v}{\partial T}$$

or

$$\left.\frac{\partial h}{\partial p}\right|_T = v - T\left.\frac{\partial v}{\partial T}\right|_p \tag{4.19}$$

This is the Clausius equation. It relates changes in thermodynamic states of a material to measurable changes in pressure and temperature.

Combining Eqs. 4.13 and 4.19 with the definition of c_p from Eq. 4.18 gives

$$dh = c_p dT + \left(v - T\left.\frac{\partial v}{\partial T}\right|_p\right) dp \tag{4.20}$$

and

$$ds = \frac{c_p}{T} dT - \left.\frac{\partial v}{\partial T}\right|_p dp \tag{4.21}$$

Had we started from Eq. 4.12, we would have arrived at

$$\left.\frac{\partial u}{\partial v}\right|_T = T\left.\frac{\partial p}{\partial T}\right|_v - p, \tag{4.22}$$

$$du = c_v dT + \left(T\left.\frac{\partial p}{\partial T}\right|_v - p\right) dv, \tag{4.23}$$

$$c_v = \left.\frac{\partial u}{\partial T}\right|_v = T\left.\frac{\partial s}{\partial T}\right|_v \tag{4.24}$$

and

$$ds = \frac{c_v}{T} dT - \left.\frac{\partial p}{\partial T}\right|_v dv. \tag{4.25}$$

4.4 Specific Heat Capacities

The specific heat capacity of a substance is defined as the ratio of its heat content to its temperature. The reference temperature is that corresponding to the definition of the heat content. In the preceding section, we introduced the two specific heat capacities

- For a constant volume process: c_v defined as $du = c_v\, dT$
- For a constant pressure process: c_p defined as $dh = c_p\, dT$

Equations 4.21 and 4.25 defined the increment ds in entropy in terms of changes in (T, p) and (T, v). Since in both cases the ds increments are the same, we may equate the two equations to give

$$\frac{c_p - c_v}{T} dT = \left.\frac{\partial v}{\partial T}\right|_p dp + \left.\frac{\partial p}{\partial T}\right|_v dv$$

Again, since T is a function of only the two states (p, v), the total differential dT may be written as

$$dT = \left.\frac{\partial T}{\partial p}\right|_v dp + \left.\frac{\partial T}{\partial v}\right|_p dv.$$

Equating like coefficients gives

$$c_p - c_v = T\left.\frac{\partial v}{\partial T}\right|_p \left.\frac{\partial p}{\partial T}\right|_v \tag{4.26}$$

The various partial gradients in these relationships can be determined experimentally as functions of measured pressures and temperatures.[3]

It was noted earlier that the ratio of the specific heat capacities is γ. This relationship can now be generalised to

$$\gamma = \frac{c_p}{c_v} = \left.\frac{\partial v}{\partial p}\right|_T \left.\frac{\partial p}{\partial v}\right|_s. \tag{4.27}$$

4.5 Property Relationships for a Perfect Gas

From the definition of enthalpy we have

$$\mathrm{d}h = du + d(pv).$$

For a perfect gas, $pv = RT$, and we have

$$d(pv) = R\mathrm{d}T.$$

For a constant volume process we have

$$du = c_v \mathrm{d}T.$$

Combining these gives

$$\mathrm{d}h = c_p \mathrm{d}T = (c_v + R)\mathrm{d}T$$

from which it follows that

$$c_p - c_v = R \tag{4.28}$$

that is, the quantitative difference between the two specific heats is numerically equal to the gas constant of the substance.

Combining this relationship with that of Eq. 4.27 gives the result

$$c_p = \left(\frac{\gamma}{\gamma - 1}\right) R. \tag{4.29}$$

[3] A physical reasoning process can be followed to show that c_p is always greater than c_v. From $\mathrm{d}q = \mathrm{d}u + p\mathrm{d}v$ we see that the addition of heat to a constant volume process will increase only the internal energy ($\mathrm{d}v = 0$), but the addition of heat to a constant pressure process will increase both internal energy and do work by causing a displacement of the boundaries of the process in order to hold pressure constant ($\mathrm{d}h = \mathrm{d}u + p\mathrm{d}v + v\mathrm{d}p$ with $\mathrm{d}p = 0$). For a given temperature change $\mathrm{d}T$, $\mathrm{d}h$ for the constant pressure process will exceed $\mathrm{d}u$ for the constant volume process by $p\mathrm{d}v$ and hence $c_p > c_v$.

This relates the specific heat capacity of a substance to its gas constant R and the adiabatic index γ. Knowing any two of these we can determine the third. For example, the specific heat capacity of dry air at normal atmospheric pressure and 20 °C is 1.012 kJ/(kg K). Given $R = 0.2869$ kJ/(kg K) for dry air the adiabatic index for air under these same conditions is calculated to be 1.39.

From Eq. 4.20, we can say that

$$c_p = \frac{\mathrm{d}h}{\mathrm{d}T}. \tag{4.30}$$

only if $\mathrm{d}p = 0$ (constant pressure process) or

$$\left.\frac{\partial v}{\partial T}\right|_p = \frac{v}{T}.$$

For a perfect gas

$$\left.\frac{\partial v}{\partial T}\right|_p = \frac{R}{p} = \frac{v}{T}$$

and thus, for a perfect gas, Eq. 4.30 is true. It will not be true for gases which deviate from the perfect gas law though it will be an adequate approximation in most cases.

Integration of

$$\mathrm{d}h = c_p\,\mathrm{d}T$$

between two points 1 and 2 yields

$$h_2 - h_1 = c_p(T_2)T_2 - c_p(T_1)T_1$$

or

$$h_2 = c_p(T_2)T_2 + (h_1 - c_p(T_1)T_1).$$

If temperature T_1 is the reference temperature used to define $h = 0$, then it follows that

$$h = c_p T \tag{4.31}$$

that is, the enthalpy of a substance (liquid or gas) can be related to its temperature via its specific heat capacity c_p. For liquids and gases, enthalpy is conventionally defined to be zero at 0 °C, and Eq. 4.31 is valid for those substances. The enthalpy of steam is not so defined, and Eq. 4.31 is not valid for steam, saturated or superheated. The specific heat capacity for these substances must be defined in terms of the enthalpy content of the steam (or vapour) phase above the liquid (or water) phase. For superheated steam with enthalpy h_s kJ/kg at temperature T_s°C, the average c_{pm} for the range of temperatures between saturation and the given temperature can be calculated from

$$c_{pm} = \frac{h_s - h''}{T_s - T_{sat}}. \tag{4.32}$$

Note that this does not equal the specific heat capacity $c_p(h_s, T_s)$ for the given steam conditions. This must be calculated from an accurate formula or extracted from a table.

For axial flow in the direction of an ordinate z, differentiating Eq. 4.4 with respect to the ordinate z yields

$$\gamma\, p\, v^{\gamma-1} \frac{\partial v}{\partial z} + v^{\gamma} \frac{\partial p}{\partial z} = R \frac{\partial T}{\partial z}.$$

For a gas for which we can write

$$h = c_p T$$

this can be written as

$$\gamma\, p\, v^{\gamma-1} \frac{\partial v}{\partial z} + v^{\gamma} \frac{\partial p}{\partial z} = \frac{R}{c_p} \frac{\partial h}{\partial z}.$$

Then,

$$\frac{\partial v}{\partial z} = -\left(\frac{v}{\gamma p}\right) \frac{\partial p}{\partial z} + \left(\frac{R}{c_p} \frac{v^{1-\gamma}}{\gamma\, p}\right) \frac{\partial h}{\partial z}. \tag{4.33}$$

Comparing this with

$$\frac{\partial v}{\partial z} = \frac{\partial v}{\partial p} \frac{\partial p}{\partial z} + \frac{\partial v}{\partial h} \frac{\partial h}{\partial z}.$$

we can equate

$$\frac{\partial v}{\partial p} = -\frac{v}{\gamma p} \tag{4.34}$$

and

$$\frac{\partial v}{\partial h} = \frac{R\, v^{1-\gamma}}{c_p\, \gamma\, p}$$

$$= \frac{v}{c_p \gamma\, T}. \tag{4.35}$$

4.5.1 Extension to Liquids

For a liquid, $\partial v / \partial p$ is a measure of volumetric compressibility. For a liquid treated as incompressible $\partial v / \partial p = 0$, otherwise $\partial v / \partial p = K_v$, the index of volumetric compressibility. Also, for a liquid

$$\frac{\partial v}{\partial h} = \frac{1}{c_p} \frac{\partial v}{\partial T}$$

where $\partial v/\partial T$ is the index of volumetric thermal expansion [m^3K^{-1}]. For small changes in temperature, this can usually be ignored. For larger temperature changes, such as encountered during the start-up phase of a thermal plant, volumetric expansion can be significant. For purposes of calculation this can be accommodated by updating the specific volume of the liquid phase in response to changes in fluid temperature.

4.6 Thermodynamic Properties of Steam

Water can exist in any of three states, solid (ice), liquid (water) or gas (vapour and steam). The solid state will not interest us here, nor will the process of conversion from the solid to the liquid phase or back again. However, the liquid and gaseous phases are of great interest, as is the process of conversion from one to the other, in both directions. By way of introduction to the subject, consider Gibbs' phase rule which states

$$F = C - P + 2,$$

where C is the number of components present in a mixture, P is the number of phases which can exist together in equilibrium and F is the number of "degrees of freedom" allowed. In the thermodynamic sense F is the number of independent variables which define the thermodynamic state. In the case of *pure* H_2O we have a single component ($C = 1$) and can identify:

State	Possible phases present	Value of F
Solid	1 (ice)	2
Sub-cooled liquid	1 (water)	2
Saturated liquid	2 (saturated water and vapour)	1
Saturated vapour	2 (saturated water and vapour)	1
Superheated steam	1 (gas)	2

At and beyond the critical point (221.29 bar or 22.417 MPa) the liquid and vapour states are no longer distinguishable as separate phases and $F = 2$.

It follows that, for any state of H_2O with the exception of the saturated state, two thermodynamic variables are required to completely identify the thermodynamic state, but only a single variable is sufficient to define the saturated state. In the saturated state both the liquid and vapour phases can exist in equilibrium.

For **wet saturated steam** the thermodynamic state is a function of the two variables temperature T and steam mass fraction x, which can take values in the range (0,1). The mass fraction x of steam present in wet saturated steam is defined as

$$x = \frac{M_v}{M_w + M_v} = \frac{M_v}{M_s},$$

where M_w and M_v are the masses of liquid and vapour, respectively, within the mass of steam M_s. The mixture properties v_s and h_s can be defined in terms of the saturation properties.

$$v_s = x\,v'' + (1-x)v',$$
$$h_s = x\,h'' + (1-x)h',$$

where a single prime $'$ indicates the saturation condition for the liquid phase and the double prime $''$ indicates that of the vapour phase.

For **dry saturated steam** the mass fraction $x = 1$ and all properties may be expressed as a function of a single variable, most conveniently pressure.

$$v'' = v(p), \;\; h'' = h(p), \;\; T'' = T(p).$$

4.6.1 Partial Derivatives of Wet Saturated Steam Properties

From the definition of steam mass quality x

$$x = \frac{h-h'}{h''-h'} = \frac{v-v'}{v''-v'}$$

we may write

$$h = h' + \left(\frac{h''-h'}{v''-v'}\right)(v-v')$$

and

$$v = v' + \left(\frac{v''-v'}{h''-h'}\right)(h-h').$$

This last equation leads to

$$\left.\frac{\partial v}{\partial h}\right|_{p=const} = \frac{v''-v'}{h''-h'} \tag{4.36}$$

for saturated vapour.

Differentiating the same equation with respect to p gives

$$\left.\frac{\partial v}{\partial p}\right|_h = \frac{\mathrm{d}v'}{\mathrm{d}p} - (h-h')\frac{\mathrm{d}}{\mathrm{d}p}\left(\frac{v''-v'}{h''-h'}\right) + \left(\frac{v''-v'}{h''-h'}\right)\frac{\mathrm{d}}{\mathrm{d}p}(h-h')$$

which reduces to

$$\left.\frac{\partial v}{\partial p}\right|_h = h\frac{\mathrm{d}}{\mathrm{d}p}\left(\frac{v''-v'}{h''-h'}\right) + \frac{\mathrm{d}}{\mathrm{d}p}\left(\frac{v'h''-h'v''}{h''-h'}\right).$$

If we neglect $\mathrm{d}v'/\mathrm{d}p$ as being negligibly small (true for most purposes) this becomes

$$\left.\frac{\partial v}{\partial p}\right|_h = h\frac{\mathrm{d}}{\mathrm{d}p}\left(\frac{v''-v'}{h''-h'}\right) - \frac{\mathrm{d}}{\mathrm{d}p}\left(h'\frac{v''-v'}{h''-h'}\right)$$

which, after further manipulation, leads to

$$\begin{aligned}
\left.\frac{\partial v}{\partial p}\right|_h &= x\left[\frac{\mathrm{d}v''}{\mathrm{d}p} - \left(\frac{v''-v'}{h''-h'}\right)\frac{\mathrm{d}}{\mathrm{d}p}(h''-h')\right] - \left(\frac{v''-v'}{h''-h'}\right)\frac{\mathrm{d}h'}{\mathrm{d}p} \\
&= x\frac{\mathrm{d}v''}{\mathrm{d}p} - x\left(\frac{v''-v'}{h''-h'}\right)\frac{\mathrm{d}h''}{\mathrm{d}p} + x\left(\frac{v''-v'}{h''-h'}\right)\frac{\mathrm{d}h'}{\mathrm{d}p} - \left(\frac{v''-v'}{h''-h'}\right)\frac{\mathrm{d}h'}{\mathrm{d}p} \\
&= x\frac{\mathrm{d}v''}{\mathrm{d}p} - \left[x\left(\frac{v''-v'}{h''-h'}\right)\frac{\mathrm{d}h''}{\mathrm{d}p} + (1-x)\left(\frac{v''-v'}{h''-h'}\right)\frac{\mathrm{d}h'}{\mathrm{d}p}\right] \\
&= x\frac{\mathrm{d}v''}{\mathrm{d}p} - \left(\frac{v''-v'}{h''-h'}\right)\frac{\mathrm{d}h}{\mathrm{d}p} \\
&= x\frac{\mathrm{d}v''}{\mathrm{d}p} - \left.\frac{\partial v}{\partial h}\right|_p\frac{\mathrm{d}h}{\mathrm{d}p}
\end{aligned}$$

which, for most purposes, can be simplified to

$$\left.\frac{\partial v}{\partial p}\right|_h = x\frac{\mathrm{d}v''}{\mathrm{d}p}. \tag{4.37}$$

4.6.2 The Equation of Clapeyron

For wet steam we may write

$$\begin{aligned}
h(T) &= (1-x)\,h'(T) + x\,h''(T) \\
&= h'(T) - x\,h'(T) + x\,h''(T) \\
&= h'(T) + x\,(h''-h') \\
&= h'(T) + x\,\mathbf{r}(T),
\end{aligned}$$

where $\mathbf{r} = h'' - h'$ and is the latent heat of evaporation. As before, the single prime indicates the saturation condition for the liquid phase while the double prime indicates the vapour phase. Similarly,

$$\begin{aligned}
v &= (1-x)\,v'(T) + x\,v''(T) \\
&= v' + x\,(v''-v')
\end{aligned}$$

with the tacit assumption that both v' and v'' are single-valued functions of T.

Proceeding as for the single-phase case, we can write the entropy differential as

$$\mathrm{d}s = \frac{1}{T}\left[\frac{\mathrm{d}h'}{\mathrm{d}T} + x\frac{\mathrm{d}\mathbf{r}}{\mathrm{d}T} - (v' + x\,(v'' - v'))\,\frac{\mathrm{d}p}{\mathrm{d}T}\right]\mathrm{d}T + \frac{\mathbf{r}}{T}\mathrm{d}x.$$

Equating the coefficients of $\mathrm{d}T$ and $\mathrm{d}x$ in this and the following entropy equation,

$$\mathrm{d}s = \left.\frac{\partial s}{\partial T}\right|_x \mathrm{d}T + \left.\frac{\partial s}{\partial x}\right|_T \mathrm{d}x$$

we can write

$$\left.\frac{\partial s}{\partial T}\right|_x = \frac{1}{T}\left(\frac{\mathrm{d}h'}{\mathrm{d}T} + x\frac{\mathrm{d}\mathbf{r}}{\mathrm{d}T}\right) - \left(\frac{v' + x\,(v'' - v')}{T}\right)\frac{\mathrm{d}p}{\mathrm{d}T}$$

and

$$\left.\frac{\partial s}{\partial x}\right|_T = \frac{\mathbf{r}}{T}.$$

Noting again that the total differential $\mathrm{d}s$ allows us to set

$$\frac{\partial}{\partial x}\left(\frac{\partial s}{\partial T}\right) = \frac{\partial}{\partial T}\left(\frac{\partial s}{\partial x}\right)$$

we can write

$$\frac{\partial}{\partial x}\left(\frac{\partial s}{\partial T}\right) = \frac{1}{T}\frac{\mathrm{d}\mathbf{r}}{\mathrm{d}T} - (v'' - v')\frac{\mathrm{d}p}{\mathrm{d}T}\frac{1}{T},$$

$$\frac{\partial}{\partial T}\left(\frac{\partial s}{\partial x}\right) = -\frac{\mathbf{r}}{T^2} + \frac{1}{T}\frac{\mathrm{d}\mathbf{r}}{\mathrm{d}T}.$$

It follows that

$$\mathbf{r} = (v'' - v')\,T\,\frac{\mathrm{d}p}{\mathrm{d}T} \tag{4.38}$$

This is the equation of Clapeyron. It relates the heat of evaporation to the saturation temperature and the gradient of the local saturation temperature/saturation pressure line. It can also be written as

$$\frac{\mathrm{d}p}{\mathrm{d}T} = \frac{1}{T}\frac{h'' - h'}{v'' - v'}. \tag{4.39}$$

4.6.3 *Calculation of Property Values*

Access to a high-quality set of property calculations or tabulations is essential to simulation of any water–steam systems, or indeed of any other substance. Some models calculate the most fundamental state variables of internal energy and specific volume (or density) and derive everything else from these two. A common approach—and the one adopted throughout this work—is to use pressure and specific enthalpy as the independent variables. All other properties—sub-cooled water, saturated and superheated vapour phases—are derived from these two. Mechanical and transport properties—thermal conductivity, absolute viscosity and surface tension—used for the calculation of the key Reynolds, Prandtl and Nusselt numbers are derived from temperatures and/or pressures.

Published tabulations of all thermodynamic and transport properties of interest are available. In most cases table lookup is the best way to access dependent properties, given the model-calculated independent variables. While formulae are available by which accurate values of all properties of interest may be calculated, many formulae have been derived for the production of those very tables or for use in offline calculations requiring high accuracy, such as contractual thermal performance determination, where material property calculations must meet international Standards. One such example is the water–steam property formulae published by the International Association for the Properties of Water and Steam (IAPWS-97). While library packages implementing these formulae are now commercially available as plug-ins for some widely used calculation packages, it is the author's view that they are too complex and too computationally intensive to be seen as a practical alternative for real-time simulation applications. Water and steam properties can best be accessed via table lookup.

Saturation Temperature of Steam

Accurate knowledge of the saturation temperature is important for simulation of steam systems and can readily be obtained from a table. However, for some purposes, a formula might be appropriate for its calculation. One such is quoted by Riemenschneider [86].

$$T_{sat} = \sum_{i=0}^{i=11} a_i \ln(p_r)^i, \qquad (4.40)$$

where $p_r = p/99.34$ with pressure p expressed in kPa. Accuracy is quoted to be $|\epsilon| < 0.0034\,\%$. The coefficients a_i are listed in the following table.

i	a_i	i	a_i
0	99.09271199	6	−3.7393484250e−4
1	27.85424215	7	−1.7417751900e−5
2	2.375357647	8	2.2071711790e−5
3	0.2107780463	9	1.5343731340e−6
4	2.1296820111e−2	10	−4.2685685100e−7
5	1.3283772900e−3	11	−4.2924602910e−8

Absolute Viscosity

Fluid viscosity is a significant physical property of gases and fluids. Its importance stems not only from its role in characterising fluid flow properties but also from its appearance in several of the most significant dimensionless similitude numbers of fluid mechanics (the Reynolds, Prandtl and Grashof numbers).

Viscosity is commonly used in either of two forms. The absolute or dynamic viscosity is a physical property defined in terms of the internal shear forces acting within a fluid or gas. Thus, if a force F [Newtons] is applied to a layer of a fluid of area A [m^2] at a distance d [m] from a surface held at zero velocity, the force will impart a lateral laminar velocity ν [m/s] to the fluid layer. The absolute viscosity μ is defined as

$$\mu = \frac{F\,d}{\nu\,A}.$$

The dimensions of μ are [N m^{-2} s] and the units are Pascal.seconds, Pa s. This unit is extremely large and the practical unit for fluids is Pa s e-3 (mPa s) and Pa s e-6 (μPa s) for gases. Absolute viscosity is frequently expressed as *poise* or *centipoise*, where 1 *centipoise* = 1 mPa s.

Wagner and Kurse (1998) [10] have published formulations from which water viscosity can be calculated, given temperature and pressure.

$$\mu = \mu^* \exp\left[\delta \sum_{i=0}^{i=19} n_i(\delta - 1)^{Ii}(\tau - 1)^{Ji}\right], \tag{4.41}$$

where $\delta = \rho/\rho^*$ and $\tau = T^*/T$. $\rho^* = 317.763\,\text{kg/m}^3$ and $T^* = 647.226\,\text{K}$. $\mu^* = 5.07\text{e}{-6}$ Pa s. The coefficients n^i are

i	Ii	Ji	n^i	i	Ii	Ji	n^i
1	0	0	0.5132047	11	2	2	−1.263184
2	0	1	0.3205656	12	2	0	0.1778064
3	0	4	−0.7782567	13	3	1	0.4605040
4	0	5	0.1885447	14	3	2	0.2340379
5	1	0	0.2151778	15	3	3	−0.4924179
6	1	1	0.7317883	16	4	0	−0.0417661
7	1	2	1.241044	17	4	3	0.1600435
8	1	3	1.476783	18	5	1	−0.01578386
9	2	0	−0.2818107	19	6	3	−0.0036295
10	2	1	−1.070786				

Scheffler et al. [11] quotes a similar but different formulation.

$$\mu = \mu^* \exp\left[\delta \sum_{i=0}^{5}\sum_{j=0}^{4} b_{ij}(\delta - 1)^i(\tau - 1)^j\right]. \tag{4.42}$$

In this case

$$\mu^* = (T/T^*)^{0.5}\left[\sum_{0}^{3} a_k (T^*/T)^k\right]^{-1} \cdot 10^{-6} \quad \text{kg/m}^3$$

The constants T^* and v^* are approximately equal to the corresponding critical values: T^* = 647.27 K and v^* = 3.147e−3. The coefficients a_{ij} are:

k	0	1	2	3
a_k	0.0181583	0.0177624	0.0105287	−0.0036744

and b_{ij} are listed in the next table.

i→ j↓	0	1	2	3	4	5
0	0.501938	0.162888	−0.130356	0.907919	−0.551119	0.146543
1	0.235622	0.789393	0.673665	1.207552	0.0670665	−0.084337
2	−0.274637	−0.743539	−0.959456	−0.687343	−0.497089	0.195286
3	0.145831	0.263129	0.347247	0.213486	0.100754	−0.032932
4	−0.0270448	−0.0253093	−0.0267758	−0.0822904	0.0602253	−0.0202595

Kinematic Viscosity

A second definition of viscosity is *kinematic* viscosity. This arises as a consequence of methods used to measure viscosity in a laboratory apparatus. A widely used method notes the time taken by a given volume of the fluid to pass under gravity through a calibrated orifice. Clearly, a more dense fluid (heavier) will be subject to a greater gravitational force than one less dense. Therefore, the flow times of two fluids of equal *absolute* viscosity will be different, the heavier fluid taking less time than the lighter.

The kinematic viscosity, which is measured on the basis of this flow time, is related to the absolute viscosity via the fluid density.

kinematic viscosity = absolute viscosity/density.

The dimensions of kinematic viscosity are $[m^2\,s^{-1}]$. If the length unit is taken as centimetre, the unit of kinematic viscosity is the *stoke*, with dimensions $[cm^2\,s^{-1}]$.

Note that in most applications of formulae involving viscosity, absolute or dynamic viscosity is intended. When applying these and other formulae, care must be exercised to ensure that the correct form and units of dynamic viscosity are being used.

4.6.4 Some Useful Approximations for Property Calculations

For many applications, such as control system design and implementation, a simple approximation can be very useful for calculation of the saturation temperature of steam.

$$T_{sat} = k_T\, p_{sat}^{0.25}. \tag{4.43}$$

p_{sat} [kPa]	k_T
100	31.33
500	32.01
1,000	31.98
5,000	31.28
10,000	30.99
16,000	30.65

The constant k_T varies slightly with pressure, as listed by the preceding table. Variation in k_T can be approximated by a polynomial function of $\bar{p} = \log_{10}(p)$, with p expressed in kPa.

$$k_T = 17.02 + 13.57\,\bar{p} - 3.893\,\bar{p}^2 + 0.3434\,\bar{p}^3$$

with errors $\pm$ 0.8 °C.

This is valid for positive (> atmospheric) but is not valid in the vacuum range. For that range, the following expression provides a good approximation to T_{sat}:

$$k_T = 7.27055 + 25.003\bar{p} - 6.57\bar{p}^2.$$

Other thermal and transport properties such as saturation conditions, thermal conductivity and dynamic viscosity can be calculated using simple approximating, frequently polynomial, functions. It is usually straightforward to construct such functions covering a limited range of validity (accuracy) but not always easy to construct *simple* approximating functions valid over large ranges of their argument variables. For example, the saturation enthalpy of steam h'' increases with increasing pressure for pressures less than approximately 30 bar and decreases for pressures above this value. h'' can be calculated from two polynomials, one valid for each range, with the argument pressure p expressed in bars.

For $p < 30$ bar,

$$h'' \approx 2654 + 27.67\,p - 2.229\,p^2 + 0.08164\,p^3 - 0.001086\,p^4$$

with errors less than 15 kJ/kg, and for $p > 30$ bar,

$$h'' \approx 2794 + 0.5946\,p - 0.0122\,p^2$$

with errors less than 5 kJ/kg. By comparison the saturation enthalpy for the liquid phase h' can be calculated over the complete pressure range from 10 kPa to the critical pressure from, again with the argument pressure p in bars,

$$442.38 + 30.5027\,p - 0.4283\,p^2 + 0.002783\,p^3 - 6.15776 \cdot 10^{-6}\,p^4$$

with errors <1 %.

Chapter 5
Conservation Equations for Flow Systems

The mathematical basis for the physical modelling of mass and thermal flow systems is the set of three conservation equations of mass, energy and momentum. These equations are coupled through their common process variables and working fluid properties. Since most physical processes have both time and spatial dependencies, the conservation equations describing their unsteady behaviour are partial differential equations in both time and the three space coordinates. In this work attention will be restricted, in the main, to the axial coordinate only, that is, spatial effects will be considered only in the direction of flow.

5.1 The General Form of the Conservation Equations

The general form of the "advection" equation in a single spatial dimension is

$$\frac{\partial \phi}{\partial t} + \nu \frac{\partial \phi}{\partial z} = f(\phi, \ldots),$$

where ν is the flow velocity and ϕ is the advected fluid property.

Limiting our treatment to flow in one dimension along an axis parallel to the direction of flow, the conservation equations may be written:

Conservation of mass:

$$\frac{\partial \rho}{\partial t} = -\frac{\partial (\rho \nu)}{\partial z}. \tag{5.1}$$

Conservation of energy:

$$\frac{1}{A}\frac{\mathrm{d}\dot{q}}{\mathrm{d}z} = -\nu \left.\frac{\partial p}{\partial z}\right|_{loss} + \rho\left(\frac{\partial h}{\partial t} + \nu\frac{\partial h}{\partial z}\right) - \left(\frac{\partial p}{\partial t} + \nu\frac{\partial p}{\partial z}\right). \tag{5.2}$$

P. O'Kelly, *Computer Simulation of Thermal Plant Operations*,
DOI 10.1007/978-1-4614-4256-1_5,

Conservation of momentum:

$$\frac{\partial p}{\partial z} = - \left.\frac{\partial p}{\partial z}\right|_{loss} - \rho\left(\frac{\partial v}{\partial t} + v\frac{\partial v}{\partial z}\right) - g\rho \sin\beta, \tag{5.3}$$

where $\begin{cases} p & \text{is pressure [kg/(s m}^2)], \\ \rho & \text{is density [kg/m}^3], \\ h & \text{is specific enthalpy [kJ/kg]}, \\ \dot{q} & \text{is heat flow [kJ/s]}, \\ A & \text{is the flow cross-sectional area [m}^2], \\ \beta & \text{is the angle of inclination of the flow to the horizontal.} \end{cases}$

The term $\left.\frac{\partial p}{\partial z}\right|_{loss}$ represents the irreversible pressure loss due to energy dissipation. Energy is lost through shear forces acting within the fluid bulk and at the interface between the flow and its containing wall. The lumped effect of these forces can be expressed as

$$\left.\frac{\partial p}{\partial z}\right|_{loss} = \frac{1}{2}\zeta\rho v^2, \tag{5.4}$$

where ζ is the local friction loss coefficient equal to ξ/d_H and d_H is the hydraulic diameter [m]. ξ is the friction factor and is discussed in more detail in Sect. 6.2. Noting that the flow velocity $v = (\dot{m}\, v)/A = \phi\, v$ this can be written as

$$\left.\frac{\partial p}{\partial z}\right|_{loss} = \frac{1}{2}\frac{\xi v}{d_H}\phi^2, \tag{5.5}$$

where $\phi = \dot{m}/A$ is the specific mass flow or mass flow density.

The mass conservation equation can be restated in terms of ϕ and density as

$$\frac{\partial \rho}{\partial t} = -\frac{\partial \phi}{\partial z}$$

or, using specific volume v, as

$$\frac{1}{v^2}\frac{\partial v}{\partial t} = \frac{\partial \phi}{\partial z}.$$

For a path of constant cross-sectional area,

$$\frac{1}{v^2}\frac{\partial v}{\partial t} = \frac{1}{A}\frac{\partial \dot{m}}{\partial z}. \tag{5.6}$$

The energy conservation equation can be restated as

$$\frac{1}{A}\frac{\mathrm{d}\dot{q}}{\mathrm{d}z} = \frac{1}{v}\left(\frac{\partial h}{\partial t} + v\frac{\partial h}{\partial z}\right) - \frac{\partial p}{\partial t} + \phi\left(\frac{\partial v}{\partial t} + v\frac{\partial v}{\partial z} + g\,\sin\beta\right) \tag{5.7}$$

after substitution for $\partial p/\partial z$ from the momentum equation.

Equations 5.6, 5.7 and 5.3 are the basic conservation equations which describe the time-varying spatial distribution of a compressible flow through a cylindrical pipe or rectangular pipe or duct inclined at an angle β to the horizontal.

The assumption that fluid properties and thermodynamic states vary only in the direction of flow is an approximation and implies uniformity or homogeneity in all other directions. It will not be necessary to assume homogeneity along the direction of flow and later, when looking at spatial discretisation of these equations, it will be expedient to assume a linear variation of fluid properties along the path through a cell.[1]

These three partial differential equations interrelate the four state variables

$$\begin{cases} p \text{ pressure } [\mathrm{kg/(m\,s^2)}] \\ \phi \text{ specific mass flow } [\mathrm{kg/(s\,m^2)}] \\ v \text{ specific volume } [\mathrm{m^3/kg}] \\ h \text{ specific enthalpy } [\mathrm{kJ/kg}] \end{cases}$$

and include the boundary heat flow $\dot{q}$ [kJ/s].

They are supplemented by the fluid properties equations

$$F_v(p,h,v) = 0$$

and

$$F_T(p,h,T) = 0$$

and by the heat transfer equation

$$\frac{\mathrm{d}\dot{q}}{\mathrm{d}z} = \alpha_{hx}A_{hx}(T - T_{ref}), \tag{5.8}$$

where T is the temperature of the working fluid (°C). α_{hx} and A_{hx} are the heat transfer coefficient and heat transfer area across which heat is exchanged between the fluid and its surroundings, at temperature T_{ref}.

Each of the conservation equations is of first order. The equations are non-linear as some coefficients are functions of the dependent variables, if only through fluid or material properties. The equations are continuous and have continuous derivatives.

[1]Fundamental to this and all derivations presented in this book is the assumption that all pipes are filled with the working medium. This excludes open channels and horizontal pipes which might not be full.

They reduce to a particularly simple form for the zero-flow condition $\dot{m} = \phi = 0$.

- Mass: $\frac{dv}{dt} = 0$ which implies no change in mass distribution.
- Energy: $\frac{1}{A}\frac{d\dot{q}}{dt} = \frac{1}{v}\frac{dh}{dt} - \frac{dp}{dt}$; the system continues to exchange heat locally with its surroundings, but there is no advection of energy to other parts of the system.
- Momentum: $\frac{dp}{dz} = -\frac{g}{v}\sin\beta$; the pressure gradient is defined by gravity only.

The conservation equations may be rearranged to remove the explicit reference to the derivatives of specific volume, using the first properties relationship. This reduces the equations to a set of three in the state variables pressure, enthalpy and specific mass flow. Then

Mass conservation:

$$\frac{\partial v}{\partial p}\frac{\partial p}{\partial t} + \frac{\partial v}{\partial h}\frac{\partial h}{\partial t} = v^2\frac{\partial \phi}{\partial z}. \tag{5.9}$$

Energy conservation:

$$\frac{\partial h}{\partial t} + v\frac{\partial h}{\partial z} = v\left(\frac{\partial p}{\partial t} + v\frac{\partial p}{\partial z}\right) + \frac{v}{A}\frac{d\dot{q}}{dz} + vv\left.\frac{\partial p}{\partial z}\right|_{loss}. \tag{5.10}$$

Momentum conservation: After some further manipulation the momentum equation is transformed to

$$\frac{\partial \phi}{\partial t} = -\left(1 + \phi^2\frac{\partial v}{\partial p}\right)\frac{\partial p}{\partial z} - \phi^2\frac{\partial v}{\partial h}\frac{\partial h}{\partial z} - 2\frac{\phi}{v}\frac{\partial v}{\partial t} - \left.\frac{\partial p}{\partial z}\right|_{loss} - \rho g\sin\beta. \tag{5.11}$$

5.1.1 Critical Flow Velocity

Noting from Eq. 4.34 that $\partial v/\partial p$ is negative (increasing pressure reduces the specific volume or increases density), the momentum equation may be written as

$$\frac{\partial p}{\partial z} = \frac{-\frac{\partial \phi}{\partial t} - \phi^2\frac{\partial v}{\partial h}\frac{\partial h}{\partial z} - 2\frac{\phi}{v}\frac{\partial v}{\partial t} - \left.\frac{\partial p}{\partial z}\right|_{loss} - \rho g\sin\beta}{\left(1 - \phi^2\left|\frac{\partial v}{\partial p}\right|\right)}. \tag{5.12}$$

It is apparent that as the denominator approaches zero, the local pressure gradient approaches an infinite value for

$$\phi = \frac{1}{\sqrt{\left|\frac{\partial v}{\partial p}\right|}}.$$

Noting that $\phi v = v$, the fluid velocity

$$v_{cr} = \frac{v}{\sqrt{\left|\frac{\partial v}{\partial p}\right|}} \tag{5.13}$$

is the maximum attainable or *critical* flow velocity and is a consequence of the compressibility of the gas or fluid. In order to increase the mass flow rate of a compressible fluid or gas once the critical velocity has been attained it is necessary to increase its density ρ (decrease v) by increasing pressure.

v_{cr} is also called the *sonic* velocity, being the speed of sound c [m/s] in the medium. We then have from Eq. 5.13

$$\frac{\partial v}{\partial p} = -\frac{v^2}{c^2}. \tag{5.14}$$

With the previous expression for $\partial v/\partial p$ this leads to

$$c^2 = \gamma p v. \tag{5.15}$$

Note that c is not a constant and will vary with pressure and temperature of the working medium. Note also that, for consistent units, pressure in Eq. 5.15 must be expressed in Pascals.

Substituting $\partial v/\partial p$ from Eq. 5.14 in Eq. 5.12, the divisor may be written as $(1 - M_a^2)$ where $M_a = v/c$ and is the Mach number, being the ratio of the local working medium flow speed to that of the local speed of sound.

The following table gives some indicative values for the velocity of sound, calculated from Eq. 5.15, for air and steam at thermodynamic conditions relevant to industrial plant.

Gas	Pressure kPa	Temperature °C	Spec. volume m^3/kg	γ	c m/s
Air	101.3	20	1.0	1.139	344.0
Air	101.3	500	2.2	1.139	510.2
Steam	1,000	240	0.206	1.3	517.5
Steam	16,000	540	0.0202	1.27	637.5

5.2 Analytical Solution of the Conservation Equations

The analytical solution of the conservation equations is not possible for general boundary conditions. However, it is possible to identify certain structural characteristics of the solution and, with some simplifying limitations, to develop analytical solutions for selected cases. These are not particularly useful renderings of the equations for our purposes, but they reveal some of the underlying structure of the equations and of the physical processes they describe.

5.2.1 An Isothermal System Without Flow Inertia

Consider a medium flowing in a conduit at a uniform steady temperature, exchanging no heat with its surroundings and performing no work. The energy conservation equation can be ignored as it will always be satisfied under these conditions, assuming a consistent thermodynamic state of the system. If we ignore inertial effects we can neglect terms involving the rate of change of flow. The mass conservation equation then reduces to

$$\frac{\partial v}{\partial t} = v^2 \frac{\partial \phi}{\partial z}$$

or, using the earlier expression for the speed of sound c,

$$\frac{\partial p}{\partial t} = -c^2 \frac{\partial \phi}{\partial z}. \tag{5.16}$$

If we restrict our attention to horizontal flow ($\sin \beta = 0$) the momentum equation reduces to

$$\frac{\partial p}{\partial z} = -\left.\frac{\partial p}{\partial z}\right|_{loss} = -\frac{1}{2}\left(\frac{\xi\, v}{d_H}\right)\phi^2. \tag{5.17}$$

Differentiating this equation with respect to z gives

$$\frac{\partial^2 p}{\partial z^2} = -\left(\frac{\xi\, v}{d_H}\right)\phi \frac{\partial \phi}{\partial z}.$$

Substituting for $\partial \phi / \partial z$ from the mass equation gives

$$\begin{aligned} \frac{\partial^2 p}{\partial z^2} &= \left(\frac{\xi\, v \phi}{d_H}\right)\frac{1}{c^2}\frac{\partial p}{\partial t} \\ &= \left(\frac{\xi \phi}{d_H \gamma p}\right)\frac{\partial p}{\partial t}. \end{aligned} \tag{5.18}$$

This has the form of the standard diffusion equation

$$\frac{\partial^2 u}{\partial z^2} = \frac{1}{D}\frac{\partial u}{\partial t}.$$

The "diffusion constant" $D = \left(\frac{\xi \phi}{d_H \gamma p}\right)^{-1}$ with dimensions [m^2/s].

Using the method of separation of variables the general solution of this equation is readily obtained as

$$p(z,t,k) = \mathrm{e}^{-(k^2 D)t}(A\ \sin kz + B\ \cos kz), \tag{5.19}$$

where k is any positive integer. A and B are constants chosen to match boundary conditions.

The steady-state profile of pressure $\bar{p}(z)$ is given by setting the time derivative to zero in Eq. 5.18. The resulting equation

$$\frac{\partial^2 \bar{p}}{\partial z^2} = 0$$

may be integrated twice, giving

$$\bar{p}(z) = a_0 z + a_1,$$

where the coefficients a_0 and a_1 are selected to match boundary conditions. If at time $t = 0$ the two end point pressures are p_1 and p_2 we have $a_0 = (p_2 - p_1)/L$ and $a_1 = p_1$. Then,

$$\bar{p}(z) = \frac{p_2 - p_1}{L} z + p_1.$$

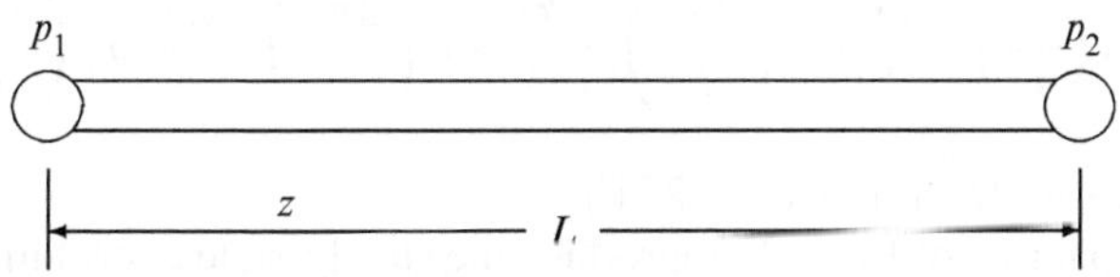

The dynamic behaviour of pressure at any point z between p_1 and p_2 will satisfy Eq. 5.18 in response to a change of either p_1 or p_2, starting from the distribution $\bar{p}(z)$.

Define a new pressure $\hat{p}(z,t) = (p(z,t) - p_2)$ which also satisfies Eq. 5.18. Then

$$\hat{p}(z,0) = \frac{p_2 - p_1}{L} z + (p_1 - p_2) \tag{5.20}$$

and

$$\hat{p}(L,t) = 0 \text{ for } t > 0.$$

This condition will be satisfied if we select $k = n\pi/L$ in Eq. 5.19 for all positive integer n and $B = 0$. The particular solution of Eq. 5.18 may then be written for any $n > 0$

$$\hat{p}(z,t,n) = \mathrm{e}^{-(n^2\pi^2 D/L^2)t} A_n \sin\left(\frac{n\pi z}{L}\right). \tag{5.21}$$

Since Eq. 5.18 is linear, the sum of all solutions for all n is also a solution, and the complete solution is

$$\hat{p}(z,t) = \hat{p}(z,0) + \sum_{n=1}^{\infty} \mathrm{e}^{-(n^2\pi^2 D/L^2)t} A_n \sin\left(\frac{n\pi z}{L}\right).$$

Using the initial profile of $\hat{p}(z,t)$ from Eq. 5.20 we have

$$\hat{p}(z,t) = \frac{p_2 - p_1}{L} z + p_1 - p_2 + \sum_{n=1}^{\infty} \mathrm{e}^{-(n^2\pi^2 D/L^2)t} A_n \sin\left(\frac{n\pi z}{L}\right)$$

from which we have, after substitution for $\hat{p}(z,t)$,

$$p(z,t) = p_1 + \frac{p_2 - p_1}{L} z + \sum_{n=1}^{\infty} \mathrm{e}^{-(n^2\pi^2 D/L^2)t} A_n \sin\left(\frac{n\pi z}{L}\right). \tag{5.22}$$

The A_n coefficients in Eq. 5.22 are those of the Fourier expansion of $\hat{p}(z,0)$ over the range $(0, L)$. Their calculation is facilitated by the Fourier series expansion of $f(z) = C_1 z + C_2$ over the range $(0, L)$,

$$C_1 z + C_2 = \frac{1}{\pi}\left(k_1 \sin\left(\frac{\pi z}{L}\right) - \frac{k_2}{2} \sin\left(\frac{2\pi z}{L}\right) + \frac{k_1}{3} \sin\left(\frac{3\pi z}{L}\right) - \frac{k_2}{4} \sin\left(\frac{4\pi z}{L}\right) + \cdots\right),$$

where $k_1 = 4C_2 + 2LC_1$ and $k_2 = 2LC_1$.

The first two terms of Eq. 5.22 reproduce the steady-state distribution of pressure from the two boundary pressures. The summated term calculates the transient departure from this steady-state distribution. Each term of the summated series is weighted by an exponential time function and decays to zero as time increases.

For air at standard pressure and temperature, $c = 344\,\mathrm{m/s}$ and $v = 1.0\,\mathrm{m^3/kg}$. A flow rate of 10 kg/s in a pipe with an internal diameter (ID) of 0.5 m gives $\phi = 51\,\mathrm{kg/(s\,m^2)}$. A typical value of $\xi = 0.02$ gives a value for D of 58,000. Then, for a pipe length of $L = 30\,\mathrm{m}$, we have for the first three exponents in this series

$$\begin{aligned} \mathrm{n} = 1 &\quad \exp = -636\,t, \\ \mathrm{n} = 2 &\quad \exp = -2{,}544\,t, \\ \mathrm{n} = 3 &\quad \exp = -5{,}724\,t. \end{aligned}$$

For steam at typical power plant operating conditions (pressure of 16 MPa and 540 °C), a flow of 100 kg/s in a pipe 0.3 m ID and a typical $\xi = 0.04$ gives $D = 80{,}000$.

For increasing n, the time exponent increases rapidly negatively, and only the first one or two terms of the series are significant. The pressure transients decay rapidly, and the new steady-state profile is achieved quickly. If flow inertia is ignored the rapid disappearance of pressure transients allows the pressure dynamics to be neglected in favour of a simple treatment of the time variation of pressure as a succession of steady-state profiles.

The diffusion equation produced by the inertialess model does not admit any wave structure in the solution and cannot be used to treat the known wave behaviour of pressure dynamics in compressible systems, particularly those involving high-speed phenomena.

Finally, we note that, had we differentiated the mass conservation equation 5.16 with respect to t and the momentum equation with respect to z, we would have obtained the equivalent equation for ϕ:

$$\frac{\partial^2 \phi}{\partial z^2} = \left(\frac{\xi \phi}{d_H \gamma p}\right) \frac{\partial \phi}{\partial t} \tag{5.23}$$

which is of exactly the same form as the pressure equation 5.18. The conclusion regarding pressure dynamics applies equally well to mass flow dynamics of inertialess systems, which may also be treated as a succession of steady-state flow profiles.

5.2.2 *An Adiabatic System with Flow Inertia*

We again ignore the energy conservation equation. The mass and momentum equations for this case may be written as

$$\frac{\partial p}{\partial t} = -c^2 \frac{\partial \phi}{\partial z}$$

and

$$\frac{\partial p}{\partial z} = -\frac{\partial \phi}{\partial t} - \frac{1}{2}\left(\frac{\xi v}{d_H}\right)\phi^2.$$

As before, differentiate the mass equation with respect to t and the momentum equation with respect to z. Then

$$\frac{\partial^2 p}{\partial t^2} = -c^2 \frac{\partial^2 \phi}{\partial t \partial z}$$

and

$$\frac{\partial^2 p}{\partial z^2} = -\frac{\partial^2 \phi}{\partial z \partial t} - \left(\frac{\xi v}{d_H}\right)\phi \frac{\partial \phi}{\partial z}.$$

Combining these two by eliminating $\partial^2\phi/\partial z\partial t$ yields

$$c^2\frac{\partial^2 p}{\partial z^2} = \frac{\partial^2 p}{\partial t^2} + \left(\frac{\xi\phi v}{d_H}\right)\frac{\partial p}{\partial t}. \tag{5.24}$$

This is the damped wave equation for which solutions for simple boundary conditions can be developed.

Alternatively, we can eliminate pressure from the mass and momentum equations by a procedure which yields an identical equation in ϕ.

$$c^2\frac{\partial^2 \phi}{\partial z^2} = \frac{\partial^2 \phi}{\partial t^2} + \left(\frac{\xi\phi v}{d_H}\right)\frac{\partial \phi}{\partial t}. \tag{5.25}$$

In terms of the variable u, which can be either p or ϕ, the damped wave equation admits the general solution, obtained readily by the method of separation of variables,

$$u(z,t) = \mathrm{e}^{-\frac{b}{2}t}(a_1 \sin kz + a_2 \cos kz)(a_3 \sin \beta t + a_4 \cos \beta t) \tag{5.26}$$

for any real value of $k > 0$. $\beta = kv'$ where $v' = \sqrt{c^2 - (b/2k)^2}$ and $b = \xi\phi v/d_H$.

The arbitrary coefficients $a_1 \cdots a_4$ are again determined from the boundary conditions by setting a_2 to 0 and a_4 to 1. We can also define $k = n\pi/L$ for all positive integers n. Equation 5.26 may then be written as

$$u(z,t) = \mathrm{e}^{-\frac{b}{2}t} a_1 \sin\left(\frac{n\pi}{L}z\right)\left[a_3 \sin\left(\frac{n\pi}{L}v't\right) + \cos\left(\frac{n\pi}{L}v't\right)\right]. \tag{5.27}$$

Being a linear equation, the general solution of Eq. 5.27 is the sum of the individual solutions for all n.

$$u(z,t) = u(z,0) + \mathrm{e}^{-\frac{b}{2}t}\sum_{n=1}^{\infty} a_n \sin\left(\frac{n\pi}{L}z\right)\left[A_n \sin\left(\frac{n\pi}{L}v't\right) + \cos\left(\frac{n\pi}{L}v't\right)\right]. \tag{5.28}$$

The coefficients a_n are determined as the coefficients of a Fourier series expansion of the initial profile of $u(z,0)$ at time $t = 0$.

$$a_n = \int_0^L u(z,0) \sin\left(\frac{n\pi}{L}\right) \mathrm{d}z.$$

This can be simplified by setting the initial profile of $u(z,0)$ to match the previous steady-state profile obtained by setting the time derivatives in Eq. 5.27 to zero. Again, the steady-state solution may be written as

$$u(z) = C_1 z + C_2$$

and the coefficients C_1 and C_2 matched to the known boundary conditions.

Complete definition of the solution expressed by Eq. 5.28 requires calculation of the coefficients A_n and for this a further boundary condition is required. This can be provided by specification of the initial rate of change of $u(z,t)$. For $u(z,t)$ given by Eq. 5.28 the rate of change is

$$\frac{\partial u(z,t)}{\partial t} = \mathrm{e}^{-\frac{b}{2}t} \sum_{n=1}^{\infty} a_n \sin\left(\frac{n\pi}{L}z\right) \times \left[-\frac{b}{2}\left(A_n \sin\beta_n t + \cos\beta_n t\right) + \beta_n\left(A_n \cos\beta_n t - \sin\beta_n t\right)\right],$$

where the subscripted β_n indicates that β is a function of n.

Setting the gradient to zero for $t = 0$ yields $A_n = b/(2\beta_n)$. The general solution to Eq. 5.28 for these boundary conditions is then given by

$$u(z,t) = u(z,0) + \mathrm{e}^{-\frac{b}{2}t} \sum_{n=1}^{\infty} a_n \sin\left(\frac{n\pi}{L}z\right)\left[\frac{b}{2\beta_n} \sin\frac{n\pi}{L}v't + \cos\frac{n\pi}{L}v't\right]. \quad (5.29)$$

Using the equalities

$$2\sin A \cos B = \sin(A+B) + \sin(A-B)$$

and

$$2\sin A \sin B = \cos(A-B) - \cos(A+B)$$

Eq. 5.29 may be expanded into

$$u(z,t) = u(z,0) + \mathrm{e}^{-\frac{b}{2}t} \sum_{n=1}^{\infty} \frac{a_n}{2}\left[\sin\frac{n\pi}{L}(z - v't) + \sin\frac{n\pi}{L}(z + v't) + \frac{b}{2\beta_n}\left(\cos\frac{n\pi}{L}(z - v't) - \cos\frac{n\pi}{L}(z + v't)\right)\right]. \quad (5.30)$$

This shows that the solution to the damped wave equation consists of two pairs of two travelling waves of differing amplitudes. One pair travels from left to right and the other from right to left. Within each pair, the two waves are phase shifted $\pi/2$ radians to each other. The speed of each wave is v' and is slightly retarded compared with the undamped speed of sound. The wave amplitudes are damped by the term $\exp(-bt/2)$. Using the same examples as the inertialess case, typical values of $b = \xi\phi v/d_H$ are around 2.04 for air at atmospheric pressure and temperature and around 1.6 for high-pressure superheated steam. In contrast to the inertialess case, the waves are not heavily damped, and several terms of the series in Eq. 5.29 may need to be considered.

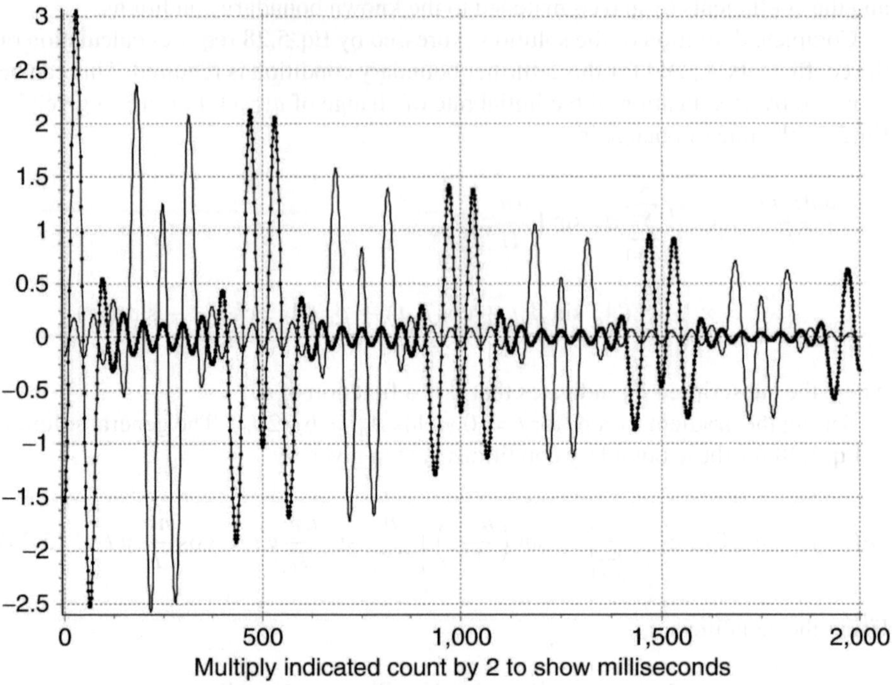

Fig. 5.1 Damped wave equation solution at 2 points in a steam line

Figure 5.1 shows the transient solution (deviation from the steady state) of Eq. 5.30 for two points 20 and 80 m along a 100 m steam line following a small but abrupt change in both end point pressures. The pipe and flow parameters are as quoted earlier. The complex combination of travelling waves gives rise to moving regions of cancellation and reinforcement of the local pressure, creating significant pressure troughs and peaks and giving rise to "beats". The transit time of the waves along the pipe is revealed by the interval between successive peaks and by the displacement in time of the momentary pressure peaks and troughs at the two spatially separated measurement points. Under some operational circumstances the amplitude of the peak pressures occurring at points of reinforcement can reach significant values and can be audibly detected as "water hammer". Their collisions with obstructions in piping systems (closed valves, pumps, etc.) or even sharp bends can be damaging to the equipment.

5.2.3 Analytical Solution of the Energy Conservation Equation

The energy conservation equation is written in terms of enthalpy and cannot treat heat transfer explicitly as this involves temperature. The equation may be expressed in terms of a single variable (temperature) when it is possible to replace the enthalpy h by $c_p T$. The specific heat capacity coefficient is tacitly assumed locally constant. Equation 5.7 may then be rewritten as

$$c_p \frac{\partial T}{\partial t} + c_p v \frac{\partial T}{\partial z} = \nu \left(\frac{\partial p}{\partial t} + \frac{\partial p}{\partial z} \right) - \frac{\nu}{A} \alpha_{hx} A_{hx} (T - T_{ref}) + \nu v \left. \frac{\partial p}{\partial z} \right|_{loss}, \quad (5.31)$$

where T_{ref} is the local temperature of the interface with which the working fluid is exchanging heat. The pressure gradient contribution to the net energy balance is negligible. If the friction heat is treated as insignificant compared to the heat transfer term, the last term in the above equation can be ignored. For an incompressible fluid the compression energy term $\nu \frac{\partial p}{\partial t}$ is not significant and in many cases even for a compressible fluid can be ignored. Define

$$\tau_t = \frac{c_p}{\alpha_{hx} A_{hx}} \frac{A_x}{\nu} \quad (5.32)$$

$$\begin{aligned} \tau_z &= v \frac{c_p}{\alpha_{hx} A_{hx}} \frac{A_x}{\nu} \\ &= v \tau_t, \end{aligned} \quad (5.33)$$

where A_x is the cross-sectional area [m^2] normal to the flow and A_{hx} is the heat transfer area per unit length of flow path. The energy balance equation can then be written as

$$\tau_t \frac{\partial T}{\partial t} + \tau_z \frac{\partial T}{\partial z} + T = T_{ref}. \quad (5.34)$$

Boundary conditions are defined by the initial temperature profile $T(z, 0)$ and the inlet temperature $T(0, 0)$.

Assuming that T_{ref} is constant, independent of z, the analytic solution $T(z, t)$ of Eq. 5.34 with the given boundary conditions may be derived using the Laplace transform method. Then

$$\begin{aligned} T(z,t) = T(z,0)\mathrm{e}^{-t/\tau_t} + \quad & T_{ref}(1 - \mathrm{e}^{-t/\tau_t}) \\ & + (T(0,0) - T_{ref})(\mathrm{e}^{-z/\tau_z} - \mathrm{e}^{-t/\tau_t}) H(t - z/v), \end{aligned} \quad (5.35)$$

where $H(t - z/v)$, the Heaviside function, is defined as $H(t - z/v) = 0$ for $t < z/v$ and $= 1$ for $t > z/v$. The detailed derivation of the solution of Eq. 5.34 is given in Appendix A.

The first two terms of Eq. 5.35 represent the space-independent time variation of the influence of the spatial boundary conditions. The third term represents the wave-front propagation of changes in inlet conditions in the direction of flow. The front travels along the flow path with velocity v where

$$v = \frac{\tau_t}{\tau_z} = \phi \, v.$$

The following diagrams depict the solution of Eq. 5.35 for flow in a pipe, with the following parameters.

Length:	40 m
Flow:	150 kg/s
Internal diameter:	300 mm
Specific flow rate $\phi = \dot{m}/A$:	2,120 kg/(s m^2)
Flow velocity:	2.122 m/s
Water/pipe heat transfer coefficient α_w:	14 kW/(m^{2}K)

The water is assumed to be at an initial uniform temperature of 330 °C over its complete length. The tube wall temperature T_{ref} is not calculated and remains constant at 330 °C. The diagrams show the spatial and time response of the water temperature following a step increase of inlet temperature to the pipe from 330 °C to 340 °C.

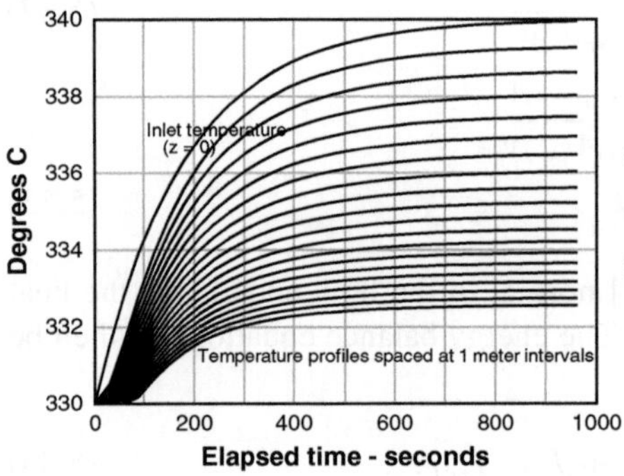

Temperatures at selected zcoords

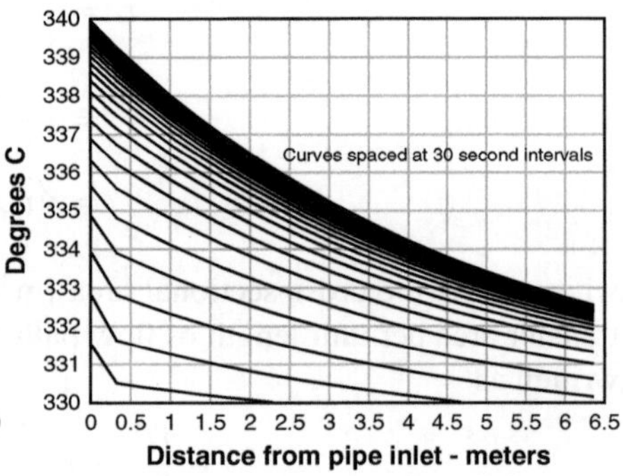

Temperatures at selected times

As $t \to \infty$ the spatial temperature profile trends to the steady-state solution of Eq. 5.35.

The transport delay, clearly evident in both diagrams, illustrates the propagation of the temperature front moving with and at the speed of the flow.

The calculation of fluid temperatures at the outlet of each of two pipes, calculated for each using Eq. 12.14, is shown by Fig. 5.2. The pipes are connected in series, with the inflow temperature of the second being the outlet temperature of the first. Both pipes have the same dimensions and flow, as listed below. The time variations of both pipes' temperatures are shown against the same time base.

Length of each pipe	20 m
Flow per pipe	250 kg/s
Pipe internal diameter	500 mm
Specific flow rate $\phi = \dot{m}/A$	1,273 kg/s m^2
Flow velocity ν	1.32 m/s
Water/pipe heat transfer coefficient α_w	5.3 kW/m^2/K

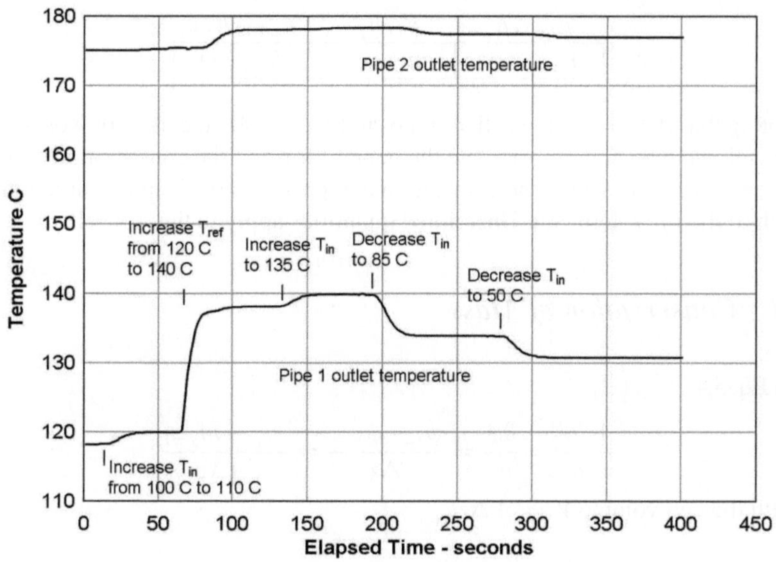

Fig. 5.2 Pipe temperatures—distributed model

The wall temperatures of the two pipes are initially set to 120 °C, and 185 °C respectively. The fluid inlet and reference temperatures of the first pipe are varied during the calculation, as shown by the annotations on the figure.

Being uniform over the whole length of the pipe, a change in wall temperature will cause an immediate movement in outlet temperature while a change in fluid inlet temperature must wait until the transport time has elapsed to cause movement.

5.3 Spatial Discretisation

While of great interest in identifying structural properties of the solutions, analytical methods of solution of the conservation equations will prove of little use to the online computational of solutions in real-time simulation systems. For this purpose it will be necessary to restructure the equations into more computationally tractable forms. Since the equations are expressed in terms of time and a spatial ordinate, discretisation of one or both of these is a natural path to follow. This section will consider discretisation of the spatial ordinate z.

Consider a flow path element divided into a series of discrete spatial cells. Each cell is filled with a working fluid whose properties are defined either at the spatial centre of the cell or at the intercell boundaries.

p_{j-1}, h_{j-1}	p_j, h_j	p_{j+1}, h_{j+1}
$\dot{m}_{j-1} \leftrightarrow$	$\dot{m}_j \leftrightarrow$	$\dot{m}_{j+1} \leftrightarrow$

Noting that $A\Delta z = V$, the cell volume, and $V/v = M$, the mass of working fluid in the cell, we may now reformulate the general form of the conservation equations as follows. We will assume the convention that positive flow is in the direction from the jth to the $(j+1)$th cell. Unscripted quantities apply to the jth cell.

5.3.1 Conservation of Mass

From Eq. 5.1

$$\frac{1}{v^2}\frac{\partial v}{\partial t} = \frac{\partial \phi}{\partial z} = \frac{\phi_j - \phi_{j-1}}{\Delta z} = \frac{\dot{m}_j - \dot{m}_{j-1}}{A\Delta z}$$

or, with the cell volume $V = A\,\Delta z$

$$\left(\frac{V}{v^2}\right)\frac{\partial v}{\partial t} = \dot{m}_j - \dot{m}_{j-1}.$$

This same result can be obtained by differentiating the cell mass $M = V/v$ for the jth cell with respect to time giving

$$\frac{dM}{dt} = -\left(\frac{V}{v^2}\right)\frac{dv}{dt} = -(\dot{m}_{j-1} - \dot{m}_j) \tag{5.36}$$

or, with $M = \rho V$,

$$\frac{dM}{dt} = V\frac{d\rho}{dt} = -(\dot{m}_{j-1} - \dot{m}_j). \tag{5.37}$$

5.3.2 Conservation of Momentum

From Eq. 5.11,

$$\frac{\partial \phi}{\partial t} + 2\frac{\phi}{v}\frac{\partial v}{\partial t} = -\left(1 + \phi^2\frac{\partial v}{\partial p}\right)\frac{\partial p}{\partial z} - \phi^2\frac{\partial v}{\partial h}\frac{\partial h}{\partial z} - \left.\frac{\partial p}{\partial z}\right|_{loss} - \rho g \sin\beta.$$

Replacing $\frac{\partial p}{\partial z}$ and $\frac{\partial h}{\partial z}$ by their first-order spatial approximations gives

$$\frac{d\phi}{dt} + 2\frac{\phi}{v}\frac{\partial v}{\partial t} = -\left(1 + \phi^2\frac{\partial v}{\partial p}\right)\left(\frac{p_j - p_{j-1}}{dz}\right) - \phi^2\frac{\partial v}{\partial h}\left(\frac{h_j - h_{j-1}}{dz}\right)$$
$$- \left.\frac{\partial p}{\partial z}\right|_{loss} - \rho g \sin\beta$$

or

$$\left(1+\phi^2\frac{\partial v}{\partial p}\right)p_{j-1}+\phi^2\frac{\partial v}{\partial h}h_{j-1}=\left(1+\phi^2\frac{\partial v}{\partial p}\right)p_j+\phi^2\frac{\partial v}{\partial h}h_j$$
$$+\left(\frac{\mathrm{d}\phi}{\mathrm{d}t}+2\frac{\phi}{v}\frac{\partial v}{\partial t}+\left.\frac{\partial p}{\partial z}\right|_{loss}+\rho g\sin\beta\right)\mathrm{d}z.$$

5.3.3 Conservation of Energy

Replacing the spatial partial derivatives in the energy conservation equation 5.7 by their first-order difference approximations gives

$$\frac{1}{A}\frac{\Delta\dot{q}}{\Delta z}=\frac{1}{v}\frac{\mathrm{d}h}{\mathrm{d}t}+\phi\frac{(h_j-h_{j-1})}{\Delta z}-\frac{\mathrm{d}p}{\mathrm{d}t}+\phi\left(\frac{\mathrm{d}v}{\mathrm{d}t}+v\frac{(v_j-v_{j-1})}{\Delta z}+g\sin\beta\right).$$

Multiply both sides of this equation by $A\Delta z=V$. Then for the jth cell,

$$\Delta\dot{q}=M\frac{\mathrm{d}h}{\mathrm{d}t}+\dot{m}_jh_j-\dot{m}_{j-1}h_{j-1}-V\frac{\mathrm{d}p}{\mathrm{d}t}+\Delta z\,\dot{m}\frac{\mathrm{d}v}{\mathrm{d}t}$$
$$+v_j\,\dot{m}_jv_j-v_{j-1}\,\dot{m}_jv_{j-1}+\dot{m}\Delta z\,g\sin\beta.$$

Given that

$$\begin{aligned}\frac{\mathrm{d}v}{\mathrm{d}t}&=\frac{\mathrm{d}\overline{\phi v}}{\mathrm{d}t}\\&=\phi\frac{\mathrm{d}v}{\mathrm{d}t}+v\frac{\mathrm{d}\phi}{\mathrm{d}t}\\&=\phi v^2\frac{\mathrm{d}\phi}{\mathrm{d}z}+v\frac{\mathrm{d}\phi}{\mathrm{d}t}\\&=\phi v^2\frac{\phi_j-\phi_{j-1}}{\Delta z}+v\frac{\mathrm{d}\phi}{\mathrm{d}t}.\end{aligned}$$

we can substitute for $\mathrm{d}v/\mathrm{d}t$ to give

$$\Delta\dot{q}=M\frac{\mathrm{d}h}{\mathrm{d}t}+\dot{m}_jh_j-\dot{m}_jh_{j-1}-V\frac{\mathrm{d}p}{\mathrm{d}t}+v^2(\dot{m}_j-\dot{m}_{j-1})+v\Delta z\frac{\mathrm{d}\dot{m}}{\mathrm{d}t}$$
$$+\frac{1}{v}v^2(v_j-v_{j-1})+\dot{m}\Delta z\,g\sin\beta.$$

Setting $v_m = v_j = v_{j-1}$ to be the mean fluid specific volume in the cell, the two terms multiplied by v^2 are equal. With Δz being the length of the cell we have for the energy equation

$$M\frac{\mathrm{d}h}{\mathrm{d}t} = (\dot{m}_j h_j - \dot{m}_{j-1}h_{j-1}) + 2v^2(\dot{m}_j - \dot{m}_{j-1}) + V\frac{\mathrm{d}p}{\mathrm{d}t} - \Delta\dot{q}$$
$$-v\Delta z\frac{\mathrm{d}\dot{m}}{\mathrm{d}t} - \dot{m}\,\Delta z\, g\,\sin\beta. \tag{5.38}$$

The individual terms of this equation have the following significance:

- $M\frac{\mathrm{d}h}{\mathrm{d}t}$ is the rate of change of energy in the cell.
- $(\dot{m}_j h_j - \dot{m}_{j-1}h_{j-1})$ is the sensible energy differential between in- and outflows.
- $V\frac{\mathrm{d}p}{\mathrm{d}t}$ is the rate of change of the cell compression energy.
- $\Delta\dot{q}$ is the heat energy exchanged across the lateral cell boundaries with external systems.
- $\dot{m}\,\Delta z\, g\,\sin\beta$ is the potential energy differential across the cell due to gravity.
- $v\Delta z\frac{\mathrm{d}\dot{m}}{\mathrm{d}t}$ denotes the flow acceleration energy.
- $v^2(\dot{m}_j - \dot{m}_{j+1})$ denotes the change in kinetic energy across the cell.

For most industrial plants the potential energy due to geodetic height differences is negligible in comparison with the other terms and may be omitted from the energy equation. The last two terms may be grouped together and denoted W, the mechanical work done by the fluid in the cell. This reveals that mechanical work done by (or to) the fluid results in a loss (or gain) of flow inertia due to acceleration (pumps) or a change in flow velocity either as a result of expansion (in turbines) or compression (in fans and compressors). The energy equation may then be written as

$$M\frac{\mathrm{d}h}{\mathrm{d}t} = (\dot{m}_j h_j - \dot{m}_{j-1}h_{j-1}) + V\frac{\mathrm{d}p}{\mathrm{d}t} - \Delta\dot{q} + W. \tag{5.39}$$

In Sect. 12.3 dealing with steam generation, we will see that the volume changes induced by phase change create a flow acceleration/deceleration term in the absence of mechanical input. Although significant to the momentum balance, this effect will be omitted from the energy balance as being insignificant when compared to the other terms.

5.4 The Serial Method: Discrete Time, Continuous Space

Before proceeding to the development of a solution algorithm for the conservation equations using the serial method, consider the solution of the equation

$$\frac{\partial u}{\partial t} + a\frac{\partial u}{\partial z} = g(z,t).$$

Replacing the time derivative by its first-order discrete equivalent gives

$$\frac{u^{n+1} - u^n}{\Delta t} + a\frac{\mathrm{d}u}{\mathrm{d}z} = g(z, n\Delta t).$$

Now replace $u^i(z)$ by a weighted mean value taken over the duration of the ith time step, with weighting parameter $\theta \in (0, 1)$.

$$u(z) \approx \theta u^{i+1} + (1-\theta)u^i$$

and

$$g(z, i\,\Delta t) \approx \theta g^{i+1} + (1-\theta)g^i.$$

The original equation can be now be written as

$$\frac{u^{n+1} - u^n}{\Delta t} + a\frac{\mathrm{d}}{\mathrm{d}z}(\theta u^{n+1} + (1-\theta)u^n) = \theta g^{n+1} + (1-\theta)g^n.$$

Grouping terms denominated at times $(n + 1)$ and n gives

$$u^{n+1} + a\theta\Delta t\frac{\mathrm{d}u^{n+1}}{\mathrm{d}z} - a\theta\Delta t\; g^{n+1} = u^n - a(1-\theta\Delta t)\frac{\mathrm{d}u^n}{\mathrm{d}z} + (1-\theta)\Delta t\; g^n = S_z^n.$$

From S_z^n we can write

$$S_z^{n+1} = u^{n+1} - a(1-\theta)\Delta t\frac{\mathrm{d}u^{n+1}}{\mathrm{d}z} + (1-\theta)\Delta t\; g^{n+1}$$

and, since

$$u^{n+1} = S_z^n - a\theta\Delta t\frac{\mathrm{d}u^{n+1}}{\mathrm{d}z} - \theta\Delta t\; g^{n+1}$$

we have the recurrence relationship for S_z^{n+1}

$$S_z^{n+1} = \frac{1}{\theta}u^{n+1} - \frac{1-\theta}{\theta}S_z^n.$$

We may now write an *ordinary* differential equation for $u^{n+1}(z)$

$$a\theta\Delta t\frac{\mathrm{d}u^{n+1}}{\mathrm{d}z} + u^{n+1} = \theta\Delta t g^{n+1} + S_z^n. \tag{5.40}$$

The solution of this equation delivers the variable $u^{n+1}(z)$ at the time step $(n+1)$ for the entire span of z, given the boundary conditions $u^{n+1}(0)$ and S_z^n from the preceding time step.

The following points should be noted:

(a) The solution is self-starting, needing only knowledge of g^{n+1} at $z = 0$ (a boundary condition normally available from interfacing systems) and S_z^n, known from the preceding time step.
(b) It requires knowledge of g^{n+1} over the complete span of z, normally not available as it will usually depend on the solution u^{n+1}.
(c) If a reasonably reliable estimate of g^{n+1} is available the equation admits the analytical solution $u^{n+1}(z) =$

$$\mathrm{e}^{-z/(a\theta\Delta t)}\left[u^{n+1}(0)+\frac{1}{a\theta\Delta t}\int^{z}\mathrm{e}^{z'/(a\theta\Delta t)}\left(S_z^n+\theta\Delta t g^{n+1}(z')\right)\mathrm{d}z'\right].$$

Direct integration using this formulation offers the advantage that the accuracy of the solution will not depend on the relative sizes of Δt and Δz as would be the case were a numerical space and time discretisation procedure to be used.
(d) This approach is valid only for non-zero values of a. In many physical problems, this coefficient is flow velocity v. When applied to the simulation of flow systems, the method must be modified to accommodate the zero-flow condition.
(e) This method is very similar to the semi-implicit method of solution of equations, described in Sect. 3.4.

The weighting parameter θ can take any value between 0 and 1. $\theta = 0$ corresponds to explicit Euler and is only conditionally stable. $\theta = 1$ corresponds to implicit Euler and is unconditionally stable but offers no computational advantage over the fully implicit method. $\theta = 0.5$ corresponds to a mid-point integration routine and is marginally stable. For most purposes a value of θ of around 0.65–0.8 will be found to be a good compromise between stability and accuracy.

We will now apply the serial method to the development of a solution procedure for the conservation equations.

5.4.1 Serial Solution of the Conservation Equations

The mass balance equation 5.1 is

$$\frac{\partial\rho}{\partial t}=-\frac{\partial\phi}{\partial z}.$$

With the time derivative replaced by its discrete time equivalent, this becomes

$$\rho^{n+1}-\rho^n=-\Delta t\frac{\mathrm{d}}{\mathrm{d}z}\left[\theta\phi^{n+1}+(1-\theta)\phi^n\right].$$

Grouping terms denominated at the $(n+1)$th and nth time steps gives

$$\rho^{n+1} + \theta\Delta t \frac{\mathrm{d}\phi^{n+1}}{\mathrm{d}z} = \rho^n - (1-\theta)\Delta t \frac{\mathrm{d}\phi^n}{\mathrm{d}z} = S_\rho^n.$$

We can then write

$$\theta\Delta t \frac{\mathrm{d}\phi^{n+1}}{\mathrm{d}z} = S_\rho^n - \rho^{n+1} \tag{5.41}$$

S_ρ^n is defined by the recursive relation

$$S_\rho^{n+1} = \frac{1}{\theta}\rho^{n+1} - \frac{1-\theta}{\theta} S_\rho^n \tag{5.42}$$

with the boundary condition

$$S_\rho^0 = \rho^0(z).$$

The momentum equation 5.3

$$\frac{\partial p}{\partial z} = -\left.\frac{\partial p}{\partial z}\right|_{loss} - \rho\left(\frac{\partial v}{\partial t} + v\frac{\partial v}{\partial z}\right) - g\rho \sin\beta$$

can be rewritten as

$$\frac{\partial \phi}{\partial t} = v\frac{\partial \rho}{\partial t} - \left(1 + \phi^2 \frac{\partial v}{\partial p}\right)\frac{\partial p}{\partial z} - \phi^2 \frac{\partial v}{\partial p}\frac{\partial h}{\partial z} - \left.\frac{\partial p}{\partial z}\right|_{loss} - g\rho \sin\beta.$$

Replacing the time derivatives by their discrete time equivalents allows this equation to be written as

$$\begin{aligned}
&(\phi^{n+1} - \phi^n) - v^n(\rho^{n+1} - \rho^n) \\
&= -\theta\Delta t \left[\left(1 + \phi^2 \frac{\partial v}{\partial p}\right)\frac{\mathrm{d}p^{n+1}}{\mathrm{d}z} + \phi^2 \frac{\partial v}{\partial h}\frac{\mathrm{d}h^{n+1}}{\mathrm{d}z} - \left.\frac{\partial p}{\partial z}\right|_{loss}^{n+1} - g\rho^{n+1}\sin\beta\right] \\
&\quad -(1-\theta)\Delta t \left[\left(1 + \phi^2 \frac{\partial v}{\partial p}\right)\frac{\mathrm{d}p^n}{\mathrm{d}z} + \phi^2 \frac{\partial v}{\partial h}\frac{\mathrm{d}h^n}{\mathrm{d}z} - \left.\frac{\partial p}{\partial z}\right|_{loss}^{n} - g\rho^n \sin\beta\right].
\end{aligned}$$

Again, group terms at the nth and $(n+1)$th time steps to give

$$\begin{aligned}
\phi^{n+1} &+ \theta\Delta t\left(1 + \phi^2 \tfrac{\partial v}{\partial p}\right)\tfrac{\mathrm{d}p^{n+1}}{\mathrm{d}z} - \theta\Delta t\left(\phi^2 \tfrac{\partial v}{\partial h}\right)\tfrac{\mathrm{d}h^{n+1}}{\mathrm{d}z} \\
&+ \theta\Delta t\left(\left.\tfrac{\partial p}{\partial z}\right|_{loss}^{n+1} - g\rho^{n+1}\sin\beta\right) = S_\phi^n,
\end{aligned} \tag{5.43}$$

where

$$S_\phi^n = \phi^n + (1-\theta)\Delta t\left(1+\phi^2\frac{\partial v}{\partial p}\right)\frac{\mathrm{d}p^n}{\mathrm{d}z} - (1-\theta)\Delta t\left(\phi^2\frac{\partial v}{\partial h}\right)\frac{\mathrm{d}h^n}{\mathrm{d}z}$$
$$+(1-\theta)\Delta t\left(\left.\frac{\partial p}{\partial z}\right|_{loss}^{n} - g\rho^n \sin\beta\right)$$

S_ϕ^{n+1} is defined by the recursion relationship

$$S_\phi^{n+1} = \frac{1}{\theta}\phi^{n+1} - \frac{1-\theta}{\theta}S_\phi^n \tag{5.44}$$

with the boundary condition

$$S_\phi^0(z) = \phi^0(z). \tag{5.45}$$

Following the same procedure the energy equation 5.7

$$\frac{1}{A}\frac{\mathrm{d}\dot{q}}{\mathrm{d}z} = -v\left.\frac{\partial p}{\partial z}\right|_{loss} + \rho\left(\frac{\partial h}{\partial t} + v\frac{\partial h}{\partial z}\right) - \left(\frac{\partial p}{\partial t} + v\frac{\partial p}{\partial z}\right)$$

becomes

$$\rho\frac{\partial h}{\partial t} + \phi\frac{\partial h}{\partial z} = \left(\frac{\partial p}{\partial t} + v\frac{\partial p}{\partial z}\right) + \frac{1}{A}\frac{\mathrm{d}\dot{q}}{\mathrm{d}z} + v\left.\frac{\partial p}{\partial z}\right|_{loss}$$

which can be written as

$$h^{n+1} - v^{n+1}p^{n+1} - v\theta\Delta t\left[v\frac{\mathrm{d}p^{n+1}}{\mathrm{d}z} - \phi\frac{\mathrm{d}h^{n+1}}{\mathrm{d}z} + \left(\frac{1}{A}\frac{\mathrm{d}\dot{q}}{\mathrm{d}z} + v\Delta p_{loss}\right)^{n+1}\right] = S_h^n, \tag{5.46}$$

where

$$S_h^n = h^n - v^n p^n - v(1-\theta)\Delta t\left[v\frac{\mathrm{d}p^n}{\mathrm{d}z} + \phi\frac{\mathrm{d}h^n}{\mathrm{d}z} + \left(\frac{1}{A}\frac{\mathrm{d}\dot{q}}{\mathrm{d}z} + v\Delta p_{loss}\right)^n\right]$$

and satisfies the recursion relationship

$$S_h^{n+1} = \frac{1}{\theta}\left[h^{n+1} - v^{n+1}p^{n+1}\right] - \frac{1-\theta}{\theta}S_h^n \tag{5.47}$$

with the boundary condition

$$S_h^0(z) = h^0(z) - v^0 p^0(z). \tag{5.48}$$

Equations 5.43 and 5.46, being two equations in two unknowns $\frac{\mathrm{d}p^{n+1}}{\mathrm{d}z}$ and $\frac{\mathrm{d}h^{n+1}}{\mathrm{d}z}$, together with their auxiliary equations 5.44 and 5.47 known from the preceding time step, may be solved to yield the spatial profiles of p^{n+1} and h^{n+1} for all $z \in (0, L)$. They can be combined as

$$D\begin{bmatrix} \frac{\mathrm{d}p^{n+1}}{\mathrm{d}z} \\ \frac{\mathrm{d}h^{n+1}}{\mathrm{d}z} \end{bmatrix} = \begin{bmatrix} \phi & \phi^2 \frac{\partial v}{\partial h} \\ v & \left(1+\phi^2\frac{\partial v}{\partial p}\right) \end{bmatrix} \times \begin{bmatrix} \frac{1}{\theta\Delta t}\left[S_\phi^n - \phi^{n+1}\right] + (\Delta p_{loss} - g\,\rho\sin\beta)^n \\ \frac{1}{v\theta\Delta t}\left[S_h^n - h^{n+1} + v^{n+1}p^{n+1}\right] + \left(\frac{1}{A}\frac{\mathrm{d}\dot{q}}{\mathrm{d}z} + v\Delta p_{loss}\right)^n \end{bmatrix}, \tag{5.49}$$

where

$$D = \phi\left[1 + \phi^2\left(\frac{\partial v}{\partial p} - v\frac{\partial v}{\partial h}\right)\right].$$

For most practical purposes the second bracketed term in D is $\ll 1$ and $D \approx \phi$. Also, if we limit consideration to non-choked flow, we can assume that $\phi^2(\partial v/\partial p) \ll 1$. Equation 5.49 can then be expanded to the two equations

$$\frac{\mathrm{d}p^{n+1}}{\mathrm{d}z} = f_1(z) + \phi\frac{\partial v}{\partial h}f_2(z) \tag{5.50}$$

and

$$\frac{\mathrm{d}h^{n+1}}{\mathrm{d}z} = v f_1(z) + f_2(z), \tag{5.51}$$

where

$$f_1(z) = \frac{1}{\theta\Delta t}\left[S_\phi^n - \phi^{n+1}\right] + (\Delta p_{loss} - g\,\rho\sin\beta)^n\,,$$

$$f_2(z) = \frac{1}{v\theta\Delta t}\left[S_h^n - h^{n+1} + v^{n+1}p^{n+1}\right] + \frac{1}{\phi A}\frac{\mathrm{d}\dot{q}}{\mathrm{d}z} + v\Delta p_{loss}.$$

The solution of these two ordinary differential equations in the spatial ordinate z yields p^{n+1} and h^{n+1} for all z along the flow path. The specific volume v^{n+1} is calculated from the properties relationship

$$v^{n+1} = v(p^{n+1}, h^{n+1})$$

given p^{n+1} and h^{n+1} and

$$\rho^{n+1} = 1/v^{n+1}.$$

The specific mass flow rate ϕ^{n+1} is calculated by direct integration of Eq. 5.41, using ρ^{n+1} and S^n_ρ.

These equations are implicitly coupled via the right side terms defined at the $(n+1)$th time step, ϕ^{n+1} and p^{n+1}. It will therefore be necessary to solve the equations iteratively. However, if the pressure equation is solved first, the updated p^{n+1} will be available for use in the solution of the second for h^{n+1}.

The profile of $p^{n+1}(z)$ is obtained by direct integration of Eq. 5.50, starting from the inlet pressure $p^{n+1}(0)$.

$$p^{n+1}(z) = p^{n+1}(0) + \int^{z} \left[f_1(z') + \phi \frac{\partial v}{\partial h} f_2(z') \right] \mathrm{d}z'.$$

Similarly $h^{n+1}(z)$ is obtained by integration of Eq. 5.51 in the direction of flow

$$h^{n+1}(z) = h^{n+1}(0) + \int^{z} \left[v f_1(z') + f_2(z') \right] \mathrm{d}z'.$$

In most cases of practical application, the solutions $p^{n+1}(z)$, $h^{n+1}(z)$ and $\phi^{n+1}(z)$ will be sought for a discrete set of z-points interior to the path $(0, L)$. Then each of these equations, now including Eq. 5.41, can be replaced by summations over a set of equal-spaced z intervals.

$$p^{n+1}(z) = p^{n+1}(0) + \sum_{j=1}^{N_z} \left[f_{1,j} + \phi \frac{\partial v}{\partial h} f_{2,j} \right] \Delta z,$$

$$h^{n+1}(z) = h^{n+1}(0) + \sum_{j=1}^{N_z} \left[v f_{1,j} + f_{2,j} \right] \Delta z,$$

$$\phi^{n+1}(z) = \phi^{n+1}(0) + \frac{1}{\theta \Delta t} \sum_{j=1}^{N_z} \left[S^n_\rho - \rho^{n+1}_j \right] \Delta z$$

and N_z is the number of steps into which the range $(0, z)$ is divided.

A recursive form is better suited to actual calculations. Each variable can be calculated at the jth z-step in terms of the known conditions at the preceding $(j-1)$th step and the three equations solved as follows.

$$p^{n+1}_j = p^{n+1}_{j-1} + \left[f_{1,j} + \phi \frac{\partial v}{\partial h} f_{2,j} \right] \Delta z, \tag{5.52}$$

$$h^{n+1}_j = h^{n+1}_{j-1} + \left[v f_{1,j} + f_{2,j} \right] \Delta z, \tag{5.53}$$

$$\phi^{n+1}_j = \phi^{n+1}_{j-1} + \frac{1}{\theta \Delta t} \left[S^n_\rho - \rho^{n+1}_j \right] \Delta z. \tag{5.54}$$

Equations 5.50 and 5.52 are valid for zero flow and show that, for this condition,

$$\frac{\mathrm{d}p}{\mathrm{d}z} = -g\,\rho \sin\beta$$

that is, the pressure gradient is defined solely by gravity.

Equations 5.51 and 5.53 are not valid for zero flow due to divisions by ν and ϕ in $f_2(z)$. However, we note that

$$\frac{1}{\phi A}\frac{\mathrm{d}\dot{q}}{\mathrm{d}z} = \frac{1}{\dot{m}}\frac{\mathrm{d}\dot{q}}{\mathrm{d}z} = \frac{1}{M}\frac{\mathrm{d}\dot{q}}{\mathrm{d}z}\Delta t,$$

where M is the mass of the working medium per unit length. We can also assume that, for zero flow, the term $(S_h^n - h^{n+1} + v^n p^{n+1})$ will be zero and can be ignored. Equation 5.51 then indicates that for zero flow,

$$\frac{\mathrm{d}h}{\mathrm{d}z} = \frac{1}{M}\frac{\mathrm{d}\dot{q}}{\mathrm{d}z}\Delta t$$

that is, the local enthalpy gradient is defined by heat transferred to or from the medium per unit time.

Being based solely on advection this equation makes no provision for thermal diffusion through the working medium. Were diffusion to be included, in this case restricted to the axial flow coordinate, a term of the form $(c\partial^2 h/\partial z^2)$ must be added to the original energy equation. This will allow heat to move between cells independently of the flow condition and can assist in the reproduction of the correct cooldown behaviour of systems. This matter is discussed in more detail in Sect. 17.2.2 dealing with network simulation.

5.5 Closing Comment

It is apparent from the evolution of this chapter that there are several possible formulations and solutions of the basic conservation equations. Which is best suited to a given simulation application cannot be defined in advance. In Sect. 12.3.5, the serial method is applied directly to the treatment of a distributed evaporator (boiler riser) while in Chap. 17 complex networks, both incompressible and compressible, are treated using conventional spatial and time discretisation.

Chapter 6
Static Components

This chapter deals with the calculation of flows through and pressure drops across pipes, ducts, valves and dampers. These are described as "static" because, compared to rotating devices such as pumps and fans, their operation does not involve the introduction of mechanical work. They are the most common building blocks of networks, and valves and dampers are the most common flow control devices.

Pipes and ducts store significant mass and energy. In large, physically extensive networks they and the working fluid they carry can store more mass and energy than the components to which they connect, and thus they play a significant role in the accurate simulation of network dynamics. The time taken by the fluid to transit pipes and ducts introduces pure transport times between their inputs and outputs, the duration of which is dependent on flow velocities and whose effects are seen by measurements of transported quantities such as temperatures and measurements related to species concentrations, including conductivity, oxygen content and concentrations of dissolved solids.

Later chapters will consider the implications of these issues for the simulation of network dynamics. This chapter considers only the characterisation of the individual components.

6.1 Bernoulli's Equation

The generalised unsteady conservation of momentum equation in one space dimension z, being the axis of flow, is given in Chap. 5 as

$$\frac{\partial p}{\partial z} = - \left.\frac{\partial p}{\partial z}\right|_{loss} - \rho\left(\frac{\partial v}{\partial t} + v\frac{\partial v}{\partial z}\right) - g\rho \sin\beta,$$

P. O'Kelly, *Computer Simulation of Thermal Plant Operations*,
DOI 10.1007/978-1-4614-4256-1_6,

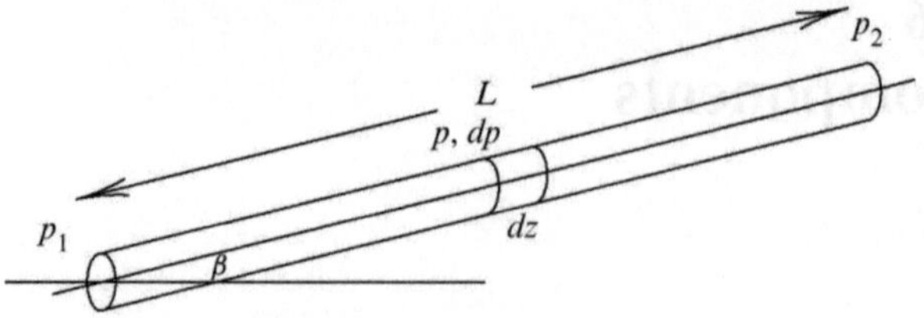

where $\begin{cases} p & \text{is pressure [kg/(s m}^2)], \\ \rho & \text{is the fluid density [kg/m}^3], \\ v & \text{is fluid flow velocity [m/s]}, \\ \beta & \text{is the angle of inclination of the flow to the horizontal.} \end{cases}$

Setting the time derivative to zero yields the steady-state equation, expressed now as a function of the spatial ordinate z only:

$$\frac{\partial p}{\partial z} = -\left.\frac{\partial p}{\partial z}\right|_{loss} - \rho v \frac{\partial v}{\partial z} - g\rho \sin\beta.$$

Having derivatives of z only, this equation may be integrated directly to give

$$p(z) = -p(z)|_{loss} - \frac{1}{2}\rho v^2 - g\, z\, \rho \sin\beta$$

or

$$\frac{pv}{g} + \frac{1}{2}\frac{v^2}{g} + \delta p_{loss} = z \sin\beta, \tag{6.1}$$

where $v = 1/\rho$ is the fluid specific volume [kg/m^3]. This is Bernoulli's equation for flow momentum conservation.

Consider a cell of length dz along the flow path. The pressure differential across the cell is dp. The velocity change is dv and the difference in geodetic height is dh. If compressibility effects are ignored over the distance dz there is no change in the specific volume v. From Bernoulli's equation we have, for conditions at the inlet and outlet of the cell,

$$\frac{pv}{g} + \frac{1}{2g}v^2 + \sin\beta\, h = \frac{(p + \mathrm{d}p)v}{g} + \frac{1}{2g}(v + \mathrm{d}v)^2 + \delta p_{loss} + \sin\beta\,(h + \mathrm{d}h).$$

The friction pressure loss gradient δp_{loss} is given by

$$\delta p_{loss} = \frac{4 f_f}{d_h}\frac{v^2}{2g}\mathrm{d}z,$$

where f_f is the (dimensionless) Darcy friction factor and d_h is the hydraulic diameter [m]. This equation can be rewritten as

$$-\frac{v}{g}\mathrm{d}p = \frac{1}{2g}\left(2v\,\mathrm{d}v + (\mathrm{d}v)^2\right) + \frac{4f_f}{d_h}\frac{v^2}{2g}\,\mathrm{d}z + \sin\beta\,\mathrm{d}h.$$

The terms in $\mathrm{d}v$ are negligible, and this equation simplifies to

$$-\frac{v\,\mathrm{d}p}{g} = \frac{1}{g}\frac{2f_f}{d_h}\frac{v^2}{g}\,\mathrm{d}z + \sin\beta\,\mathrm{d}h$$

or

$$-v\,\mathrm{d}p = \frac{2f_f}{d_h}v^2\mathrm{d}z + g\ \sin\beta\,\mathrm{d}h.$$

Multiplying both sides of this equation by p/v gives

$$-p\,\mathrm{d}p = \frac{2f_f}{d_h}\frac{p}{v}v^2\mathrm{d}z + \frac{p}{v}g\ \sin\beta\,\mathrm{d}h.$$

For a perfect gas, $pv = ZRT$ where the symbols have their usual meanings. Introducing $\phi = \dot{m}/A_x$ as the mass flow density, being the mass flow $\dot{m}$ crossing a flow cross-sectional area A_x, we may write this equation as

$$\begin{aligned}-p\,\mathrm{d}p &= \frac{2f_f}{d_h}ZRT\phi^2\mathrm{d}z + \frac{p}{v}g\ \sin\beta\mathrm{d}h\\ &= \frac{2f_f}{d_h}ZRT\frac{\dot{m}^2}{A_x^2}\mathrm{d}z + \frac{p}{v}g\ \sin\beta\,\mathrm{d}h.\end{aligned}$$

For a round pipe or conduit of equivalent area, $A_x = \pi/4\ d_h^2$, where d_h is the hydraulic diameter. The second of these equations may be written as

$$-p\,\mathrm{d}p = \frac{32f_f}{\pi^2 d_h^5}ZRT\ \dot{m}^2\mathrm{d}z + \frac{p}{v}g\ \sin\beta\,\mathrm{d}h.$$

Integrating this equation between two points 1 and 2 along a path length L gives

$$-\int_{(1)}^{(2)} p\,\mathrm{d}p = \frac{32f_f}{\pi^2 d_h^5}ZRT\ \dot{m}^2\int_0^L \mathrm{d}z + \frac{p}{v}g\ \sin\beta\int_{(1)}^{(2)}\mathrm{d}h$$

from which we obtain

$$-\frac{1}{2}(p_2^2 - p_1^2) = \frac{32f_f}{\pi^2 d_h^5}ZRT\ \dot{m}^2L + \frac{p_{av}}{v_{av}}g\ \sin\beta\,\Delta h$$

or

$$p_1^2 - p_2^2 = \frac{64 f_f}{\pi^2 d_h^5} ZRT\, \dot{m}^2 L + 2\frac{p_{av}}{v_{av}} g\, \sin\beta\, \Delta h.$$

This can be rearranged to give mass flow in terms of the difference in pressures at each end of the flow path as

$$\dot{m}^2 = \frac{\pi^2}{64} \frac{d_h^5}{f_f L\, ZRT} \left[(p_1^2 - p_2^2) - 2\frac{p_{av}}{v_{av}} g\, \sin\beta\, \Delta h \right]. \tag{6.2}$$

It is assumed here that the pressure ratio across the flow path never approaches that at which the flow velocity will approach its maximum, or critical, value (refer to Chap. 5).

The terms $p_{av} = (p_1 + p_2)/2$ and $v_{av} = (v_1 + v_2)/2$ are the average values of pressure and specific volume over the geodetic height Δh. In many cases the geodetic height difference can be ignored. Then, with A_x as defined earlier, Eq. 6.2 becomes

$$\dot{m}^2 = \frac{1}{4} \frac{d_i A_x^2}{f_f L\, ZRT_{av}} (p_1^2 - p_2^2). \tag{6.3}$$

This is the general steady-state flow equation for a compressible medium.

For a component for which $(p_1 - p_2)/p_1 < 0.1$ (used as a condition for incompressibility[1]) this equation can be simplified as follows. Noting that

$$(p_1^2 - p_2^2) = (p_1 + p_2)(p_1 - p_2) = 2p_{av}(p_1 - p_2)$$

we have, with $p_1 v_1 = ZRT_1$ and $p_{av} v_{av} = ZRT_{av}$,

$$\begin{aligned} \dot{m}^2 &= \frac{1}{2} \frac{d_i A_x^2 p_{av}}{f_f L\, ZRT_{av}} (p_1 - p_2) \\ &= \frac{1}{2} \frac{d_i A_x^2}{f_f L\, v_{av}} (p_1 - p_2) \\ &= \mathcal{A}(p_1 - p_2). \end{aligned} \tag{6.4}$$

The admittance $\mathcal{A}$ is given from

$$\mathcal{A} = \frac{1}{2} \frac{d_i A_x^2}{f_f L\, v_{av}}.$$

[1]Compressible flow is distinguished from incompressible flow by the variation in flow medium density along the flow path. Conventionally, a compressible flow medium may be treated as incompressible if the pressure change along the considered section of flow is less than 10 % of the operating pressure (absolute) of the medium. The ASME uses 7 % to distinguish fans from compressors.

This form is suitable for incompressible (hydraulic) and for low-pressure gas networks. It is not appropriate for high-pressure gas networks nor for steam networks (particularly including turbine expansion stages or turbine throttle valves working at high pressure drops). For these the compressible form $(p_1^2 - p_2^2)$ must be retained or at least approximated.

Equation 6.3 can be reworked in terms of $\mathcal{A}$. Then,

$$\begin{aligned}
\dot{m}^2 &= \frac{1}{2}\mathcal{A}\frac{v_1}{ZRT_{av}}(p_1^2 - p_2^2) \\
&= \frac{1}{2}\mathcal{A}\frac{1}{p_1}(p_1^2 - p_2^2) \\
&= \frac{\mathcal{A}}{2p_1}(p_1 + p_2)(p_1 - p_2) \\
&= \frac{1}{2}\mathcal{A}\left(1 + \frac{p_2}{p_1}\right)(p_1 - p_2)
\end{aligned}$$

for $p_2/p_1 = \Pi \gg \Pi_{cr}$, the critical pressure ratio. This shows that the effect of compressibility on a piping section is to reduce the apparent admittance as the pressure drop across the element increases. The 10 % limit mentioned earlier can be applied as below this limit the error incurred by neglecting the influence of compressibility is less than the uncertainty in prediction of the friction factor.

6.2 Piping Section

The pressure loss due to friction in a piping section of length L [m] can be calculated from the standard handbook formula [9]

$$\begin{aligned}
\Delta p &= \zeta p_s \\
&= \zeta \frac{v}{2}\phi^2 \\
&= \zeta \frac{v}{2}\frac{\dot{m}^2}{A_x^2} \qquad \text{[Pascals]},
\end{aligned}$$

where ζ is the resistance coefficient and p_s is the velocity head. With

$$\zeta = \xi L/d_h \tag{6.5}$$

we may write

$$\Delta p = \xi \frac{L}{d_h} \frac{1}{2} \frac{v}{A_x^2} \dot{m}^2$$

$$\dot{m}^2 = \mathcal{A} \Delta p \tag{6.6}$$

and

$$\mathcal{A} = \frac{2\, d_h A_x^2}{\xi L\, v}, \tag{6.7}$$

where $\begin{cases} d_h & \text{is the hydraulic diameter of the piping section [m],} \\ v & \text{is the fluid specific volume [m}^3\text{/kg],} \\ A_x & \text{is the flow cross-sectional area normal to the flow direction [m}^2\text{].} \end{cases}$

This is the same formula as derived in the preceding section, with one difference. Two friction factors are in common use. ξ is the Fanning friction factor. It is 4 times the Darcy friction factor f_f used in the preceding section

$$\xi = 4\, f_f .$$

The Fanning friction factor is used throughout the remainder of this book and is given from the following formula, due to Prandtl,

$$\xi = \left(1.14 + 2.0 \log\left(\frac{d_i}{K}\right)\right)^{-2}, \tag{6.8}$$

where d_i is the pipe section internal diameter [m] and K is the hydraulic roughness of the surface [mm]. The hydraulic roughness is quoted in handbooks for a variety of materials. Typical values are (quoting from Steinmueller [9])

Non-acid-cleaned steel tubes with mill scale	K = 0.02 − 0.06 mm,
Acid-cleaned steel tubes	K = 0.03 − 0.04 mm,
Moderately rusted steel tubes	K = 0.10 − 0.20 mm.

Values for other substances are to be found in, for example, Perry [17].

Also, for a pipe, quoting Swamee and Jain [16], Basu et al. [15] give the following formula for ξ:

$$\xi = 0.25 \left[\log\left(\frac{K}{3.7\, d_h} + \frac{5.74}{Re^{0.9}}\right)\right]^{-2} \tag{6.9}$$

for $5000 < Re < 10^8$ and $10^{-6} < (K/d_h) < 0.01$.

6.3 Valves and Associated Piping

A valve is never found in isolation. It is always connected into a network or attached to another component by a section of pipe at each end of the valve. It is therefore appropriate to treat a valve as if always joined to a section of piping on both sides and to consider the combined pressure loss in both valve and associated piping. That having been said, some valves do connect directly to the body of a component via a piping stub too short to be considered as a separate piping element. This is common practice for tanks, pumps, headers and valves that dump to waste.

The mass flow through a valve is conventionally defined in terms of the valve flow coefficient k_v and the differential pressure Δp_{vlv} across it:

$$\dot{m}^2 = k_v^2 \, v \, \Delta p_{vlv}, \tag{6.10}$$

where v is the specific volume of the working medium. The flow coefficient of the valve is a function of the cross-sectional area A_x at the valve throat, which in turn depends on valve stem position.

Consider the following section of piping containing a pump, a regulating valve and other components which may be treated as lumped admittances upstream and downstream of the valve.

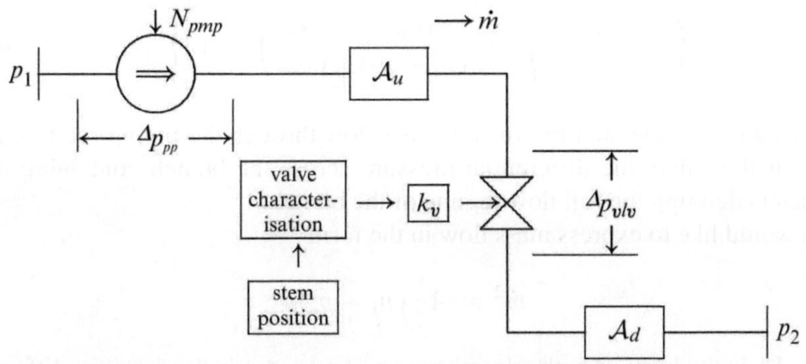

The pressure drop across the pipe sections excluding the valve is $(p_1 - p_2 - \Delta p_{vlv})$. The mass flow through the piping is equal to that through the valve and is given by

$$\dot{m}^2 = \mathcal{A}_{pp} \, (p_1 - p_2 - \Delta p_{vlv}), \tag{6.11}$$

where $\mathcal{A}_{pp}$ is the combined admittance of the upstream and downstream piping and other pressure loss elements. The maximum flow $\dot{m}_{100}$ through the valve for a given p_1 and p_2 is

$$\dot{m}_{100}^2 = k_{vs}^2 \, v \, \Delta p_{100}, \tag{6.12}$$

where Δp_{100} is the pressure differential across the fully open valve. The maximum value of k_v, denoted k_{vs}, is attained with the valve fully open. We may therefore write

$$\left(\frac{\dot{m}}{\dot{m}_{100}}\right)^2 = \left(\frac{k_v}{k_{vs}}\right)^2 \left(\frac{\Delta p_{vlv}}{\Delta p_{100}}\right) \tag{6.13}$$

from which follows

$$\dot{m}^2 = \left(\frac{\dot{m}_{100}^2}{\Delta p_{100}}\right) \left(\frac{k_v}{k_{vs}}\right)^2 \Delta p_{vlv}.$$

From Eq. 6.11 we may write

$$\left(\frac{\dot{m}}{\dot{m}_{100}}\right)^2 = \frac{(p_1 - p_2 - \Delta p_{vlv})}{(p_1 - p_2 - \Delta p_{100})}$$

from which

$$\Delta p_{vlv} = (p_1 - p_2)\left[1 - \left(\frac{\dot{m}}{\dot{m}_{100}}\right)^2\right] + \Delta p_{100}\left(\frac{\dot{m}}{\dot{m}_{100}}\right)^2.$$

After substitution of this expression into the equation for $\dot{m}^2$ and some rearrangement of terms, we arrive at

$$\left(\frac{\dot{m}}{\dot{m}_{100}}\right)^2 = \left\{1 + \frac{\Delta p_{100}}{(p_1 - p_2)}\left[\left(\frac{k_{vs}}{k_v}\right)^2 - 1\right]\right\}^{-1}. \tag{6.14}$$

This equation is exact and returns the mass flow through the piping section, given the k_v of the valve, the differential pressure across the branch containing it and parameters defining the full flow regime of the branch.

We would like to express mass flow in the form

$$\dot{m}^2 = \mathcal{A}_\Sigma\,(p_1 - p_2),$$

where $\mathcal{A}_\Sigma$ is the lumped path admittance and $(p_1 - p_2)$ is the pressure differential across the path. A minor rearrangement of terms in Eq. 6.14 produces

$$\left(\frac{\dot{m}}{\dot{m}_{100}}\right)^2 = \frac{\left(\frac{k_v}{k_{vs}}\right)^2}{(p_1 - p_2 - \Delta p_{100})\left(\frac{k_v}{k_{vs}}\right)^2 + \Delta p_{100}}(p_1 - p_2) \tag{6.15}$$

which is of the required form but now has the differential term $(p_1 - p_2)$ as part of the admittance term itself. This implicit dependence on $(p_1 - p_2)$ can be eliminated as follows. Define

$$Dp_{12} = |p_1 - p_2|_{100},$$

where Dp_{12} is the branch pressure differential under full rated flow conditions and will be known from plant operating data. From Eq. 6.11,

$$\dot{m}_{100}^2 = \mathcal{A}_{pp}(Dp_{12} - \Delta p_{100})$$

giving

$$\mathcal{A}_{pp} = \frac{\dot{m}_{100}^2}{Dp_{12} - \Delta p_{100}}.$$

Substitution for this expression back in Eq. 6.15 gives

$$\left(\frac{\dot{m}}{\dot{m}_{100}}\right)^2 = \frac{\left(\frac{k_v}{k_{vs}}\right)^2}{\Delta p_{100} + \left(\frac{k_v}{k_{vs}}\right)^2 (Dp_{12} - \Delta p_{100})}(p_1 - p_2). \tag{6.16}$$

The combined piping-valve admittance is calculated from,

$$\begin{aligned}\mathcal{A}_{\Sigma} &= \frac{\left(\frac{k_v}{k_{vs}}\right)^2}{\Delta p_{100} + \left(\frac{k_v}{k_{vs}}\right)^2 (Dp_{12} - \Delta p_{100})} \\ &= \frac{\left(\frac{k_v}{k_{vs}}\right)^2}{\Delta p_{100}\left(1 + \left(\frac{k_v}{k_{vs}}\right)^2 \left(\frac{Dp_{12}}{\Delta p_{100}} - 1\right)\right)}.\end{aligned} \tag{6.17}$$

We will denote the ratio $\frac{Dp_{12}}{\Delta p_{100}}$ as the "DP ratio". It is the ratio of the pressure drop across the flow branch, including the valve, to that across the valve alone under full flow conditions. It is a parameter of the branch/valve combination and can take any value greater than unity. A value of unity implies zero pressure drop across the piping as all of the drop is then across the valve. A value of, say, ten implies that only 10 % of the branch pressure drop occurs across the valve.

In principle the same reasoning applies to a damper located in a duct. However, in almost all cases, the pressure drop along the duct is much less than across the damper, and for most practical purposes, the DPR for a damper can be treated as unity.

6.3.1 Flow Characteristics of Various Valve Types

Comparing the flow equation for a valve, Eq. 6.10, with that of the pressure loss through a piping section reveals that a valve may be treated as a piping section of variable flow cross section and that the k_v of the valve is simply

$$k_v^2 = \frac{2\,A_x^2}{\zeta},$$

where A_x is the flow cross-sectional area of the valve throat and $k_{v,\max} = k_v$ for the fully open valve. Differing valve types have differing characteristics as a result of

(i) Differing relationships between the stem position and the throat area
(ii) The increased pressure drop in the valve over that given by the pipe analogy caused by the internal flow path and obstructions introduced by its construction

The first of these is a matter of the valve design, and a specific characteristic is achieved by appropriate design of the valve drive linkages or by shaping the valve cone, seat or porting. Typical characteristics are, where θ is the relative valve stem position $0 \leq \theta \leq 1$,

Linear	θ,	
Gate or globe valve	$\sqrt{\theta}$,	
Control type valve	θ^2,	
Butterfly valve	$\left(\frac{1.4}{1-\frac{\sin(\theta)}{\sin(\theta_0)}} - 1\right)$	$\theta_o =$ diaphragm angle when shut.

Figure 6.1 shows the normalised flow characteristics of each of these types for variation of relative opening from 0 % to 100 % and for a range of DP ratios from 1 to 10.

The second factor (ii) depends on the specific valve and type. Typical values of the resistance coefficient ζ are given in Steinmueller as

Gate or globe type valve	3.6 – 4.0,
Angle type valve	2.0 – 2.8,
Swing type check valve	0.8 – 1.9.

6.3.2 An Operational Note About Valves

The preceding development and the illustrations of Fig. 6.1 reveal a subtle but often forgotten principle of the action of valves as flow regulating devices. Before a valve can influence the flow through it significantly, the pressure drop across the valve must be significant compared to the total pressure drop across the branch. As Fig. 6.1 reveals, as the DP ratio increases, the valve must close further before any significant reduction in flow is observed. The valve characteristics also show that as the branch DP ratio increases each valve tends towards a similar shape and progressively loses its own characteristic shape. It is for this reason that control valve throat sizing is usually of a smaller diameter than the connecting piping and that a control valve must be matched, not only to its test head/flow characteristic but also to the piping section into which it will be installed.

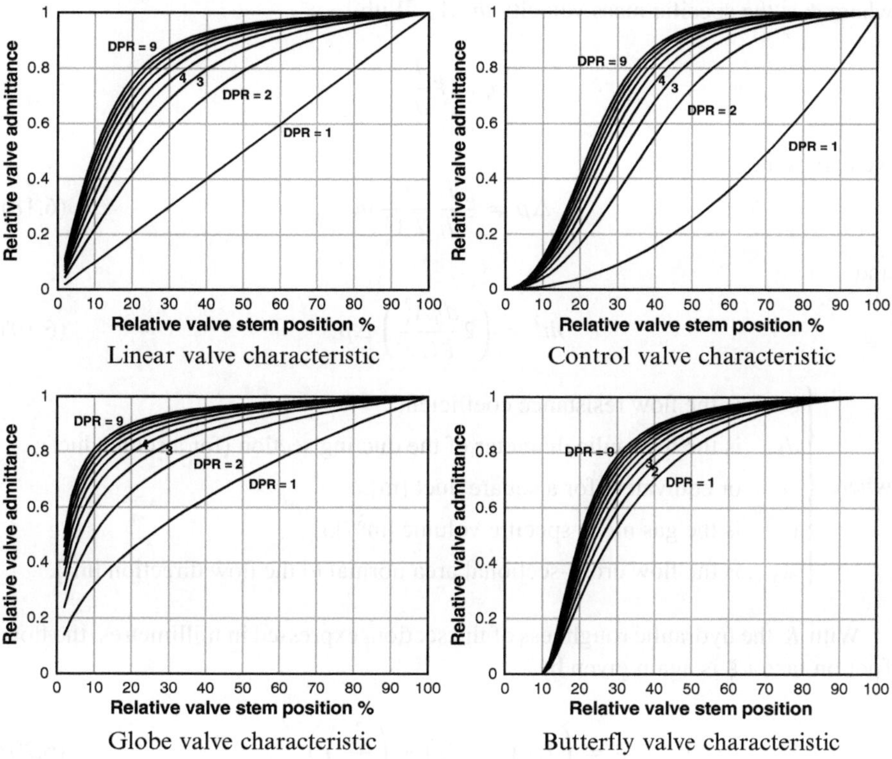

Fig. 6.1 Valve flow characteristics as function of type and DP ratio

6.4 Ducts and Dampers

6.4.1 Ducts

Sections of ducting are widely used to carry gases of various compositions around the plant. They are usually operated at or near atmospheric pressure (generally less than say 30 mbar gauge) and therefore have thin walls. They vary in length from a metre or two up to several hundreds of metres and in cross-section from a few square centimetres up to several square metres. Pressure gradients are small but significant to the distribution of flows among several interconnected branches.

The pressure loss due to friction in a ducting section of length L [m] is calculated from the standard formulae (Eck [19] p. 389).

$$\Delta p = \zeta \frac{v}{2} \phi^2 = \zeta \frac{v}{2} \frac{\dot{m}^2}{A_x^2},$$

where ϕ is the specific mass velocity $\dot{m}/A_x$. With,

$$\zeta = \xi \frac{L}{d_h}$$

we may write

$$\Delta p = \xi \frac{L}{d_h} \frac{v}{2A_x^2} \dot{m}^2 \tag{6.18}$$

and

$$\dot{m}^2 = \left(2 \frac{d_h A_x^2}{\xi L v} \right) \Delta p, \tag{6.19}$$

$$\text{where} \begin{cases} \zeta & \text{is the flow resistance coefficient,} \\ d_h & \text{is the hydraulic diameter of the ducting section (for a round duct),} \\ & \text{or equivalent for a square duct [m],} \\ v & \text{is the gas mean specific volume [m}^3\text{/kg],} \\ A_x & \text{is the flow cross-sectional area normal to the flow direction [m}^2\text{].} \end{cases}$$

With K the hydraulic roughness of the section, expressed in millimetres, the flow friction factor ξ is again given by

$$\xi = \left(1.14 + 2.0 \log \left(\frac{d_h}{K} \right) \right)^{-2}. \tag{6.20}$$

The hydraulic diameter for a round duct is its internal diameter [m]. For a rectangular duct $d_h = 4A_x/U$ where A_x is the duct cross-sectional area perpendicular to the flow direction and U is the duct circumference [m]. Alternatively, for a rectangular duct with cross-sectional dimensions a, b expressed in millimetres, Basu et al. [15] give the following formula for calculation of the hydraulic diameter d_h:

$$d_h = 1.3 \frac{ab^{0.625}}{(a+b)^{0.25}}. \tag{6.21}$$

For large ducts ($d_h > 0.8$ m) the flow friction factor is substantially independent of d_h and hydraulic roughness as the velocity profile is constant across most of the duct section. For these ducts (the majority of combustion air and flue gas ducts) ξ may be assigned the value of 0.0023.

For ducts containing tube banks (superheaters, reheaters and economisers) due allowance is to be made in the calculation of A_x for the area occupied by the tubing. The effects of tube and bank spacing are considered via bank factors, as described in Sect. 9.3.

Fig. 6.2 Utility boiler louvre control dampers during manufacture (with permission of Everlee Engineering, Johannesburg, South Africa)

The admittance of a duct section free of obstructions is calculated from

$$\mathcal{A} = 2\frac{d_h A_x^2}{\xi L v} \tag{6.22}$$

and the symbols have their usual meanings.

6.4.2 *Dampers*

Dampers are usually connected into a duct network or are attached to another component by a short section of duct at each end of the damper. Some dampers are integrated with the body of an associated component. This is common practice for fans, pulverisers, heat exchangers, etc. It is sufficient to consider such dampers as integral to the associated component.

The most commonly used dampers are:

(a) Louvre dampers with parallel action
(b) Louvre dampers with counter-rotating action
(c) Butterfly dampers

Figure 6.2 illustrates large louvre dampers as used in a large utility boiler, in differing stages of manufacture and erection. The multiple parallel operating louvres are intended to provide restriction to the gas flow while introducing minimal turbulence to the flow.

The mass flow through a damper is related to the pressure differential across it by

$$\dot{m}^2 = k_v^2 \, v \, \Delta p_{dpr}, \tag{6.23}$$

where Δp_{dpr} is the differential pressure across the damper [kPa] and v is the gas specific volume [m^3/kg]. k_v is the flow coefficient of the damper and is a function of damper drive shaft angle of rotation. Note that

(a) For vanes positioned linearly by an electrically driven actuator, the angle of rotation is proportional to motor runtime.
(b) For vanes positioned by an hydraulic or pneumatic actuator, the angle of rotation is proportional to actuator position.

6.5 A Note to Practical Applications

Pressure drops and control damper and valve characteristics will differ, sometimes significantly, from the theoretical values predicted by the formulae presented in this chapter. Practical pipes include bends, irregularities, abrupt changes in cross-sectional area, internal scaling, corrosion and damage, any or all of which will increase pressure losses beyond the theoretical values. While the standard formulae will in general give good agreement with actual plant data it will almost always be necessary to modify the theoretical coefficients to give a better match to reality, if this is important.

In some applications, such as high-pressure gas networks, special formulae have been developed to suit the applications. The interested reader is referred to the specialised literature if the methods described here are found wanting. For example, the book of Osiadacz [21] provides a detailed treatment of the simulation of high-pressure gas networks.

Chapter 7
Turbomachines

Rotating machines, specifically that group of rotating machines known as turbomachines, are widely used throughout the manufacturing, process and power generation industries for the forced movement of air, gases, fluids and steam. They encompass fans, pumps, compressors and turbines. Each includes a small number of sub-types which cover a large range of configurations, capacities and operating conditions.

A fan, pump or compressor induces movement of a fluid or gas by the forced rotation of its rotary component, the impeller. Usually driven by an electric motor or sometimes, in the case of large pumps, fans and compressors, by a steam turbine, these machines force the working fluid (liquid or gas) along a path from its inlet (intake) to its outlet (discharge). The flow path between inlet and outlet is defined by the shape, orientation and spacing of fixed and rotating blades.

The flow through the impeller can be resolved into three principal directions:

- Axial, being parallel to the axis of rotation of the impeller
- Radial, also called centrifugal, directed outwards towards the blade tip along the local blade's surface
- Circumferential or tangential, in the direction of the blade's forward movement

The flow between adjacent pairs of blades exhibits a velocity profile normal to the flow. A thin boundary layer is attached to each blade surface. The bulk of the medium moves between these layers as a mixture of linear and circulatory flow, with a velocity profile which is maximum along a line close to the centreline of the inter-blade space. The design of the impeller ensures that, under design conditions, the circulatory component is minimal, and the flow may be considered to be linear and smooth. Under some operating conditions, this will not be the case, and the machine's performance will deteriorate as a result. In this treatment we will ignore the velocity profile normal to the flow path and consider all flow streamlines to be of equal velocity, generally aligned with one or other of the blade surfaces.

At a radial position r from the shaft centreline the blade has a forward or tangential speed of $r\omega$ where ω is the rotational speed of the impeller [rad/s].

P. O'Kelly, *Computer Simulation of Thermal Plant Operations*,
DOI 10.1007/978-1-4614-4256-1_7,

The flow into the impeller may be unconstrained or may be directed along a specific path by guide vanes ahead of the rotating blades which introduce the flow to the impeller at some desired angle. If the guide vanes introduce the flow at any angle other than normal to the plane of rotation of the impeller they are said to introduce "pre-rotation" which can be positive or negative. The angle between the inflow direction and the angle of the blade leading edge is the incidence angle. It is not uncommon in large fans and compressors to use fixed or adjustable guide vanes to induce some desired incidence angle.

Two types of turbomachines are in common usage:

- Centrifugal flow, in which the gas flow is propelled radially through the rotor
- Axial flow, in which the direction of flow is aligned predominantly along the rotor shaft, with no or negligible radial component

A third type, known as mixed flow, incorporates a mixture of both axial and radial flow. The impeller blades of such machines frequently resemble those of an axial flow fan but with an enlarging conical hub to induce the radial component.

Turton [23] provides a general coverage of the underlying principles of analysis of turbomachines.

7.1 Fans

Fans are distinguished from compressors by the small pressure ratio between outlet and inlet (<7 %). These small compression ratios allow the neglect of density changes of the gas flowing through the fans and the simple treatment of interchangeable volumetric and mass flows using an average gas density.

The flow leaving the impeller may be passed to the pipe or duct (the diffuser) along a free path in which any swirl induced by the rotary impeller will persist. In larger axial flow fans the diffuser may be preceded by a set of fixed blades which straighten the flow and convert the swirl energy into static pressure.

Unducted single-stage axial flow fans, with or without external shrouding, are frequently found in industrial applications as the source of air flow for heat exchangers (such as fin-fan coolers) and air-cooled condensers. [18] provides a detailed coverage of the industrial applications of fans of all types.

7.1.1 Centrifugal Fans

The following diagram depicts the velocity vectors at both root and periphery (discharge) for a centrifugal fan with forwards-curving blades. Inlet conditions and velocities at the blade root are identified by the subscript "1" while those at the outer blade edge (outlet conditions) are subscripted "2".

Fig. 7.1 A large forced draft centrifugal fan rotor (Courtesy of Millmerran Operating Company, Australia)

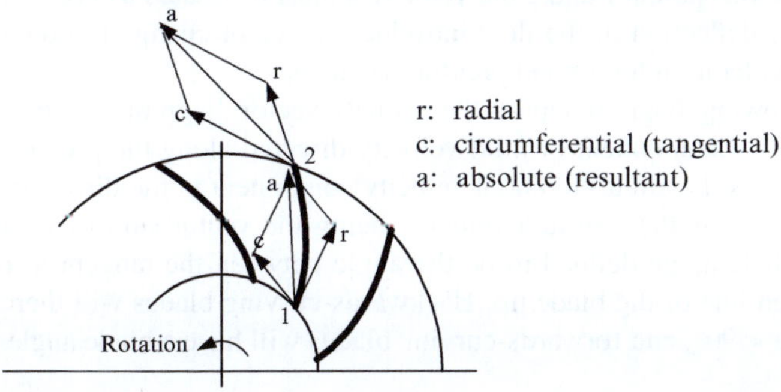

The following two diagrams show the same information for a centrifugal fan with (a) backwards-curving blades and (b) straight blades (no curvature at the tip).

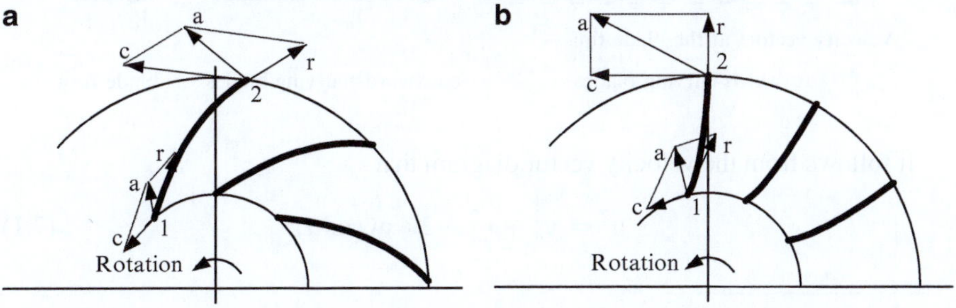

A key performance parameter of a fan (or pump or compressor) is the mass or volumetric flow of the working medium. The relationship between the pressure differential across the fan and the volumetric or mass flow through it is the *head/flow characteristic*. The actual flow achieved in a specific installation will depend on the match between the head developed by the fan and the system resistance characteristic. Since most flow resistance arises as friction losses these are adequately represented as proportional to the square of flow velocity[1] and the system resistance characteristic will be a quadratic function of mass (or volumetric) flow, passing through the zero-flow origin.

Flow Through the Impeller

If we neglect the irreversible pressure loss term in the Bernoulli equation, Eq. 6.1, we can write, for two points 1 and 2 along a flow path,

$$p_1 v_1 + \frac{1}{2} v_1^2 = p_2 v_2 + \frac{1}{2} v_2^2.$$

For a straight flow path the velocity v is aligned with the flow direction and at the two points can be treated as a scalar. If the flow undergoes a change of direction between the two points 1 and 2 the velocities must be treated as vectors. As we will now see the deflection of the flow introduces a vector change in flow momentum, creating acceleration forces and pressure changes.

The following diagram depicts the velocity vector diagram for a centrifugal fan. v_r is the radial component of fluid velocity directed along the pressure surface of the blade, v_c is the circumferential velocity component in the direction of impeller rotation and v_a is the absolute velocity, being the vector sum of v_r and v_c. β is the local blade angle defined to be the angle between the tangent to rotation and the extended line of the blade tip. Backwards-curving blades will therefore have a blade angle < 90°, and forwards-curving blades will have a blade angle > 90°.

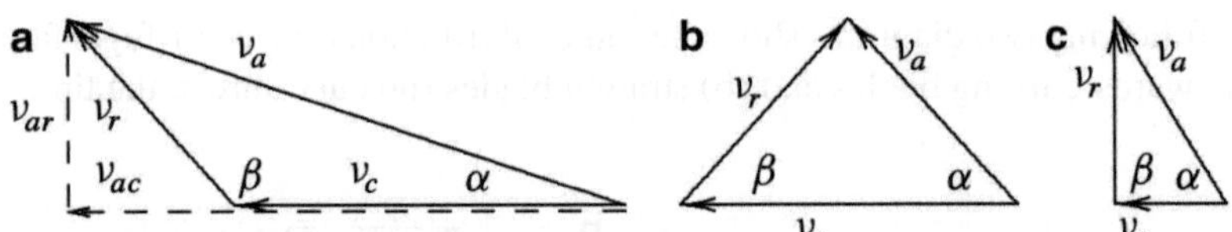

Velocity vectors at the blade tip:

forwards-curving blades backwards-curving blades blade root

It follows from the velocity vector diagram that

$$v_c^2 = v_r^2 + v_a^2 - 2 v_r v_a \cos\alpha, \tag{7.1}$$

[1]For fans or pumps discharging to some forms of filters, including liquid baths, the system resistance will deviate significantly from v^2 and may even be constant.

$$\cos\alpha = \frac{v_r^2 + v_a^2 - v_c^2}{2v_c\, v_a}, \tag{7.2}$$

$$v_c = v_a \cos\alpha \frac{\tan\beta}{\tan\alpha + \tan\beta}, \tag{7.3}$$

$$v_{ac} = v_a \cos\alpha. \tag{7.4}$$

This last equation defines v_{ac} to be the component of absolute velocity v_a in the circumferential (rotational) direction.

We will later refer to v_{ar} being the component of absolute velocity in the radial direction, calculated from

$$\tan(\pi - \beta) = -\tan\beta = \frac{v_{ar}}{v_{ac} - v_c}$$

from which

$$v_{ac} = v_c - \frac{v_{ar}}{\tan\beta}. \tag{7.5}$$

The energy imparted to the fluid by the impeller appears as an increase in kinetic energy (as a result of an increase in absolute velocity) and an increase in static pressure caused by centrifugal force. In general the radial fluid speed at the tip is less than at the root, giving rise to an increase of static pressure at the tip.

The Euler Equation for a Fan

The application of Bernoulli's equation at the inlet and outlet edges produces the equation

$$p_1 v_1 - p_2 v_2 = \left[(v_{a2}^2 - v_{a1}^2) + (v_{c2}^2 - v_{c1}^2) + (v_{r1}^2 - v_{r2}^2)\right]. \tag{7.6}$$

If the pressure differential is sufficiently small to allow neglect of compression of the gas so that we can assume $v_1 = v_2 = v$ we can write

$$\Delta p_{th} = \frac{1}{v}\left[(v_{a2}^2 - v_{a1}^2) + (v_{c2}^2 - v_{c1}^2) + (v_{r1}^2 - v_{r2}^2)\right]. \tag{7.7}$$

This is referred to as the "theoretical" pressure rise as it takes no account of internal losses or of the fact that the radial and circumferential velocities do not line up exactly with the blade angles, either at the inlet or outlet of the stage.

Equation 7.1 allows us to eliminate v_{r1} and v_{r2} from Eq. 7.7 to give the usual form of the Euler equation for a fan

$$\Delta p_{th} = \frac{1}{v}(v_{c2}\, v_{a2c} - v_{c1}\, v_{a1c}). \tag{7.8}$$

Given that local compressibility effects can be ignored in a fan, volumetric and mass flow can be used interchangeably if the local gas density ρ or specific volume v is known.

$$\dot{m} = Q\,\rho = \frac{Q}{v}. \tag{7.9}$$

The assumption of incompressibility also means that the in and out volumetric and mass flows are equal; hence, $Q_1 = Q_2 = Q$ and $\dot{m}_1 = \dot{m}_2 = \dot{m}$. The total volumetric flow leaving the fan (the sum of the individual blade channel flows) can be calculated from

$$Q = \nu_{a2r} A_2 = \nu_{a2r} 2\pi r_2 b_{c2} \quad [\mathrm{m}^3/\mathrm{s}],$$

where $A_2 = 2\pi r_2 b_{c2}$ is the discharge area of the fan disk and r_2 and b_2 are the outer radius of the fan disk and the chord width of the fan blade at the tip, respectively.[2] This equation can be rearranged into

$$\nu_{a2r} = \frac{Q}{2\pi\, r_2\, b_{c2}}.$$

Replacing ν_{a2r} in Eq. 7.5 gives, for the stage discharge flow,

$$\nu_{a2c} = \nu_{c2} - \frac{Q}{2\pi\ r_2\, b_{c2}\ \tan\beta_2} \tag{7.10}$$

and, at the inlet,

$$\nu_{a1c} = \nu_{c1} - \frac{Q}{2\pi\ r_1\, b_{c1}\ \tan\beta_1}. \tag{7.11}$$

With these substituted into Eq. 7.8, the theoretical pressure developed in the fan becomes

$$\Delta p_{th} = \frac{1}{v_2}\nu_{c2}^2 - \frac{1}{v_1}\nu_{c1}^2 - \frac{Q}{v\,2\pi}\left(\frac{\nu_{c2}}{r_2\, b_{c2}\ \tan\beta_2} - \frac{\nu_{c1}}{r_1\, b_{c1}\ \tan\beta_1}\right).$$

The two circumferential velocities ν_{c1} and ν_{c2} are equal to the fan rotational speed ω [rad/s] multiplied by the respective radius r. This equation can then be written in terms of the fan speed and the relevant dimensions.

$$\Delta p_{th} = \omega^2 \frac{1}{v} r_2^2 \left(1 - \frac{r_1^2}{r_2^2}\right) - \dot{m}\frac{\omega}{2\pi}\left(\frac{1}{b_{c2}\ \tan\beta_2} - \frac{1}{b_{c1}\ \tan\beta_1}\right). \tag{7.12}$$

In many cases, the blade entry (root) angle is close to 90°, and $\tan\beta_1$ is much larger than $\tan\beta_2$. If additionally the outer radius is some multiples greater than the root radius, Eq. 7.12 may be simplified to

---

[2]This ignores the area occupied by the blades seen edge on; this is small, and little error is incurred.

$$\Delta p_{th} = \frac{1}{v}\omega^2 r_2^2 - \dot{m}\left(\frac{\omega}{2\pi\, b_{c2}\, \tan\beta_2}\right). \tag{7.13}$$

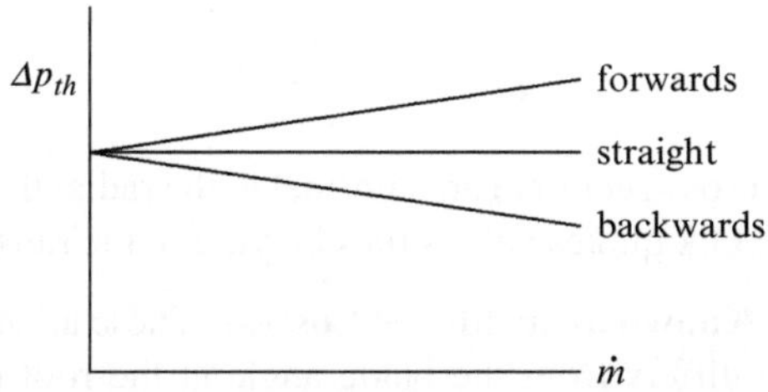

Equation 7.13 reveals some significant features of the centrifugal fan developed head/flow characteristic, in the absence of losses.

1. The zero-flow developed head is proportional to the square of the blade tip speed, being the product of the rotational speed and impeller outer radius.
2. As flow increases through the fan the developed head changes linearly with the mass flow and fan speed. The direction of change is set by the construction of the fan via the sign of $\tan\beta_2$, that is, whether the fan blades are inclined forwards (in the direction of rotation) or backwards:
   - The head reduces for backwards-curving blades.
   - The head increases for forwards-curving blades.
   - There is no change for straight-tipped blades.

Pressure Losses

Pressure losses arise in the fan as friction and shock losses.

Friction Losses Within the Impeller These constitute the largest contribution to fan losses at higher flows. They are difficult to predict theoretically using the standard formulae which work well for pipes and ducts. The combination of complex flow patterns and geometries and a mixture of laminar and turbulent flow regions within the inter-blade space render the assumptions behind the standard correlations inappropriate to this case. Moreover, what might be a true picture of things at one flow is unlikely to be valid at any other throughput.

Eck [19][3] provides an attempt at consolidation of a theoretical basis for the prediction of frictional losses in the impeller and arrives at the simple formula:

$$\Delta p_{loss} = C_w\, A_s \frac{\rho}{2} w_1^2 \frac{w_1}{Q}, \tag{7.14}$$

[3] Also available in English [20].

where A_s is the total internal surface area of the fan, including blades and the casing sidewalls, and w_1 is the maximum gas velocity. If we take w_1 as the radial component of the inlet gas velocity v_{r1}, this formula may be written as

$$\Delta p_{loss} = C_w \frac{A_s}{A_1^2} \frac{v}{2} \dot{m}^2, \tag{7.15}$$

where A_1 is the inlet cross-sectional area normal to the radial flow and $\dot{m}$ is the mass flow through the fan. Eck quotes values for C_w lying in the range 0.0065–0.007.

Shock Losses (Also Known as Incidence Losses) These arise when the incoming flow approach angle differs from the blade angle at the root (non-zero incidence angle), and the flow must be turned through that angle. The fan design will ensure optimal operation of the fan at its nominal operating point, and any departures from the design point will result in a loss of performance, either as a reduction in efficiency or a decrease in the developed head. It should be noted that a fan's optimum efficiency and maximum developed head will not occur at the same flow.

A formula relating the pressure change due to shock loss and flow is difficult to develop. Assuming that the design of the fan results in minimum shock loss at the nominal operating condition, any departure from this condition will cause the shock loss to increase. The contribution to the total losses at higher flows is not significant and can be ignored in the formulation of the head/flow characteristic. At low flows, for which viscous friction losses are reduced, shock loss will be significant. A workable approximation can be made which can be fitted to a fan's known efficiency characteristic. Since shock losses are minimum at the design operating point flow, we might assume a loss proportional to the square of the difference between this and any other mass flow.

$$\Delta p_{shock} = \Delta p_{shock,min} + C_{shock} (\dot{m} - \dot{m}_{OP})^2, \tag{7.16}$$

where $\dot{m}_{OP}$ is the mass flow at the nominal operating point. The constants $\Delta p_{shock,min}$ and C_{shock} are selected to give the best match to the fan's known efficiency characteristic (refer to Sect. 7.1.7).

7.1.2 Calculation of the Head/Flow Characteristic

Equations 7.13 and 7.15 can be combined to give the equation of the theoretical head/flow characteristic of a centrifugal fan, neglecting the shock losses.

$$\Delta p_{th} = \left(\frac{r_2^2}{v}\right) \omega^2 - \dot{m} \left(\frac{\omega}{2\pi\, b_{c2} \tan \beta_2}\right) - \left(C_w \frac{A_s}{A_2^2} \frac{v}{2}\right) \dot{m}^2. \tag{7.17}$$

The first term is the zero-flow head and is proportional to the square of the impeller rotational speed ω. This can be expressed in terms of the more usual revolutions per minute (rpm) by noting that ω [rad/s] = $2\pi N$ [rpm]/60. The second and third terms show the variation of the head across the fan as a function of the mass flow [kg/s]. Equation 7.17 assumes zero whirl at the fan inlet ($\beta_1 \simeq 90°$).

The following diagram shows the calculated Δp_{th} for the three possible blade orientations of a centrifugal fan. The characteristics have been adjusted to pass through the same nominal operating point. This diagram suggests that only fans with forwards-curving blades will show a peak in the head/flow characteristic, that is, a section to the left of the peak showing a positive slope. In real fans, this is not the case, and most fans will show some peak in the characteristic, with a loss of developed head at lesser flows. This is caused by departure of the flow pattern from optimal which, to a first approximation, can be included by a shock loss term of the form of Eq. 7.16.

7.1.3 The Head/Flow Characteristic from Fan Data

The head/flow characteristic can be predicted using a fan's known physical design parameters and physical dimensions. However, for simulation purposes, since the fan is known, it is more convenient and more accurate to fit an analytic function to the fan's test head/flow characteristic. The preceding theoretical analysis has given insight into the specific form of this fitting function and, equally importantly, the manner in which its coefficients might be varied to accommodate varying plant operational conditions.

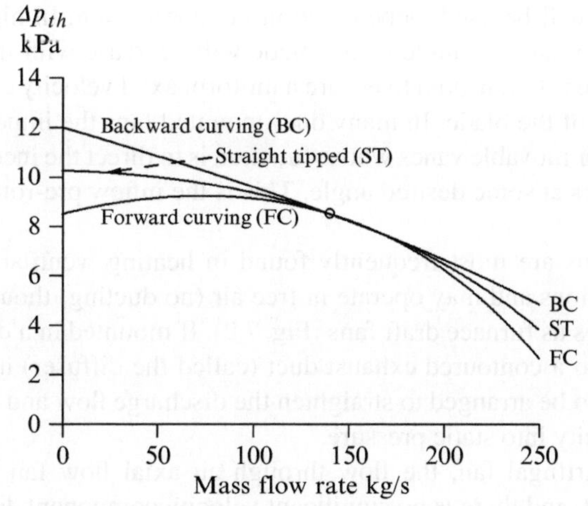

The discussion has not provided any information of the selection of the fan design parameters and dimensions to meet a specific performance objective as that is the business of the fan designer. The purpose here is to provide a well-founded basis for using the known parameters of a given fan to calculate its performance under varying operational conditions.

Units

Using the consistent SI system of units adopted throughout this book—$\dot{m}$ in kg/s and lengths in metres—Eq. 7.17 will calculate Δp_{th} in Pascals (Pa). The usual practical unit of pressure is kiloPascals (kPa). This is significant when using calculated values but will be absorbed when fitting to actual measured data if pressures are quoted in kPa.

Other units are often used when quoting fan test data. In customary US and British units developed pressure or head is usually quoted in inches of mercury (in Hg) (1 technical atmosphere (atm) = 101.3 kPa (absolute) = 29.98 in Hg). Therefore multiply head in inches by 2.6442 to obtain head in kPa. Volumetric flows are usually quoted in SI units as m^3/s and in customary US units and British units as cubic feet per minute (CFM) (1 cu. foot = 0.028317 cu.m). Multiply CFM by 4.7195E-4 to obtain m^3/s.

Note that test pressures are often quoted in gauge (above atmospheric) units rather than absolute (above zero).

7.1.4 Axial Flow Fans

Axial flow fans use an arrangement of blading to propel the gas predominantly parallel to the impeller axis of rotation. The fan consists of a bladed impeller rotating in free air or within an enclosure or duct. The impeller is provided with a number of radially mounted blades inclined to the axis of rotation. Typically anywhere between 6 and 32 blades will be used, depending on the application. In high-capacity and high-performance fans the angle of the blade will decrease with increasing radius (distance from the rotation axis) to ensure a uniform axial velocity component along the radial length of the blade. In many duct-mounted fans the impeller is preceded by a set of fixed or movable vanes whose function is to direct the incoming flow onto the rotating blades at some desired angle. This is the inflow pre-rotation mentioned earlier.

Axial flow fans are most frequently found in heating, ventilation and cooling (HVAC) applications and may operate in free air (no ducting) though they are also used in large sizes as furnace draft fans (Fig. 7.2). If mounted in a duct, the fan will often exhaust into a contoured exhaust duct (called the diffuser) in which a set of fixed blades are to be arranged to straighten the discharge flow and convert some of the exhaust velocity into static pressure.

Unlike a centrifugal fan, the flow through an axial flow fan has little or no radial component, and there is no significant velocity component directed outwards towards the blade tip.

Figure 7.3 shows the velocity vector diagrams for the inlet and discharge from an axial blade row. The blades are shown stretched out as an extended row, an arrangement which permits the treatment of each pair of adjacent blades as if isolated from the others. The blade row is shown preceded by a row of inlet guide

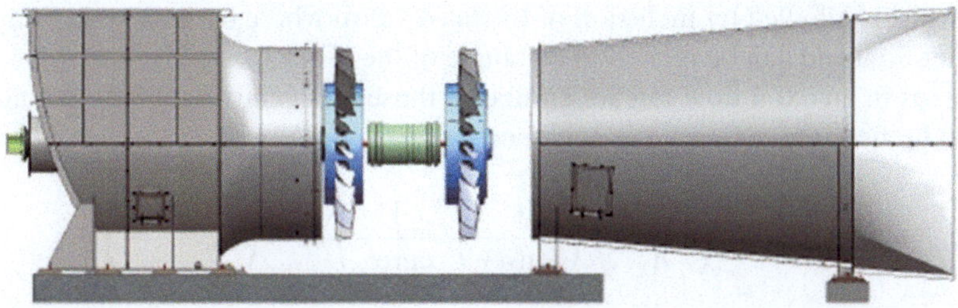

Fig. 7.2 Schematic view of a typical two-stage axial flow fan (Courtesy of TLT-Babcock Fans Inc. USA)

Fig. 7.3 Axial fan velocity vectors

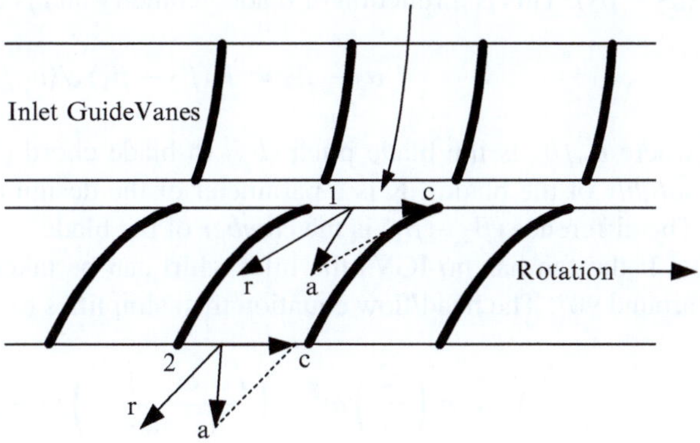

vanes (IGV). In the absence of a radial velocity component all velocity vectors lie in the plane of the page. The Euler equation developed for the centrifugal fan is valid for the axial flow fan. The relationship between the head across the fan and mass flow has the same form as Eq. 7.8 with the modifications that:

- The circumferential speed of the impeller is calculated using a radius r_M defined at the root mean square radius of the blade

$$r_M^2 = \frac{r_1^2 + r_2^2}{2}$$

  and is therefore the same for all points on the blade.
- Each blade has a constant chord ($b_{c1} = b_{c2} = b_c$).
- A_x is the free cross-sectional area of the inter-blade space, normal to the flow path, and is the annular area between the root (or hub) and the blade tip:

$$A_x = 2\pi\,(r_2^2 - r_1^2)$$

  reduced by an allowance for the blade thickness.

Inlet whirl is allowed by inclusion of the $\tan\alpha_1$ term where α_1 is the entry angle of the inlet flow and can be related to the angle of the IGVs.

Losses in an axial flow fan are similar to those in a centrifugal flow fan and are treated in the same manner. We may then write

$$\Delta p_{th} = \left(\frac{r_2^2}{v}\right)\omega^2 - \left(\frac{r_M\,\omega}{A_x}\right)\left(\frac{1}{\tan\alpha_2} - \frac{1}{\tan\alpha_1}\right)\dot{m} - \left(C_w\,\frac{A_s}{A_x^2}\frac{v}{2}\right)\dot{m}^2. \quad (7.18)$$

The difference between the blade leading edge and fluid angles at the inlet to the impeller is called the *incidence* angle $(\beta_1 - \alpha_1)$. The difference between the trailing edge blade angle β_2 and the exit flow angle α_2 is referred to as the *deviation* $(\alpha_2 - \beta_2)$. This is a function of blade geometry and is calculated from

$$\alpha_2 - \beta_2 = K(\beta_2 - \beta_1)\surd(b_p/b_c), \quad (7.19)$$

where b_p/b_c is the blade pitch (b_p) to blade chord (b_c) ratio, also known as the *solidity* of the blades. K is a parameter of the design and is typically around 0.26. The difference $(\beta_2 - \beta_1)$ is the *camber* of the blade.

If the fan has no IGVs the inlet whirl can be taken to be zero, and α_1 will be around 90°. The head/flow equation then simplifies to

$$\Delta p_{th} = \left(\frac{r_2^2}{v}\right)\omega^2 - \left(\frac{r_M\,\omega}{A_x}\frac{1}{\tan\alpha_2}\right)\dot{m} - \left(C_w\,\frac{A_s}{A_x^2}\frac{v}{2}\right)\dot{m}^2. \quad (7.20)$$

The Head/Flow Characteristic from Plant Data

The following figure shows the characteristics calculated from Eq. 7.18 for a fictitious but realistic multistage axial flow fan for a variety of inlet vane angles. The fan blade angles are $\beta_1 = 30°$ and $\beta_2 = 50°$. The inflow angle α_1 in Eq. 7.18 is taken as equal to the guide vane angle. The individual characteristics are calculated for various incidence angles.

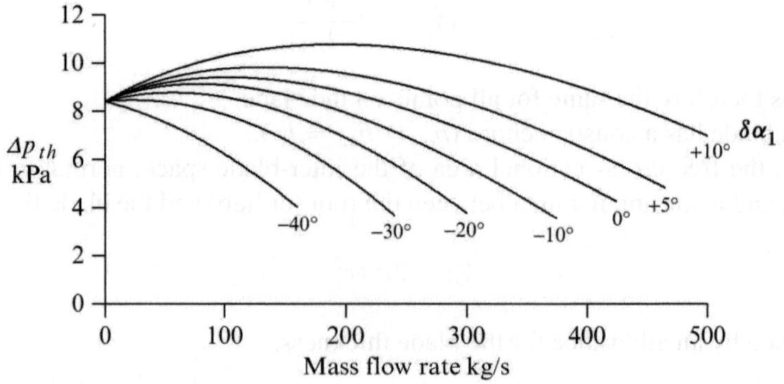

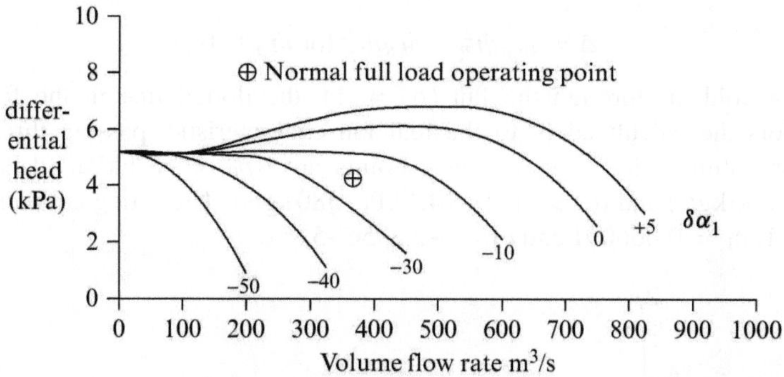

Fig. 7.4 Head/flow characteristic of a large axial flow fan

Figure 7.4 shows the head/flow characteristic of an actual axial flow induced draft fan for a large power station unit. The horizontal axis is volumetric flow but can be replaced by mass flow by application of the appropriate gas density.[4] Flow through the fan is controlled by adjustment of the orientation of the IGVs.

Comparison of these two sets of characteristics—the simplified theoretical and the actual—suggests that Eq. 7.18 can be used to develop a satisfactory approximation, at least over the part of the characteristic of most operational interest. Then

$$\Delta p = \mathrm{d}p_0 + c_1\,\dot{m} + c_2\,\dot{m}^2,$$

where the coefficients $\mathrm{d}p_0$, c_1 and c_2 are selected to match the real characteristic. The following three points are convenient references since they are usually easily identifiable and span the normal operating range:

- $\mathrm{d}p_0$, the zero-flow head ($\dot{m} = 0$)
- $\mathrm{d}p_p, \dot{m}_p$ at the point of maximum fan differential pressure
- $\mathrm{d}p_R, \dot{m}_R$ at the nominal operating point (rated conditions)

"Rated conditions" imply rated speed and rated position of any control devices such as control vanes.

The fitting coefficients c_1 and c_2 are calculated from

$$c_1 = \left(\dot{m}_R^2(\mathrm{d}p_p - \mathrm{d}p_0) - \dot{m}_p^2(\mathrm{d}p_R - \mathrm{d}p_0)\right)/\Delta,$$

$$c_2 = \left(-\dot{m}_R(\mathrm{d}p_p - \mathrm{d}p_0) + \dot{m}_p(\mathrm{d}p_R - \mathrm{d}p_0)\right)/\Delta,$$

[4]Since this is an induced draft fan it will be carrying hot flue gas at around 120–130°C. The gas density will be around 0.86 kg/m³.

where

$$\Delta = \dot{m}_p \dot{m}_R^2 - \dot{m}_R \dot{m}_p^2 \text{ for } \dot{m}_p > 0.$$

Assuming cold air through the fan ($\rho = 1$), the dotted line in the following figure plots the calculated fit to the real fan characteristic, passing through the design operating point. The reference points are $\mathrm{d}p_0 = 5.1\,\mathrm{kPa}$, $(\mathrm{d}p_p, \dot{m}_p) = (5.4\,\mathrm{kPa}, 200\,\mathrm{kg/s})$ and $(\mathrm{d}p_R, \dot{m}_R) = (4.1\,\mathrm{kPa}, 380\,\mathrm{kg/s})$. The fitting coefficients are $\mathrm{d}p_0 = 5.1$, $c_1 = 0.006091$ and $c_2 = -2.295\mathrm{e}{-5}$.

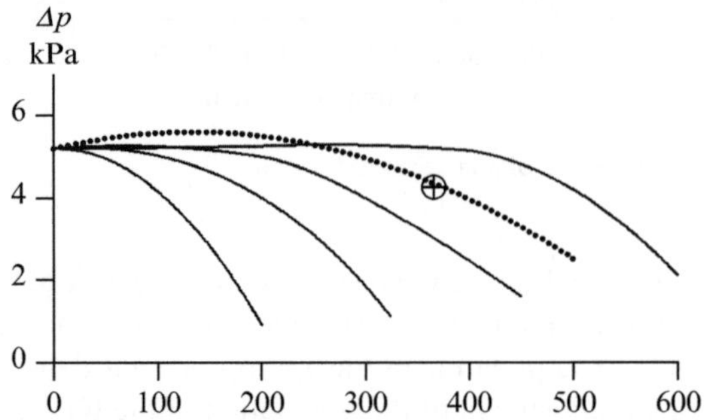

Any characteristic may be represented by a polynomial in powers of $\dot{m}$ using a standard fitting procedure such as least squares. Applying this method to fit a quadratic function to the same fan produces the coefficients $\mathrm{d}p_0 = 5.093$, $c_1 = 0.005531$ and $c_2 = -2.815\mathrm{e}{-5}$, very close to the coefficients calculated using specific points.

Some fans such as centrifugal fans with straight or backwards-curving blades exhibit no "peak" in the characteristic. For such fans, the linear term in Eq. 7.18 falls away, and $c_1 = 0$. The quadratic term is then given simply as

$$c_2 = -\frac{\mathrm{d}p_0 - \mathrm{d}p_R}{\dot{m}_R^2}.$$

7.1.5 *Fan Flow Control*

Control of the flow through centrifugal and axial flow fans is most frequently implemented using one of two methods:

(a) A regulating damper either upstream or downstream of the fan
(b) Variation of fan speed

For axial fans, but less frequently for centrifugal fans, a third alternative uses adjustable inlet vanes.

Damper Control

A flow control damper changes the system flow resistance characteristic. Since the flow through the fan is defined by the point of intersection of the fan and system characteristics, moving the system characteristic changes the flow, as shown by the following diagram for three values of system resistance.

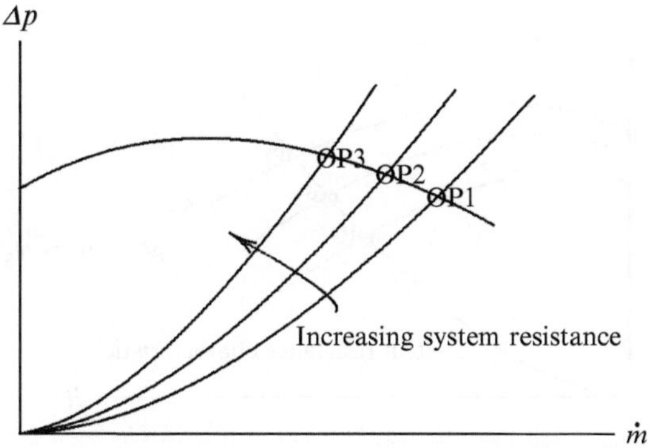

The three operating points are designated OP1, OP2 and OP3. Increasing the system resistance by closing the damper—which increases the pressure drop across the damper—will move the resistance characteristic upwards in the direction of reducing flow. In similar fashion opening the damper will move the resistance characteristic downwards in the direction of increasing flow. The system characteristic changes, but the fan characteristic remains unchanged.

Damper control is the most widely used form of flow control, particularly for centrifugal fans, and requires no special construction features of the fan. It is simple to implement and is the cheapest alternative to install and maintain. It suffers from greater losses than the other alternatives and penalises overall plant efficiency, but this disadvantage can be minimised by good damper design.

Speed Control

Variation of fan speed produces the same result but via a different mechanism. The fan characteristic is moved up by increasing speed or down by reducing speed while the system characteristic remains unchanged. As before, the flow through the fan is defined by the intersection of the fan and system characteristics.

As speed changes, the individual fan characteristics do not simply shift up or down in proportion to the square of the speed. As indicated by Eqs. 7.17 for the centrifugal fan and 7.18 for the axial fan, while the zero-flow developed head is proportional to the square of the fan speed, the characteristic includes a term proportional to the product of fan speed and mass (or volumetric) flow. This term causes the fan characteristics to diverge as speed is reduced.

Speed control is the most expensive form of flow control and is most frequently used for large-capacity axial flow fans. The fan is usually directly coupled to its prime mover—usually an electric motor—and heavy current electronic motor speed control is more expensive than, say, simple damper control. However, speed control offers operational advantages of high efficiencies and precision control. Reducing costs and increasing conversion efficiencies are now making this alternative more attractive.

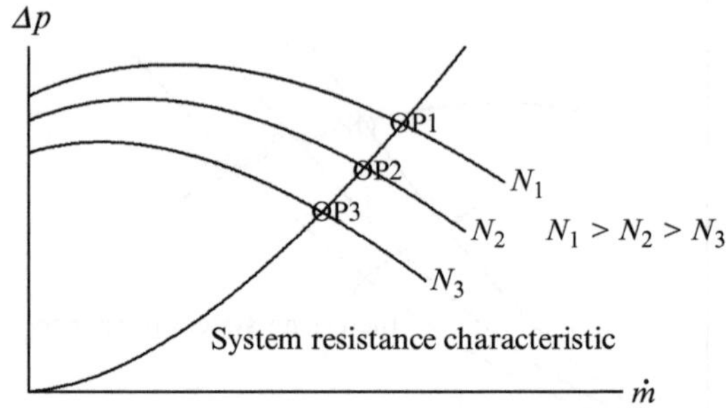

The fan speed N enters the calculation of the fan characteristic via the coefficients $\mathrm{d}p_o$ and c_1. These, determined for a specific case at rated speed N_R, are designated $\mathrm{d}p_{0,R}$ and $c_{1,R}$. Thereafter, for any other speed N,

$$\mathrm{d}p_0(N) = \mathrm{d}p_{0,R}\left(\frac{N}{N_R}\right)^2$$

$$c_1(N) = c_{1,R}\left(\frac{N}{N_R}\right).$$

Inlet Guide Vane Control

IGV control is widely use for large, constant speed, axial flow fans. The vanes themselves are usually integrated with and part of the fan casing though some manufacturers offer them as an add-on extra. As with the speed variation method, variation of the inlet vanes moves the fan characteristic by rotating it about the zero developed head point rather than by shifting it up and down. The system characteristic remains unchanged.

The design and orientation of the vanes will have been optimised to suit the fan and its duty and can be expected to return good efficiency and utilisation. As already discussed, variation of vane position varies the angle α_1 at which the incoming air flow meets the leading edge of the fan blade. The added whirl or pre-rotation improves or detracts from the internal developed head. The effect is not unlike a variation of entry shock loss though it is more a change of "utilisation" of the airflow than a loss.

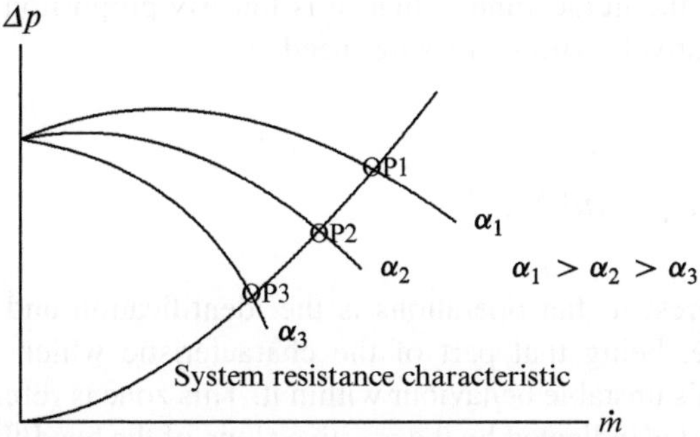

The influence of guide vane angle, assumed equal to the inlet flow incidence angle, is included in the coefficient c_1 since from Eq. 7.18,

$$c_1 = \frac{r_M\,\omega}{A_x}\left(\frac{1}{\tan\alpha_2} - \frac{1}{\tan\alpha_1}\right).$$

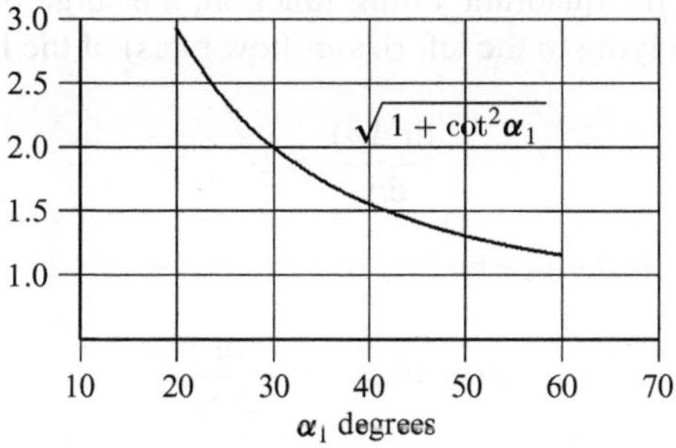

The inlet flow angle α_1 appears only in the second bracketed term. We can therefore write, with $c_{1,R}$ being the value of c_1 determined from rated conditions and $\delta\alpha_1$ the difference between α_1 and the rated value α_{1R},

$$\delta\alpha_1 = \alpha_1 - \alpha_{1R},$$

$$c_1 = c_{1,R} + \delta\alpha_1 \frac{\mathrm{d}c_1}{\mathrm{d}\alpha_1}(\alpha_1).$$

The gradient of c_1 with respect to α_1 is obtained by differentiating c_1,

$$\frac{\mathrm{d}c_1}{\mathrm{d}\alpha_1} = -\frac{r_M\,\omega}{A_x}\sqrt{1+\cot^2\alpha_1}.$$

The plot of the function $\sqrt{1+\cot^2\alpha_1}$ indicates that the guide vane influence is greatest at smaller angles, increasingly rapidly for $\alpha_1 < 30°$. It decreases uniformly at a decreasing rate as α_1 increases.

For a given α_1 the guide vane influence is linearly proportional to fan speed ω and increases negatively with increasing speed.

7.1.6 Fan Surge and Stall

Of particular interest to fan operations is the identification and exclusion of the "prohibited" zone, being that part of the characteristic which is to be avoided because of the fan's unstable behaviour within it. This zone is referred to as the stall or surge region and is indicated by the positive slope of the head/flow characteristic.

While arising from quite different physical bases, surging and stalling are now recognised as related. This matter is described in some detail when dealing with compressors (Sect. 7.3.4). Fan surging arises as a consequence of the interaction of the fan with its integrated system and can arise as a normal consequence of maloperation if the fan's operating point is allowed to enter the prohibited zone. Both centrifugal and axial flow fans can exhibit surge and stall.

With reference to the quadratic fitting function, the surge or stall region can be identified as the zone lying to the left (lesser flow rates) of the line

$$\frac{\mathrm{d}(\Delta p)}{\mathrm{d}\dot{m}} = 0$$

which, from Eq. 7.17, is defined by

$$\dot{m} \leq \dot{m}_{lim} = -\frac{c_1}{2\,c_2}. \tag{7.21}$$

Axial flow fans show behaviour in the surge/stall regions which deviates markedly from the simple quadratic characteristic. The next figure illustrates an axial fan head/flow characteristic exhibiting this behaviour.

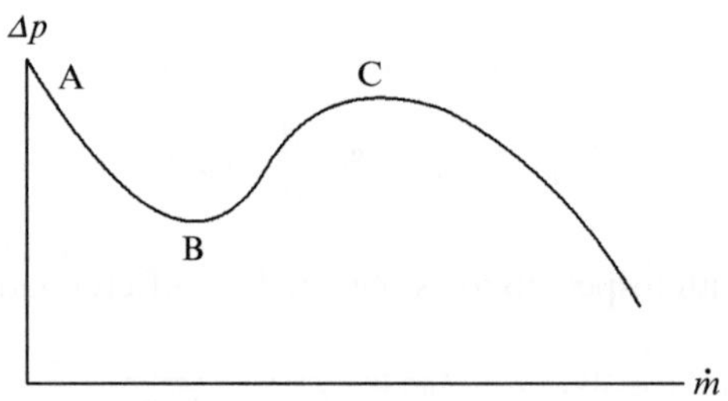

In the region to the right of C the fan will be developing usually more than 50 % of its rated capacity, and the flow through the fan will be close to its design optimum (normal operating point). The fan cannot operate stably within the B–C region—the surge zone—in which flow and developed pressure will oscillate. Flow oscillations may cause flow reversal, and the fan will exhibit reduced and oscillating pressures, reduced power consumption, increased vibration and increasing temperatures. Although stable operation in the region A–B, a region exhibiting a negative slope of the head/flow characteristic, is possible, it is unlikely due to the probable appearance of stall at such low flows.

Fan stall is essentially aerodynamic in origin and originates within the inter-blade space of the fan. It is a complex phenomenon, essentially the result of centrifugal force, which tends to separate the flow from the blade surface, exceeding viscous forces which tend to keep it attached. The picture is made more complex by local changes in static pressure associated with local velocity profiles—amplified when the blades are not aerofoil shapes—and when the static pressure, particularly at the trailing edge of the blades, is influenced by conditions downstream from the fan.

7.1.7 Power Requirements

The power developed by a fan is given by the product of the developed head and volumetric flow.

$$P_F = \Delta p_{th}\, \dot{m}/\rho.$$

Using the fan of Fig. 7.4 as an example, the following figure shows the calculated fan power against volumetric flow for a range of vane positions.

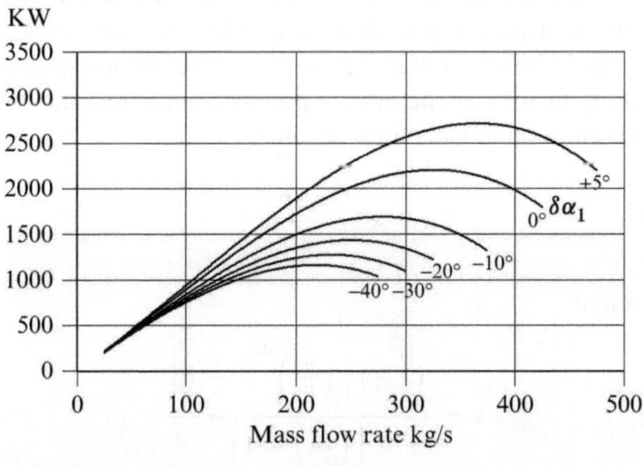

The power supplied by the prime mover is related to the fan power via the fan efficiency η_F calculated from

$$\eta_F = \frac{P_F - P_{loss}}{P_F} \times 100\,\%. \tag{7.22}$$

Power losses are calculated from the relevant pressure losses using the definitions of Δp_{loss} discussed in Sect. 7.1.1. The following coefficients were developed there for the friction and shock losses.

Friction loss from Eq. 7.15:

$$k_{loss} = C_w \frac{A_s}{A_2^2}\frac{v}{2}.$$

Shock loss from Eq. 7.16:

$$k_{shock} = C_{shock,min} + C_{shock}(\dot{m} - \dot{m}_{OP})^2.$$

In each case, the flow-dependent head loss is calculated by multiplying the pressure loss coefficient by the square of the mass flow. Then

$$\Delta p_{loss} = (k_{loss} + k_{shock})\,\dot{m}^2.$$

The power lost by irreversible friction and shock losses is then the product of the head loss and volumetric flow.

$$P_{loss} = \Delta p_{loss}\dot{m}/\rho. \tag{7.23}$$

With Δp quoted in kPa, $\dot{m}$ in kg/s and density ρ in kg/m^3, fan power P_F is calculated in kJ/s or kW.

It is usual to base fan power calculations on inlet conditions and to ignore compressibility of the gas flowing through the fan. It may then be assumed that both inlet and discharge volumetric flows are equal. In practice this is not strictly correct, and even the slight compression through the fan will increase the discharge volumetric flow rate. The effect of compression can be accommodated by definition of a compressibility factor κ as the ratio of pressure rise across the fan without (Δp) and with ($\Delta p'$) compression taken into account.

$$\kappa = \frac{\Delta p}{\Delta p'}.$$

It can be shown that

$$\kappa = \frac{\left(\frac{n}{n-1}\right)\left[\left(\frac{p_2}{p_1}\right)^{\frac{n-1}{n}} - 1\right]}{\left[\left(\frac{p_2}{p_1}\right) - 1\right]}, \tag{7.24}$$

where n is the polytropic index.

For polytropic efficiencies in the range 0.9–1.0 and fan compression ratios p_2/p_1 less than 1.1 (10 % compression), κ will have values $\geq$0.96, and little error will be incurred if compression is ignored.

The following diagram shows the variation of fan efficiency as a function of mass flow, with the losses calculated using Eq. 7.23. The sharp decrease in efficiency at lower flows is a direct result of the increase in shock losses and other effects, such as flow separation, which manifest themselves at low flows.

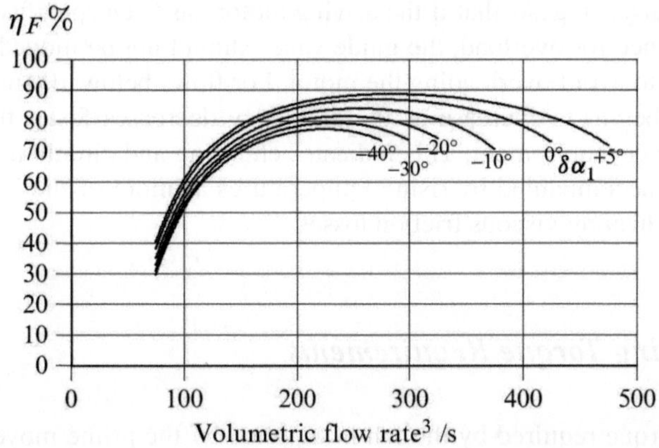

The approach described here is heuristic and is not based on a detailed analysis of the physical processes occurring at such low flows. Despite this, it offers a reasonably accurate reproduction of a fan's performance, whatever the physical basis may be.

The power P_{PM} supplied to the fan by its prime mover is shown by the next figure, calculated from $P_{PM} = P_F/\eta_F$, again for a range of inlet vane positions. Figure 7.4 shows the fan's operating point at around 380 m³/s for a guide vane relative position of around −20°.

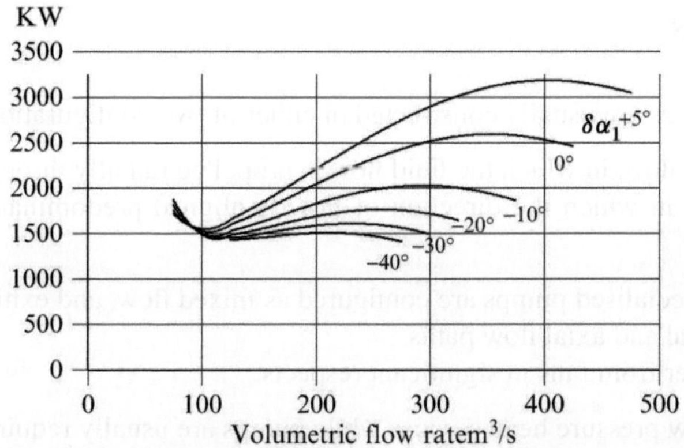

Assuming it is carrying flue gas at 125°C and density 0.86 kg/m³, this indicates a mass flow of around 330 kg/s or 1,176 t/h and a motor power of some 1,700 kW. The actual motor power is calculated from the fan required power by using the motor efficiency, which can usually be assumed to be around 90 %. The actual motor power under these conditions will then be around 1,900 kW or 1.9 MW.

The motor power characteristics for a relative vane position less than −30° (−40, −50, etc.) are substantially flat, and motor power shows relatively little change with gas flow. The reduced power required to move a lower flow rate of gas is offset by the decrease in efficiency, keeping the power requirement up. The power characteristic also suggests that if the driving motor has been specified for full load plus an allowance for overload, the guide vanes should not be moved beyond, say, $-10°$ in order to avoid overloading the motor. For flows below $100\,m^3/s$ the power characteristic begins to increase as the efficiency decreases faster than the flow-dependent power requirement. This indicates churning and circulating flow within the fan and is accompanied by rising temperatures as more of the input energy is converted into heat by viscous friction losses.

7.1.8 Driving Torque Requirements

The driving torque required by the fan is supplied by the prime mover. For a non-zero speed ω, the torque required by the fan is calculated from

$$T_{qF} = P_{PM}/\omega.$$

With P_{PM} in kW and ω in rad/s, the torque is calculated in newton metres [Nm].

The general question of driving torque, speed calculation and prime mover matching is discussed later in Sect. 7.4.

7.2 Pumps

Like fans, pumps are usually constructed in either of two configurations:

- Centrifugal flow, in which the fluid flow is propelled radially through the rotor
- Axial flow, in which the direction of flow is aligned predominantly along the rotor shaft

Some more specialised pumps are configured as mixed flow, and exhibit aspects of both centrifugal and axial flow paths.

Pumps differ from fans in significant respects:

1. Fans are low pressure head devices while pumps are usually required to develop a high head, several orders of magnitude greater than a fan.[5]
2. Pumps handle incompressible fluids at high velocities and must be constructed to withstand substantially larger forces than fans.

[5]There are of course instances of pumps which develop very small heads, such as metering pumps, medical device pumps, fuel pumps for petrol engines—but not diesels which develop substantial pressures—and vacuum pumps. However, in this book, we will deal with larger pumps in industrial scale applications.

3. Pumps tend to be physically smaller than fans for a given mass flow in consequence of the higher densities of their working media.

Dimensional analysis can be used to develop general relationships among variables which, in a general fashion, link a pump or fan's design parameters and its design objectives. The following list sets out the dimensional relationships between volumetric flow Q, mass flow $\dot{m}$, total head differential H, required torque T and power consumption P on the one side and a machine's rotational speed N, its diameter (or other dimensions) D and the working medium density ρ on the other.

$$
\begin{aligned}
Q &\mapsto N\ D^3 \\
\dot{m} &\mapsto \rho\ N\ D^3 \\
H &\mapsto \rho\ N^2 D^2 \\
T &\mapsto \rho\ N^2 D^5 \\
P &\mapsto \rho\ N^3 D^5 \\
S &\mapsto \rho\ N^2 D^4 / b_t^2.
\end{aligned}
$$

The last item in the list relates mechanical stress S acting on the blades to density, speed, dimension and blade thickness b_t.

The importance of density ρ in differentiating fans and pumps is now clear. For a fan handling air at atmospheric pressure and 20°C, ρ is 0.83 kg/m^3. For a pump handling water, ρ is close enough to 1,000 kg/m^3.

Pumps and fans typically work within similar mass flow regimes meaning that ND^3 for a pump must be around 1,000 times less than a fan. If all of this were taken up by dimension alone, a pump's dimensions (say, impeller diameter) will be smaller than a fan by around a factor of 10 ($\sqrt[3]{1,000}$). A pump's developed head can be expected to be between 100 and 10,000 times that of a fan which, given the reduction in dimension, will require either some increase in speed, since most of the increase will be provided by the higher density, or an increase in dimension, or both. Of greatest significance is the blade stress which, being proportional to the density of the working medium, must be offset by a reduction in the product of dimension and speed or by an increase in blade thickness. There is a practical limit to the blade thickness in order to maximise the flow cross-sectional area through the pump. The optimal design of a pump to balance these competing influences is complex. Suffice to say that, compared to a fan of similar mass flow capacity, a pump will be smaller, faster and thicker.

Fig. 7.5 Centrifugal pump impeller with double-sided entry (Courtesy of Millmerran Operating Company, Australia)

7.2.1 *Calculation of the Head/Flow Characteristic*

Brennen [24] gives a development of the calculation of the head/flow characteristic for non-cavitating pumps which is applicable to all types of pumps. Though developed specifically for pumps, it parallels the development presented in Sect. 7.1 for fans and results in an equation for pumps very similar to that developed for fans. This illustrates the essential commonality which exists throughout the turbomachinery family.

Following Brennen, but retaining the nomenclature of Sect. 7.1, we introduce a dimensionless head coefficient[6]

$$\psi = \Delta p_{th}/(\rho r_2^2 \omega^2) \tag{7.25}$$

and a flow coefficient

$$\phi = Q/(A\, r\, \omega), \tag{7.26}$$

where ϕ, A and r may be defined at either the inlet (1) or discharge (2) of the pump. Conservation of mass requires that the mass flow rate $\dot{m}$ be the same at inlet and outlet. Then we can write

$$\dot{m} = A\, \rho\, \nu_{a2r},$$

where r_2 is the blade tip radius, b_{c2} is the blade chord at the tip and ν_{a2r} is the flow velocity normal to the exit perimeter. For a centrifugal pump the flow discharge area $A_2 = 2\pi\, r_2 b_{c2}$, and for an axial flow pump, $A_2 = \pi(r_2^2 - r_1^2)$. In each case a more

[6]The subscript $_{th}$ is retained as this is the theoretical head which will differ from the practical head.

exact treatment requires an allowance for the area occupied by the blades, but this will be ignored here.

The Euler Eq. 7.17 derived for fans can be reformulated for non-cavitating flow in terms of ψ and ϕ. Assuming no inlet swirl—the case with most pumps—we have

$$\psi = 1 - \phi_2 \cot \beta_2. \tag{7.27}$$

A more exact analysis would treat the variable velocity profile across the flow stream. This equation would apply to each streamline, and total effects would be integrated across all streamlines. For our purposes, however, we will assume that all stream lines move with the same velocity and treat only a representative line.

The derivation of Eqs. 7.17 and 7.27 made no allowance for the effects of viscous friction losses. Brennen includes these effects by defining a friction loss factor f as

$$\Delta p_{loss} = f \frac{\rho}{2} \bar{v}_M^2,$$

where $\bar{v}_M^2$ is the vector mean of the absolute velocities v_{a1} and v_{a2} at the inlet and outlet of the pump. If β_M is the angle of v_M and $\bar{v}_M$ is its magnitude, then

$$\cot \beta_M = \frac{1}{2}(\cot \beta_1 + \cot \beta_2)$$

and

$$\bar{v}_M = \frac{v_{a2r}}{\sin \beta_M}.$$

In the absence of inlet swirl, the analysis produces the following equation for the head/flow characteristic in terms of the non-dimensional head and flow coefficients ψ and ϕ.

$$\psi = 1 - \phi \cot \beta_2 - f \left[\phi^2 + \frac{1}{4}(1 + \phi \cot \beta_2)^2 \right] \tag{7.28}$$

which reduces to Eq. 7.27 for $f = 0$.

The figure below plots ψ against ϕ, calculated from Eq. 7.28, for $\beta_2 = 45°$ and values for f of 0.0, 0.05, 0.1 and 0.15.

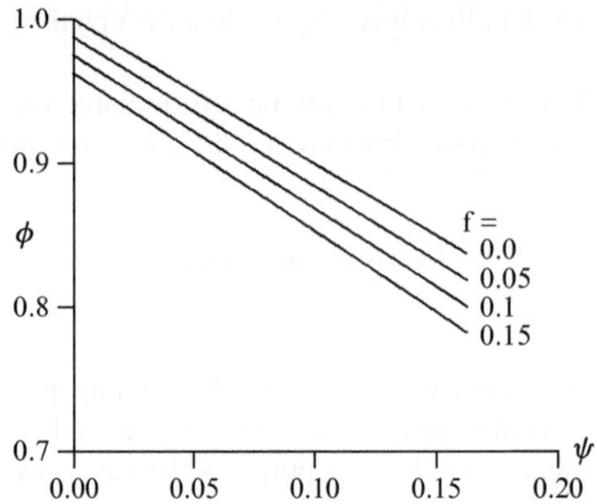

If we now replace ψ and ϕ from Eqs. 7.25 and 7.26, we obtain

$$\Delta p_{th} = (1 - f/4)\frac{r_2^2\omega^2}{v} - \frac{r_2\omega}{A}\cot\beta_2\,\dot{m} - f\frac{v}{A^2}\left(1 + \frac{1}{4}\cot^2\beta_2\right)\dot{m}^2. \quad (7.29)$$

With the inclusion of a shock loss term, similar to that used in the derivation of the fan characteristic, Eq. 7.29 becomes

$$\Delta p_{th} = (1 - f/4)\frac{r_2^2\omega^2}{v} - \frac{r_2\omega}{A}\cot\beta_2\,\dot{m} - f\frac{v}{A^2}\left(1 + \frac{1}{4}\cot^2\beta_2\right)\dot{m}^2$$
$$+ \left(C_{shk,min} + C_{shk}(\dot{m} - \dot{m}_{OP})^2\right). \quad (7.30)$$

Like Eq. 7.17, Eq. 7.30 is quadratic in $\dot{m}$. Its coefficients have the same dependence on the plant's physical dimensions as the fan, with some differences in detail. In contrast to the fan equation, Eq. 7.30 includes a friction term in the zero-flow head coefficient, and the friction loss term uses Brennen's formulation in place of that of Eck.

Recirculation or Backflow

The high pressure differential across the pump induces a flow from the discharge side to the inlet through the gap between the impeller blade tips and the casing. This recirculation or backflow mixes with the inlet stream and causes a deflection of that stream from its ideal inflow angle. This is a form of shock loss and results in a deterioration of the pump's performance. It can also contribute to the onset of cavitation since it disturbs flow conditions at the pump inlet. It is particularly significant at low flows when the discharge pressure is highest and inlet conditions are marginal.

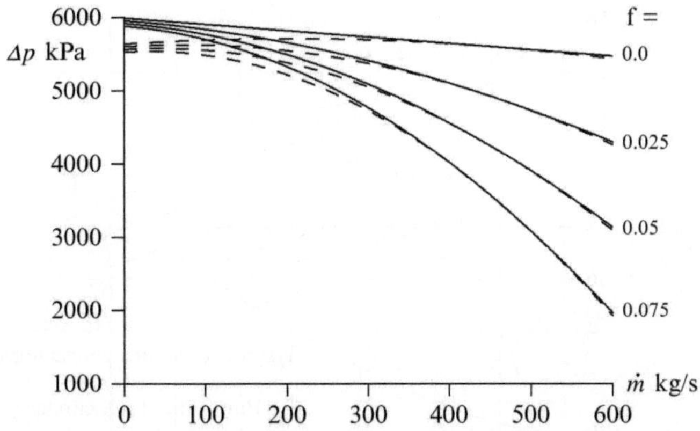

Fig. 7.6 Calculated pump head/flow characteristic

7.2.2 *The Head/Flow Characteristic from Pump Data*

Figure 7.6 shows a family of head/flow characteristics calculated from Eq. 7.30 for a range of values of the viscous friction loss factor f.

The pump is single-stage centrifugal with the following specifications:

Impeller blade tip radius r_2	18.0 cm
Mean blade chord b_c	12.0 cm
Rotational speed N	5,000 rpm
Blade tip angle β_2	45°

The full lines show the characteristic with no shock losses ($C_{shk,min}$, $C_{shk} = 0$). The dashed lines show the characteristic with $C_{shk} = 0.00175$. This shows clearly the increased shock loss effect at low flows due to flow disturbances such as separation and backflow. The loss of developed head causes rollover of the characteristic as flow approaches zero. The effect at high flows is much less pronounced.

Figure 7.7 shows the head/flow characteristics of a typical large variable speed centrifugal flow pump. This pump has two stages, and it can be assumed that the stages are equal, each developing 1/2 of the total pressure ratio. The test characteristic shows no peak, therefore $c_1 = 0$, and can be extrapolated back to the zero-flow axis to determine the zero-flow head. This yields a $\mathrm{d}p_0$ of 6,750 kPa at 5,000 rpm. Using the quadratic curve-fitting procedure outlined for fans we determine c_2 to be −0.0139.

Assuming the pump has backwards-curving blades with a tip angle of 45°, a mean chord of 10.0 cm and a rated rotational speed of 5,000 rpm, these parameters would indicate that the pump has a blade radius of 18.4 cm. We can also determine the friction parameter f to be 0.056, assuming a root-to-tip radius ratio of 0.5.

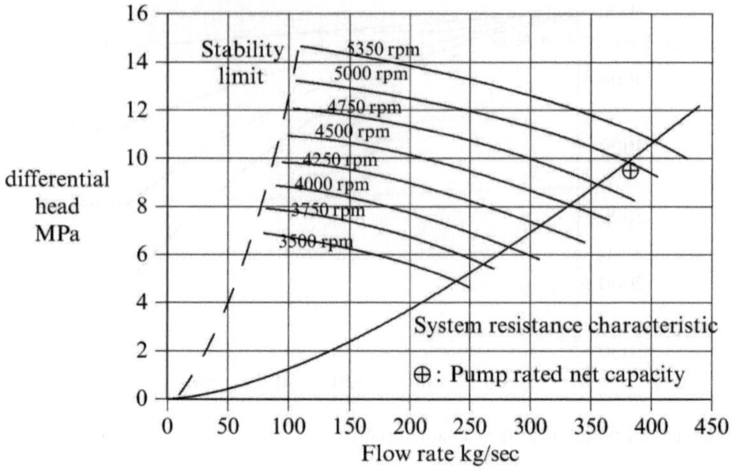

Fig. 7.7 Typical pump head/flow characteristic

Units

In customary US and British units, developed pressure or head is usually quoted in feet of water (1 atmosphere (atm) = 101.3 kPa (absolute) = 33.89 ft). Therefore, multiply head (ft) by 2.9891 to obtain head in kPa. Volumetric flows are quoted in SI units as m^3/s and in customary US units as US gallons per minute (GPM) or in British units as imperial gallons per minute, also GPM or sometimes IGPM (1 US gallon = 4.54609e–3 m^3 and 1 imperial gallon = 3.785435e–3 m^3).

7.2.3 Pump Efficiency

The power developed by a pump is given by the product of the developed head and volumetric flow:

$$P_P = \Delta p_{th}\, \dot{m}/\rho.$$

Pump efficiency is calculated as described for fans. Using the pump of Fig. 7.6 as an example, the following figure shows the calculated pump power against mass flow. The power supplied by the prime mover is related to the pump power via the pump efficiency η_P calculated from

$$\eta_P = \frac{P_P - P_{loss}}{P_P} \times 100\,\%. \tag{7.31}$$

Power losses are calculated from the relevant pressure head loss by multiplying $\Delta p_{..}$ by the volumetric flow. Then

$$P_{loss} = (\Delta p_{loss} + \Delta p_{shock})\,\dot{m}/\rho. \quad (7.32)$$

With $\Delta p_{..}$ quoted in kPa (absolute), $\dot{m}$ in kg/s and density ρ in m^3/kg, pump power P_P is calculated in kJ/s or kW.

The following diagram shows the variation of pump efficiency as a function of mass flow, with the losses calculated using Eq. 7.32.

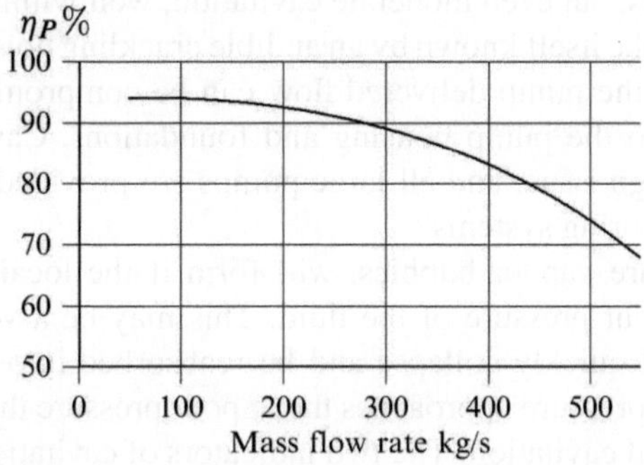

The power P_{PM} supplied to the pump by its prime mover is shown by the next figure, calculated from $P_{PM} = P_P/\eta_P$.

Figure 7.6 shows the pump's operating point at around 380 kg/s or 1,368 t/h. The motor power is some 1,900 kW. With a calculated efficiency of 85 %, the actual motor (prime mover) power is calculated to be 2,235 kW.

Pump efficiency decreases rapidly for flows much in excess of the design or rated flow as friction losses within the pump become significant. Efficiency at lower flows depends strongly on shock losses which are relatively minor here.

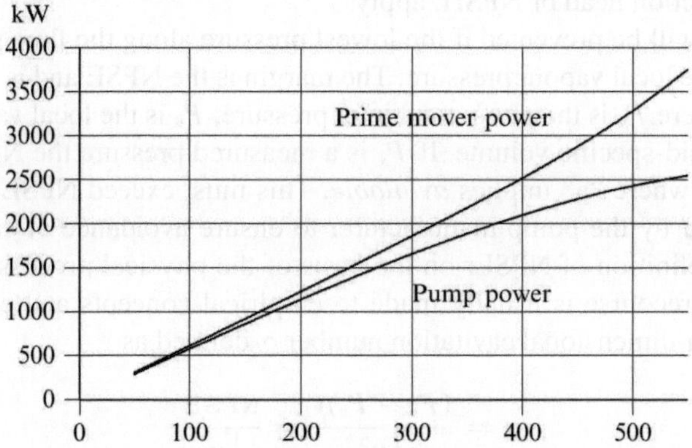

7.2.4 Pump Cavitation

Cavitation is the term used to describe the formation and subsequent collapse of vapour bubbles within the fluid body in a centrifugal pump. Bubbles form in a low-pressure region of the flow and are transported to a region of higher pressure in which they collapse. Formation and collapse of bubbles create pressure shocks which can be destructively high, reaching 2,000 bar under worst circumstances. The impact of these repetitive shocks causes pitting and consequent corrosion of and damage to the impeller surfaces. Minor cavitation may occur without observable symptoms, but even moderate cavitation, well within the pump's ability to withstand, will make itself known by an audible crackling noise. Under conditions of severe cavitation the pump delivered flow can be compromised and significant vibrations induced to the pump housing and foundations. Cavitation is a serious operational and design issue, and all large pumps are provided with some form of protection and monitoring systems.

Cavities, which are vapour bubbles, will form if the local pressure decreases below the local vapour pressure of the fluid. This may be a very localised effect, and the bubbles will quickly collapse and be reabsorbed into the fluid body. The closer the bulk fluid pressure approaches the vapour pressure the greater will be the degree and severity of cavitation. The two indicators of cavitation are the local fluid pressure and its vapour pressure, which is a function of temperature. Cavities are most likely to form at the point of lowest pressure which, for a pump, is the inlet throat. The throat pressure is calculated from the source pressure plus any static column pressure less any flow friction pressure loss less the dynamic velocity head.

A second form of cavitation can occur within the impeller blade passages if the flow breaks away from the blade-guiding surfaces, that is, if the pump stalls. This is a low-flow phenomenon and ideally is avoided by minimum flow protection. The present discussion is focused on the first form of cavitation for which the concepts of net positive suction energy (NPSE), formerly and still in some places known as net positive suction head or NPSH, apply.

Cavitation will be prevented if the lowest pressure along the flow path is always greater than the local vapour pressure. The margin is the NPSE and is the difference $P_s v - P_v v$, where P_s is the pump total inlet pressure, P_v is the local vapour pressure and v is the fluid-specific volume. If P_s is a measured pressure the NPSE is known as the NPSEa where "a" implies *available*. This must exceed NPSEr which is the NPSE *required* by the pump manufacturer to ensure avoidance of cavitation. The quantitative definition of NPSEr on the basis of the physical processes involved is complex, and recourse is usually made to empirical concepts and constants. One such is the non-dimensional cavitation number σ defined as

$$\sigma = \frac{(P_s - P_v)v}{\frac{1}{2}\nu^2} = \frac{\mathtt{NPSE}}{\frac{1}{2}\nu^2}$$

where ν is some reference velocity such as the impeller tip velocity. Empirical relationships are available relating the cavitation number to a characteristic number of a given pump. The cavitation number is used to determine the NPSEr for a given pump duty.

A second empirical non-dimensional number used within the industry is the "suction-specific speed" S_s. Similar to the "specific speed" used to characterise a pump's performance related to its geometry, S_s is defined as

$$S_s = \frac{\omega\, \dot{V}^{1/2}}{\mathrm{NPSE}^{3/4}}$$

where ω is the angular rotational speed of the impeller and $\dot{V}$ is the volumetric flow rate through the pump. Like the cavitation number, the suction-specific speed can be related to a characteristic of a given pump to determine the NPSEr.

Most pumps can continue to operate with some degree of cavitation with minimal loss of performance. The characteristic crackling noise will be audible, but pump performance will decrease noticeably only if σ has decreased significantly. The value of σ at which performance begins to deteriorate is termed the "breakdown cavitation number", and a loss of developed head $> 5\,\%$ can be expected, as well as a noticeable increase in vibration. Should the suction inlet bulk pressure decrease below the fluid vapour pressure the resulting large-scale flashing of the fluid can completely block the pump inlet, causing flow to stop. This is a transient effect as the sudden decrease in flow will increase the inlet pressure with the consequent collapse of the vapour bubbles. The process is unstable and can result in serious equipment damage.

From the simulation point of view cavitation is an operational event that can arise at any time as a result of malfunction or maloperation. It requires calculation of the parameters which indicate the presence of cavitation and of the performance changes within the pump.

7.3 Compressors

Compressors are distinguished from fans by the larger pressure differential they develop. Unlike fans, the larger pressure head developed by compressors requires inclusion in their analysis of the effects of compression of the working medium. Continuity of mass allows us to assume equal mass flows into and out of the compressor stage, but volumetric flows will differ.

Compressors are widely used in the power and process industries in a wide variety of roles, as blowers supplying air to steel blast furnaces, on offshore oil and gas platforms compressing gases for transport to land stations, as the compression units for high-pressure gas distribution networks, for on-site compression of gas products for storage, in compressed air systems,[7] and so on.

Like fans, compressors are built in either centrifugal (also called radial) or axial flow configurations. Some salient constructional and operating differences between the two types are summarised in the following table.

Centrifugal	Axial
Large frontal area	Smaller frontal area
High pressure rise per stage	Low pressure rise per stage
Fewer stages for a given head	More stages for a given head
Higher developed head, smaller throughput	Lower developed head, higher throughput
Pressure ratios per stage: 1.5–3.5	1.05–1.2
Lower efficiency 77 %–82 %	Higher efficiency 85 %–92 %
Compact, robust construction	Long axial length, sensitive alignment
Large margin to surge and stall	Small margin to surge and stall
Wide operating flexibility	Narrow operating flexibility

Centrifugal compressors have a wide range of operating speeds, much wider than axial flow compressors. This difference is a direct consequence of their larger margin between the surge, stall and choking limits. For applications requiring wide speed variation, such as aircraft jet engines, axial compressors are provided with special constructional features such as stage bypasses to extend the operating speed range.

By way of introduction let us return to the basic momentum conservation equation

$$\frac{\partial p}{\partial z} = -\left.\frac{\partial p}{\partial z}\right|_{loss} - \rho\left(\frac{\partial v}{\partial t} + v\frac{\partial v}{\partial z}\right) - g\rho \sin\beta.$$

In Chap. 6 this was restructured into the steady-state Bernoulli's equation

$$\frac{pv}{g} + \frac{1}{2}\frac{v^2}{g} + \delta p_{loss} = \sin\beta\, z.$$

This may be simplified by ignoring irreversible friction losses (δp_{loss}) and potential energy contributed by geodetic height variations. It may then be written in the differential form

$$-v\,\mathrm{d}p = v\,dv.$$

[7] We will not consider reciprocating compressors here although these are widely used in smaller compressed air and HVAC systems.

This equation can be integrated directly along a flow path between two points 1 and 2, 1 being on the upstream side of 2. Then

$$-\int_{(1)}^{(2)} v\,\mathrm{d}p = \int_{(1)}^{(2)} v\,\mathrm{d}v = \frac{1}{2}\left(v_2^2 - v_1^2\right).$$

The right-hand side denotes the change in flow kinetic energy between 1 and 2. The left-hand integral can be evaluated once the pv process is defined. For an adiabatic process

$$\frac{p}{p_1} = \left(\frac{v}{v_1}\right)^{\gamma}$$

and

$$\int_{(1)}^{(2)} v\,\mathrm{d}p = \left(\frac{\gamma}{\gamma-1}\right) p_1 v_1 \left[\left(\frac{p_2}{p_1}\right)^{\frac{\gamma-1}{\gamma}} - 1\right]$$

$$= \left(\frac{\gamma}{\gamma-1}\right) Z\,R\,\mathcal{T}_1 \left[\left(\frac{p_2}{p_1}\right)^{\frac{\gamma-1}{\gamma}} - 1\right],$$

where R is the gas constant and $\mathcal{T}_1$ is the gas temperature (K) at the inlet. Z is the gas compressibility constant, usually close to 1.

This is the compression work W_{12} needed to compress the gas from its state (p_1, v_1) to (p_2, v_2) and is equal to the gain in enthalpy Δh_{12} when moving from state 1 to state 2.

$$\int_{(1)}^{(2)} v\,\mathrm{d}p = \Delta h_{12}.$$

Thus the enthalpy gain in isentropic compression is equal to the gain in flow kinetic energy.

$$\Delta h_{12} = \frac{1}{2}\left(v_2^2 - v_1^2\right).$$

This can be written in terms of the enthalpy conditions at the flow points 1 and 2, given that, in compression, $\Delta h_{12} = h_2 - h_1$. Then,

$$h_1 - \frac{1}{2}v_1^2 = h_2 - \frac{1}{2}v_2^2.$$

A practical compressor process will be better approximated as a polytropic process. With n as the polytropic index, the enthalpy gain in compression now includes polytropic warming and

$$\Delta h_{12} = \left(\frac{n}{n-1}\right) Z\,R\,\mathcal{T}_1 \left[\left(\frac{p_2}{p_1}\right)^{\frac{n-1}{n}} - 1\right] \tag{7.33}$$

noting the relationship between n and γ via the η_p is the polytropic efficiency (Eq. 4.8 for compression). Equation 7.33 may be inverted to define the pressure ratio Π in terms of the polytropic enthalpy increase.

$$\Pi = \frac{p_2}{p_1} = \left[\left(\frac{n-1}{n}\right)\frac{1}{Z R T_1}\Delta h_{12} + 1\right]^{\frac{n}{n-1}}. \tag{7.34}$$

An expression for the theoretical isentropic enthalpy change or "head" Δ_{th} can be obtained from the Euler equation for the stage. The Euler equation was derived for fans and pumps in the form

$$\Delta p_{th} = \frac{1}{v}(v_{c2}v_{a2c} - v_{c1}v_{a1c}).$$

By neglecting compressibility of the working medium as it passed through the fan it could be assumed that $v_1 = v_2 = v$, and the Euler equation produced a calculation of the theoretical pressure rise across the device. With compressibility taken into account $v_1 \neq v_2$ and the Euler equation defines the isentropic enthalpy change to be

$$\Delta h_{th} = v_{c2}v_{a2c} - v_{c1}v_{a1c}.$$

While having much in common, centrifugal and axial flow compressors are distinctly different and will be discussed separately.

7.3.1 Centrifugal Flow Compressors

Figure 7.8 shows a typical impeller for a small centrifugal compressor. The sloping and narrowing profile of the blades from inlet to discharge is clearly visible.

Making some rough assumptions about flow cross-sectional areas, we can write expressions for the volumetric flows Q [m^2/s] entering and leaving the compressor blading. The subscript 1 denotes inlet (hub), and 2 denotes outlet (blade tip):

$$Q_1 = v_{a1r}A_1 = v_{a1r}2\pi\, r_1\, b_1 \tag{7.35}$$

and

$$Q_2 = v_{a2r}A_2 = v_{a2r}2\pi\, r_2\, b_2, \tag{7.36}$$

where v_{ar} is the radial component of the absolute velocity at inlet and outlet and r and b denote the radius and blade chord respectively.

Conservation of mass through the compressor means that the entry and exit mass flows will be equal ($\dot{m}$). The differing specific volumes at entry (v_1) and exit (v_2) mean that the entry Q_1 and exit Q_2 volumetric flows will differ, with $Q_1 > Q_2$. It should be noted that, when in manufacturer's data the compressor head/flow

Fig. 7.8 A centrifugal compressor (turbocharger) cutaway view

characteristic is quoted against volumetric flow, the inlet or suction flow is usually intended.

Noting that $Q = \dot{m}\, v$, Eqs. 7.35 and 7.36 may be rearranged to yield expressions for the radial velocity components

$$v_{a1r} = \frac{\dot{m}\, v_1}{2\pi\, r_1\, b_1} \tag{7.37}$$

and

$$v_{a2r} = \frac{\dot{m}\, v_2}{2\pi\, r_2\, b_2}. \tag{7.38}$$

From the velocity triangle we can write

$$v_{ac} = v_c - \frac{v_{ar}}{\tan\beta}$$

At the compressor inlet,

$$v_{ac1} = v_{c1} - \frac{1}{v_1}\frac{\dot{m}}{2\pi r_1 b_1 \tan\beta_1}$$

and at the outlet

$$v_{ac2} = v_{c2} - \frac{1}{v_2}\frac{\dot{m}}{2\pi r_2 b_2 \tan\beta_2}.$$

Substituting these back into the Euler equation gives, after some reduction,

$$\Delta h_{th} = v_{c2}^2 - v_{c1}^2 - \dot{m}\frac{v_{c2}}{2\pi\, v_2 r_2 b_2}\left(\cot\beta_2 - \frac{v_{c1}}{v_{c2}}\frac{v_2}{v_1}\frac{r_2}{r_1}\frac{b_2}{b_1}\cot\beta_1\right). \tag{7.39}$$

With $v_{c1} = \omega\, r_1$ and $v_{c2} = \omega\, r_2$, Eq. 7.39 may be written as

$$\Delta h_{th} = \omega^2 r_2^2\left(1 - \frac{r_1^2}{r_2^2}\right) - \dot{m}\frac{\omega}{2\pi v_2\, b_1}(\cot\beta_2 - k_v \cot\beta_1) \tag{7.40}$$

where $k_v = (v_2/v_1)(b_2/b_1)$. Since $v_2 < v_1$ and in most centrifugal compressors $b_2 < b_1$, k_v will be <1. Compared to the incompressible case this reduces the influence of the inlet blade angle on the developed head.

It is common practice in the compressor literature to introduce non-dimensional quantities in place of the developed head and mass flow. For example, we might introduce the head coefficient ψ and the flow coefficient ϕ, defined as

$$\psi = \frac{\Delta h_{th}}{v_{c2}^2} \text{ and } \phi = \frac{\dot{m}v}{2\pi\omega_2^2 b_2}.$$

Equation 7.40 may then be written in dimensionless form as

$$\psi = 1 - \phi(\cot\beta_2 - k_v\cot\beta_1)/(1 - r_1^2/r_2^2) \tag{7.41}$$

or, ignoring the influence of the hub-to-tip radius as second order,

$$\psi = 1 - \phi(\cot\beta_2 - k_v\cot\beta_1). \tag{7.42}$$

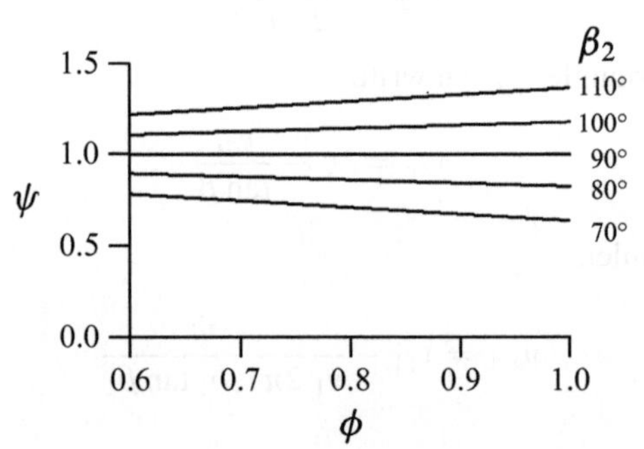

The adjacent figure shows the relationship between ψ and ϕ for a variety of discharge blade angles β_2. For simplicity, it is assumed that the flow enters the impeller with zero swirl ($\beta_1 = 90°$). Angles $\beta_2 > 90°$ indicate forwards-inclined blades while $\beta_2 < 90°$ indicate backwards-inclined blades. These curves show the same form and structure as shown for incompressible fans (Sect. 7.1.1). The key difference is in the definition of ψ which for the incompressible case $= f(\Delta p)$ and for the compressible case $= f(\Pi)$.

The following diagram shows the pressure ratios for a typical single-stage centrifugal compressor, calculated using Δh_{th} from Eq. 7.40 and the pressure ratio Π from Eq. 7.34. A family of curves is shown using rotational speed as the parameter. The calculation assumes zero inlet swirl. An allowance for losses has been made in the form of a loss term proportional to the square of mass flow, as described for fans by Eq. 7.17. In a departure from usual industry practice the pressure ratios are plotted against discharge volumetric flow in order to emphasise the decrease in volumetric flow with increasing compression.

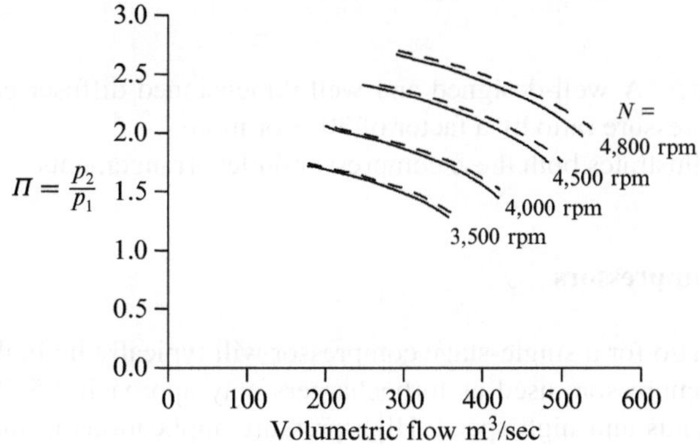

The set of dotted curves shows the same information but with an inlet whirl (inflow angle) of +10°.

Equations 7.34 and 7.40 are coupled via the ratio (v_2/v_1) in k_v since $v_2/v_1 = \Pi^{1/\gamma}$. However, the coupling is weak, and, at least for the simulation purpose, little error is incurred by using a single-shot calculation (no iteration).

Influence of the Discharge Diffuser

The discharge flow leaving the impeller enters a spiral plenum called the volute. The inner surface of the volute is circular and is separated from the impeller by a small gap which ensures clearance between the rotating impeller and the stationary casing. The outer edge of the volute is usually formed as an Archimedean spiral which terminates at the compressor discharge flange.

The performance of a compressor can be improved, sometimes significantly, by interposing a series of divergent channels, defined by curved fixed vanes, between the impeller discharge and the volute. Being divergent, the flow cross-sectional area increases along the flow path, causing a deceleration of the flow. In accord with Bernoulli's equation this causes some of the kinetic energy of the flow, indicated by velocity, to be transformed into potential energy, indicated by an increase in static pressure. This improves both the compressor's pressure ratio

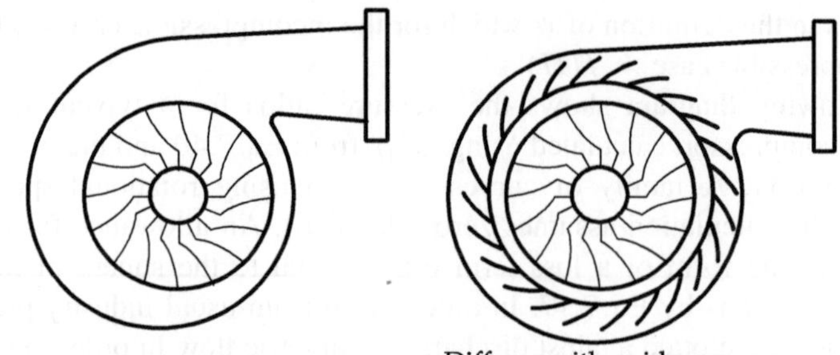

Fig. 7.9 Centrifugal compressor discharge arrangements

and its efficiency. A well-designed and well-dimensioned diffuser can improve a compressor's pressure ratio by a factor of 30 % or more.

Figure 7.9 illustrates both these compressor outlet arrangements.

Multistage Compressors

The pressure ratio for a single-stage compressor will typically lie in the range 1.8–2.5, though compressors used as turbochargers may approach 3.5. These tend to be of small radius and high speed. High-pressure applications in the oil and gas industries can use gas pressures of the order of several MPa which implies pressure ratios of multiple decades. The higher ratios require the cascading of several stages. For example, assuming a series of stages having equal pressure ratios of, say, 2:1, an overall pressure ratio of 16 would require 4 stages. Compressors with 6 or 7 stages are not uncommon.

Equation 7.34 indicates that the stage pressure ratio depends inversely on the inlet absolute temperature. In a cascade arrangement the inlet temperature to a stage is the discharge temperature from the preceding stage. The stage temperature ratio, assuming a polytropic process, is given from

$$\frac{T_2}{T_1} = \left(\frac{p_2}{p_1}\right)^{\frac{n-1}{n}}. \tag{7.43}$$

For example, a compression ratio of 2:1 with a polytropic efficiency of 94 % and assuming $\gamma = 1.4$ would give a temperature ratio of 1.235. Assuming an inlet temperature of 300 K (27°C) for the first stage, its outlet temperature and therefore the inlet temperature to the second stage would be 370.5 K or 97.1°C and so on for subsequent stages. It is for this reason that some compressors and turbochargers are fitted with "intercoolers" or heat exchangers between stages to reduce inlet temperatures to subsequent stages and maximise stage pressure ratios.

Flow-Limiting Mechanisms

Compressors suffer from three flow-limiting mechanisms:

- Surging, which limits the maximum developed pressure ratio for a given speed
- Stalling, which limits the minimum flow permitted
- Choking, which limits the maximum flow achievable

Bearing in mind that surge and stall are related but different phenomena, we can summarise each as follows.

Surging is an operational issue for all compressors and arises as a consequence of the interaction between the compressor and the flow system in which it is located. Large industrial compressors, particularly those delivering high flows and high pressures, are almost invariably fitted with anti-surging control systems.

Stalling has been discussed in Sect. 7.1.6 in connection with fans. The same principle applies to compressors. The higher blade channel pressure gradients in compressors tend to initiate blade stall earlier than in fans, but otherwise, the two cases are very similar, and prediction of the stall line follows similar guidelines. It is interesting to note that a "rotating stall" was first observed by Whittle [22] in a centrifugal compressor during the early development of gas turbines. A rotating stall is caused by the transient breakaway of flow from the suction side of a blade in one inter-blade channel which causes that blade to stall. Being a transient effect, the normal flow regime re-establishes itself but not before it disturbs the flow in the preceding channel which then stalls. This sequence continues, and the stall propagates itself slowly around the rotor, opposite to the direction of impeller rotation.

Choking is a compressor phenomenon and is a consequence of compressibility effects as flow velocities within the compressor approach the local sonic velocity (Mach number approaches unity). Since flow is induced by rotation of the impeller, the maximum flow velocities can be expected in the vicinity of the highest impeller speeds at the blade tips.[8] Since the local speed of sound c is proportional to $\sqrt{T}$ it increases with increasing compression, reaching its maximum at the blade tip. For the example quoted earlier, a stage pressure ratio of 2:1 produces a temperature ratio of 1.235:1 and an increase in c of approximately 11 % compared to the hub. Once a choke point is reached the developed pressure ratio decreases rapidly with no change in throughput flow [26]. Given that the tip speed is the product of blade rotational speed and radius, the Mach number limitation means that high-speed compressors will have small radii. For example, a compressor running at a nominal 20,000 rpm (2,094 rad/s) and a temperature ratio of 1.235:1 will reach Mach 1 tip speed for a tip radius of 0.184 m.

---

[8] Axial flow compressors tend to use aerofoil-shaped blades. These can cause a narrowing of the channel width causing locally higher velocities which can lead to choking within the inter-blade space.

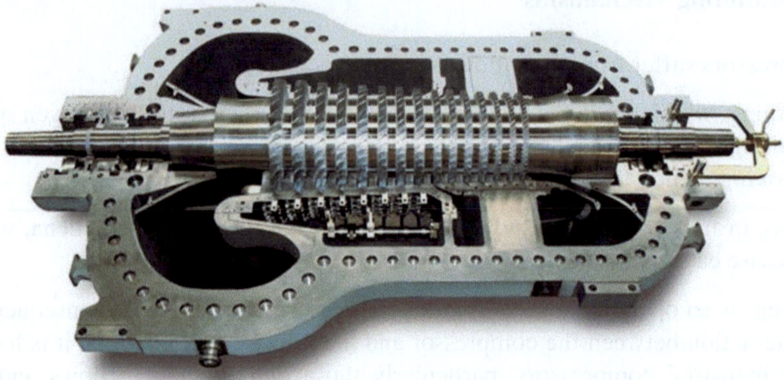

Fig. 7.10 A large industrial axial flow compressor (Photo courtesy of GE Oil & Gas)

The impeller inlet also presents a possible choke point and, for a given maximum flow rate, must be sized to ensure a maximum inlet flow velocity less than or approaching sonic speed. The incoming stream is introduced to the eye of the impeller through a convergent nozzle whose throat area roughly equals that of the eye, defined by the hub radius. This area must be sized to ensure that at the maximum mass flow rate the flow velocity remains sufficiently below the sonic speed.

7.3.2 Axial Flow Compressors

The axial flow compressor is designed to provide flow through the device which is aligned with the axis of rotation of the rotor. In the following analysis, we will assume a zero radial component of flow, although in reality, there will be some radial flow component induced by the radial pressure gradient along the blade.

Figure 7.10 shows the internal construction arrangement of a large industrial axial compressor as might be found in, for example, the oil and gas industries used for the compression and transmission of large volumes of gas. Some axial compressors, known as axi-radial or axi-centrifugal, are fitted with one or more centrifugal stages after the axial stages, employed to take advantage of the centrifugal compressor's higher pressure ratio capabilities to increase developed head.

A stage of an axial flow compressor consists of a row of rotating blades followed by a row of stationary blades. These correspond functionally to the impeller and diffuser of a centrifugal compressor but are so arranged that the flow enters the stage from the preceding stage with zero or minimal "whirl" (radial component) and leaves the stage with zero whirl.

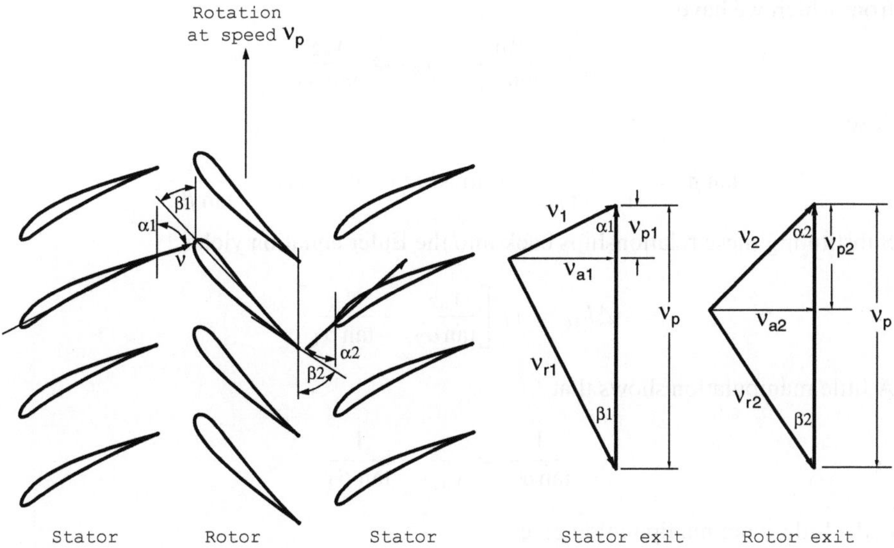

Fig. 7.11 Axial flow compressor velocity triangles

Figure 7.11 depicts three extended lines of blades, two stationary (stator rows) and one moving (the rotor). Unless otherwise noted all angles are referenced to the plane of rotation of the rotor. Other reference planes sometimes used include the axis of rotation (normal to the plane of rotation) and the blade chord line.

The rotor moves with rotational velocity ω rad/s and induces a flow velocity whose component in the direction of rotation equals that of the rotating blades. The induced flow enters the first stator row from the left (in the diagram) with velocity v_1 at an angle α_1. The vector combination of the forwards blade movement and the axial flow through the inter-blade channel produces a resultant flow velocity v_{r2} relative to the moving blade and v_2 relative to the following stator row. Flow leaves the rotor at an angle β_2, being the angle of the rotor blade trailing edge. The following stator row will ideally be aligned with this flow and will turn the flow to meet the next rotor blade row at an optimum angle.

Referring to the two velocity triangles, one at the rotor inlet (stator exit) and the other at the rotor exit, we may derive the equations describing the enthalpy rise and pressure ratio across the stage.

The Euler equation for an axial flow compressor reduces to

$$\Delta h_{th} = v_b(v_{b2} - v_{b1})$$

because we can assume negligible radial flow. From the velocity triangles, we can write

$$\tan\alpha_1 = \frac{v_{a1}}{v_{b1}}; \quad \tan\alpha_2 = \frac{v_{a2}}{v_{b2}}$$

from which we have

$$v_{b1} = \frac{v_{a1}}{\tan\alpha_1}; \quad v_{b2} = \frac{v_{a2}}{\tan\alpha_2}.$$

Also,

$$\tan\beta_1 = \frac{v_{a1}}{v_b - v_{b1}} \quad \text{from which} \quad v_b - v_{b1} = \frac{v_{a1}}{\tan\beta_1}.$$

Substituting these relationships back into the Euler equation yields

$$\Delta h_{th} = v_b \left[\frac{v_{a2}}{\tan\alpha_2} - \frac{v_{a1}}{\tan\alpha_1} \right].$$

A little manipulation shows that

$$\frac{1}{\tan\alpha_2} = \frac{v_b}{v_{a2}} - \frac{1}{\tan\beta_2}$$

and a little more manipulation gives

$$\Delta h_{th} = v_b^2 - v_b v_{a1} \left[\frac{v_{a2}}{v_{a1}} \frac{1}{\tan\beta_2} + \frac{1}{\tan\alpha_1} \right].$$

The flow velocities at inlet and outlet are related to the mass flow $\dot{m}$ through the compressor via the flow-specific volume v and the annular area A. Assuming equal mass flows at inlet and outlet, $\dot{m}_1 = \dot{m}_2 = \dot{m}$, we can write

$$v_{a1} = \frac{\dot{m} v_1}{A_1} \quad \text{and} \quad v_{a2} = \frac{\dot{m} v_2}{A_2}.$$

With these substitutions,

$$\Delta h_{th} = v_b^2 - v_b \dot{m} \frac{v_1}{A_1} \left[\frac{v_{a2}}{v_{a1}} \frac{1}{\tan\beta_2} + \frac{1}{\tan\alpha_1} \right].$$

Little error is incurred by assuming equal axial velocity components through the compressor, and this equation can be simplified to

$$\Delta h_{th} = v_b^2 - v_b \dot{m} \frac{v_1}{A_1} \left[\frac{1}{\tan\beta_2} + \frac{1}{\tan\alpha_1} \right]. \tag{7.44}$$

Define a mean radial position along the blade r_m, where

$$r_m = \sqrt{(r_t^2 + r_h^2)/2}$$

and r_t and r_h are the tip and hub radii (measured from the shaft centreline). Then,

$$v_b = \omega\, r_m \quad \text{and} \quad A_1 = 2\pi (r_t^2 - r_h^2)$$

giving

$$\Delta h_{th} = \omega^2 r_m^2 - \omega \dot{m} \frac{r_m v_1}{A_1} \left[\frac{1}{\tan\alpha_2} + \frac{1}{\tan\alpha_1} \right]. \tag{7.45}$$

With the pressure ratio given from Eq. 7.34, we may then write, from Eq. 7.45,

$$\frac{p_2}{p_1} = \left\{ 1 + \left(\frac{\gamma - 1}{\gamma} \right) \frac{\omega^2 r_m^2}{Z\,R\,T_1} \left[1 - \left(\frac{\dot{m} v_1}{\omega\, r_m A_1} \right) \left(\frac{1}{\tan\alpha_2} + \frac{1}{\tan\alpha_1} \right) \right] \right\}^{\frac{\gamma}{\gamma-1}}. \tag{7.46}$$

With the sonic velocity $c = \sqrt{\gamma Z R T}$, this equation may be written as

$$\frac{p_2}{p_1} = \left\{ 1 + (\gamma - 1)\, M_a^2 \left[1 - \left(\frac{\dot{m} v_1}{\omega\, r_m A_1} \right) \left(\frac{1}{\tan\alpha_2} + \frac{1}{\tan\alpha_1} \right) \right] \right\}^{\frac{\gamma}{\gamma-1}}. \tag{7.47}$$

since the ratio $\omega\, r_m / c$ is the Mach number M_a.

Alternatively, given that

$$c_p = \left(\frac{\gamma}{\gamma - 1} \right) R.$$

Eq. 7.46 can be written as

$$\frac{p_2}{p_1} = \left\{ 1 + \left(\frac{\omega^2 r_m^2}{Z c_p T_1} \right) \left[1 - \left(\frac{\dot{m} v_1}{\omega\, r_m A_1} \right) \left(\frac{1}{\tan\alpha_2} + \frac{1}{\tan\alpha_1} \right) \right] \right\}^{\frac{\gamma}{\gamma-1}}. \tag{7.48}$$

A similar result is quoted by Boyce [25],

$$\frac{p_2}{p_1} = \left\{ 1 + \left(\frac{\omega^2 r_2^2}{c_p T_1} \right) \left[1 - \left(\frac{v_{a2}}{\omega\, r_2} \right) \left(\tan\beta_2 + \frac{v_{a1} r_1}{v_{a2} r_2} \tan\alpha_2 \right) \right] \right\}^{\frac{\gamma}{\gamma-1}},$$

where α_2 is the absolute flow angle at the stator exit and β_2 is the relative flow angle at the rotor exit referred to the axial flow direction, that is, normal to the rotational plane. r_2 is the impeller tip radius, and c_p is the specific heat capacity of the working medium. Referred to the circumferential plane the angles may be replaced by their complements and the tangent replaced by cotangent. With

$$v_{a2} = \frac{\dot{m} v}{A_2}$$

and with an adjustment made for the angles, this may be written as

$$\frac{p_2}{p_1} = \left\{ 1 + \left(\frac{\omega^2 r_2^2}{c_p T_1} \right) \left[1 - \left(\frac{\dot{m} v}{\omega\, r_2 A_1} \right) \left(\frac{1}{\tan\beta_2} + \frac{v_{a1} r_1}{v_{a2} r_2} \frac{1}{\tan\alpha_2} \right) \right] \right\}^{\frac{\gamma}{\gamma-1}}. \tag{7.49}$$

The flow characteristics of a single stage of an axial flow compressor are displayed by the following diagram, calculated using Eq. 7.46. A typical set of design parameters has been assumed for the purpose of calculation.

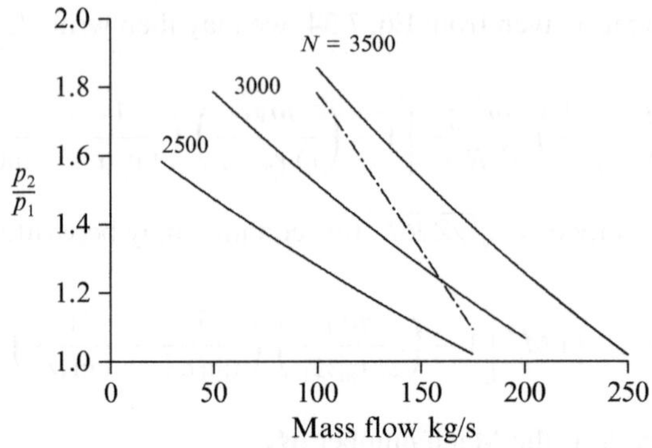

The figure illustrates the steep characteristic typical of an axial flow compressor. Characteristics are shown for only a limited range of developed heads as the maximum pressure ratio will be limited by the surge or stall line.

The full-line characteristic is idealised as it includes no allowance for losses. Including a quadratic allowance for losses as described earlier for the centrifugal machines produces the dash-dot curve for $N = 3{,}500$ rpm. Inclusion of losses steepens the characteristic significantly.

Multistage Axial Flow Compressors

As mentioned in the introduction to this section, a single stage of an axial flow machine will produce only a relatively small pressure ratio. Industrial axial flow compressors are therefore built as multistage machines to develop the often large ratios needed by applications. The performance of an axial flow stage is sensitively dependent on its design parameters and the optimum design of a compressor is a complex task. For the simulation task, if it can be assumed that the design of the compressor permits the assumption that all stages of the compressor develop equal pressure ratios, the overall developed head of a compressor having "n" stages will be that of the single-stage characteristic raised to the power "n". For example, a compressor with atmospheric suction and with a single-stage ratio of 1.25 must deliver air against a head of 4,000 kPa. This requires an overall pressure ratio of around 40 and 17 stages for the compressor.

The Influence of Inlet Guide Vanes

Many compressors are fitted with IGV as flow regulation devices. When operating within a range that does not obstruct the compressor inlet—otherwise they would be acting as dampers—the IGVs vary the angle at which the incoming flow strikes the rotor blade leading edge and so vary the angle α_1. The following diagram shows the effect of variation of the IGV position. The compressor speed is held constant at 3,500 rpm, and again, losses are not included. Negative IGV angles indicate angles less than the blade inlet angle and direct the flow against the rotation of the blades.

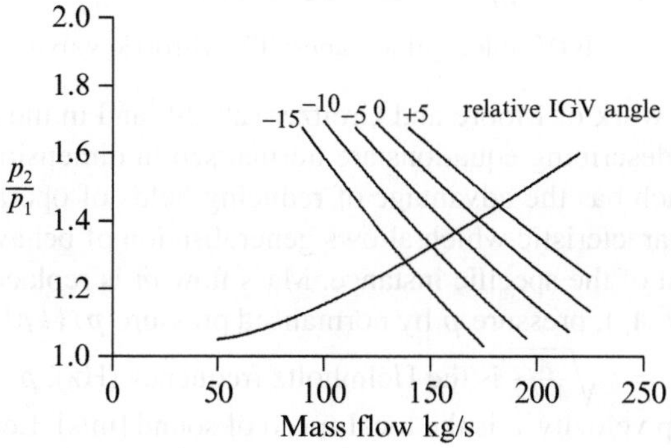

7.3.3 *Fitting to Plant Data*

This section has used a simplified theoretical analysis to derive the typical head/flow characteristics of axial flow compressors in terms of their physical construction details. This procedure is useful only as an approximation to the actual characteristic which, as mentioned, is sensitive to design parameters. However, the analysis does indicate the structure of suitable forms of approximation which can be fitted to a known compressor's characteristic. In the simulation context, this means fitting an equation of the form of Eq. 7.46, 7.47 or 7.48, with or without explicit inclusion of losses, to a known compressor characteristic while retaining access to compressor speed and IGV blade angle for the inclusion of control effects. Fitting can best be accomplished by estimating the parameters used in these equations from known physical construction parameters and then adjusting them to give an optimum fit.

7.3.4 *Modelling of Surge and Stall Effects*

As mentioned previously, surge and stall during compressor operation are serious operational problems. A significant body of literature has been developed dealing with the prediction and prevention of compressor surge and stall. The field is summarised by Willems [27]. The compact model proposed originally by Greitzer [28] in 1976 and subsequently updated by Moore and Greitzer [29] in 1986 was derived by treating a compressor feeding a lumped volume (the plenum) in the form shown by the following figure.

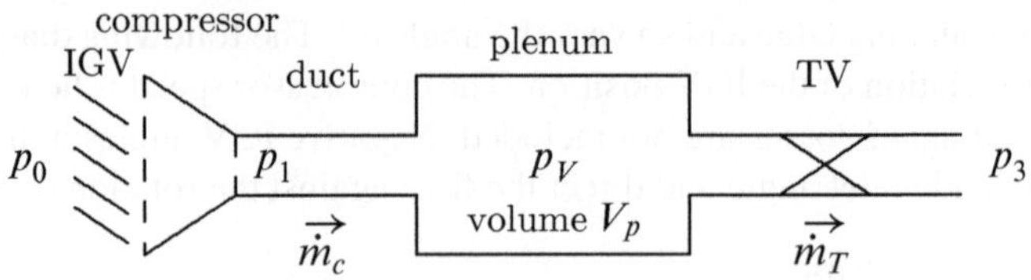

IGV: inlet guide vanes; TV: throttle valve

In the original work of Moore and Greitzer [28, 29] and in the current literature such as [30], the describing equations are normalised in dimensions, variables and time. This approach has the advantage of reducing fields of operating characteristics to a single characteristic which allows generalisation of behavioural aspects in terms independent of the specific instance. Mass flow $\dot{m}$ is replaced by normalised mass flow $\dot{m}/(\rho U A_c)$, pressure p by normalised pressure $p/(\frac{1}{2}\rho U^2)$ and time t by $t \cdot \omega_H$, where $\omega_H = c\sqrt{\frac{A_c}{V L_c}}$ is the Helmholtz frequency (Hz). ρ is the gas density, and U the rotor tip velocity. c is the local speed of sound [m/s]. Length is expressed as multiples of rotor mean radius.

The model to be described here can treat only axial flow effects and is thus limited to the treatment of surge, which is an axial flow phenomenon. Rotating stall, on the other hand, is a radial/circumferential phenomenon which, although related to surge, has a quite different physical basis. The literature cited previously treats both surge and rotating stall, but here, we will confine interest to surge.

The model consists of three ordinary differential equations. The first states the mass balance for the plenum

$$\left(\frac{V}{\gamma p\, v}\right)\frac{\mathrm{d}p_V}{\mathrm{d}t} = \dot{m}_c - \dot{m}_T = \dot{m}_c - k_T\sqrt{(p_V - p_3)}$$

where the throttle valve characteristic is approximated by the parameter k_T. If choked flow through the valve is assumed, the last term can be simplified to $k_T\sqrt{p_r}$. Writing

$$\tau = \frac{V}{\gamma p\, v} \quad \text{and} \quad k'_T = \frac{k_T}{|p_V - p_3| > 0}$$

this equation can be written as

$$\tau\frac{\mathrm{d}p_V}{\mathrm{d}t} = \dot{m}_c - k'_T(p_V - p_3). \tag{7.50}$$

From Eq. 5.15, we have

$$c^2 = \gamma\, p\, v$$

giving

$$\tau = \frac{V_p}{c^2},$$

where c is again the speed of sound.

The second differential equation describes the momentum balance for the flow through the compressor.

$$\frac{\mathrm{d}\dot{m}_c}{\mathrm{d}t} = \frac{A_c}{L_c}(p_1 - p_V).$$

Writing

$$\mathcal{I} = \frac{L_c}{A_c}$$

this becomes

$$\mathcal{I}\frac{\mathrm{d}\dot{m}_c}{\mathrm{d}t} = p_1 - p_V, \tag{7.51}$$

where L_c and A_c are the length and flow cross section of the duct joining the compressor discharge and the plenum. Flow inertia $\mathcal{I}$ is related to the *inertance*, discussed later in Chap. 17.

Neglecting flow dynamics within the compressor, we can use an algebraic equation to define the duct (= compressor discharge) pressure p_1 in terms of the compressor static head/flow characteristic. Assuming polytropic compression from

$$\Pi^{\frac{n-1}{n}} = \frac{n-1}{n}\frac{1}{ZRT_1}\Delta h_{12} + 1$$

we obtain

$$p_1 = p_0\left[\left(\frac{n-1}{n}\right)\frac{\Delta h_{12}}{ZRT_1} + 1\right]^{\frac{n-1}{n}}$$

which, when substituted back into Eq. 7.51, gives

$$\frac{\mathrm{d}\dot{m}_c}{\mathrm{d}t} = \frac{A}{L}\left\{p_0\left[\left(\frac{n-1}{n}\right)\frac{\Delta h_{12}}{ZRT_1} + 1\right]^{\frac{n-1}{n}} - p_V\right\}. \tag{7.52}$$

The real enthalpy change Δh_{12} from compressor inlet to outlet includes losses Δh_{loss}

$$\Delta h_{12} = \Delta h_{th} + \Delta h_{loss}$$

which themselves depend on the operating point. Indeed it is these losses which cause the roll-over of the head/flow characteristic from a negative to a positive slope at low flows, and it is, as we shall see, this reversal of slope which creates the conditions for flow instability. Therefore, if the head losses are properly included into the calculation of Δh_{12}, Eq. 7.52 can be used as it stands.

An alternative approach, and the one used here, is to approximate the head/flow characteristic by a suitable approximation, preferably of polynomial form, to the known head/flow characteristic. Following Greitzer [28] a suitable cubic is

$$\Psi_c(\Phi) = \Psi_{co} + H\left[1 + \frac{3}{2}\left(\frac{\Phi - \Phi_0}{W} - 1\right) - \frac{1}{2}\left(\frac{\Phi - \Phi_0}{W} - 1\right)^3\right] \tag{7.53}$$

where $\Psi_c(\Phi)$ is the normalised head developed by the compressor and Φ is the normalised flow through it. H and W are defined by the following diagram, drawn in normalised form using the parameters indicated.

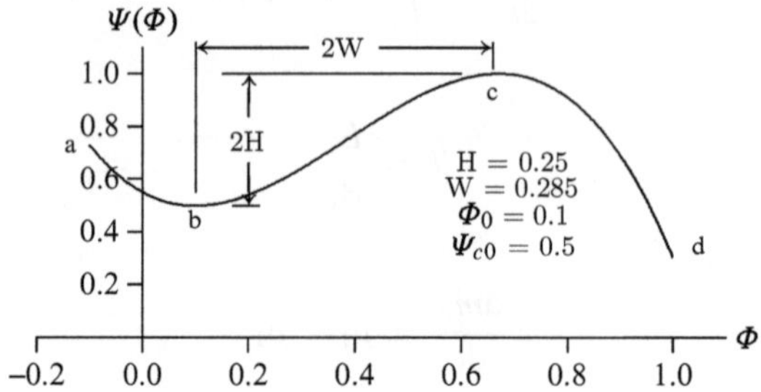

Φ_o is used to shift the characteristic along the Φ-axis.

The third equation describes the variation of compressor speed ω [rad/s] from a simple torque balance

$$I\frac{\mathrm{d}\omega}{\mathrm{d}t} = T_{drv} - T_{cpr}, \tag{7.54}$$

where T_{drv} is the torque developed by the driving device (electric motor, steam or gas turbine) and T_{cpr} is the counter-torque imposed by the compressor. The compressor torque is a non-linear function of speed and flow, as discussed earlier.

The use of a polynomial to represent the compressor head/flow characteristic relieves the analysis of the need to consider the underlying physical effects, as would have been the case had we used the enthalpy calculation of Eq. 7.52. It also opens the analysis to a consideration of the underlying mathematics of the describing equations. For this purpose we will assume that the compressor speed remains constant and combine the two remaining Equations 7.50 and 7.51 into the single second-order equation

$$\frac{\mathrm{d}^2\dot{m}_c}{\mathrm{d}t} + \frac{k'_T}{\tau}\left[1 - \frac{\tau}{\mathcal{I}}\frac{\partial F_p}{\partial \dot{m}_c}\frac{1}{k'_T}\right]\frac{\mathrm{d}\dot{m}_c}{\mathrm{d}t} + \frac{1}{\tau\mathcal{I}}\dot{m}_c - \frac{k'_T}{\tau\mathcal{I}}F_p(\dot{m}_c) = -\frac{k'_T}{\tau\mathcal{I}}p_3, \tag{7.55}$$

where $F_p(\dot{m}_c)$ is the polynomial approximation to the compressor characteristic[9] and $\partial F_p/\partial \dot{m}_c$ is its local gradient.

This equation is of the general form

$$\frac{\mathrm{d}^2x}{\mathrm{d}t} + f\left(x, \frac{\mathrm{d}x}{\mathrm{d}t}\right) = g(t), \tag{7.56}$$

where the forcing function $g(t)$ may be zero, yielding the autonomous form, may be a constant or may be any other function of t. For a stable solution x remains bounded as $t \to \infty$.

---

[9]This may be expressed in terms of pressure ratio or developed head. Here, we use developed head.

A well-known form of Eq. 7.56 is the damped harmonic oscillator

$$\frac{\mathrm{d}^2 x}{\mathrm{d}t} + 2\zeta\omega\frac{\mathrm{d}x}{\mathrm{d}t} + \omega^2 x = g(t), \tag{7.57}$$

where ζ is the damping factor and ω the harmonic frequency. If both ζ and ω are constants this equation is linear with constant coefficients and has the general solution, for $g(t) = 0$,

$$x(t) = C\mathrm{e}^{-\zeta t}\sin(\omega t + \theta),$$

where C and θ are constants selected to satisfy boundary conditions. This solution is stable ($x \to 0$ as $t \to \infty$) if ζ is positive but diverges if ζ is negative.

Two non-linear forms of Eq. 7.56 are of interest to the present discussion. Van der Pol's equation is

$$\frac{\mathrm{d}^2 x}{\mathrm{d}t} - \mu(1 - x^2)\frac{\mathrm{d}x}{\mathrm{d}t} + x = g(t), \tag{7.58}$$

where $\mu > 0$ is a constant, usually ≤ 1, and Duffing's equation

$$\frac{\mathrm{d}^2 x}{\mathrm{d}t} + a\frac{\mathrm{d}x}{\mathrm{d}t} + f(x) = g(t), \tag{7.59}$$

where a is a constant and $f(x)$ is a third-order polynomial in x with no constant term.

A more general form combining features of both these equations is

$$\frac{\mathrm{d}^2 x}{\mathrm{d}t} + f_d(x)\frac{\mathrm{d}x}{\mathrm{d}t} + f_r(x) = g(t), \tag{7.60}$$

where $f_d(x)$ is a non-linear *damping* function and $f_r(x)$ is a non-linear *restorative* function. We will limit discussion to the homogeneous case $g(t) = 0$. Hayashi [33] provides a comprehensive treatment of the analysis of these and other non-linear systems.

Van der Pol's equation describes a self-exciting oscillator and exhibits *limit cycle* behaviour. This behaviour is not possible with linear systems and comes about via the non-linear damping term $\mu(1 - x^2)$. For $x < 1$ damping is negative (equivalent to negative ζ in Eq. 7.57) and the solution diverges, but for $x > 1$, damping changes to positive and the solution converges. If $\dot{x}$ were to be plotted against x—the so-called "phase space"—the resulting trajectory would form a squared-off ellipse rotating around the point $\dot{x} = 0, x = 1$. Solutions which start within the ellipse will converge onto it from within while those which start outside will converge onto it from without. The rate of convergence is defined by the restorative function $f_r(x)$ which for van der Pol's equation is simply x.

By comparing Eq. 7.55 with each of these equations, we observe the following.

Helmholtz Frequency Equating the coefficients of the linear terms of Eqs. 7.55 and 7.57, we can set $1/(\tau\mathcal{I}) = \omega^2$. But $\tau = V_p/c^2$, giving

$$\omega = c\sqrt{\frac{L_c}{V_p A_c}}$$

which we recognise as the Helmholtz frequency defined above.

Damping The coefficient of the rate term is the damping coefficient. For Eq. 7.55 this is

$$\frac{k_T'}{\tau}\left[1 - \frac{\tau}{\mathcal{I}}\frac{\partial F_p}{\partial \dot{m}_c}\frac{1}{k_T'}\right].$$

Comparing this to the van der Pol equation, we can identify k_T'/τ with μ. The damping coefficient will change sign when the bracketed term passes through zero. Like van der Pol's equation, the solution of Eq. 7.55 will become unstable if the damping term becomes negative which happens when

$$1 < \frac{\tau}{\mathcal{I}}\frac{\partial F_p}{\partial \dot{m}_c}\frac{1}{k_T'}$$

or

$$\frac{\partial F_p}{\partial \dot{m}_c} > k_T'\frac{\mathcal{I}}{\tau}$$

τ, k_T' and $\mathcal{I}$ will be small numbers, and for all practical purposes, the solution will become unstable where the gradient of the compressor characteristic becomes slightly positive.

This behaviour is illustrated by the following computational results. Equation 7.55 is decomposed into two first-order DEs by defining

$$\frac{d\dot{m}_c}{dt} = y \tag{7.61}$$

and writing Eq. 7.55 as

$$\frac{dy}{dt} = -\frac{k_T'}{\tau}\left[1 - \frac{\tau}{\mathcal{I}}\frac{\partial F_p}{\partial \dot{m}_c}\frac{1}{k_T'}\right]y - \frac{1}{\tau\mathcal{I}}\dot{m}_c + \frac{k_T'}{\tau\mathcal{I}}\left(F_p(\dot{m}_c) - p_3\right). \tag{7.62}$$

The results depicted by Fig. 7.12 were computed by solving Eqs. 7.61 and 7.62 using the compressor whose head/flow characteristic is shown earlier. Flow variations were induced by varying the back pressure p_3. Initialised to a value for which the flow $\dot{m}_c$ was greater than the positive slope threshold (the region c–d of the characteristic), back pressure was progressively increased until the flow reduced below the stability threshold ($\Phi \approx 0.66$, point "c" on the characteristic), whereupon the flow abruptly transitioned into a stable limit cycle. After some 12 cycles the

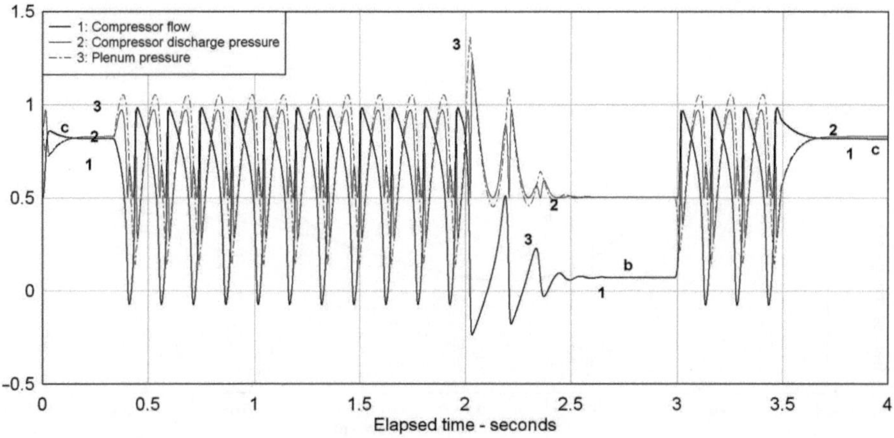

Fig. 7.12 Compressor entering and leaving the unstable region

back pressure was increased further to bring the flow below the point marked "b" whereupon the operating point entered the second stable range on the characteristic (the region a–b). The limit cycle was replaced by a stable but much reduced flow. The back pressure was again increased to bring the operating point into the unstable range and reactivate the limit cycle. Finally the back pressure was reduced to bring the flow back into the normal stable operating region (c–d).

It is noted that for a brief interval during each limit cycle the flow reversed. If this were to happen in an actual compressor, the compressor would have entered "deep stall".

This example shows that the flow calculation can be caused to enter and leave the unstable region at will, simply by bringing the flow into the unstable region.

Being offset slightly along the Φ-axis, this characteristic provides a second stable region close to zero flow. In practice, however, this is unlikely to be stable, given the turbulence and resulting stall that develops in the inter-blade space of the rotor at such low flows.

Computational Issues Equations 7.61 and 7.62 were solved using fourth Runge–Kutta, and the results are shown by Fig. 7.12. Individual traces on Fig. 7.12 are:

1. Compressor flow 2. Compressor discharge pressure 3. Plenum pressure

The model parameters were selected to give $\tau = 0.025$ s and $\mathcal{I}$ 0.01, producing a limit cycle period of around 100 ms, close to the theoretical Helmholtz frequency of 10 Hz. Numerical stability limited the discrete time step to less than 1.7 ms.[10] Although the limit cycle period would suggest a much larger time step, the cycle

[10]Published actual compressor test results [31, 32] indicate that compressor surge frequencies are to be expected in the vicinity of 20 Hz. Computation at this frequency would require the use of a significantly smaller time step.

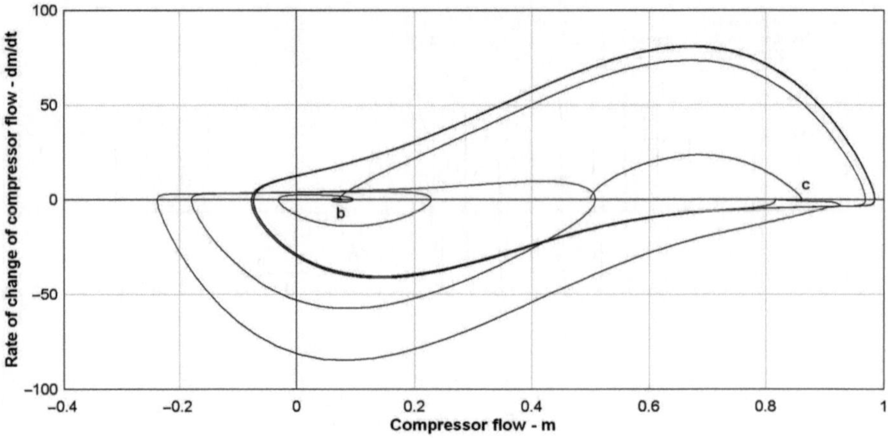

Fig. 7.13 State-space trajectories for compressor surge

waveform exhibits steeper leading and trailing edges which induce high-frequency components into the limit cycle spectrum. It is these components which determine and reduce, accordingly, the requisite time step. This is an aspect of the solution of non-linear equations which is absent from linear treatments.

These very short time steps cannot be used in real-time simulation nor do they need to be. It may be of interest to identify the onset of surge but not necessarily the details of the resulting oscillations. Assuming a minimum time step of, say, 50 ms, the onset and general behaviour of surge oscillations can be stably reproduced but at a lower oscillation frequency, maybe 2 Hz.

Figure 7.13 shows the state-space plot of these oscillations, $\dot{m}_c$ vs $d\dot{m}_c/dt$. The time for one circuit of the loop is 0.1 s. Small damped but stable oscillations are visible around both stable limit points—the points marked "b" and "c" on the characteristic—which converge to a point on the zero rate axis. The centre of the closed trajectories can be located at a flow rate of around 0.45, being the center of the unstable branch of the characteristic. It requires only a small movement of flow away from either of these points into the positive slope region of the characteristic to initiate the limit cycle.

Inclusion of Rotor Speed The model can be completed by adding Eq. 7.54 to the mix. The driving torque is supplied by the driving device. The compressor retarding torque is that given by Eq. 7.66 which is a linear function of ω.

7.4 Speed Calculation for Rotating Drives

The rate of change of speed ω [rad/s] is given by

$$I_C \frac{d\omega}{dt} = T_{drv} - T_{drvn}, \tag{7.63}$$

where T_{drv} is the torque developed by the driving device, T_{drvn} is the retarding torque presented by the driven device and I_C is the combined moment of inertia of the coupled set. If we have analytical expressions for the two torque terms as functions of ω and other relevant process variables, this equation can be solved to give the time-varying speed of the set.

7.4.1 *Treatment of Geared Drives*

Gearing is used where the operating speeds of the driving and driven devices are different. Although drives are frequently directly coupled to the driven device there are many instances where gearing is required to match the torque/speed characteristics of the drives. Given that power is the product of torque and speed a high-torque device will usually be driven at a low speed, and *vice versa* a low-torque device may require a high speed. In each case the driven device must be matched to the torque/speed characteristic of the driving device, for a given power rating.

A similar case is that of the variable speed drive for which the driving and driven devices are linked by a variable speed coupling, such as a fluid coupling. This is equivalent to a gearbox with a continuously variable speed ratio.

For these cases we may write Eq. 7.63 as

$$I_{drv}\frac{d\omega_m}{dt} + I_{drvn}\frac{d\omega_d}{dt} = T_{PM} - T_d .$$

With the speed ratio $G = \omega_m/\omega_d$, this becomes

$$(I_M + I_d/G)\frac{d\omega_m}{dt} = T_{PM} - T_d/G \qquad (7.64)$$

that is, both the driven device MoI and retarding torque have been reduced by the gear ratio, as seen by the driving device.

7.4.2 *Retarding Torque*

Any rotating machine will exhibit two forms of retarding torques in the bearings:

(a) *Static Friction*: This is the resistance to turning offered by the shaft at standstill in its bearings. At standstill the oil film between the shaft and the bearing surface is very thin or non-existent. The coefficient of static friction is high, and a substantial torque may be needed to rotate the shaft. For roll and ball bearings this is less of a problem than for journal bearings—the majority in large industrial-sized rotors. Heavy components are often provided with a source of high-pressure oil for lifting (or "jacking") the shaft off the bearing surface prior to first rotation.

In the absence of a high-pressure lubricating or lifting oil supply, static friction is high at zero or low speed but decreases quickly once the shaft begins to turn. To ensure continuity of the speed calculation the static friction torque can be weighted by a term such as $e^{-k\omega}$, where the constant k is selected to ensure a very rapid reduction in the torque component as the drive accelerates away from standstill. This term has the opposite effect as the drive decelerates close to zero speed, at which time the weighting term again becomes effective and brings the speed smoothly to zero.

(b) *Rolling Friction*: As shown by Eq. 7.76 (Sect. 7.5.2), torque due to rolling friction is proportional to ω. The coefficient of proportionality is small and, assuming normal operation of the lubrication system and once static friction has been overcome, can generally be ignored. Even mechanical damage, short of massive bearing failure, is unlikely to affect the motor speed dynamics but will influence bearing temperatures.

Fans, pumps, compressors and similar devices develop retarding torques from the movement of their working media, giving rise to the following.

(a) *Fluid Drag*: Rotation of the impeller creates static pressure in the casing as the blades exert force on the working fluid. It is this action which develops the zero-flow static head dp_0. From Eq. 7.17 for a centrifugal flow and Eq. 7.18 for an axial flow fan, we have, for the zero-flow head,

$$dp_0 = \rho\, r_2^2[1 - (r_1/r_2)^2]\,\omega^2$$

or

$$dp_0 = \rho\, r_2^2\, \omega^2$$

if we ignore the significance of the bracketed term ($r_1 \ll r_2$). The equation for pumps is similar.

Since pressure is force acting over an area, the head can be treated as a drag force F_{dr} acting over the blade area A_{bl} and exerting a torque calculated as if the force acts at the mid-point of the blade, at a distance $(r_1 + r_2)/2$ from the shaft centreline. Then,

$$dp_0 = F_{dr}/A_{bl}$$

from which

$$F_{dr} = dp_0 A_{bl}.$$

For a centrifugal fan or pump, $A_{bl} = (r_2 - r_1)b_c$, with the terms defined as in Sect. 7.1.1. Then,

$$F_{dr} = \rho r_2^2 \omega^2 (r_2 - r_1) b_c.$$

The drag torque is calculated as the drag force F_{dr} acting at the mid-point of the blade, at a moment arm $(r_1 + r_2)/2$. Then

$$
\begin{aligned}
T_{dr} &= F_{dr}(r_1 + r_2)/2 \\
&= \rho r_2^2 \omega^2 (r_2 - r_1)(r_2 + r_1) b_c/2 \\
&= \rho r_2^2 \omega^2 (r_2^2 - r_1^2) b_c/2 \\
&= \rho r_2^4 (1 - (r_1/r_2)^2)(b_c/2)\, \omega^2
\end{aligned}
$$

that is, the drag torque is proportional to the square of the rotational speed with the coefficient of proportionality k_{drag} given by

$$
k_{drag} = k_O N_{bl}\, \rho\, r_2^4 (1 - (r_1/r_2)^2) b_c/2 \tag{7.65}
$$

where N_{bl} is the number of blades and k_O is an adjustment parameter to fit observed or expected performance.

(b) *Flow Torque*: Torque is defined to be equal to the rate of change of angular momentum. If v_1 is the velocity of the fluid (or gas) into the impeller and v_2 is its velocity leaving the impeller ($v_2 > v_1$) and $\dot{m}$ is the fluid mass flow rate which, assuming conservation of mass, is the same at inlet and outlet, then the developed torque T is

$$
\begin{aligned}
T &= \dot{m}(v_2 r_2 - v_1 r_1) \\
&= \dot{m}\left(v_2 r_2 - v_2 \frac{r_1}{r_2} r_1\right) \\
&= \dot{m} v_2 \frac{1}{r_2}(r_2^2 - r_1^2).
\end{aligned}
$$

Assuming that the medium velocity leaving the tip equals the tip linear velocity U (no slip), we may write

$$
\begin{aligned}
T &= \dot{m} U \frac{1}{r_2}(r_2^2 - r_1^2) \\
&= 2\pi\, \omega\, \dot{m}\, r_2^2 \left[1 - \left(\frac{r_1}{r_2}\right)^2\right].
\end{aligned} \tag{7.66}
$$

An alternative formulation may be developed by noting that, for mass flows greater than zero, the motor must deliver a torque based on the power used to generate that flow. Given that power = developed head (dp) × volumetric flow ($\dot{v}$) and torque = power ÷ speed, we may write, for $\omega > 0$,

$$
\begin{aligned}
T_{\dot{v}} &= P_{drv}(\dot{v})/\,\omega \\
&= dp\, \dot{v}/\,\omega.
\end{aligned} \tag{7.67}
$$

It is common operating practice to start fans and pumps against a closed discharge which opens only when some threshold speed or pressure has been exceeded. This means that $T_{\dot{v}}$, effective only if flow has been established, could be zero until some time into the start.

Comminution Equipment

Crushing and grinding of rock and coal are size-reduction activities described generically as comminution. Originally this term is applied in the mining industry to the reduction of the size of rock extracted, particularly by blasting, from mines and quarries to facilitate its transport as the first step towards separation of minerals from waste product. The term is now widely applied to all forms of size reduction in the processing or preprocessing of minerals and solid fuels. Rotary crushing and grinding equipment might be described as "friction controlled". Tube mills are usually physically large with a high moment of inertia and operate at a relatively low speed of rotation, usually in the range 10–15 rpm.[11] The high retarding torques arise from the internal friction within and lifting forces needed to rotate the contents of the machine. Once the high static friction has been overcome and the device is rotating, the load torque is only weakly dependent on load, that is, throughput of ground product, as this is frequently determined by airflow through the mill while the charge weight remains relatively constant. A significant fraction of the power supplied to a comminution machine goes into friction and other losses rather than into the size-reduction process. This is usually described as the grinding efficiency of the mill which is the ratio of the breakage energy to the total energy supplied to the mill. Tube mills typically exhibit grinding efficiencies in the 5–8 % range while roller-table mills exhibit around 7–15 % efficiency [34]. Primarily for this reason roller-table mills are widely used for the grinding of coal in power stations and are replacing the more traditional tube mills in other applications such as cement raw mills. Tube mills continue to be used for hard abrasive coals.

The fluid drag and flow torque terms described above for fans, pumps, etc. are not relevant to comminution equipment. The retarding torque can be developed heuristically as the sum of (1) a stiction term $k_0 e^{-k_{st}\omega}$ applying when the mill is stationary or nearly stationary, (2) a term defining the internal friction losses and proportional to the product of the internal mass of the mill's contents and rotational speed (this appears as the no-load torque of the charged mill) and is independent of the ground throughput of the mill and (3) a term derived from a breakage law via which the specific energy E_{brk} expended by the size-reduction process may be calculated.

$$T_L = k_0 e^{-k_{st}\omega} + k_1 M\omega + k_2 M\, E_{brk}. \qquad (7.68)$$

The coefficients k_0, k_1 and k_2 are selected to match the known power consumption of the equipment.

[11] The speed of rotation is selected to be some fraction of the rotating tube's critical speed, itself primarily a function of the diameter of the tube.

7.4.3 Driving Torque

Electric Motor Drives

With some few exceptions, such as those driven by steam turbines, most fans, pumps and compressors in industrial applications are driven by electric motors. As well as meeting the rated load requirements, selection of the motors must ensure smooth acceleration and deceleration during starting, stopping and load transients. It is necessary, for example, to ensure the motor's ability to overcome static torque at standstill and that the current taken by the motor during starting does not exceed the maximum rating for the motor or lead to excessive winding temperatures.

The plant automation systems may monitor the run-up time of the motor by checking that within a certain monitoring time the motor has passed some preset speed limit, typically 60 % of rated speed. Failure to meet this criterion will trip the motor circuit breaker and raise an alarm. In some critical systems a pump may be required to accelerate to rated speed in the shortest possible time.

For simulation purposes it is important, particularly for large motor drives, to reproduce the run-up and rundown characteristics closely, both to satisfy any automation supervisory requirements and to ensure the correct response of the process to the associated flow and pressure changes.

Most large electric motor drives used in industrial applications are induction motors. Unlike synchronous motors, which operate at a constant "synchronous" speed defined by the supply frequency, induction motors run at a variable speed somewhat less than that of the supply. If N_s is the synchronous speed (usually stated in rpm but can be quoted in rad/s in which case it is denoted ω_s), N_d is the speed of the drive (also normally quoted in rpm but denoted ω_d if in rad/s), f_s is the frequency of the supply (Hz) and n_p the number of poles in the motor, then

$$N_s = \frac{120\, f_s}{n_p}$$

and

$$s = \frac{N_s - N_d}{N_s} = 1 - \omega_d/\omega_s$$

is the *slip*.

The synchronous speed of a 2-pole induction motor connected to a 50 Hz supply is 3,000 rpm and 3,600 rpm if connected to 60 Hz. When running at 1,430 rpm on a 50 Hz supply, for example, a 4-pole motor has a slip $s = 0.0467$ or 4.67 %.

The following figure shows typical torque/speed characteristics for an induction motor and a load with stiction. While only representative, this characteristic shape applies generally to all induction motors, whether squirrel cage or wound rotor. The latter type may be provided with means for varying the rotor resistance during starting to control the maximum starting current.

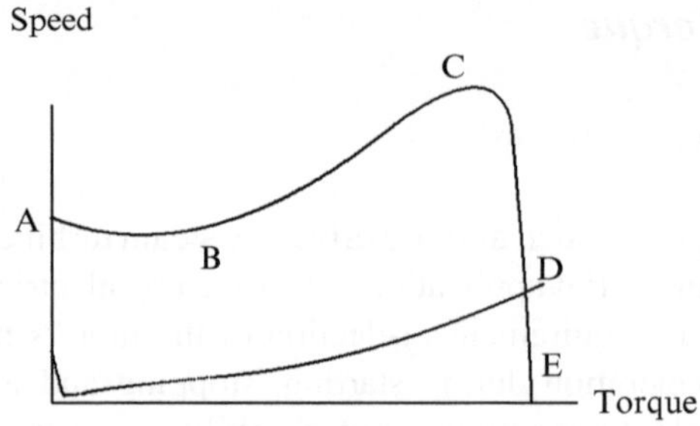

At A the motor is at standstill, and the torque T_0 is the no-load or blocked rotor torque. The torque at B is the "pull-up" torque and must exceed the load torque if the run-up is to be successful. Point C is the maximum torque developed by the motor at slip s_M and speed N_M. D designates the point at which the motor torque equals the load torque and is the stable operating point. At E the motor speed equals synchronous speed, and the developed torque is zero. Windage and friction losses will prevent the motor reaching this speed, even in the no-load state.

The difference between the two curves is the acceleration torque.

Induction Motor Torque/Speed Characteristic

The relationship between the speed of an induction motor and its developed torque can be derived in terms of the motor's physical parameters by using its equivalent circuit in which some inductive and resistive elements are expressed as functions of slip. A detailed treatment of even the single-phase lumped equivalent circuit is outside the scope of this book.[12] We will simply note that, with reference to the parameters of the equivalent circuit, the torque developed by an induction motor for any slip s is

$$T(s) = \frac{1.5}{\pi}\frac{n_p}{f}V_T^2 \frac{R_r/s}{\sqrt{(R_s + R_r/s)^2 + X_l^2}}, \tag{7.69}$$

where

- R_r is the rotor resistance [ohms]
- R_s is the stator resistance [ohms]
- X_l is the total leakage reactance (rotor plus stator) [ohms]
- V_T is the stator RMS phasor terminal volts
- n_p is the number of poles
- f is the supply frequency [Hz]

These parameters are usually available, or can be readily calculated, from the motor faceplate details.

[12] The interested reader is referred to [36] for a comprehensive treatment of this topic.

The maximum torque and the slip at which it occurs are obtained by differentiating Eq. 7.69 with respect to s and setting the derivative to zero. Then,

$$T_M = \frac{0.75}{\pi}\frac{n_p}{f}\frac{V_T^2}{R_s + \sqrt{R_s^2 + X_l^2}} \tag{7.70}$$

and

$$s_M = \frac{R_r}{\sqrt{R_s^2 + X_l^2}}. \tag{7.71}$$

As the rotor resistance increases, the peak torque occurs at a greater value of slip (lower speed), and the available starting torque increases.

As is often the case in simulation, an approximating function can be found which fits a given static characteristic sufficiently closely to avoid the use of a complex physical model. This approach can be computationally faster and reduces the detailed physical properties needed for each drive. The following formula gives a good approximation to the typical torque/speed characteristic and provides a set of parameters which can be selected to match a given characteristic.

$$\frac{T_{DR}(\omega)}{T_M} = 2\left[\left(\frac{s}{s_M}\right)^2 + \left(\frac{s_M}{s}\right)^2\right]^{-k} \quad \text{for } s > 0, \tag{7.72}$$

where k is some index which can take any value in the range $0 < k < 1$. In practice best results are obtained with k values in the range $0.2 < k < 0.5$. As shown by the accompanying figure variation of k simulates variation of the rotor resistance which moves the peak value of the characteristic along the slip axis.

That Eq. 7.72 ensures the maximum torque T_M occurs at $s = s_M$ is shown by differentiating with respect to s. Then

$$\frac{\mathrm{d}}{\mathrm{d}s}\left(\frac{T_{DR}(\omega)}{T_M}\right) = \frac{4k}{\omega_s}\frac{1}{s}\left[\left(\frac{s}{s_M}\right)^2 - \left(\frac{s_M}{s}\right)^2\right]\left[\left(\frac{s}{s_M}\right)^2 + \left(\frac{s_M}{s}\right)^2\right]^{-(k+1)}$$

which is zero for $s = s_M$.

The following figure shows torque/speed characteristics $T_{DR}(\omega)/T_M$ as calculated from Eq. 7.72 for the three combinations of values of s_M and k: (0.15, 0.25), (0.15, 0.50) and (0.2, 0.25). A typical induction motor will develop maximum torque at around $s = 0.15$ and will exhibit a ratio of standstill to maximum torque in the range 2:1–5:1.

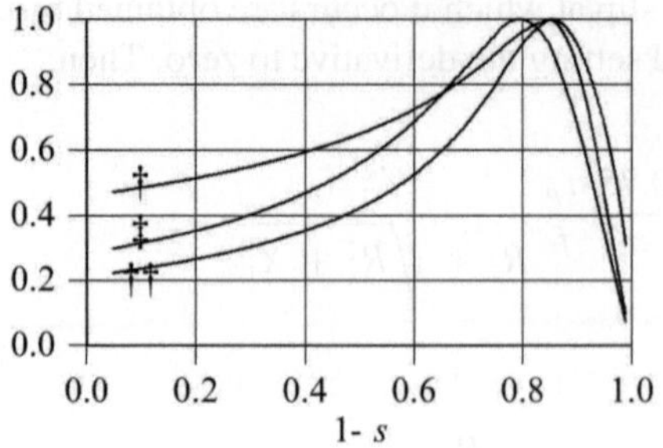

Steam Turbines as Drives

When used to drive fans, pumps or compressors, steam turbines are usually operated as variable speed drives. Unlike their role when driving synchronous power generators, when used as utility drives, the speed of the turbine/driven device combination varies with load on the driven device. Accurate speed control within the simulation requires accurate and stable modelling of the turbine/drive torque balance.

The power developed by the turbine is the product of the real enthalpy drop across it and the steam mass flow through it,

$$P_{trb} = \Delta h\, \dot{m}_s$$

and is discussed in detail in Chap. 14.

7.4.4 Speed Calculation from the Torque Balance

Multiplying both sides of 7.63 by ω gives

$$\begin{aligned} I_C \omega \frac{\mathrm{d}\omega}{\mathrm{d}t} &= T_{drv}\omega - T_{drvn}\omega \\ &= P_{drv} - P_{drvn}, \end{aligned}$$

where P_{drv} and P_{drvn} are the *powers* developed by and consumed by the drive and driven device respectively. Noting that

$$\frac{\mathrm{d}\,\omega^2}{\mathrm{d}t} = 2\omega \frac{\mathrm{d}\omega}{\mathrm{d}t}$$

this can be written in terms of ω^2 as

$$\frac{I_C}{2} \frac{\mathrm{d}\,\omega^2}{\mathrm{d}t} = P_{drv} - P_{drvn}. \tag{7.73}$$

The power required by the driven device is the product of its developed head and the volumetric flow through it.

$$P_{drv} = \mathrm{d}p\,\dot{v}.$$

If we treat dp as the internal developed head (better than differential pressure as this includes internal losses), we can write $\mathrm{d}p = (\mathrm{d}p_0/\omega_0^2)\,\omega^2$, where $\mathrm{d}p_0$ is the zero-flow head coefficient from the drive's test head/flow characteristic and ω_0 is the rated speed of the driven device [rads/s]. The torque balance then becomes

$$\frac{I_C}{2}\frac{\mathrm{d}\omega^2}{\mathrm{d}t} = P_{drv} - \frac{\mathrm{d}p_0}{\omega_0^2}\,\dot{v}\,\omega^2 - F_{drag}(\omega) \tag{7.74}$$

or in normalised form

$$\frac{I_C}{2}\frac{\mathrm{d}N^2}{\mathrm{d}t} = \frac{1}{\omega_0^2}P_{drv} - \frac{\mathrm{d}p_0}{\omega_0^2}\,\dot{v}\,N^2 - F_{drag}(N) \tag{7.75}$$

where the normalised speed $N = \omega/\omega_0$ and ω_0 is the rated drive speed (rad/s). $F_{drag}(N)$ is derived from internal bearing friction and fluid drag losses.

Since Eq. 7.74 is expressed in terms of ω with units of rad/s the moment of inertia I is expressed in consistent units of kg m^2. With ω_0 expressed in rad/s, Eq. 7.75 is also consistent.

Once P_{drv} is known, Eqs. 7.74 and 7.75 are linear in ω^2 or N^2 and may be solved using standard techniques. The preferred method is fourth-order Runge–Kutta because of its fourth-order accuracy in time.

Figure 7.14 shows the start-up speed transient calculated using Eq. 7.74 for a large centrifugal fan.

In many applications the rundown behaviour is of more interest than the run-up. The rate of decay of pressure and flow following trip of the driving motor is significant to the control of water level in drum boilers, of pressure and boiler feedflow in once-through boilers, of air flow to or flue gas flow from a furnace, of airflow through a coal mill and so on.

Once the circuit breaker has tripped, only retarding torques remain. Without additional breaking, the rundown time of large motors is much longer than run-up, as the following figures demonstrate. These, calculated for a typical fan and pump, show that the start-up transient completes within 20 s or so. For demonstration purposes, it is then assumed that, at 30 s, the driving motor circuit breaker trips open for some unspecified reason (Fig. 7.15).

In both cases, as the speed passes through 1,000 rpm on the way down, the discharge damper or valve closes, reducing flow to zero. The drive continues to coast down, slowed by fluid drag and bearing losses. Because of the larger fluid drag, pumps slow more rapidly than fans. The drive might be equipped with a brake to bring the shaft to a full stop from a low speed.

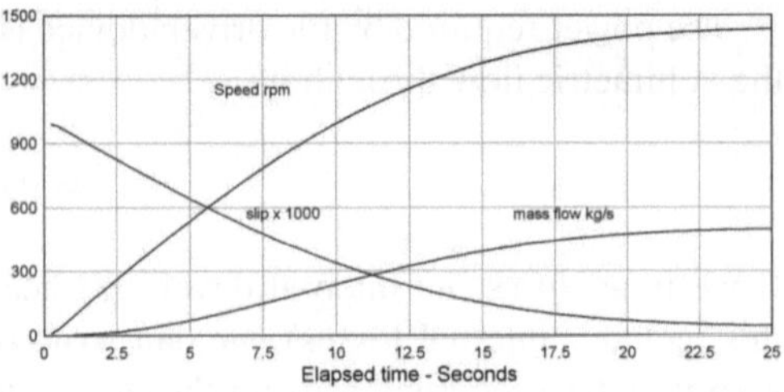

Fig. 7.14 Centrifugal fan start-up

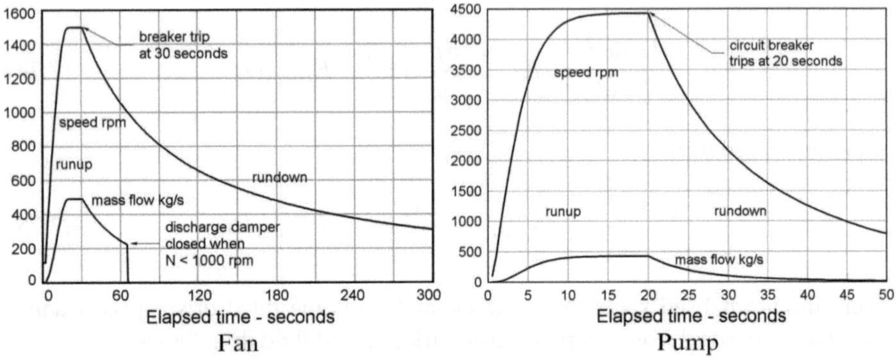

Fig. 7.15 Motor rundown following breaker trip

Run-up time for a motor drive is defined variously as the time to reach some defined speed, typically 60 % or 90 % of rated speed. As the calculated results show, these times are unaffected by the position of discharge dampers or valves as, during the initial development of flow, the significant retarding torques are not flow dependent.

These speed curves have been calculated using the following fan and pump data.

This form of speed calculation is appropriate for large motors or for those for which start-up current transients are of interest. These generally are supplied from higher-voltage buses—3.3 kV and higher—and draw substantial power. A first-order lag approximation to the start-up and shutdown transients will usually suffice for smaller motors.

	Fan	Pump
Type of device	Centrifugal	Centrifugal
Working medium	Air	Water
Nominal speed rpm	1,500	5,000
Root radius mm	300	100
Tip radius mm	1,000	200
Blade chord mm	500	100
Nominal mass flow kg/s	380	400
Nominal volumetric flow m^3/s	380	0.4

7.4.5 Calculation of the Moment of Inertia

Exact calculation of the moment of inertia (MoI) of a complex shape is not straightforward and, in most cases dealt with here, not necessary. A good approximation can usually be arrived at by treating the rotating shape as built up from a set of fundamental simple shapes, each having an easily computed MoI. The usual set of shapes includes a cylinder, rotating about its longitudinal axis; a hollow disk, also rotating about its longitudinal axis; and a flat plate rotating about a longitudinal axis offset from and parallel to the plate's central axis; the offset is the distance between the axis of rotation and the centreline of the plate.

	Mass	Radius of rotation	MoI
Cylinder	$\pi r^2 L\rho$	$r/\sqrt{2}$	$(\pi/2)\, r^4 L\, \rho$
Radius r			
Length L			
Hollow disk	$\pi(r_o^2 - r_i^2)t\rho$	$\sqrt{r_o^2 + r_i^2}$	$\pi(r_o^4 - r_i^4)t\rho$
Inner radius r_i			
Outer radius r_o			
Thickness t			
Offset plate	$w\,t\,L\rho$	$\sqrt{\frac{w^2+t^2}{12} + \delta^2}$	$\left(\frac{w\,t^3+t\,w^3}{12} + w\,t\,\delta^2\right) L\rho$
Width w			
Thickness t			
Length L			
Offset δ			

The MoI is calculated as $M\, r_{rot}^2$, where M is the rotating mass [kg] and r_{rot} is the distance from the axis of rotation [m]. In each case ρ is the density [kg/m^3] of the material. All dimensions are metres.

7.5 Bearings

Bearings used in all rotating machinery carry the weight of the shaft and rotor, resist the out-of-balance forces developed by rotation and maintain the central alignment of the rotating shaft. Friction developed between the stationary bearing surface and the rotating shaft is reduced by lubrication via which a thin film of lubricant—usually oil but sometimes water[13]—is maintained between the moving surfaces. Shear forces within the lubricant film will develop heat within the lubricant, some of which will be transferred to the shaft and bearing housing, raising their

[13]We will not consider non-fluid lubricants such as graphite.

temperatures, while the rest will heat the lubricant. If the heat developed is small it can be dissipated by natural convection from the bearing housing, as is the case for small, sealed bearing units. However, for all larger rotating machines, since the lubricant is the heat sink for the metal parts, it must be circulated through a suitable cooling system to avoid excessive temperatures within the shaft or structure. Where it is provided, the lubrication system provides an important cooling role in addition to its primary lubrication role.

There are three principal types of oil-lubricated bearings used in industrial plant. Sealed bearings are usually packed with grease or other lubricant. Rotation of the shaft develops heat which liquefies part of the grease and provides the lubrication film. The bearing develops little heat which is dissipated from the casing by natural convection. Splash bearings, including oil-bath and splash-ring types, provide their own irregular circulation path from and back to an internal sump. Again, the bearing develops little heat which dissipates from the casing. The type most commonly found in larger plant components is the pressurised bearing (Fig. 7.16). These are supplied with oil from an external pump, with the oil returned to its storage tank through one or more water-cooled heat exchangers, usually of the shell-tube type.

Pressurised bearings appear most frequently as journal, ball or roller, and tilting pad bearings. The journal bearing is widely used for industrial applications, given its robust construction, easy maintenance and relative insensitivity to dirty and dusty environments. Tilting pad bearings are frequently used as thrust bearings in steam and gas turbines and to support heavy rotating loads such as rotary air heaters.

Figure 7.17 shows a typical oil supply and cooling system. The lubricant is injected under pressure into the bearing through inlet ports and removed via drain ports. The flow path between the two ports is complex. The incoming lubricant distributes itself into a thin film which fills the space between the bearing and the shaft surfaces. Rotation of the shaft causes the film to rotate. For an unweighted shaft, such as a vertically aligned shaft, the oil film maintains a uniform thickness around the shaft. For a load-carrying shaft, the load causes a displacement of the

Fig. 7.16 High-pressure journal bearing showing oil entries (Courtesy of Millmerran Operating Company, Australia)

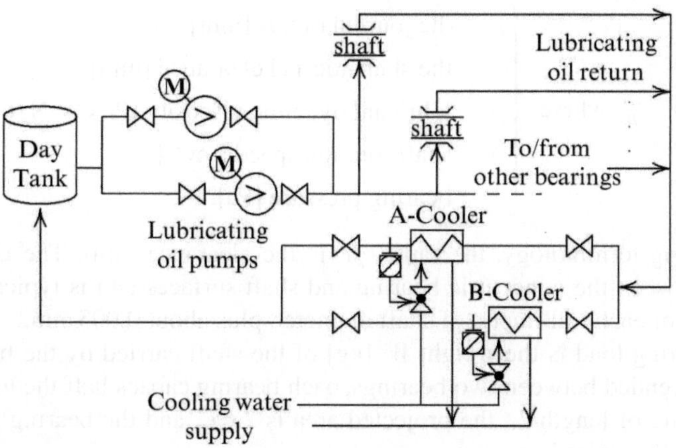

Fig. 7.17 Typical lubricating oil system: general arrangement

shaft away from the normal centreline of rotation, and a wedge is formed in the oil film. A region of local high pressure forms in the vicinity of the wedge, ahead (in the sense of the direction of shaft rotation) of the point of minimum clearance. As the film rotates, the off-centre axis rotates with it, inducing vibration. The coefficient of rotating friction, and therefore the heat developed within the bearing, is a function of film thickness, rotational speed and lubricant viscosity, itself a function of lubricant temperature. This last effect is auto-stabilising since viscosity and therefore generated heat decrease with increasing temperature. On the other hand, should the temperature increase be excessive as a result of, say, inadequate cooling, the reduced lubricant shear strength will be insufficient to withstand the weight of the rotating shaft, which may then contact the bearing surface, resulting in bearing and possibly shaft mechanical damage.

The following formulae are used for the calculation of a variety of bearing parameters. They are derived from a simplified analysis of the hydrodynamics of journal bearings, are widely used for bearing design and have been rearranged somewhat to suit the requirements of simulation analysis of plant for which the design details are known.

7.5.1 The Sommerfeld Number

An important parameter in bearing analysis is the dimensionless Sommerfeld number S:

$$S = \left(\frac{r}{c}\right)^2 \frac{\mu N}{P} \text{ where } \begin{cases} r & \text{the journal radius [mm]} \\ c & \text{the shaft-journal clearance [mm]} \\ \mu & \text{lubricant dynamic viscosity [Pa s} = \text{N/m}^2\text{s],} \\ N & \text{shaft rotation speed [rev/s]} \\ P & \text{bearing pressure [Pa].} \end{cases}$$

In bearing terminology, the ratio r/c is the clearance ratio. The clearance is the gap between the concentric bearing and shaft surfaces and is typically around 0.001 mm for each millimetre of shaft diameter, plus about 0.005 mm.

The bearing load is the weight W [kg] of the shaft carried by the bearing. For a shaft suspended between two bearings, each bearing carries half the total weight. For a bearing of length L, the projected area is $2\,r\,L$, and the bearing pressure = bearing load/projected area is

$$P = \frac{W}{2\,r\,L}.$$

The maximum pressure developed in the oil wedge exceeds this pressure by some factor, dependent on the specific design of the bearing. Design charts are available for the estimation of this factor [35]. Typically it will be in range of 2 or 4 times the bearing pressure P.

There is no advantage in designing a bearing whose length exceeds the shaft diameter. We may therefore usually assume that $L \leq 2\,r$.

7.5.2 Derived Bearing Quantities

Bearing Power Loss

The viscous friction factor of the oil film is given by Petroff's equation

$$f_f = 2\pi^2 \left(\frac{\mu N}{P}\right) \frac{r}{c}.$$

This drag creates a friction torque T_f needed to overcome it. The torque is calculated as the drag force $f_f\,W$ acting at a moment distance r.

$$T_f = f_f\,W\,r$$

The power required of the prime mover to overcome bearing friction is then given from

$$P_{wr} = T_f \cdot 2\pi N \quad [J\ s^{-1} \text{ or Watts}]. \tag{7.76}$$

Bearing Oil Flow

A bearing is designed for a given bearing load and a minimum oil film thickness at a rated shaft speed. Using the Sommerfeld number as ordinate, design charts (Juvinall [35]) are used to determine the maximum pressure in the bearing and the requisite volumetric oil flow to the wedge.

For dynamic analysis purposes a simple approximation can be used based on a single admittance coefficient defined by the bearing's rated mass flow $\dot{m}_{oil_{100}}$ and rated pressure drop $\Delta p_{brg_{100}}$ across it.

$$\mathcal{A}_{brg} = F(\mu, N) \frac{\dot{m}_{oil_{100}}}{\sqrt{\Delta p_{brg_{100}}}}.$$

$F(\mu, N)$ is a function of lubricant viscosity μ and rotational speed N, obtained as an approximation to the design charts.

Bearing Temperature Rise

From the point of view of the bearing designer, the conservative assumption is that all of the heat generated in the oil is transferred to the oil, raising its temperature. This assumption guarantees that if the oil flow is sufficient to remove all of this heat, nothing will get too hot. In practice the heat is generated in the oil but part is transferred to the shaft and the casing. For the moment let us accept the designer's assumption and calculate the oil temperature rise as it passes through the bearing from the equation

$$P_{wr} = \dot{m}_{oil} c_{p,oil} \Delta T. \tag{7.77}$$

The dynamic variation of the oil temperature can be calculated from the simplified bearing heat balance

$$M_{brg} c_{p,brg} \frac{dT_{brg}}{dt} = \dot{m}_{oil} (c_{p,oil} T_{oil,in} - c_{p,brg} T_{brg}) + P_{wr}. \tag{7.78}$$

In writing this equation we have lumped the oil, shaft and casing masses together and tacitly assumed that:

- The oil, shaft and bearing casing are all at the same temperature T_{brg} which is the temperature of the oil leaving the bearing.
- A representative mass M_{brg} and specific heat capacity $c_{p,brg}$ can be defined for this agglomeration.
- The in- and outflows of oil are equal.

These assumptions are not unreasonable. The mass of oil in the bearing is very small and contributes little to the net energy storage in the bearing. Heat transfer from the oil to both shaft and casing is very good, and little temperature differential is expected to be maintained between them and the oil.

The consistency of these calculations can be confirmed by verifying that the calculated oil flow can remove the calculated heat from the bearing and that the calculated bearing oil temperature rise is reasonable. For example, a pump operating at 1,500 rpm (25 s^{-1}) has a shaft weight of 1,000 kg and a shaft diameter of 100 mm. The rated flow of lubricating oil is assumed to be 10 l/min (0.14 kg/s for an oil density of 858 kg/m^3) per bearing. Assuming an initial mean bearing oil temperature of 38°C and a supply temperature of 30°C and representative values at 40°C for the oil's specific heat capacity (1.95 kJ/kg C) and viscosity μ (6.81 mPa s or 0.00681 Ns/m^2), we calculate the following:

Bearing clearance c	$= 0.001 \cdot 100 + 0.005$	$= 0.105$ mm
Clearance ratio r/c	$= 250/0.105$	$= 2,381$
Bearing pressure P	$= 500/(2 \cdot 0.05 \cdot 0.1)$	$= 50,000$ Pa
Sommerfeld number S	$= 2,381^2 \cdot \mu \cdot 25/P$	$= 19.304$
Viscous friction factor f_f	$= 2\pi^2(\mu \cdot 25)/P \cdot 2,381$	$= 0.16$
Bearing power loss P_{wr}	$= f_f \cdot 500 \cdot 0.25 \cdot 2\pi \cdot 25$	$= 3,141$ watts
Oil temperature rise ΔT	$= P_{wr}/(\dot{m}\, c_p)$	$= 11.5$°C

7.5.3 Bearing Vibration

A healthy bearing with normal lubrication exhibits a "normal" level of vibration, and several factors and influences contribute to this vibration. These include minor out-of-balance rotational forces, irregularities in movement of the lubricating oil film, random variations in the forces driving the shaft, irregularities in the bearing surfaces, torsional oscillations of the shaft, interactions with supporting foundations and so on. Departures from the normal level of vibration can indicate a deterioration of any of the several factors including the mechanical condition of the bearing or shaft surfaces, the cleanliness or effectiveness of the lubricating oil supply, an out-of-balance condition of the shaft or a shaft misalignment. As the bearing surface wears the bearing clearance increases, and changes to the oil film flow pattern will change the observed vibration level or spectral composition. The change in oil film behaviour may become so extreme as to allow contact between the bearing and shaft surfaces with almost certain mechanical damage to both. A change in flow rate or incoming temperature, high or low, of the lubricating oil will change the oil temperature in the bearing and hence its viscosity. Again the behaviour of the oil film will change and with it the vibration characteristics. Should the shaft be a fan, compressor or turbine, blade damage or fracture will create an out-of-balance condition, and the bearing vibration will change to reflect this.

Being related to the rotational speed of the shaft, bearing vibration will reflect the resonances of the shaft. Leaving aside reciprocating engines and compressors, whose bearings are arranged as closely spaced load-bearing components resisting significant off-centre forces, the shaft in a turbomachine is freely suspended between two bearings. The shaft length will be several times greater than the shaft diameter and will exhibit one or more resonant or critical speeds. Around these speeds the shaft will flex away from its true centreline and cause a slight misalignment (eccentricity) within each bearing, with a corresponding increase in vibration level. Each critical speed will be marked by a characteristic frequency in the vibration spectral plot.

From this discussion it is apparent that bearing vibration measurements can provide important diagnostic information relevant to condition monitoring and early fault detection. Large and mission-critical rotating machines are routinely equipped with monitoring systems which deliver online analysis of these vibration characteristics. By assisting the identification of the probable cause of an abnormal vibration characteristic this information can be used to anticipate and prevent equipment failure.

Being a complex periodic waveform, usually possessing some stochastic component, a vibration measurement must be reduced to a representative value in order to be suitable for control room display. This may be the simple mean, a peak or peak-to-peak value or an RMS (*root mean square*) mean. It may indicate amplitude or offset (microns), velocity (mm/s) or acceleration (G or m/s^2). It may be displayed as a scalar *equivalent* value or resolved into its x-(lateral) and y-(vertical) components.

Vibration is measured using a variety of devices, suited to the location, type of measurement required and post-measurement processing to be performed. The simplest and most widely used is an accelerometer which delivers an electrical signal proportional to the instantaneous acceleration in the axis direction of the sensor. A single integration of the output yields velocity, and a second integration yields amplitude. Shaft displacement and eccentricity are usually measured using proximity metres based on eddy currents induced in the head of the measuring device. Permissible values of vibration levels have been established in various national and international Standards such as ISO 10816 (previously ISO 2372) for machines operating in the 600–12,000 rpm range. These are given for different sizes and speed ranges of machines and generally lie in the ranges of (quoting RMS velocities) 0.45–2.80 mm/s for small machines going up to 0.71–4.5 mm/s for large machines. Tolerable levels lie in the ranges of 2.8–7.1 mm/s for small machines and 4.5–11.2 mm/s for large machines. Not-permissible ranges are anything greater than these. Allowable and tolerable ranges are quoted for both rigid and flexible foundations.

For the simulation purpose vibration measurements can be reproduced empirically, conveniently as a sequence of base functions combined via a series of weighting coefficients which themselves can be functions of related process variables. The base functions are chosen to reproduce known characteristics such as critical speeds. A useful representation is, where v is the per unit vibration measure and V is the required scaled vibration measure,

$$v = \sum_{i=1}^{n_c} a_i \, \mathrm{e}^{-\left(\frac{N-N_i}{\sqrt{2}w_i}\right)^2} + a_0 N^2$$

$$V = K_{sc}\, v, \tag{7.79}$$

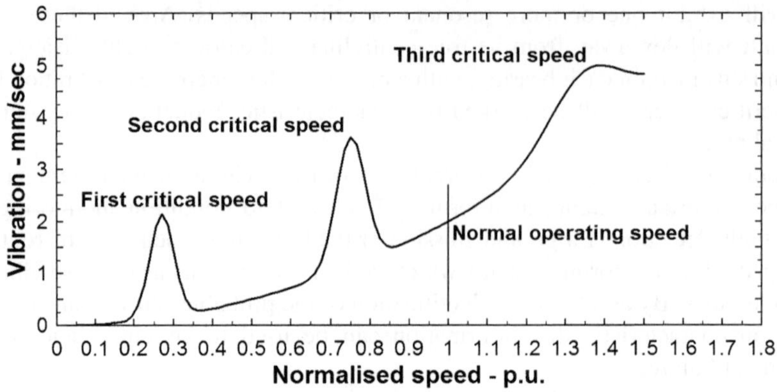

Fig. 7.18 Typical turbine vibration/speed characteristic

where $\begin{cases} N \text{ is the per unit shaft speed} \\ N_i \ (\text{i} = 1, n_c) \text{ are the } n_c \text{ critical speeds (per unit)} \\ w_i \ (\text{i} = 1, n_c) \text{ are the widths of speed variation about } N_i \\ a_i \ (\text{i} = 1, n_c) \text{ are amplitude constants for each critical speed} \\ a_0 \text{ is a constant.} \end{cases}$

This function represents the superposition of a quadratic speed dependency on the sum of a series of bell-shaped curves, each of width w_i and amplitude a_i about a critical speed N_i. The scaling factor K_{sc} is used to scale the output to suit the desired measurement display. Equation 7.79 is capable of adjustment to reproduce most vibration/speed characteristics of power generation and industrial turbine drives. This is illustrated by Fig. 7.18 which shows a typical vibration/speed characteristic generated using Eq. 7.79.

Large turbomachines such as power-generating turbines exhibit two or three critical speeds. The first is typically around 800 rpm, the second around 2,400 rpm while the third will be well above (but not always) the normal operating speed of 3,000 rpm or 3,600 rpm. Smaller turbines used to drive pumps, fans or blowers in industrial applications will exhibit other critical speeds but rarely will more than three be considered and frequently only two.

Other sources of vibration can appear as a result of abnormal operation. For example, a turbine being accelerated with a high eccentricity will experience high vibrations during the start. This condition can be replicated by replacing a_0 in Eq. 7.79 by $a_0(1 + X/X_0)$ where X is the current eccentricity and X_0 is some appropriate reference value.

Chapter 8
Heat Transfer

In accordance with the second law of thermodynamics heat is transferred unaided from a higher temperature to a lower temperature, but not in the opposite direction. Heat may be transferred unaided by any of three processes: conduction, convection or radiation.

Heat transfer by conduction causes heat to flow with no relative motion of the conducting medium. Thus, heat is conducted along shafts, across tube walls and through stationary fluids. Heat transfer by convection involves the movement (displacement) of the transfer medium involved. Thus, heat is transferred by convection through and with rising gases and vapours, through moving fluids, through rotating oil and water films and throughout the atmosphere. It is convective currents which create and drive the weather. Heat transfer by radiation is heat transfer "at a distance" without physical contact between the heat source, the emitter, and the heat sink, the receiver. Radiative heat transfer is an electromagnetic phenomenon and follows the same physical laws as other electromagnetic radiation.

The modelling of heat transfer requires the consideration of the physical process, the correlation of the physically and heuristically defined parameters with the mechanical and thermodynamic state of the process or media and the physical construction and operation of the heat transfer equipment. As we will see from the development of the material of this chapter, the heat transfer coefficient, expressed in a particular form (and there are several forms from which to choose), plays a central role.

Following the well-established precedent we will consider the three heat transfer mechanisms in the order quoted in the first paragraph.

P. O'Kelly, *Computer Simulation of Thermal Plant Operations*,
DOI 10.1007/978-1-4614-4256-1_8,

8.1 Conduction

8.1.1 Conduction Along a Straight Path

The rate of heat flow $\dot{q}$ along a linear conducting path is described by the Fourier equation in one dimension.

$$\frac{\mathrm{d}\,q(z)}{\mathrm{d}t} = -\lambda A \frac{\mathrm{d}T}{\mathrm{d}z}, \tag{8.1}$$

where A [m^2] is the area of the conduction path normal to the heat flow direction and λ is the coefficient of thermal conduction [J/(s m K)]. For two points separated by a distance Δz [m] along the path and having a temperature differential ΔT [K] between them, this becomes

$$\frac{\mathrm{d}\,q(z)}{\mathrm{d}t} = -\lambda A \frac{\Delta T}{\Delta z}. \tag{8.2}$$

8.1.2 Conduction Through a Cylindrical Wall

The steady-state radial distribution of temperature T through a cylindrical wall satisfies the following equation in the radial dimension r:

$$\frac{1}{r}\frac{\mathrm{d}}{\mathrm{d}r}\left(r\frac{\mathrm{d}T}{\mathrm{d}r}\right) = 0.$$

Integration of this equation yields the steady-state radial distribution of temperature.

$$T(r) = c_1 \ln(r) + c_2.$$

For a cylindrical wall with outer surface temperature T_o at radius r_o and inner surface temperature T_i at radius r_i,

$$c_1 = \frac{\Delta T}{\ln\left(\frac{r_o}{r_i}\right)},$$

where $\Delta T = T_o - T_i$, the temperature differential across the wall. Differentiate $T(r)$ with respect to r. Then

$$\frac{\mathrm{d}T(r)}{\mathrm{d}r} = \frac{c_1}{r} = \frac{\Delta T}{r\,\ln(\frac{r_o}{r_i})}.$$

The general formulation of Fourier's law of heat conduction in the cylindrical radial dimension is

$$-\lambda\, A(r)\, \frac{\mathrm{d}T(r)}{\mathrm{d}r} = \frac{\mathrm{d}q(r)}{\mathrm{d}t}.$$

Noting that for a cylinder of length l, $A(r) = 2\pi r l$ and substituting for $\mathrm{d}T(r)/\mathrm{d}r$ from the preceding equation, we arrive at the well-known equation for heat conduction through a cylindrical wall

$$\frac{\mathrm{d}q}{\mathrm{d}t} = -2\pi\lambda\, l \frac{\Delta T}{\ln(\frac{r_o}{r_i})}.$$

Define

$$r_{log} = \frac{r_w}{\ln(\frac{r_o}{r_i})},$$

where the wall thickness $r_w = (d_o - d_i)/2$. Then the conduction equation may be written

$$\frac{\mathrm{d}q}{\mathrm{d}t} = -\lambda\, \frac{A_{log}}{r_w}\, \Delta T$$

and $A_{log} = \pi r_{log} l$.

8.2 Convection

8.2.1 Forced Convective Heat Transfer Within Tubes and Ducts

The heat transfer for single phase convective heat transfer from a moving fluid or gas to its containing surface—here assumed to be metal—is computed from

$$\dot{q}_{hx} = \alpha_{hx}\, A_{hx}\, (T_f - T_m), \tag{8.3}$$

where T_m is the notional temperature at the inner surface of the metal interface. This form of the equation defines the heat transfer coefficient in terms of the temperature differential across the heat transfer interface and is valid for both laminar and turbulent flow regimes. For the moment we will consider turbulent flow (Reynolds Number >2,000). The laminar flow situation (low flow velocity) will be discussed a little later.

Various methods are available for predicting the heat transfer coefficient α_{hx} for particular geometries of the heat transfer surface, its orientation to the flow, the flow regime and the nature and properties of the working medium. Unlike electrical circuits and components whose electrical properties can be accurately predicted, heat transfer and pressure drop coefficients cannot be predicted accurately, and their determination depends on extensive accumulation of test data and its use to identify a best fit by a heuristic formula. Such formulae are called "correlations" and are frequently expressed in terms of one or more of the dimensionless "similitude numbers" of Reynolds Re, Nusselt Nu, Prandtl Pr and Graetz Gr. These are defined as

$$Re = \frac{\phi L}{\eta} \quad Nu = \frac{\alpha L}{\lambda} \quad Pr = \frac{c_p \eta}{\lambda} \quad Gr = \frac{\phi c_p d}{\lambda},$$

where
- η is the dynamic viscosity of the fluid [Pa·s],
- λ is the thermal conductivity of the fluid [W/(m K)],
- c_p is the specific heat of the fluid [kJ/(kg K)],
- L, d is a representative distance [m].

Note that the wall metal conductivity λ_w is used in the Biot Number $Bi = \frac{\alpha L}{\lambda_w}$.

For closed pipes and ducts—the most widely met fluid conduits—a popular correlation for fluids, including superheated steam, is that of Sieder–Tate

$$Nu = 0.023 Re^{0.8} Pr^{0.33} (\mu_b / \mu_w),$$

where μ_b and μ_w are the fluid viscosities evaluated at the temperature of the bulk fluid (subscript b) and the heat transfer surface (the wall, subscript w), respectively. For gases and vapours the Hausen form applies:

$$Nu = 0.341 Re^{0.6} Pr^{0.36}$$

d_i is the pipe inner diameter [m]. The specific mass flow $\phi = \dot{m}/A_x$ where A_x is the flow cross-sectional area.

The heat transfer coefficient α can be extracted to give the standard Colburn correlation for transfer between a fluid (not gases or steam) and a circular pipe:

$$\alpha_{hxm} = 0.023\, \lambda^{0.58}\, \eta^{0.38}\, c_p^{0.42}\, \phi^{0.8}\, d_i^{-0.2} \qquad [\text{kW}/(\text{m}^2\,\text{K})].$$

Other correlations derived from the Hausen form are available for gases and steam.

The product $(\lambda^{0.58}\, \eta^{0.38}\, c_p^{0.42})$ can be represented as a simple function of the interfacial (boundary) temperature $\mathcal{F}(T)$. Then

$$\alpha_{hxm} = 0.023\, \mathcal{F}(T)\, \phi^{0.8}\, d_i^{-0.2} \qquad [\text{kW}/(\text{m}^2\,\text{K})]. \tag{8.4}$$

The following section presents useful approximations to $\mathcal{F}(T)$ and α_{hxm} for a variety of working fluids. Unless otherwise noted dynamic viscosity is quoted in mPa·s for fluids and μ Pa·s for gases.

For water, saturated and superheated steam, the predicted values of α_{hxm} have been compared with the corresponding values extracted from the nomograms on pages 116 and 118 of Steinmuller [9]. Agreement is within 10–30 % over the full range of the working medium pressures, temperatures and specific mass flow rates, quite adequate for most purposes.

8.2.2 Heat Transfer Correlations for Various Media

Water

The heat transfer coefficient α_{hx} for water can be calculated from

$$\alpha_{hx} = 0.023\,\mathcal{F}(T)\,\phi^{0.8} d_i^{-0.2} \qquad [\text{kW}/(\text{m}^2\,\text{K})]. \tag{8.5}$$

The following table lists values of $\mathcal{F}(T)$ for the range of water temperatures 0–350 °C. This can be adequately fitted by the quadratic approximation

$$\mathcal{F}(T) = 1.3159 - 0.4557\left(\frac{T}{100}\right) + 0.0725\left(\frac{T}{100}\right)^2.$$

Errors are within ±3 % for $T > 30\,°\text{C}$. The errors in the vicinity of 0 °C are caused by the rapid (>quadratic) increase in dynamic viscosity of water at low temperatures.

All property values used for this calculation apply to saturation conditions. Some errors (>30 %) will be incurred when used at pressures for which the boundary temperatures are less than saturation.

T (C)	λ	η	cp	True value	$\mathcal{F}(T)$	Error %
0	0.558	1.794	4.226	1.631	1.316	−19.3
50	0.647	0.555	4.178	1.132	1.106	−2.31
100	0.690	0.278	4.191	0.907	0.933	+2.86
150	0.692	0.185	4.215	0.783	0.796	+1.64
200	0.672	0.139	4.262	0.707	0.695	−1.58
250	0.636	0.110	4.343	0.646	0.630	−2.46
300	0.542	0.092	4.483	0.588	0.601	+2.30
350	0.456	0.073	4.790	0.598	0.609	+1.81

Lubricating Oil

The following table lists values of $\mathcal{F}(T)$ for a standard lubricating oil (SAE 20) for a range of oil temperatures from 0 °C to 120 °C and the errors incurred using the quadratic approximation,

$$\mathcal{F}(T) = 1.458 - 1.65\left(\frac{T}{100}\right) + 0.746\left(\frac{T}{100}\right)^2,$$

$$\alpha_{hx} = 0.023\mathcal{F}(T)\,\phi^{0.8} d_i^{-0.2} \quad [\text{kW}/(\text{m}^2\,\text{K})].$$

T (C)	λ	η	cp	True value	$\mathcal{F}(T)$	Error %
0	0.144	33.0	1.77	1.560	1.458	−6.5
20	0.144	13.05	1.851	1.117	1.158	+3.67
40	0.143	6.81	1.934	0.885	0.918	+3.68
60	0.142	4.18	2.018	0.745	0.737	−1.16
80	0.141	2.83	2.102	0.651	0.616	−5.45
100	0.140	2.0	2.186	0.578	0.554	−4.07
120	0.138	1.54	2.269	0.527	0.553	+4.84

Light Fuel Oil

These include petroleum products with kinematic viscosities less than around eight centistokes (mm^2/s) of which the most common is diesel fuel.

Density ρ is calculated from a reference value ρ_{15}, quoted at 15 °C (hence the subscript), by the following formula.

$$\rho = \rho_{15} - 0.0064(T - 15),$$

where $\rho_{15} = 0.940\,\mathrm{kg/dm^3}$. Note that this equation produces density in $\mathrm{kg/dm^3}$ and density is used in these units in the following properties calculations.

Dynamic Viscosity η is calculated from kinematic viscosity ν as $\eta = \nu/\rho$.

Kinematic Viscosity ν follows a log–log relationship as a function of $\log(\mathcal{T})$, $\mathcal{T}$ in degrees Kelvin. Then

$$\log\log(\nu + 0.8) = d\ \log(\mathcal{T}) + c_0$$

with $d = -3.923$ and $c_0 = 9.52392$.

Specific Heat Capacity c_p is calculated from

$$c_p = [(2.96 - 1.33\,\rho) + (0.00615 - 0.00230\,\rho)\,T]\,z.$$

The temperature T is used in degrees Celsius and density ρ in $\mathrm{kg/dm^3}$. The factor z is a function of temperature

$$z = 0.067\,k + 0.35,$$

where $k = \sqrt[3]{T_s}/\rho_{15}$ and T_s is the saturation temperature of the oil [°C]. For most purposes z can be assumed to be unity for light oils.

Thermal Conductivity λ is a function of temperature and density and is calculated from

$$\lambda = \frac{0.1172}{\rho}(1.0 - 0.00054\,T)$$

and again T is expressed in degrees Celsius and ρ in kg/dm^3.

The following table lists values of $\mathcal{F}(T)$ for the range of light fuel oil temperatures 0–120 °C and the errors incurred using for $\mathcal{F}(T)$ the quadratic approximation,

$$\mathcal{F}(T) = 0.7404 - 0.483\left(\frac{T}{100}\right) + 0.279\left(\frac{T}{100}\right)^2,$$

T (C)	λ	η	cp	True value	$\mathcal{F}(T)$	Error %
0	0.113	7.815	1.582	0.748	0.740	−1.07
20	0.128	3.805	1.829	0.649	0.655	+0.92
40	0.147	2.073	2.077	0.590	0.592	+0.03
60	0.174	1.203	2.324	0.554	0.551	−0.54
80	0.214	0.717	2.571	0.536	0.532	−0.75
100	0.280	0.422	2.818	0.532	0.536	+0.75
120	0.409	0.231	3.066	0.546	0.562	+2.93

$$\alpha_{hx} = 0.023\mathcal{F}(T)\,\phi^{0.8} d_i^{-0.2} \quad [\text{kW/m}^2\,\text{K})].$$

Heavy Fuel Oil

Heavy fuel oil is the "bottoms" product of the crude oil distillation process. Officially designated as no. 6 fuel oil it is available in many grades, also known as Bunker B or C in the English/US system or as S in the German classification system. The density and viscosity of every grade of heavy fuel oil are strongly temperature dependent, and below some temperature (usually around 40 °C or so and denoted the "pour point"), the oil is non-Newtonian and only semi-fluid. This temperature is also referred to as the lower limit of "pumpability".

Density ρ may be calculated from a reference value ρ_{15}, usually quoted at 15 °C, by the following formula:

$$\rho = \rho_{15} - 0.0067(T - 15),$$

where $\rho_{15} = 0.940$ kg/dm^3.

Dynamic Viscosity η is again calculated from kinematic viscosity ν as $\eta = \nu/\rho$.

Kinematic Viscosity is calculated as shown for light fuel oil but with the coefficients:

$$d = -3.431541 \text{ and } c_0 = 8.94349.$$

Specific Heat Capacity and Thermal Conductivity Coefficient are calculated as for light oil. A value of around 0.96 for z is more appropriate for heavy oil.

The following table lists values of $\mathcal{F}(T)$ for the range of oil temperatures 40–120 °C and the errors incurred using for $\mathcal{F}(T)$ the quadratic approximation,

$$\mathcal{F}(T) = 4.696 - 5.639\left(\frac{T}{100}\right) + 2.382\left(\frac{T}{100}\right)^2,$$

T (C)	λ	η	cp	True value	$\mathcal{F}(T)$	Error %
50	0.137	99.798	2.126	2.493	2.472	−0.84
60	0.151	54.996	2.248	2.149	2.170	+0.98
70	0.167	31.826	2.371	1.896	1.916	+1.05
80	0.188	19.108	2.493	1.708	1.709	0.0
90	0.215	11.765	2.616	1.568	1.550	−1.15
100	0.253	7.337	2.738	1.466	1.439	−1.84
120	0.390	2.777	2.983	1.352	1.359	+0.44

$$\alpha_{hx} = 0.023\mathcal{F}(T)\,\phi^{0.8}d_i^{-0.2} \quad [\text{kW}/(\text{m}^2\,0\text{K})].$$

Since the pour point for heavy oils is generally between 40 °C and 50 °C the range of temperatures for which these properties are calculated is limited to >50 °C.

8.2.3 *Heat Transfer at Low Flow Rates*

The heat transfer correlations quoted in the previous section are valid only for Reynolds numbers $Re > 2{,}100$ (turbulent flow). For $Re < 2{,}000$ the flow may be treated as laminar. The calculation of heat transfer in laminar flow is complicated by the close interaction of momentum and heat exchange. Although this has been the subject of extensive theoretical investigation, and the relevant mass and momentum equations have been solved for simple geometries (flat plate, horizontal and vertical tubes), the practical solution is greatly complicated by the presence of local eddies, natural convection and local diffusion. Practical correlations useful for the prediction of convective heat transfer in laminar flows are to be found in the literature such as Perry [17]. These are based on Nusselt and Grashof numbers and are specific to particular geometries. For this work we can rely on the observation that for low Reynolds numbers, heat transfer reduces with reducing flow velocities to some minimum value, below which the heat transfer coefficient becomes essentially independent of velocity, a consequence of eddies and natural convection within the boundary layer. For $Gr > 100$ the Sieder–Tate correlation is satisfactory for circular tubes of small diameters (the usual case in tubular heat exchangers) and small temperature differentials [17].

$$Nu = 1.86\,Gr^{\frac{1}{3}}\left(\frac{\mu_b}{\mu_w}\right)^{0.14}.$$

Noting that the Graetz number $Gr = Re\,Pr\,d/L$, we may write

$$Gr = \frac{\phi\,c_p\,d}{\lambda}.$$

From the definition of Nu we then have

$$\frac{\alpha L}{\lambda} = 1.86\left(\frac{\phi\,c_p\,d}{\lambda}\right)^{\frac{1}{3}}\left(\frac{\mu_b}{\mu_w}\right)^{0.14}$$

from which

$$\alpha = 0.0186\,\lambda^{2/3}c_p^{1/3}\left(\frac{\mu_b}{\mu_w}\right)^{0.14}\phi^{1/3}d^{1/3}/L \quad [\mathrm{kW/(m^2\,K)}]. \qquad (8.6)$$

The L quoted in the Nusselt and Graetz numbers is a length parameter over which heat transfer is taking place. For transfer across a fluid film, L is the thickness of the film. For transfer to a tube L is the tube diameter and for a plate, L is the plate length.

The fluid viscosities μ_w, evaluated at the temperature of the wall, and μ_b, evaluated at the fluid bulk temperature, are strongly temperature dependent for water and oil. At the operating temperatures typical of most power and utility plant fluid/fluid heat exchangers their ratio can be greater than 2:1 under low-flow conditions. If ignored this would introduce about 10 % error ($2^{0.14} = 1.102$) into the calculated value of the heat transfer coefficient. It is common practice to evaluate each at the same average temperature $(T_b + T_w)/2$ and to ignore the temperature dependence of the viscosity ratio ($\mu_w = \mu_b$).

With this approximation, the heat transfer coefficient of Eq. 8.6 for low flow rates (Re < 2,000) can be approximated for flow through a pipe of internal diameter d_i by

$$\alpha = 0.0186\mathcal{F}(T)\phi^{1/3}d_i^{-2/3} \quad [\mathrm{kW/(m^2\,K)}] \qquad (8.7)$$

with

$$\mathcal{F}(T) = 1.612 - 0.01374\left(\frac{T}{100}\right) + 0.00976\left(\frac{T}{100}\right)^2.$$

The heat transfer coefficient will not reduce to zero for zero flow but will remain above some minimum value. There are no clear guidelines as to the determination of this limit and in a specific case its selection must be matched to actual plant observation or expectations.

8.2.4 Convective Heat Transfer Between Steam and Metal Surfaces Within a Tube or Pipe

For heat transfer at the steam/metal boundary

$$\dot{q}_{hx} = f_{sc}\,\alpha_{hx}\,a_{hx}\,(T_s - T_m), \tag{8.8}$$

where

- T_s is the steam temperature within the cell,
- T_m is the inner surface temperature,
- f_{sc} is a factor which accounts for the deterioration of heat transfer caused by the surface fouling,
- α_{hx} is the local convective heat transfer coefficient,
- a_{hx} is the heat transfer area per unit length of the inner pipe surface.

The heat transfer coefficient α_{hxm} is the standard Hausen correlation (Steinmueller [9]) for an individual tube of inner diameter d_i, assuming dry conditions on the inner surface of the pipe.

$$\alpha_{hx} = 0.0224\,\lambda^{0.58}\,\eta^{0.38}\,c_p^{0.42}\,\phi^{0.8}\,d_i^{-0.2} \quad [\mathrm{kW/(m^2\,K)}],$$

where

- η is the dynamic viscosity of the steam [μ Pa·s],
- λ is the thermal conductivity of the steam [W/(m K)],
- c_p is the specific heat of steam [kJ/(kg K)].

For dry saturated steam, the product ($\lambda^{0.58}\,\eta^{0.38}\,c_p^{0.42}$) can be represented as a function $\mathcal{F}(T)$ of steam temperature. Then,

$$\alpha_{hx} = 0.0042\,\mathcal{F}(T)\phi^{0.8}\,d_i^{-0.2} \quad [\mathrm{kW/(m\,K)}]$$

using for $\mathcal{F}(T)$ the cubic approximation

$$\mathcal{F}(T) = -0.9995 + 2.429\left(\frac{T}{100}\right) - 1.135\left(\frac{T}{100}\right)^2 + 0.26\left(\frac{T}{100}\right)^3.$$

The following table lists values of these parameters and the resulting $\mathcal{F}(T)$ for saturated steam in the range of temperatures 100–366 °C.

P (kPa)	T (C)	λ	η	c_p	True value	$\mathcal{F}(T)$	Error %
101.3	100	0.0246	12.19	2.05	0.4077	0.4545	−11.48
235.1	125	0.0283	12.96	2.10	0.4573	0.4754	7.91
482.2	150	0.0318	13.86	2.22	0.5137	0.4628	9.91
904.1	175	0.0357	14.76	2.47	0.5885	0.6066	−3.08
1575.3	200	0.0404	15.66	2.72	0.6734	0.7384	−9.66
2583.6	225	0.0448	16.57	3.19	0.7810	0.8317	−6.49
4029.3	250	0.0507	17.52	3.68	0.9101	0.9023	0.86
6026.3	275	0.0585	18.55	4.51	1.1007	1.0076	8.46
8703.7	300	0.0699	19.70	5.54	1.3613	1.2472	8.38
12213.7	325	0.0895	21.25	6.80	1.7624	1.7626	−0.01
16752.0	350	0.1332	23.71	8.05	2.4834	2.7373	−10.22
20329.9	366	0.2268	27.30	11.50	4.1448	3.7046	10.62

This correlation is, as before, valid for $Re > 2{,}100$. The coefficient 0.0042 has absorbed the units used for the calculation of the fitting function $\mathcal{F}(T)$. The cubic approximation for $\mathcal{F}(T)$ gives an adequate fit over the quoted pressure range but not at the higher pressures (above around 185–190 bar) due to the rapid increase of the heat transfer coefficient as pressure approaches the critical point. For this high-pressure region the table lookup procedure described below is preferred as the table accurately straddles the saturation line.

For superheated steam the heat transfer coefficient is calculated from

$$\alpha_{hx} = 0.0082\,\mathcal{F}(T)\phi^{0.8}\,d_i^{-0.2} \quad [\mathrm{kW/(m^2\,K)}]. \tag{8.9}$$

Values of $\mathcal{F}(T)$ for the superheated region can be interpolated from the tabulation of Fig. 8.1, given pressure (in bar) and temperature (°C).

Figure 8.2 displays this table graphically. While the table is limited by the discretisation of the temperature scale, the left hand end of each individual curve will terminate at the saturation temperature below the critical point and at the critical temperature (374.2 °C) for all pressures equal to and greater than the critical pressure (221.2 bar). The table and graph have been artificially extended below these limiting temperatures for some pressures (values indicated in italics) in order to guarantee continuity of the table lookup across the saturation line or critical temperature.

The rapid changes in $\mathcal{F}(T)$ in the vicinity of the critical temperature and pressure are driven largely by changes in c_p. These changes are so rapid that accurate calculation of conditions in the near vicinity of the critical point is not possible without finer discretisation of temperature, but sufficient accuracy is achievable using values around the edges of this "near" vicinity. Compared with the Steinmuller nomograms cited earlier, Eq. 8.9 delivers heat transfer coefficients within ±20 % of the nomogram values.

	Temperature °C									
Pressure bar	250	300	350	375	400	425	450	500	550	600
1	0.606	0.685	0.768	0.811	0.855	0.900	0.955	1.032	1.123	1.217
20	0.700	0.738	0.809	0.856	0.883	0.921	0.963	1.050	1.141	1.236
40	0.887	0.841	0.866	0.903	0.926	0.955	0.993	1.076	1.167	1.262
60	*1.114*	1.010	0.949	0.978	0.984	1.000	1.030	1.109	1.199	1.194
80		1.333	1.071	1.089	1.059	1.058	1.078	1.147	1.235	1.330
100		*1.608*	1.276	1.240	1.158	1.131	1.136	1.191	1.274	1.369
125		*2.079*	1.596	1.493	1.324	1.252	1.229	1.254	1.328	1.421
150		*3.150*	2.262	1.844	1.554	1.415	1.349	1.329	1.385	1.473
175			*3.568*	1.871	1.631	1.502	1.415	1.444	1.524	1.617
200			*6.891*	2.318	1.909	1.691	1.511	1.503	1.573	1.666
225			*14.50*	9.055	3.095	2.202	1.900	1.630	1.576	1.628
250			*12.00*	7.840	4.708	2.709	2.112	1.726	1.643	1.687
275			*7.500*	7.298	6.210	3.559	2.451	1.808	1.689	1.734
300			*6.850*	6.997	6.514	4.339	3.102	2.043	1.740	1.725

For 1 bar the table entry values are:
at 100°C : 0.404; at 150°C : 0.466 and at 200°C : 0.533.

Fig. 8.1 Values of $(\lambda^{0.58}\eta^{0.38}c_p^{0.42})$ for superheated steam

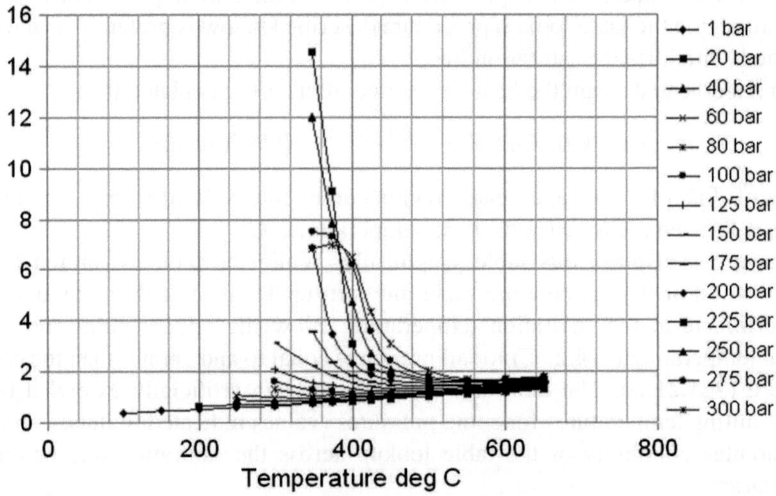

Fig. 8.2 Graph of tabulation of $\mathcal{F}(T)$, Fig. 8.1

8.2.5 *Modification of the Steam Heat Transfer Coefficient by Moisture*

Heat transfer from wet steam to a metal surface is greatly enhanced by the presence of moisture deposited on the surface. This condition can arise during initial warming of steam lines, valve chests and steam turbine internals (casings, blading, shaft). It is of operational significance as it can cause rapid warming of and therefore large temperature gradients in cold metal components at rates which can lead to high local stresses and superficial cracking. Being a consequence of local condensation it is sometimes referred to as "condensation warming".

It almost invariably occurs at low steam velocities and low pressures. Under these conditions condensation can occur in either of two forms. At low condensation rates moisture will be deposited on the metal surfaces as droplets. Depending on the geometry and inclination of the surface the droplets may remain and accumulate until they coalesce into larger droplets and finally into a small stream. If the velocity of the passing steam is sufficiently high they may be entrained and removed or propelled along the inner surface of the component until they drop off or are removed by a drainage port.

In each case the heat transfer to the metal surface is greatly enhanced beyond that of a dry surface. Heat transfer in the presence of film condensation has been analysed both theoretically, particularly in the pioneering work of Nusselt (1916), and experimentally, and useful formulae are available. No such thorough analysis has been undertaken for dropwise condensation though it is generally agreed that heat transfer rates for dropwise condensation greatly exceed those of film condensation by "6 to 18 times" (Perry [17]) and "at least a factor of 10" (Butterworth and Hewitt [63]).

The theoretical analysis is based on a number of assumptions which apply to almost all practical operations during which film condensation will occur:

- Steam flow rates are low.
- Vapour pressures are low.
- Flow at the surface interface is laminar ($Re < 2{,}100$).
- Condensation films are thin (for tubes, $\ll$ than the wall thickness).

Butterworth and Hewitt quote the effective heat transfer coefficient under these conditions for filmwise condensation on vertical plates as

$$\alpha = 0.943 \left(\frac{\lambda_l^3 \rho_l (\rho_l - \rho_v) g\, \mathbf{r}}{\mu_l (T_s - T_w) L} \right)^{1/4}, \tag{8.10}$$

where λ_l is the coefficient of thermal conductivity of the liquid, μ_l is the liquid viscosity, $\mathbf{r}$ is the latent heat at the local vapour pressure, T_s and T_w are the steam and wall temperatures, respectively, and g is the acceleration due to gravity. L is the length of plate over which condensation takes place. The coefficient of 0.943 assumes that each of these quantities is expressed in British (or customary US) units, yielding a heat transfer coefficient in Btu/(hr ft²F).[1] Perry quotes this same equation for plates and uses (after rearrangement from the original Nusselt form)

$$\alpha = 0.73 \left(\frac{\lambda_l^3 \rho_l (\rho_l - \rho_v) g \, \mathbf{r}}{\mu_l (T_s - T_w) D} \right)^{1/4} \tag{8.11}$$

for condensation on horizontal tubes where D is the outside diameter of the tube. Again the scaling coefficient assumes customary US units. For variables expressed in SI units the scaling coefficient is 4.145.

Perry provides a nomograph (expressed in customary US units) for the determination of film condensation heat transfer coefficients for both vertical and horizontal condensation ([17], pp. 10–22). For example, for water (condensate) at a film temperature of 100 °C and atmospheric pressure, a horizontal steam line of 300 mm diameter and an assumed temperature differential across the film of 50 °C (initial warming phase), the heat transfer coefficient is 4.8 kW/(m² K). Equation 8.11 yields 4.6 kW/(m² K). For steam conditions and a flow rate typical of this stage of a boiler cold start, Steinmuller (p. 116) gives a heat transfer coefficient of some 0.5 kW/(m² K) for a dry surface. This suggests that film condensation produces a ten-fold increase in the local heat transfer coefficient. Now include the "up to 10 times increase" due to dropwise condensation, and we have the figure of "possibly up to 100 times" increase in heat transfer during condensation warming sometimes claimed by plant operators (anecdotal evidence only).

Butterworth and Hewitt suggest that the deposition of moisture virtually ceases once the inner metal temperature exceeds 95 % of the local saturation temperature. The reader is referred to Sect. 15.1.1 for further discussion of this subject.

8.2.6 Convective Heat Transfer from Gases to the Outer Surfaces of Metal Tubes

With the exception of tubular gas–gas heat exchangers, gases usually contact metal surfaces on the outside surfaces of tubing in the form of tube banks. Heat transfer is calculated from the standard equation

$$\dot{q}_{hx} = \alpha_{hx} A_{hx} (T_g - T_m), \tag{8.12}$$

where T_g is the local gas temperature and T_m is the notional temperature at the outer surface of the metal interface. For smooth tubes, that is, not finned, the heat transfer coefficient α_{hxm} is the standard Brandt correlation for an individual tube of outer diameter d_o.

$$\alpha_{hx} = 0.287 \, \lambda^{0.636} \, \eta^{-0.236} \, c_p \, \phi^{0.6} \, d_o^{-0.4} \quad [\mathrm{W/(m^2\,K)}].$$

[1] To convert to W/(m²K) multiply Btu/(hr ft²F) by 5.678.

For dry air, the product ($\lambda^{0.636}\,\eta^{-0.236}\,c_p$) can be represented as a simple function of temperature $\mathcal{F}(T)$. Then,

$$\alpha_{hx} = 0.00295\,\mathcal{F}(T)\,\phi^{0.6}\,d_o^{-0.4} \quad [\text{kW}/(\text{m}^2\,\text{K})]. \tag{8.13}$$

The following table gives values of this product for the range of air temperatures 100–1,300 °C and the errors incurred, using for $\mathcal{F}(T)$ the quadratic approximation,

$$\mathcal{F}(T) = 1.2928 + 0.1596\left(\frac{T}{100}\right) - 0.00298\left(\frac{T}{100}\right)^2.$$

This approximation fits the true function with errors not exceeding ±1 % over the whole range.

The actual values are influenced by both the moisture content and CO_2 content of the gas. The stated values apply to dry air. For a mid-range CO_2 (fluegas) the coefficient in Eq. 8.13 is around 0.0035. The reader is referred to Steinmuller [9] for correction factors to be applied for other moisture (air and fluegas) and CO_2 (flue gas) contents. The table uses T in °C, λ in J/(m K), η in μ Pa·s and c_p in kJ/(kg K).

T	λ	η	cp	True value	$\mathcal{F}(T)$	Error %
100	0.0314	21.52	1.011	1.455	1.449	+0.39
300	0.0439	29.02	1.046	1.736	1.745	−0.48
500	0.0556	35.33	1.093	2.013	2.016	−0.16
700	0.0666	40.91	1.137	2.270	2.264	+0.27
900	0.0771	46.01	1.169	2.491	2.488	+0.13
1100	0.0869	50.80	1.196	2.687	2.688	−0.03
1300	0.0963	55.35	1.218	2.863	2.864	−0.03

For heat transfer to a non-tubular surface, such as the flat plates found in rotary and plate-type air heaters, the diameter d_o used in Eq. 8.13 is replaced by d_{eq}, calculated from the geometry of the flow passing the surface,

$$d_{eq} = \frac{4\,A_{xs}}{U_{fs}}, \tag{8.14}$$

where A_{xs} is the cross-sectional area of the flow passing the surface and U_{fs} is the peripheral length of the flow enclosure.

8.2.7 Non-condensing Convective Heat Transfer from Steam to the Outer Surfaces of Metal Tubes

In many heat transfer applications, heat is exchanged between steam flowing inside a tube and gas flowing across the outer tube surface. This arrangement is preserved in air-cooled condensers but reversed in water-cooled condensers in which steam flows across the outer surface of tubes cooled from within by the water flow. These cases of a condensing environment are treated in Chap. 15.

In some applications, steam flows across the outer surface of a tube or shaft without condensing. In such cases, the steam may be treated as a gas and the heat transfer treated accordingly. Using steam parameters, the function $\mathcal{F}(T)$ of Eq. 8.13 is

$$\mathcal{F}(T) = -6.506 + 15.24\left(\frac{T}{100}\right) - 8.787\left(\frac{T}{100}\right)^2 + 1.779\left(\frac{T}{100}\right)^3$$

for $100 < T < 374$°C and again,

$$\alpha_{hx} = 0.00495\,\mathcal{F}(T)\,\phi^{0.6}\,d_o^{-0.4} \quad [\mathrm{kW/(m^2\,K)}]. \tag{8.15}$$

8.2.8 Heat Transfer to Finned Tubes

Following Th. E. Schmidt, Steinmuller [9] derives the heat transfer coefficient for finned tubes in a bank located within a duct as follows.

$$\alpha = \alpha_R\left[1 - (1 - \eta_R)\left(\frac{A_R}{A}\right)\right],$$

where A is the total heat absorbing surface, being the fin area A_R plus the free tube surface area A_0. α_R is the heat transfer coefficient of the fin, extracted from

$$Nu = c\,Re^{0.675}\,Pr^{0.33}\left(\frac{A}{A_0}\right)^{-0.37},$$

$$Nu = \frac{\alpha_R d_t}{\lambda},$$

$$Pr = \frac{c_p\,\eta}{\lambda},$$

$$Re = \frac{\dot{m}_g d_t}{A_Q \eta},$$

where $\begin{cases} d_t & \text{is the outer diameter of the tube,} \\ \dot{m}_g & \text{is the gas mass flow rate,} \\ A_Q & \text{is the area of the narrowest flow cross section, equal to the cross-sectional area of the duct reduced by the projection of one tube row.} \end{cases}$

For in-line tubes $c = 0.3$; for staggered tubes, $c = 0.45$.

For circular or spiral fins

$$\frac{A_Q}{A_{duct}} = \frac{t_b - d_t}{t_b} - 2\frac{h_R S_R}{t_R t_b},$$

where $\begin{cases} t_b & \text{is the transverse pitch of the fins,} \\ t_R & \text{is the fin pitch,} \\ h_R & \text{is the fin height,} \\ S_R & \text{is the fin thickness.} \end{cases}$

A_0 is the surface area of the unfinned tube $= \pi d_t L$ where L is the tube length. We may then write

$$\left(\frac{A}{A_0}\right) = 1 + 2\frac{h_R}{t_R}\left(1 + \frac{S_R}{d_t} + \frac{h_R}{d_t}\right),$$

$$\left(\frac{A_R}{A}\right) = \frac{S_R d_t + 2h_R(d_t + S_R + h_R)}{t_R d_t + 2h_R(d_t + S_R + h_R)}.$$

The heat transfer coefficient α_R is extracted as

$$\alpha_R = c\,(\lambda^{0.345}\eta^{-0.345}c_p^{0.33})\,\phi^{0.675}d_t^{-0.325}\left(\frac{A_R}{A}\right)^{0.37}, \tag{8.16}$$

where $\phi = \dot{m}_g/A_x$ and A_x is the gas flow cross-sectional area. This can be written

$$\alpha_R = c\,\mathcal{F}(T)\,\phi^{0.675}d_t^{-0.325}\left(\frac{A_R}{A}\right)^{0.37}, \tag{8.17}$$

where $\mathcal{F}(T) = \lambda^{0.345}\eta^{-0.345}c_p^{0.33}$ is approximated by

$$\mathcal{F}(T) = 0.10302 + 0.0030432\left(\frac{T}{100}\right) - 0.00010152\left(\frac{T}{100}\right)^2$$

which fits the true value within ±0.85 % within the range 100–1,300 °C.

η_R is the fin efficiency, defined as the ratio of actual heat transfer to the maximum achievable were all of the fins at the temperature of the base tube. Complex formulae are available for the calculation of η_R for certain fin/tube geometries in terms of modified Bessel functions or alternatively by the use of various approximations [37]. For example, Steinmuller [9] quotes Th. E. Schmidt in setting

$$\eta_R = \frac{\tanh x}{x} \text{ where } x = \sqrt{\frac{2\alpha_R \cdot h_R^2}{\lambda_R S_R}}.$$

Perrotin [37] cites Hong and Webb [38] for the improved approximation

$$\eta_R = \frac{\tanh(m\, r_t \psi)}{m\, r_t \psi} \cos(m\, r_t\, \psi),$$

where r_t is the tube radius and

$$m = \sqrt{\frac{2\,\alpha_R}{\lambda_R S_r}},$$

$$\psi = \left(\frac{d_f}{d_t} - 1\right)\left(1 + 0.35 \ln\left(\frac{d_f}{d_t}\right)\right).$$

8.2.9 Treatment of Gases of Arbitrary Composition

The heat transfer approximation described above applies only to air or to gases whose thermal and transport properties approximate those of air. For gases of other composition a similar approximation may be developed in the same way using properties calculated from the given composition data.

8.2.10 Other Convective Heat Transfer Correlations

All of the heat transfer correlations quoted so far are based on the fundamental definition of the heat transfer coefficient from

$$\alpha = \frac{\dot{q}}{T_w - T_f}, \tag{8.18}$$

where $\dot{q}$ is the specific heat flux exchanged between a fluid at temperature T_f and a wall at temperature T_w. The fluid temperature can be, for example, the "bulk" or average temperature, the centre line temperature or the temperature of the fluid film in contact with the wall. An assumed form of a relationship among the principal dimensionless constants of Reynolds, Nusselt, Prandtl, etc. is then fitted to experimental data. As the heat transfer coefficient appears in at least one of these constants it can be extracted from the fitting function. A different correlation has been developed for each of these fluid temperature assumptions. Whichever assumption is made, the chosen form leads to a "correlation" which usually carries the name of the discoverer. We have made frequent use of the correlations due to Colburn, Hausen and Brandt in applications which are close to the experimental conditions under which they were developed (convective heat transfer to fluids, gases or steam without phase change). If we introduce phase change, these correlations are no longer valid, and other methods must be used.

A class of correlations dealing with two-phase heat transfer assumes that the fluid temperature of interest is not the bulk temperature but the saturation temperature of the fluid flowing in a tube. This is appropriate to the case of boiling heat transfer as the fluid will be at or close to its local saturation temperature. Since the saturation temperature is a function of local pressure p this leads to a class of heat transfer correlations of the form

$$(T_w - T_{sat}) = a\,\phi^b \exp(-p/c).$$

Various investigators have applied this form to their data and produced differing sets of the coefficients a, b and c. For example, the Jens–Lottes correlation uses $a = 25$, $b = 0.25$ and $c = 62$. This correlation is not of the form of Eq. 8.18, and the heat transfer coefficient does not appear explicitly. Another variant of this form is

$$(T_w - T_{sat}) = a\,\phi^b \exp(-p/p_{cr}),$$

where p_{cr} is the critical pressure for the fluid under test.

These issues are discussed further in Chap. 13.

8.3 Radiation

Being an electromagnetic wave phenomenon, thermal radiation is emitted within a band of wavelengths or frequencies defined by the temperature of the emitting body. The relationship between temperature and frequency is defined quantitatively by Planck's Law. Wien's law of displacement states that the frequency of maximum intensity over the emitted spectrum is inversely proportional to the absolute temperature of the emitter.

The *emissive power* of an emitting body or surface is defined to be the flux density due to the emitter throughout a hemisphere centred on the emitter. The emissive power of a perfect black-body emitting surface is defined by the Stefan–Boltzmann law to be

$$E = \sigma\,\mathcal{T}^4, \tag{8.19}$$

where $\mathcal{T}$ is the absolute temperature of the emitting surface [K]. σ is the Stefan–Boltzmann constant 5.6697×10^{-8} W/(m^2 K^4). The total heat energy radiated from a hemispherical source of area A m^2 is $A \cdot E$.

The ratio of actual emission to ideal black-body emission is ϵ, the emissivity of the emitter, noting that the emitter may be a surface—solid or liquid—or a gas. The emissive power of a non-black (= grey) body is therefore

$$E = \epsilon \sigma \mathcal{T}^4.$$

A diffuse emitter is one which emits radiation equally in all directions.

Of the emitted energy incident upon a surface, some will be absorbed and the rest reflected. A black body absorbs all of the radiation incident upon it. Practical surfaces are not perfect absorbers, and the ratio of actual to ideal absorption is the absorptivity of the surface, Λ.

According to Kirchoff's Law, the emissivity and absorptivity of a surface, under certain conditions, are numerically the same. For practical purposes these conditions can be relaxed with little error, particularly if only monochromatic radiation (corresponding to a single representative emitter temperature) is assumed.

8.3.1 Hemispherical Radiation

Consider radiation within a notional collimated tube or pencil diverging from a radiant source through an area $\mathrm{d}A$ normal to the ray tube. If this area subtends a solid angle $\mathrm{d}\Omega$ steradians to the source of the radiation, the heat flux $\mathrm{d}\dot{q}$ passing through $\mathrm{d}A$ within the ray tube inclined at angle θ to the normal to $\mathrm{d}A$ is

$$\mathrm{d}\dot{q} = I(\theta, \phi) \cos\theta \, \mathrm{d}A \, \mathrm{d}\Omega. \tag{8.20}$$

The intensity of thermal radiation or flux density I has dimensions energy (Joules) per unit time (seconds) per unit area (m^2) or power (J/s or Watts) per unit area. Dimensionally, therefore, radiation intensity I = Power/Area. The absence of wavelength λ from the argument list of I indicates that this is the *total* intensity, integrated over all frequencies.

Consider an incremental plane area $\mathrm{d}A$ at the centre of a hemisphere of unit radius, radiating in or receiving heat from all directions above it. Then the heat flux $\dot{q}$ passing along a ray tube inclined at an angle θ to the normal to $\mathrm{d}A$ and which subtends an incremental solid angle $\mathrm{d}\Omega$ to $\mathrm{d}A$ is given from

$$\dot{q} = I(\theta, \Omega) \int_{2\pi} \mathrm{d}\Omega \, \cos\theta$$

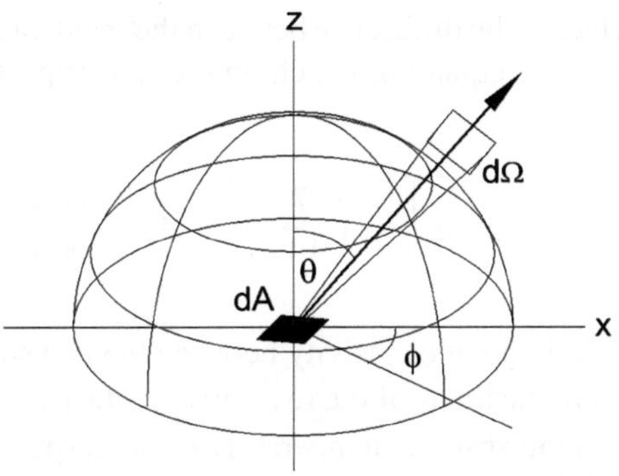

since the unit sphere subtends a solid angle of 2π steradians about dA. Given that d$\Omega = 2\pi \sin\theta \, \mathrm{d}\theta$, this equation may be integrated to give

$$E = \dot{Q} = I \int_0^{\pi/2} 2\pi \sin\theta \cos\theta \, \mathrm{d}\theta = \pi I \int_0^1 \mathrm{d}(\sin^2\theta)\mathrm{d}\theta = \pi I. \tag{8.21}$$

The emissive power of a radiating surface is therefore π times the radiation intensity. If E is the total emissive power of a black body from the Stefan–Boltzmann Law, integrated over all frequencies and in all directions, the intensity I_b of black-body radiation is E/π.

If we assume that the environment through which the individual ray tubes pass is homogeneous with refracting index = 1, the intensity will be independent of θ and the radiation is isotropic. The intensity I will then be a constant.

Emitting and receiving surface areas can be calculated for simple geometries such as parallel plates or concentric cylinders or flames of assumed geometries. If the receiving surface is sufficiently far from the source, the source can be successfully treated as a point source. This approach is found in the analysis of heat transfer from industrial fires, such as may arise as a consequence of industrial accidents, or from bushfires [14]. A radiant source which allows the assumption of radiation symmetric in both azimuthal and polar coordinates can also be treated as a point source and radiation intensity computed independent of emitter area.

Restricting attention to Lambert sources we can assume that radiation is emitted equally in all directions.

Radiation energy absorbed by a surface raises the temperature of the surface which emits energy proportional to the fourth power of its absolute temperature and its own emissivity. Heat emitted is the sum of the reflected part of the incident radiation and its own intrinsic, temperature dependent emission. The net energy transferred to the surface is the difference between the incident and emitted energies, expressed by the following equation, which ignores any repeat reflections from the second surface.

$$\dot{q}_{rad} = \epsilon_{12}\, c_R \left[A_1 \left(\frac{\mathcal{T}_1}{100} \right)^4 - A_2 \left(\frac{\mathcal{T}_2}{100} \right)^4 \right], \tag{8.22}$$

where $\begin{cases} \epsilon_{12} & \text{is the joint emissivity between the two surfaces,} \\ A_1, A_2 & \text{are the areas of the respective surfaces [m}^2\text{],} \\ \mathcal{T}_{1,2} & \text{is the absolute temperature of the respective surface.} \end{cases}$

The joint emissivity is given by

$$\frac{1}{\epsilon_{12}} = \frac{1}{\epsilon_1} + \frac{A_1}{A_2} \left(\frac{1}{\epsilon_2} - 1 \right).$$

Dividing the temperatures by 100 has eliminated the 10^{-8} multiplier in the Stefan–Boltzmann constant. The new constant $c_R = 5.6697\,\mathrm{W/(m^2\,K^4)}$ is referred to as the *radiation factor*.

It should be noted that this relationship is valid only for parallel plane surfaces as the intensity of the incident radiation depends on the angle of incidence (Lambert's cosine rule). However, if we assume that the emitting surfaces are Lambert surfaces, which absorb and emit energy independent of incident angle, then the relationship can be applied to any configuration. Most industrial engineering flames and surfaces may be treated as Lambert surfaces.

Typical values of emissivities for boiler flames and receiving surface materials are (Steinmueller [9]),

Bituminous coal	0.55–0.80
Fuel oil	0.45–0.85
Natural gas	0.40–0.60
Steel, oxidised	0.80–0.95
Aluminium, oxidised	0.08–0.12
Refractory	0.50–0.75

Noting that

$$\mathcal{T}_1^4 - \mathcal{T}_2^4 = (\mathcal{T}_1^2 + \mathcal{T}_2^2)(\mathcal{T}_1 + \mathcal{T}_2)(\mathcal{T}_1 - \mathcal{T}_2)$$

Eq. 8.22 can be written

$$\dot{q}_{rad} = \frac{\alpha_R\, A_1}{100} (\mathcal{T}_1 - \mathcal{T}_2)$$

which has the same form as the convection heat transfer equations. α_R is the heat transfer coefficient given by

$$\alpha_R = \epsilon_{12}\, c_R \left[\left(\frac{T_1}{100}\right)^2 + \left(\frac{T_2}{100}\right)^2\right]\left[\left(\frac{T_1}{100}\right) + \left(\frac{T_2}{100}\right)\right]$$

$$\approx \epsilon_{12} c_R \left(\frac{T_1}{100}\right)^3 \quad \text{if } T_1 \gg T_2.$$

In the furnace, where temperatures are very high and predictable and vary within a fairly narrow envelope, this form of representation can be used advantageously to save computation effort.

8.3.2 Luminous Gas Radiation

Flame radiation arises primarily as radiation from combusting carbon and other hot particles (soot, char and ash) and from carbon dioxide and water vapour which have been raised to the flame temperature. These sources are distributed through the flame volume and create the dependence of total emission on flame volume, shape and composition.

The flame temperature in a furnace is always much greater than the temperature of the receiving surface, and radiant heat transfer can be assumed to be independent of the receiving surface temperature.[2] In this case the total radiant energy from the flame is given as

$$\dot{q}_{rad} = \epsilon f_v\, A_{fl}\, c_R \left(\frac{T_{fl}}{100}\right)^4, \tag{8.23}$$

where A_{fl} is the effective surface area of the flame. f_v is a view factor which depends on the furnace geometry.

The emissivity of a flame depends on its composition and is formed from two parts, the luminous and the non-luminous. A flame is a stirred zone of combusting fuel mixed with high temperature combustion products—including intermediate products of combustion which undergo further transformations—and non-combusting moisture, ash and flue gases. Coal and oil flames produce substantial quantities of intermediate products such as coke and soot which are combusted within the flame zone. Soot, produced principally by thermal cracking of hydrocarbons, gives oil flames their characteristic bright appearance (high luminosity). The hydrogen produced by the same process oxidises to water vapour, producing a low luminosity blue flame. The fixed carbon component of coal burns with a bright flame. The emissivity of the ash is high and, at flame temperatures, contributes up to 60 % of the total radiative heat available from the flame. Natural gas (basically methane) burns with a pale blue, almost transparent flame, with low emissivity.

[2]For example, a large boiler furnace will have a flame temperature of around 1,500 K. With waterwall tube temperatures around 360 °C or 633 K, the fourth power of the ratio of flame to wall temperatures is around 32.

Equation 8.22 can be applied to the luminous radiation from the flame to the furnace wall tubes to give

$$\dot{Q}_R = A_w f_v c_R \left(\frac{\epsilon_w + 1}{2}\right) \left[\epsilon_g \left(\frac{T_g}{100}\right)^4 - \Lambda_{g,w} \left(\frac{T_w}{100}\right)^4\right], \quad (8.24)$$

where A_w is the receiving surface area [m^2], T_g and T_w are the absolute temperatures of the gas and receiving surface, respectively. The emissivities ϵ_g of the gas and ϵ_w of the wall and absorptivity $\Lambda_{g,w}$ are complex functions of the partial pressures of CO_2 and H_2O vapour, the total gas pressure and temperature and the temperature of the surface. Tables are available in Perry [17], Steinmueller [9] and Shvets [51] for the lookup of these quantities. The lookup parameter is the product of gas partial pressure and the beam length l. For transfer from gas to a tube bank or enclosure Shvets recommends $l = 4m\,V/A_w$ where V is the enclosure volume and m is a correction factor, around 0.9. Basu et al. [15] quotes Holman [12] in suggesting $l = 3.6\,V/A_w$.

The double subscript on Λ in Eq. 8.24 indicates that it is to be evaluated at the absorber surface temperature while the emissivity is evaluated at gas temperature. In this case, these two parameters will not be numerically equal.

The view factor f_v can take a value in the range (0,1). View factors are difficult to calculate for most practical cases but may be calculated for certain idealised configurations. Hottel and Sarofim [48], Rohsenow [44] (Chap. 7) and Perry [17] (Chap. 10) provide useful discussions of its meaning and values for differing geometries.

8.3.3 Non-luminous Gas Radiation

Different forms of thermal radiation can be characterised on the basis of their spectral composition. Luminous radiation is, fairly obviously, made up of frequency components in the visible range. Non-luminous radiation is produced in a furnace mainly in the infrared range.

Luminous radiation from the flame is the principal mechanism for heat transfer in fired furnaces. Non-luminous radiation is emitted by hot flue gas after it has left the combustion zone and is a significant contributor to total heat energy received by the radiant heating surfaces. Non-luminous radiation differs from luminous radiation in key aspects, which bear on its modelling treatment.

1. Flame radiation is dominated by radiation from hot combusting particles, mostly carbon; gas radiation consists mainly of radiation from low emissivity gases (N_2 and residual O_2) and high emissivity water vapour and CO_2. Flame radiation is composed of a mix of luminous and non-luminous radiation.

2. Beyond the furnace, fluegas temperatures can be of the same order as the temperature of the receiving surface and radiant heat transfer calculation should take account of the receiving surface temperature.
3. Non-luminous radiant heat exchange is strongly dependent on the composition of the emitting gas, which influences its emissivity and absorptivity.

In industrial plant non-luminous emissions come mainly from triatomic gases (H_2O, SO_2, CO_2). While mono- and diatomic gases (N_2, O_2, H_2) have low emissivities and absorptivities, triatomic gases have large emissivity and absorption coefficients, and their emission and absorption behaviour deviates markedly from the ideal Stefan–Boltzmann black-body model. Shvets [51] quotes the emissive powers of CO_2 and H_2O at an absolute temperature $\mathcal{T}$ as

$$E_{CO_2} = 3.5\, p^{0.33} l^{0.33} \left(\frac{\mathcal{T}}{100}\right)^{3.5} \quad [\mathrm{W/m^2}], \tag{8.25}$$

$$E_{H_2O} = 3.5\, p^{0.8} l^{0.6} \left(\frac{\mathcal{T}}{100}\right)^{3} \quad [\mathrm{W/m^2}], \tag{8.26}$$

where p is the partial pressure of the gas [bar] and l is the beam length [m].

8.3.4 Absorption Along a Ray Path

Since our discussion of radiative transfer will focus on furnaces which are generally of rectangular or sometimes cylindrical geometry, we may ignore the polar ordinate ϕ and consider transmission along paths lying in the (z, θ) plane only. That is, we will assume symmetry in the azimuthal coordinate direction. Radiation along a path

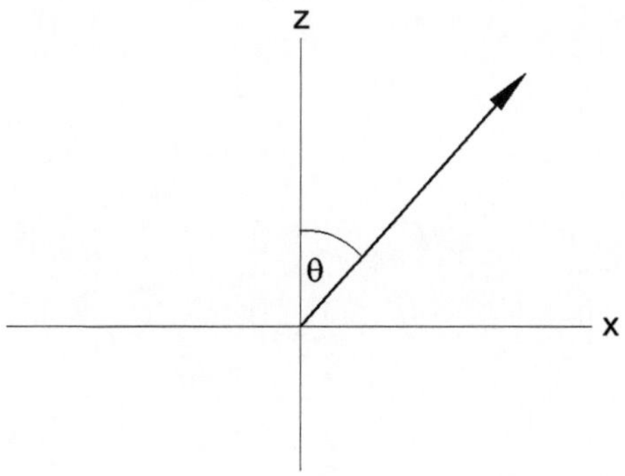

is described by the radiative transport equation (RTE) which defines the radiative intensity I at any point at a distance s along the path from the emitter. Along a single transmission ray or along a collimated ray tube, intensity remains constant unless changed by absorption or scattering [48]. The initial intensity leaving the emitter I_e is enhanced by radiation received from other sources and by radiation scattered into its path. It is diminished by absorption by the medium through which it passes and by scattering out of the path. The local rate of change of the intensity is defined by the balance of these elements. As absorption and scattering are functions of the transmission medium and radiation frequencies the coefficients of the RTE are functions of these variables. If we ignore scattering and assume that the frequency dependencies have been eliminated or totalised by an appropriate integration process the RTE can be written simply as

$$\frac{\mathrm{d}I_r}{\mathrm{d}s} = -k_{ab} I_r \text{ and } I_r(0) = I_e, \tag{8.27}$$

where the absorption coefficient k_{ab} is a complex function of the composition, temperature and pressure of the gas along the path. The solution of this equation is

$$I_r(s) = I_r \mathrm{e}^{-k_{ab} s}. \tag{8.28}$$

Computationally, an incremental form of solution can be used to advance the solution known at distance step n to the next step $(n + 1)$ along a distance increment Δs.

$$I_r^{n+1} = I_r^n \, \mathrm{e}^{-k_{ab} \Delta s} \tag{8.29}$$

which is closely approximated by

$$\Delta I_r^{n+1} = I_r^{n+1} - I_r^n = -k_{ab} \Delta s \, I_r^n \tag{8.30}$$

and is the absorption within the $(n + 1)$th zone.

The application of these equations to the practical calculation of furnace heat distribution is discussed in Sect. 11.4.2 in more detail.

Chapter 9
Heat Exchangers

Heat exchangers provide a path for the exchange of heat between fluids and fluids, fluids and gases and gases and gases. The two or more heat-exchanging media are separated physically by a metal tube or plate wall. Some applications feature three or more heat exchange media, with multiple fluid circuits exchanging heat within a common shell. Our interest here will concern two-sided heat exchange only.

Industrial heat exchangers in both fluid and gas applications include the following types:

- Shell-tube
- Tubular (crossflow, counter-flow, co-flow)
- U-tube
- Air cooled (fan assist or natural convection)

The chapter concludes with an expanded treatment of rotary air heaters as the rotating element does not lend itself to the simpler treatments used for static heat exchangers.

9.1 Heat Exchangers for Fluids

9.1.1 Shell-Tube Heat Exchangers

The most widely used heat exchanger for hydraulic applications is the shell-tube type. They are commonly used as fluid-fluid heat exchangers in hydraulic fluid circuits serving as oil coolers, closed-circuit water coolers, condensate coolers, etc.

A typical shell-tube heat exchanger is depicted in Fig. 9.1. The incoming flow on the primary (tube) side enters the inlet plenum and is distributed equally to each of the tubes opening through the tube sheet. The tube sheet at each end of the shell serves to separate the primary and secondary (shell) side flows.

P. O'Kelly, *Computer Simulation of Thermal Plant Operations*,
DOI 10.1007/978-1-4614-4256-1_9,

Fig. 9.1 Shell-tube heater exchanger—general arrangement (Courtesy of Industrial Products Marketing Pty Ltd Australia)

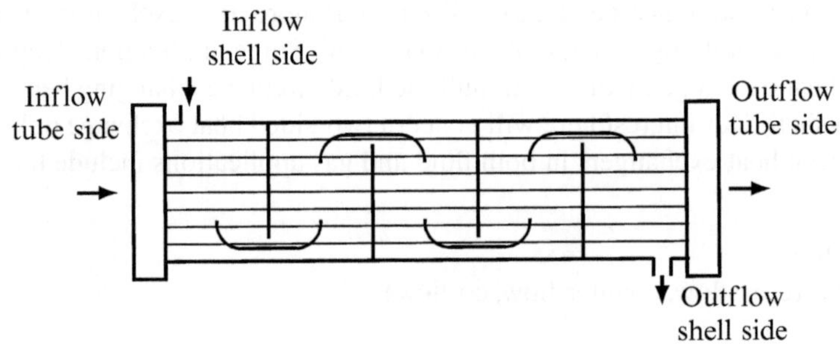

Fig. 9.2 Depiction of flow pattern, shell side and co-flow

After leaving the tubes the individual flows are recombined in the outlet plenum and leave the heater through the outlet stub. The flow on the shell side is directed across the tube bank. The path from inlet to outlet is guided by a series of baffle dividers arranged to ensure maximum exposure of the tubes to the flow while minimising the pressure loss (Fig. 9.2).

Pressure Drop Tube Side

The flow admittance of the tube side is calculated using the standard pipe formula (Eq. 6.2) for a single tube. The total admittance of the tube bank is then the admittance of the individual tube multiplied by the square of the number of parallel tubes. This follows from the observation that, if $\dot{m}$ is the total flow to a bank of n_{tube} parallel tubes, the flow through each tube is $\dot{m}_t = \dot{m}/n_{tube}$. All tubes will see the same differential pressure which is also seen by the bank. It follows that, with A the bank flow admittance, a the individual tube flow admittance and Δp the pressure differential across the bank,

$$\Delta p = \frac{\dot{m}^2}{A}$$
$$= \left(\frac{\dot{m}}{n_{tube}}\right)^2 \frac{1}{a}.$$

Therefore,

$$A = a \cdot n_{tube}^2.$$

The friction pressure drop along a single tube is calculated from

$$\Delta p = \zeta \frac{v}{2} \left(\frac{\dot{m}}{A_x}\right)^2 \quad \text{Pascals.}$$

A_x is the internal tube flow cross-sectional area $= \frac{\pi}{4} d_i^2$.

The resistance coefficient ζ is given by the standard formula equation 6.5.

Pressure Drop Shell Side

It is not possible to give a general and simple formula for the calculation of pressure drop across the shell side of a shell-tube heat exchanger because of the wide variety of types with their various tube and flow arrangements. Steinmuller [9] quotes

$$\Delta p = \zeta \frac{v}{2} \left(\frac{\dot{m}}{A_x}\right)^2 \quad \text{Pascals.}$$

A_x is the free flow cross-sectional area of the flow across the tube bank, calculated using the centreline tube row.

The resistance coefficient $\zeta = k\, f_e\, f_z\, Z$ accounts for the multi-tube, crossflow arrangement of the tube bank. The parameter f_e is the tube bank arrangement factor, and f_z is a correction factor if the number of tube rows Z is less than 10. k is a coefficient equal to 0.8 for clean tubes and higher values for fouled tubes. Values for f_e and f_z are tabulated in [9] for specific heat exchanger geometries and flow regimes.

9.1.2 U-Tube Heat Exchangers

These are commonly used for tank immersion heaters and feedwater heaters. Calculation of pressure drop and flow admittance on the primary side is the same as used for the shell-tube type, assuming that the heater is implemented as a simple bank of parallel tubes.

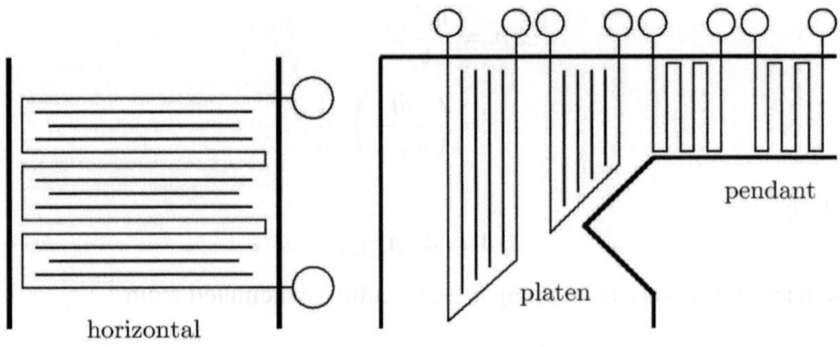

Fig. 9.3 Duct section showing various tube bank arrangements

There is no defined flow regime on the secondary side of an immersion heater, the tubes simply being immersed in a pool of fluid which may or may not be moving. A feedwater heater is in fact a condenser in which the tubes are either horizontally or vertically arranged. It is treated in detail in Chap. 16.

9.2 Tubular Heat Exchangers in Ducts

9.2.1 *Pressure Drop Calculations*

A tube bank is arranged as a set of tube bundles either across or parallel to (pendant within) the gas flow. Various configurations are possible, as shown by Fig. 9.3.

These configurations are commonly used for superheaters, reheaters and economisers in which heat is transferred from the flue gas stream to the secondary flow through the tube walls. It is safe to assume that the bulk of the pressure drop occurs across the tube bank, with the duct loss adding a small contribution.

Duct Side

The pressure loss across a tube bank is given from

$$\Delta p = \frac{k}{2\,v} \nu^2,$$

where ν is the flow velocity through the narrowest cross section normal to the flow path. This is usually the smallest lateral inter-tube space. A convenient formula is given by Steinmueller [9]:

$$\Delta p = \zeta \frac{v}{2} \left(\frac{\dot{m}}{A_x} \right)^2 \quad \text{Pascals,} \tag{9.1}$$

$$\zeta = k\, f_e\, f_z\, Z, \tag{9.2}$$

where
- ζ is the duct flow resistance factor,
- A_x is the duct free flow cross-sectional area,
- f_e is the tube bank arrangement factor,
- f_z is a correction factor for $Z < 10$,
- Z is the number of tube rows,
- k is a coefficient = 0.8 for clean tubes and higher values for fouled tubes.

Values for f_e and f_z may be found from standard handbook tables (e.g. Steinmueller [9] pp. 153–154) for specific heat exchanger geometries and flow regimes, given the details of plant equipment.

The flow admittance for the duct section containing the tube bank is

$$\mathcal{A} = \left(2 \frac{A_x^2}{\zeta\, v} \right). \tag{9.3}$$

The free cross-sectional area of the duct A_x is the duct cross-sectional area less than the area of tubing presented to the gas flow. For cross-flow heat exchangers

$$A_x = w_d\, d_d \; - l_t d_o\, n_{tp}, \tag{9.4}$$

where
- w_d is the duct width [m],
- d_d is the duct depth [m],
- l_t is the tube length across the gas path [m],
- d_o is tube outer diameter,
- n_{tp} is number of tube panels across the gas path.

For parallel flow (tubes aligned with the flow path) Basu et al. [15] quote

$$\Delta p = \xi \frac{L_t}{d_h} \frac{v^2}{2\, v} \quad \text{Pascals,} \tag{9.5}$$

where the equivalent hydraulic diameter d_h is given from

$$d_h = \frac{4 \left(a\, b - n_t \frac{\pi d_x^2}{4} \right)}{2(a + b) + n_t \pi\, d_o}$$

and
- a is the duct width [m],
- b is the duct depth [m],
- n_t is number of tubes,
- ξ is the Fanning friction factor.

Effect of Finned Tubes

The presence of fins on the outer surface of tubes increases the pressure loss across the bank. The increase is difficult to predict, and specific data appropriate to each case must be relied upon for the best estimation of the effects of the fins.

Tube Side

The heat exchanger consists of a parallel array of tubes for each of which the admittance is calculated using the parameters of the single tube. The total admittance of the tube bank is the admittance of the individual tube multiplied by the square of the number of parallel tubes. For smooth tubes (not rifled),

$$\mathcal{A} = \left(2\frac{d_H A_x^2}{\xi L v}\right) n_{tube}^2, \tag{9.6}$$

where all the terms inside the parentheses are calculated on a per-tube basis. ξ is the single-tube friction factor given by Eq. 6.20.

9.3 Heat Flows and Temperature Profiles

The heat flow arrangements for most heat exchangers are shown separately for the primary (tube) and secondary (shell or duct) sides by Figs. 9.4 and 9.5 for three adjacent cells along the flow path.

Heat exchange is treated separately at the inner and outer surfaces of the tube. The local cell tube temperature is calculated from the balance of these two heat flows. The details of this calculation are described in Sect. 10.3. There, the tube wall is treated as two zones, an inner and an outer. Heat transfer at each surface is convective, with conduction moving heat radially through and axially along the wall.

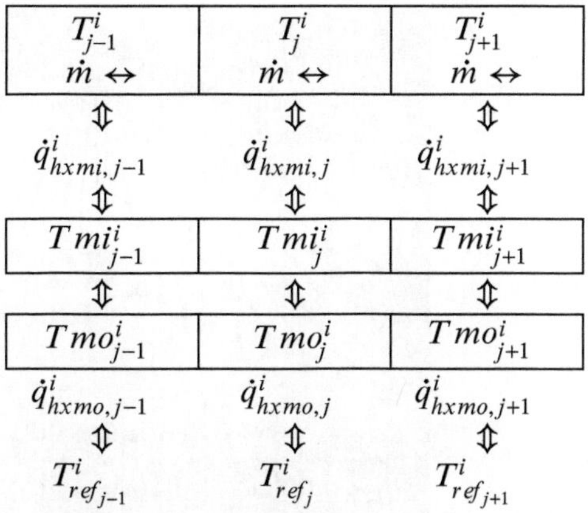

Fig. 9.4 Primary side heat flow network—inter-cell heat transfers

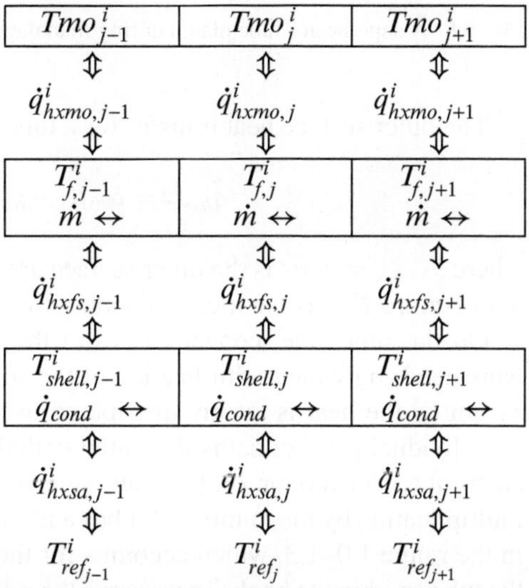

Fig. 9.5 Secondary side heat flow network—inter-cell heat transfers

9.3.1 Shell-Tube Heat Exchangers

Computationally this is the simplest heat exchanger geometry as it allows the assumption of well-defined flow regimes on both the primary (simple turbulent flow through the individual tubes) and secondary or shell side (slug flow across the tube bank). This in turn allows the use of standard heat transfer correlations.

On both the inner (i) and outer (o) surfaces a composite heat transfer coefficient $\bar{\alpha}_{hxm}$ is defined for heat transfer from the tube metal temperature to the fluid (or gas), $\bar{\alpha}_{hxmi}$ for the inner and $\bar{\alpha}_{hxmo}$ for the outer surface. The details are developed further in Sect. 10.3.

Fig. 9.6 A superheater tube platen during manufacture (Courtesy of Foster-Wheeler Corp.)

The outer surface heat transfer for a tube is similar to Eq. 8.3.

$$\hat{\dot{q}}_{hxmo} = \bar{\alpha}_{hxmo}\, a_{hxmo}\, (T_{mo} - T_{ref}), \tag{9.7}$$

where $a_{hxmo} = \pi\, d_o$ is the outer surface area per unit length per tube. The reference temperature T_{ref} is the local secondary fluid temperature.

On the tube side, conversion from the single-tube heat transfer to the whole bank needs only the multiplication of the single-tube result by the number of tubes, assuming the heat is evenly distributed over all tubes. Since the tubes are treated as individual parallel paths the outlet enthalpy from the bank is equal to that from each tube. Conversion on the shell side from the single tube to the bank requires multiplication by the number of tubes and by a bank arrangement factor f_{bk}, usually in the range 1.0–1.3, which accounts for the effects of spacing and arrangement of the tubes within the bank. To preserve the overall heat transfer balance, the shell-side heat transfer coefficient must be adjusted by the same bank factor:

$$\hat{\dot{q}}_{hxmo} = \frac{\alpha_{hxmo}\, a_{hxmo}}{f_{bk}} (T_{mo} - T_{ref}), \tag{9.8}$$

$$\dot{q}_{hxmo} = \hat{\dot{q}}_{hxmo}\, n_{tube}\, f_{bk}. \tag{9.9}$$

9.3.2 Tube Bank Heat Exchangers

This is an important category as it encompasses superheaters, reheaters, economiser and furnace tubes. The following treatment assumes gas on the outer tube side and either gas, steam or liquid inside the tubes. It assumes that the gas flows across the tubes which are aligned in rows normal to and along the gas flow path.

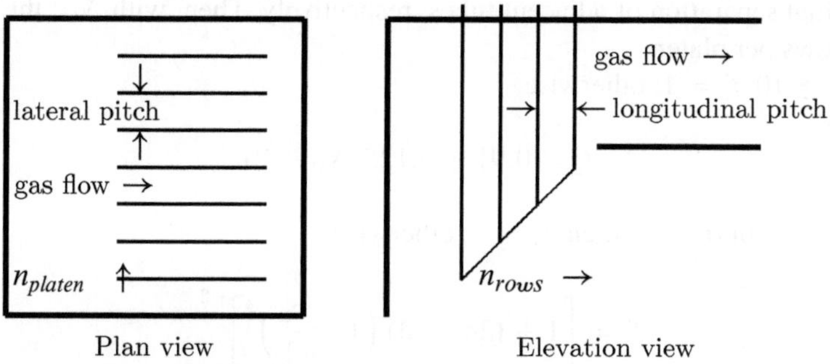

Fig. 9.7 Definition of tube pitch for tubular heat exchangers

Convective Heat Transfer to Tubes from the Outer Gas Side

Heat transfer to the tube bank is calculated from Eq. 8.12 on a per-tube basis, for which the average area a_{hxm} is the surface area of one tube exposed to the gas flow. The heat transfer coefficient for the gas flowing across a single tube is calculated from Eq. 8.13, modified by the factor f_G which accounts for the influence of gas composition on heat transfer.

$$\hat{\alpha}_{hxm} = \alpha_{hxm} \; f_G$$

f_G can be as high as 1.3 for high moisture content but can be assumed to be around 1.04–1.07 for moisture of around 5%.

The heat transfer coefficient α_{hxm} for a single tube is calculated using the standard Brandt correlation for flow across an individual tube. Total heat transferred to the tube bank in the jth cell is then

$$\dot{q}_{hxm} = \alpha_{hxm} \, a_{hxm} \, n_{tb} \; f_s f_z (T_g - T_m), \tag{9.10}$$

where n_{tb} is the number of parallel tubes in the bank. The bank influence factors f_s and f_z may be calculated using the following formulae quoted by Basu et al. [15].

f_s is the bank geometric arrangement factor and depends on the tube arrangement (staggered or in-line), and the lateral and longitudinal spacing of the tube plates and rows. f_z is the bank row factor and provides a correction which accounts for the longitudinal spacing of the tubes along the gas path if the row count is less than 10 (Fig. 9.7).

Correction Factors for In-Line Tubes, Crossflow

Define the transverse pitch $\sigma_1 = S_1/d_o$ and the longitudinal pitch $\sigma_2 = S_2/d_o$ where d_o is the outer diameter of the tube and S_1 and S_2 the transverse and longitudinal separation of adjacent tubes, respectively. Then, with N_{tr}, the number of tube rows per platen,

if $N_{tr} \geq 10$ $f_z = 1$; otherwise,

$$f_z = 0.91 + 0.125(N_{tr} - 2).$$

if $\sigma_1 \leq 1.5$ or $\sigma_2 \geq 2$, then $f_s = 1$; otherwise,

$$f_s = \left[1 + (2\sigma_1 - 3)\left(1 - \frac{\sigma_2}{2}\right)^3\right]^{-2}.$$

Correction Factors for Staggered Tubes, Crossflow

If $N_{tr} > 10$, then $f_z = 1$;
otherwise,

if $\sigma_1 < 3$, then $f_z = 3.12\, N_{tr}^{0.05} - 2.5$.
if $\sigma_1 > 3$, then $f_z = 4.0\, N_{tr}^{0.02} - 3.2$.

Define

$$\psi_0 = \frac{\sigma_1 - 1}{\sigma_2 - 1} \qquad \psi_\sigma = \sqrt{\frac{\sigma_1^2}{4} + \sigma_2^2}$$

Then,

If $0.1 < \psi_\sigma < 1.7$, then $f_s = 0.34\, \psi_0^{0.1}$.
If $1.7 < \psi_\sigma < 4.5$ and $\sigma_1 < 3.0$, then $f_s = 0.275\, \psi_0^{0.5}$.
If $1.7 < \psi_\sigma < 4.5$ and $\sigma_1 > 3.0$, then $f_s = 0.34\, \psi_0^{0.1}$.

Heat Transfer Between Tube Banks and the Tube Fluid

If the tube is carrying gas, the inner surface convective heat transfer coefficient α_{hxmi} is the same Brandt-type coefficient used for the outer surface but calculated using the inner gas conditions. For steam or hydraulic fluid in the tube, the inner surface convective heat transfer coefficient α_{hxmi} is a Colburn or Hausen-type coefficient calculated using the local steam or fluid conditions.

9.3.3 Temperature Profiles Through Heat Exchangers

Tubular heat exchangers in both shell-tube and duct-mounted applications may be connected in either co-flow or counter-flow arrangement, illustrated by Fig. 9.8 for a duct-mounted tubular heat exchanger bank:

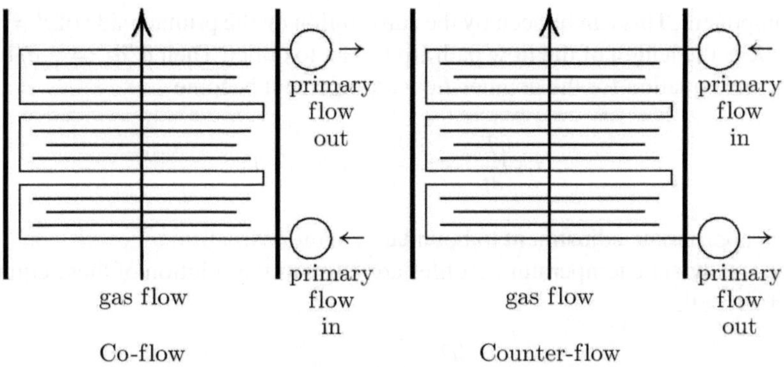

Fig. 9.8 Co- and counter-flow tube bank arrangements

In this illustration, flow within the tubes is directed across the gas flow. Throughout this work, we consider variations of fluid properties only in the axial ordinate, that is, in the direction of flow. To extend this to the crossflow or x-axis would lead to a significant increase in computation effort and to be meaningful, even if restricted to these heat exchange components, would require the simultaneous calculation of gas enthalpies and temperatures normal to the flow. A simple theoretical analysis suggests a parabolic variation of gas temperature across the flow, with its maximum value aligned with the flow centreline. Actual plant measurements consistently show major deviations from this ideal. Depending on the construction and operating modes of the plant the gas temperature profile in the superheater and reheater regions normal to the flow direction may not even be smoothly continuous and may show significant bands of well-delineated temperature zones. For these reasons the crossflow variations of gas and fluid properties are replaced by representative axial values, developed by some averaging process across the x-axis.

For the purposes of this illustration we will ignore the tube metal and assume that heat is transferred directly between the fluids.

The temperatures of the primary (T_1) and secondary (T_2) fluids or gases exchanging heat along parallel flow paths in the axial flow direction can be described as a function of time by the following equations (refer to Eq. 5.34):

$$\tau_{t1}\frac{\partial T_1}{\partial t} + \tau_{z1}\frac{\partial T_1}{\partial z} + T_1 = T_2, \tag{9.11}$$

$$\tau_{t2}\frac{\partial T_2}{\partial t} + \tau_{z2}\frac{\partial T_2}{\partial z} + T_2 = T_1. \tag{9.12}$$

These equations apply to the co-flow arrangement in which the two flows exchange heat while moving in the same direction from inlet to outlet. The incoming hot stream exchanges heat with the incoming cold stream. In the counter-flow arrangement this is reversed with the flows moving in opposite directions through the component. This can be seen by the substitution on the primary side of $z' = L - z$ where L is the length of the flow path from inlet to outlet. Then $\partial/\partial z' = -\partial/\partial z$ and the second equation for the counter-flow arrangement become

$$\tau_{t2}\frac{\partial T_2}{\partial t} - \tau_{z2}\frac{\partial T_2}{\partial z} + T_2 = T_1 \tag{9.13}$$

with an appropriate adjustment to boundary conditions.

The steady-state temperature profiles are given as the solution of these equations with $\partial/\partial t = 0$.

$$\tau_{z1}\frac{\partial T_1}{\partial z} + T_1 = T_2, \tag{9.14}$$

$$\tau_{z2}\frac{\partial T_2}{\partial z} + T_2 = T_1 \tag{9.15}$$

for co-flow, and

$$\tau_{z1}\frac{\partial T_1}{\partial z} + T_1 = T_2, \tag{9.16}$$

$$-\tau_{z2}\frac{\partial T_2}{\partial z} + T_2 = T_1 \tag{9.17}$$

for counter-flow.

Equations 9.14 and 9.15 or 9.16 and 9.17 may be combined by definition of $\mathbf{T} = [T_1\ T_2]^T$. Then,

$$\frac{\mathrm{d}\mathbf{T}}{\mathrm{d}z} = \mathbf{A}\,\mathbf{T},$$

where

$$\mathbf{A} = \begin{bmatrix} -\frac{1}{\tau_{z1}} & \frac{1}{\tau_{z1}} \\ \frac{1}{\tau_{z2}} & -\frac{1}{\tau_{z2}} \end{bmatrix}.$$

The solution of this equation can be generated at a series of points equally spaced at Δz intervals axially along the flow path, from inlet to outlet, using Eq. 3.7. Then,

$$\mathbf{T}^{j+1} = [\mathbf{I} + \mathbf{A}\boldsymbol{\Phi}]\mathbf{T}^{j},$$

where $\boldsymbol{\Phi} = \exp(\Delta z\,\mathbf{A})$ is the matrix exponential of $\mathbf{A}$. Subject to certain limitations on $||\mathbf{A}||$ the matrix exponential may be approximated by the first two terms of the exponential series:

$$\boldsymbol{\Phi} \approx \mathbf{I} + \Delta z\,\mathbf{A}$$

and

$$\mathbf{T}^{j+1} = \left[\mathbf{I} + \mathbf{A} + \Delta z\, \mathbf{A}^2\right] \mathbf{T}^j.$$

After substitution for **A** and some algebraic manipulations, we arrive at two equations from which the steady-state profiles of the two flow temperatures can be calculated. Defining

$$\begin{aligned} a_1 &= 1 - 1/\tau_{z1} & b_1 &= 1/\tau_{z1}^2 + 1/(\tau_{z1}\tau_{z2}), \\ a_2 &= \phantom{1 - {}} 1/\tau_{z1} & b_2 &= 1/\tau_{z1}^2 + 1/(\tau_{z1}\tau_{z2}), \\ a_3 &= \phantom{1 - {}} 1/\tau_{z2} & b_3 &= 1/(\tau_{z1}\tau_{z2}) + 1/\tau_{z2}^2, \\ a_4 &= 1 - 1/\tau_{z2} & b_4 &= 1/(\tau_{z1}\tau_{z2}) + 1/\tau_{z2}^2. \end{aligned}$$

we can write

$$T_1^{j+1} = (a_1 + \Delta z\, b_1) T_1^j + (a_2 - \Delta z\, b_2) T_2^j,$$
$$T_2^{j+1} = (a_3 - \Delta z\, b_3) T_1^j + (a_4 + \Delta z\, b_4) T_2^j.$$

The following figures illustrate the steady-state temperature profiles calculated from these equations. The tubular heat exchanger carries water on the tube side (flow 1) and is heated by a gas flow on the duct side (flow 2). The τ parameters were calculated from Eq. 5.33 using the following parameters:

		Flow 1	Flow 2
v	m³/kg	0.001	1.2
d_i	m	0.01	1.0
c_p	kJ/(kg K)	4.12	1.2
α_{hx}	kJ/(m² k)	1.5	0.07
A_x	m²	0.00031	3.141
ν	m/s	0.35	0.61
$\dot{m}$	kg/s	0.11	1.57
A_{hx}	m²/m	0.03142	3.142
τ		9.62	8.57

Co-flow

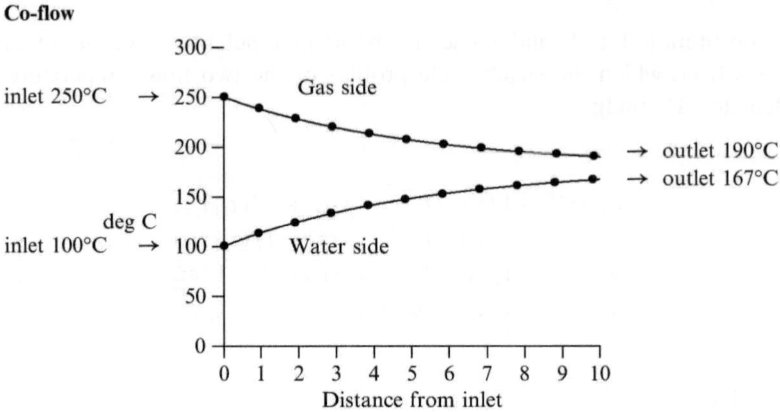

Counter-flow

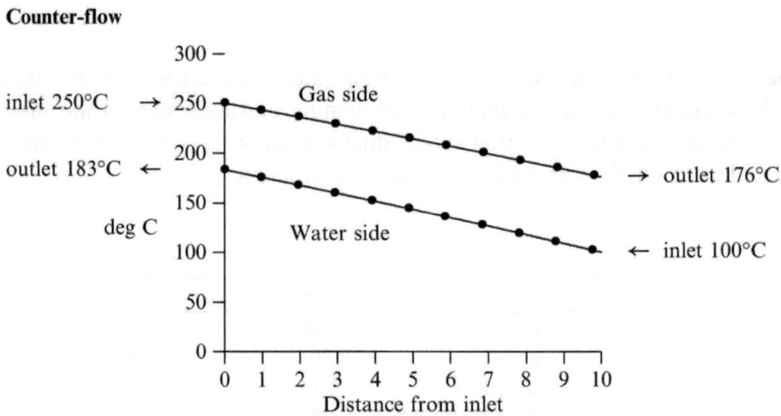

These profiles illustrate the characteristics typical of each type of heat exchanger configuration.

Co-flow

(a) The temperature differential between hot and cold sides is maximum at the inlet and reduces progressively through the component. The heat load is unevenly distributed through the component.

(b) The maximum cold-side outlet temperature is limited to be less than the hot-side *outlet* temperature.

Counter-flow

(a) The temperature differential between hot and cold sides is substantially constant along both flow paths, and the heat load is distributed more evenly over the heat exchanger.
(b) The maximum cold-side outlet temperature is limited to be less than the hot-side *inlet* temperature.

This last feature of the counter-flow heat exchanger offers some advantages for networks of heat exchangers. In a large utility boiler the flue gas passes in a single contiguous stream from the furnace exit to the air heater inlet. The ordering of the various superheater and reheater tube banks along this path is selected to maximise heat pickup by each bank. Typically, superheaters are set up in a co-flow arrangement, while reheaters are frequently arranged as counter-flow.

9.3.4 Heat Exchangers for Ambient Cooling

The most common instances of heat transfer to ambient via forced convection loss from the outside of the heat exchanger tubes are transformer cooling loops, some oil coolers and air-cooled condensers. This case also covers the widely used class of heat exchangers known as fin–fan coolers.

Forced Air Cooling

Heat flows directly from the metal parts at temperature T_m to ambient via a heat transfer mechanism based on a forced air flow (established by fans) perpendicular to a bank of tubes. The tubes may be finned or smooth. Heat transfer reverts to natural convection if the source of forced air flow (fans) is not present.

Heat transfer from the tube surface via forced convection to ambient is given by

$$\dot{q}_{amb} = \alpha_{af} a_{af} (T_m - T_{amb}), \tag{9.18}$$

where
- $\dot{q}_{amb}$ is the heat transferred from a tube bank per unit tube length to ambient,
- a_{af} is the total surface area per unit length exposed to the air flow,
- α_{af} is the heat transfer coefficient.

For finned tubes the heat transfer coefficient α_{af} is defined as

$$\alpha_{af} = k_{af} \frac{\nu_{\perp}^{0.6}}{d_{ot}^{0.4}} \left(\frac{d_{cc}}{d_{cc} - d_{ot}} \right)^{0.6},$$

where
- $\nu_{\perp}$ is the air velocity normal to the tube bank,
- d_{ot} is the outside diameter of the smooth tubes (measured at the fin root),
- d_{cc} is the centre-to-centre spacing of the tubes in a row,
- k_{af} is a scaling constant, theoretical value 0.0016 W/(m K).

The same heat transfer coefficient may be used for smooth tubes if the bare tube area is used.

Natural Air Cooling

If the element (duct, pipe, valve, pump, heat exchanger shell) is insulated, heat flows from the metal casing at temperature T_m to ambient through a series path made up of:

- Conduction through the pipe or chest wall from the inner zone to the insulation interface
- Conduction and convection across the metal-insulation interface
- Conduction through the insulation
- Convection and possibly radiation from the insulation outer surface to ambient

By far the largest temperature drop along this path is through the insulation. As the heat flux across the metal-insulation interface will be small, it can safely be assumed that T_m will closely approximate the metal interface temperature. The heat flow path can then be represented with little error by assuming that the complete drop in temperature from T_m to ambient occurs across the insulation. Define A_{insul} to be the insulation area per unit length across which heat is transferred. Then

$$\dot{q}_{hmxo} = A_{insul} \lambda_{insul} (T_m - T_{amb}) \tag{9.19}$$

λ_{insul} is the coefficient of thermal conductivity for the insulation material with a typical value of around 0.05 W/(m K).

It appears reasonable to assume that, in keeping with plant experience and with the intended purpose of insulation, the insulation surface temperature will approximate ambient, and that radiation from the insulation surface can be neglected. However, radiation from large-lagged surfaces can be appreciable. An allowance for this effect can be made in individual cases by an adjustment of the effective insulation thickness to increase ambient losses beyond the theoretical value.

Heat storage in the insulation is negligible, and no insulation temperature need be calculated. Air and gas ducts operating at or near ambient pressure have thin-containing walls in which little heat is stored. It is usually unnecessary to calculate a wall temperature different from the enclosed gas temperature.

9.3.5 Rotary Air Heaters: Ljungström and Rothemuhle

Air heaters are used to recover furnace flue gas waste heat and return it to the furnace in the form of pre-heated combustion air. For this reason they are called "recuperative"[1] heaters and contribute significant gains to overall efficiency. In power plants they are usually found in any of three configurations:

1. Rotary of the Ljungström type
2. Fixed plate of the Rothemuhle type, also called "rotary"
3. Tubular

Another form of air heater is often used in geographical regions subject to low ambient temperatures. These use a steam-heated coil to preheat combustion air before it enters the rotary air heater. This prevents excessively low temperatures in the rotary heater which might cause condensation and corrosion. Steam coil air preheaters are another form of tubular heat exchanger and are discussed elsewhere.

Rotary air heaters are built in either of two forms. In the Ljungström arrangement the heat transfer medium rotates through fixed flue gas and air streams. In the Rothemuhle type the heat transfer medium remains stationary, and the gas streams are rotated across it by a rotating hood. The Rothemuhle hood is much lighter than the Ljungström rotor, but construction and operation are more complex. In the course of a single rotation the Rothemuhle hood will traverse four streams, two gas and two air. The bi-sector Ljungström will traverse two streams and the tri-sector type three, one gas and two air streams. However, the analysis of both types is essentially the same, and the following discussion will focus on the Ljungström type which is more commonly found (Fig. 9.9).

The air heater consists of a central rotating element (the rotor) installed within a casing that is divided into two (bi-sector type) or three (tri-sector type) sectors. The casing is integrated with the gas and air duct structures with expansion joints linking it to the ducting. The rotor is the medium of heat transfer and is built up as a steel frame into which the heat transfer elements ("baskets") are inserted. This provides relatively easy access to the individual baskets for assembly, cleaning and maintenance. It is a substantial piece of equipment, weighing up to 130 tonnes in an industrial boiler and in excess of 400 tonnes in a large utility boiler. The rotor diameter depends on rating, ranging up to some 8 m in industrial boilers and more than 20 m in utility boilers.

[1]The term "regenerative" is also used. A "regenerative" heat exchanger is distinguished from a "recuperative" heater by its use of cyclic heating and cooling. In the heating phase heat is absorbed from a hot stream and given up to the cooler stream during the cooling phase. A tubular air heater, for example, is not cyclic and heat exchange is continuous. Both are "recuperative" in that both recover heat from a waste stream and return it to the process.

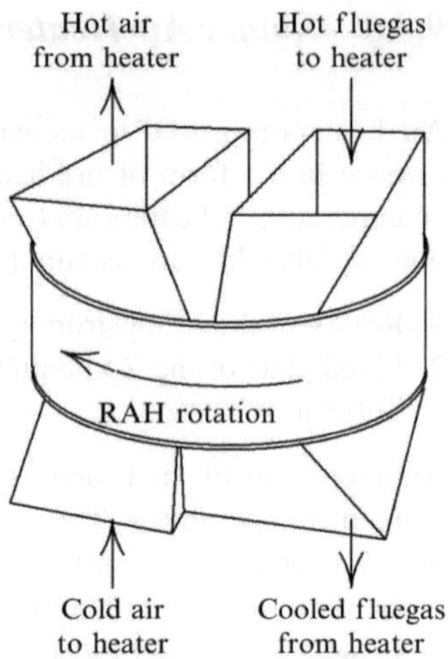

Fig. 9.9 General arrangement of a rotary air heater

Rotation of the rotor is necessary for heat transfer to take place. The rotor is carried by a vertical shaft[2] supported at its base by a single horizontal thrust bearing with its own lubrication system and at the top by a simple shaft alignment bearing which supports no vertical load. The rotor is turned slowly by an electric motor through gearing at around 1–2 rpm. It is divided into circumferential segments, the number of segments being selected so that the angle subtended by a segment is a multiple of 7.5°. Thus, typically, a small rotor might have 12 segments and up to 24 in a large one. Each segment is divided into boxes, each carrying one or more baskets mounted in individual support frames. These are grouped vertically into hot end, intermediate and cold end zones, in each of which the basket material and design is matched to the local temperature conditions and maintenance requirements for cleaning and replacement (Fig. 9.10).

Baskets are formed from a number of compact vertical steel plates with a large surface area relative to their volume. The plates are typically around 0.5–1.2 mm thickness, spaced around 5–6 mm apart. They are not flat, and a variety of bends and ripples are used to maximise the heat transfer area while minimising their thermal mass. This facilitates the rapid heating and cooling of the baskets as they move between the hot gas and cool air streams while minimising the air and gas flow pressure drops. Different manufacturers use different designs, but three or four designs are typical, each optimised for use in the cold, middle or hot zones. For example, plates in the hot zone are designed for maximum heat transfer while those in the cold zone must provide for possible ash fouling, minimisation of corrosion and simplified maintenance.

[2]In most air heaters the rotor rotates in the horizontal plane. Smaller units are sometimes mounted vertically with horizontal air and gas flows.

Fig. 9.10 Assembled baskets for installation (Courtesy of Millmerran Operating Company, Australia)

Radial and circumferential seals are provided to restrict leakage of gas and air between the sectors and between the duct and casing while the rotor is moving. Being connected close to the discharge side of the combustion air fans, the air side is at a higher pressure that the gas side which, in a balanced draft furnace, is maintained at a slight suction. Leakage of air to the flue gas is a major operational issue with air heaters and is usually around 4–8% of the air flow. Sootblowers are provided for the removal of ash which may be deposited from the flue gas.

Tri-sector types are commonly found in power generation plants with coal firing. The flue gas sector spans about 40% of the rotor. The second sector is smaller and is connected to the cold air duct from the forced draft (combustion air) fans. After heating, this air—secondary air—is distributed via the hot air duct to the furnace burner registers. The third sector is the smallest. It heats the primary air which is ducted to the coal pulverising mills for drying and transport of the fuel to the burners.

Consider the passage of a single radial segment as it moves through a complete revolution. At a rotational speed of N revolutions per minute the segment will take 60/N seconds for a complete revolution. Around 40% of this time is spent first in the gas stream and then an equal period in the air stream. The missing time is spent out of either gas stream while passing through the bulky central support structure which occupies an area equivalent to some 20–25% of the rotor area. As the segment moves through the flue gas stream it absorbs heat at a rate proportional to the local differential between gas and segment material temperatures. The total heat absorbed is proportional to the time spent in the gas stream.

Leaving the hot gas stream, the segment passes through the unheated support structure where it neither gains nor loses significant heat. It then moves into the cooler air stream, moving in the counter-flow direction, to which it loses heat at

Fig. 9.11 View from underneath the rotor (Courtesy of Millmerran Operating Company, Australia)

a rate proportional to the local differential between air and segment temperatures. The segment leaves the air stream with residual heat and, having again crossed the central support structure, re-enters the flue gas stream and the process repeats.

For a tri-sector air heater, the process is identical, but the rotor traverses two independent air streams during a single rotation.

As the segment moves from the gas to the air streams it transports a small parcel of gas or air from one stream to the other. This is trapped within the free volume of the segment and released into the opposite stream. Since the same *volume* of gas is transported in each direction but at differing temperatures, this represents a *mass* leakage from the air to the gas side. It is typically around 1–4% of the air flow.

The counter-flow arrangement of gas and air flows through the air heater ensures that a relatively constant temperature differential is maintained between the hot and cold sides along the flow paths. Flue gas enters the air heater from the economiser or reheater at around 320–360 °C. It leaves the air heater at around 130–150 °C. This avoids condensation and corrosion within the heater and gas duct and ensures that the further losses in temperature as the flue gas passes through the dust collection plant do not reduce gas temperatures going to the induced draft fans below 115–120 °C, again to avoid condensation and possible corrosion. With these gas temperatures air leaves the air heater at around 285–330 °C.

Under typical operating conditions temperature differentials across an air heater can approach or exceed 300K. Differentials of this magnitude across a physical length of some 2–3 m create substantial temperature gradients which, for adequate simulation, need an appropriate form of spatial discretisation in the flow direction.

To describe this process mathematically we will treat the continuous movement of the rotor as a series of incremental steps, each of distance Δs, and follow the path of a single segment as it moves through a complete revolution. The reference time $t = 0$ is defined to be the instant at which the segment first enters the gas stream. The segment is divided into a series of contiguous zones or cells along the gas stream, with the first cell at the gas inlet and the last cell being the gas outlet. An identical discretisation scheme is used for the air side. A uniform distribution of gas and air temperature is assumed across the duct normal to the flow at each inlet. Radial variations of gas, air and segment material temperatures are ignored.

The gradient of gas temperature along the flow is described by the equation

$$c_g \dot{m}_g \frac{\mathrm{d}T_g}{\mathrm{d}z} = \dot{q}_{gm}(z),$$

where $\dot{q}_{gm}(z)$ is the local gradient of heat exchange between the gas and the rotor segment. $\dot{m}_g$ is the mass flow of gas through the segment, and c_g is its specific heat capacity. The gradient of air temperatures is described by an identical equation, expressed in terms of air-side variables.

$$c_a \dot{m}_a \frac{\mathrm{d}T_a}{\mathrm{d}z} = \dot{q}_{am}(z).$$

Denote the rotational position of the segment by the index j and the flow-axial cell of the segment by index k. Then, $T^j_{g,k}$ is the representative gas temperature in the kth cell, and $T^j_{m,k}$ is the representative temperature of the segment material. Both are defined at the same rotational position. Similarly, $T^j_{a,k}$ is the representative air temperature in the kth cell.

The rate of heat transfer from the kth cell of the gas to the kth cell of the segment at position j is

$$\dot{q}^j_{gm,k} = \alpha^j_k \, A_k \left(T^j_{g,k} - \overline{T}^j_{m,k} \right),$$

where α^j_k is the local heat transfer coefficient, gas to metal and A_k the segment heat transfer area. $\overline{T}^j_{mk}$ is the metal temperature used for the calculation of heat transfer and is defined as

$$\overline{T}^j_{m,k} = \theta T^j_{m,k} + (1-\theta) T^j_{m,k-1},$$

where θ is a weighting parameter, selected within the range (0,1) to yield the best approximation to the non-linear variation of the metal temperature along the flow. Computation experience suggests a value around 0.8–0.9 gives best results.

If the change in segment temperature within a cell from one step $(j-1)$ to the next j is ΔT, we can write for the segment

$$T^j_{m,k} = T^{j-1}_{m,k} + \Delta T^j_{m,k}.$$

The heat transferred from gas to segment over the step length Δs is $\dot{q}^j_{gm,k} \Delta s$. We then have

$$\Delta T^j_{m,k} = \frac{\dot{q}^j_{gm,k}}{c_m M_m} \Delta s,$$

where c_m is the specific heat capacity of the segment material with mass M_m.

The same procedure produces a similar result for the air side.

After some manipulations we arrive at four recursive equations from which the gas, air and segment temperatures on both sides can be calculated, given as boundary conditions the inlet temperatures of the gas and air streams and the profile of segment temperatures entering the gas stream, assumed equal to those leaving the air stream.

On the gas side,

$$\begin{aligned} T^j_{g,k} &= \frac{1+\theta\Delta s B_m}{D_g}\left[T^j_{g,k-1} + (1-\theta)\Delta z B_g T^j_{m,k-1}\right] \\ &\quad + \frac{\theta\Delta z B_g}{D_g}\left[T^{j-1}_{m,k} - (1-\theta)\Delta z B_g T^j_{m,k-1}\right], \end{aligned} \tag{9.20}$$

$$\begin{aligned} T^j_{m,k} &= \frac{\Delta s B_m}{D_g}\left[T^j_{g,k-1} + (1-\theta)\Delta z B_g T^j_{m,k-1}\right] \\ &\quad + \frac{1+\Delta z B_g}{D_g}\left[T^{j-1}_{m,k} - (1-\theta)\Delta z B_g T^j_{m,k-1}\right], \end{aligned} \tag{9.21}$$

and on the air side,

$$\begin{aligned} T^j_{a,k} &= \frac{1+\theta\Delta s B_m}{D_a}\left[T^j_{a,k-1} + (1-\theta)\Delta z B_a T^j_{m,k-1}\right] \\ &\quad + \frac{\theta\Delta z B_a}{D_a}\left[T^{j-1}_{m,k} - (1-\theta)\Delta z B_a T^j_{m,k-1}\right], \end{aligned} \tag{9.22}$$

$$\begin{aligned} T^j_{m,k} &= \frac{\Delta s B_m}{D_a}\left[T^j_{g,k-1} + (1-\theta)\Delta z B_a T^j_{m,k-1}\right] \\ &\quad + \frac{1+\Delta z B_a}{D_a}\left[T^{j-1}_{m,k} - (1-\theta)\Delta z B_a T^j_{m,k-1}\right], \end{aligned} \tag{9.23}$$

where

$$B_g = \frac{\alpha^j A^j}{c_g \dot{m}_g} \qquad B_a = \frac{\alpha^j A^j}{c_a \dot{m}_a} \qquad B_m = \frac{\alpha^j A^j}{c_m M_m}.$$

$$D_g = 1 + \theta\Delta s B_m + \Delta z B_g,$$
$$D_a = 1 + \theta\Delta s B_m + \Delta z B_a.$$

Heat Transfer

The standard Brandt correlation for gas/metal heat exchange can be used for calculation of α in each cell and at each step. The gas and air mass flow velocities are derived from the flow through the inter-plate spaces of the individual baskets. It may be assumed that each plate is exposed to the same mass flow and steam temperature. As mentioned in Sect. 8.2.6 the application of the Brandt correlation to a flat plate, as here the case, requires the determination of an equivalent tube diameter d_{eq} whose calculation is also detailed in that section.

A number of dimensional parameters must be known for the calculation of heat transfer areas, heat transfer coefficients, mass flow velocities and pressure-loss coefficients. These are determined from the overall dimensions of the heater and from the number and dimensions of the individual baskets. Since much of this detailed design information may not be readily available, judicious assumptions can be made where necessary to substitute for missing data. Hot and cold end and intermediate baskets are not of equal size, but reasonable estimates can be made assuming they are. Typical numbers for a large rotary air heater are as follows (Fig. 9.12):

Average basket length: 1 m
Average basket height: 0.7 m
Average basket width: 0.6 m
Assuming 6 mm spacing each basket will contain around 120 plates
Each plate is 0.7 m $\times$ 0.6 m giving an average plate area of 0.42 m^2
Area per basket = area per plate x plates per basket $\approx$ 70 m^2
Mass of plates per basket, assuming 0.6 mm plate thickness, $\approx$ 560 kg
Baskets per segment, assuming six baskets in each of three zones
(hot, intermediate, cold) = 18
Total heat transfer area per segment
$= 70\,m^2 \times 18$ baskets $\times$ 2 (two sides per plate)
$= 1,890\,m^2$.

Applying this data to the calculation of the heat transfer coefficient for a single plate, we can represent the inter-plate space as a rectangle 6 mm high and 1,000 mm long, giving an inter-plate flow cross-sectional area of 0.006 m^2 and peripheral distance of 2.012 m. Applying the calculation method of Sect. 8.2.6 yields an equivalent hydraulic diameter of 0.01 m. Assuming this is the area presented to flow between any pair of plates the total flow area will be 0.006 times the number of inter-plate spaces presented to the incoming flow. From the quoted data there are 120 plates in each of six baskets in each of 12 segments giving a total flow cross section of 52 m^2. For a total gas flow of say 300 kg/s, the mass flow velocity ϕ will be around 5.5 kg/s/m^2 which, with an average gas specific volume of 1.25 m^2/kg, indicates a gas velocity in the inter-plate space of 7.2 m/s. The Brandt heat transfer coefficient for these conditions is 0.546 kJ/s/m^2/C.

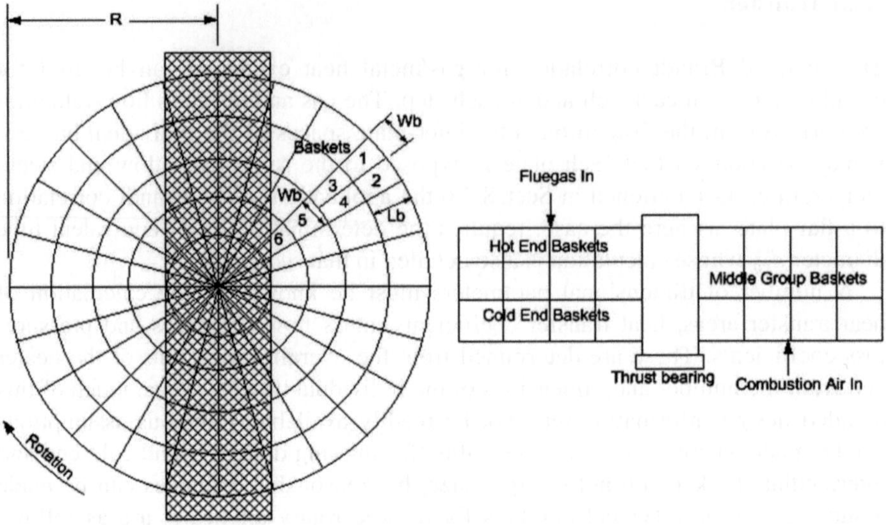

Fig. 9.12 Plan and elevation sections of a rotary air heater

With this data and a spatial division of ten zones in the direction of flow, Eqs. 9.20–9.23 produce the following profiles of temperatures on the gas and air sides of a bi-sector air heater. Temperatures are shown in degrees Celsius.

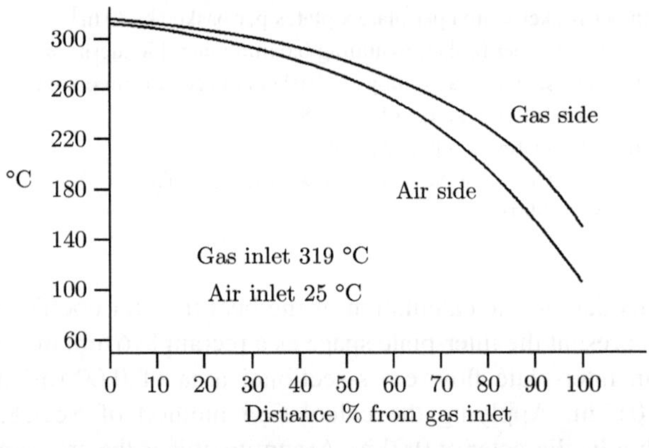

Pressures in the Air Heater

The tri-sector air heater is the more general case and will be used to illustrate a method of calculation. The air heater connects into the flue gas and air flow paths as shown by the following figure:

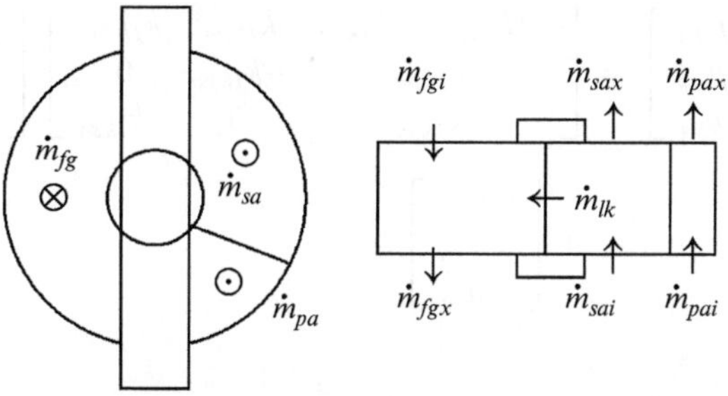

The representative pressure in any sector (p_{fg}, p_{pa}, p_{sa}) in the flue gas, primary and secondary air sectors respectively, will be determined from the balance of mass flows into and out of the sector. With normal flow rates and degrees of fouling, the pressure differential between input and outlet of any path will be around 1 kPa. For a furnace with balanced draft the flue gas pressure in the air heater will be somewhat lower than furnace pressure (suction) and the pressure on the air side somewhat lower than the combustion air fan discharge pressure. Given that both furnace and fan discharge duct pressure are controlled to set-point values by modulating control loops, pressures within the air heater will tend to remain fairly constant over wide ranges of operation and under few circumstances would be expected to move by more than 2 or 3 kPa. It is therefore permissible to treat the calculation of pressures on an incompressible basis derived from mass balances alone.

Given the small pressure differentials it is permissible to calculate the internal leakage flows as linear functions of the pressure differentials rather than the more correct $\sqrt{dp}$. Then,

$$\dot{m}_{lkpa} = k_{lkpa}(p_{pa} - p_{fg})$$

and

$$\dot{m}_{lksa} = k_{lksa}(p_{sa} - p_{fg})$$

where $\dot{m}_{lkpa}$ and $\dot{m}_{lksa}$ are the leakage flows from the primary and secondary air sectors[3] to the flue gas. k_{lkpa} and k_{lksa} are selected to give the anticipated leakage rates. Defining

$$\delta\dot{m}_{fg} = \dot{m}_{fgi} - \dot{m}_{fgx}$$
$$\delta\dot{m}_{pa} = \dot{m}_{pai} - \dot{m}_{pax}$$
$$\delta\dot{m}_{sa} = \dot{m}_{sai} - \dot{m}_{sax}$$

as the inflow/outflow differentials for each sector, the following equation can be solved to yield the three sector pressures, given the appropriate boundary conditions:

[3] There will be a primary air sector only in a tri-sector air heater.

$$\mathcal{C}\frac{\mathrm{d}}{\mathrm{d}t}\begin{bmatrix} p_{fg} \\ p_{pa} \\ p_{sa} \end{bmatrix} = \begin{bmatrix} -(k_{lkpa}+k_{lksa}) & k_{lkpa} & k_{lksa} \\ k_{lkpa} & -k_{lkpa} & 0 \\ k_{lksa} & 0 & -k_{lksa} \end{bmatrix}\begin{bmatrix} p_{fg} \\ p_{pa} \\ p_{sa} \end{bmatrix} + \begin{bmatrix} \delta\dot{m}_{fg} \\ \delta\dot{m}_{pa} \\ \delta\dot{m}_{sa} \end{bmatrix}. \tag{9.24}$$

Writing

$$\mathbf{A} = \frac{1}{\mathcal{C}}\begin{bmatrix} -(k_{lkpa}+k_{lksa}) & k_{lkpa} & k_{lksa} \\ k_{lkpa} & -k_{lkpa} & 0 \\ k_{lksa} & 0 & -k_{lksa} \end{bmatrix} \qquad \mathbf{D} = \frac{1}{\mathcal{C}}\begin{bmatrix} \delta\dot{m}_{fg} \\ \delta\dot{m}_{pa} \\ \delta\dot{m}_{sa} \end{bmatrix} \tag{9.25}$$

and

$$\mathbf{p} = \begin{bmatrix} p_{fg} \\ p_{pa} \\ p_{sa} \end{bmatrix}$$

this equation can be written compactly as

$$\dot{\mathbf{p}} = \mathbf{A}\mathbf{p} + \mathbf{D}. \tag{9.26}$$

One possible discrete-time solution sequence is

$$\mathbf{p}^{n+1} = (\mathbf{I} + \mathbf{A}\Phi)\mathbf{p}^{n} + \Delta t \Phi \mathbf{D}, \tag{9.27}$$

where $\Phi = \exp(\Delta t \mathbf{A})$ is the matrix exponential. The constant $\mathcal{C}$ is selected to ensure stability of the calculation and has the form of a "compressibility" coefficient.

Pressure Losses

The pressure differentials across the individual sectors are important indicators of the air heater thermal performance and mechanical condition. A high differential indicates excessive fouling of or mechanical damage to the baskets and may initiate sootblowing or maintenance. The pressure losses may be calculated from the usual Fanning formula equation 6.18.

$$\begin{aligned} \Delta p &= \xi \frac{L}{d_h}\frac{v}{2A_x^2}\dot{m}^2 \\ &= \xi \frac{L}{d_h}\frac{v}{2}\phi^2. \end{aligned}$$

The friction factor ξ is quoted by Habbitts et al. [39] in the range of 0.023–0.03 as a function of Reynolds number for a variety of plate geometries. For simulation purposes it is sufficient to use a mid-range value as the actual value in operations will be determined by the mechanical condition of the baskets and their degree of fouling.

Chapter 10
Temperatures Through Component Walls

Pressure-bearing components are usually constructed of steel with containing walls ranging in thickness between 5 and 80 mm. Boiler tubes (superheaters, reheaters, economisers and evaporators) typically have an outside diameter of 50–60 mm and a wall thickness of around 4–10 mm. Boiler drums can have wall thicknesses up to 22 cm. Valve chests have casing thicknesses of around 10–20 mm with significantly thicker flanges. HP turbine casings can be up to 30 mm thick and LP turbines around 5 mm. Again, flanges will be of greater thickness.

Throughout this work all pressurised flow path components, including the casings of turbines, are treated as some form of hollow cylinder.[1] Since most analytical methods provide tractable solutions only when applied to simple shapes and boundary conditions this discussion will be limited to these cylindrical shapes.

We will now look at the calculation of the temperatures within the walls of these components.

10.1 Analytical Solution of the Heat Conduction Equation

The distribution of temperature throughout a conducting material of a given geometry is given at any spatial point and time by the solution of the heat conduction equation

$$\nabla^2 T = \frac{1}{D}\frac{\partial T}{\partial t},$$

where D is the heat diffusion coefficient

$$D = \frac{k}{\rho c_p}, \tag{10.1}$$

[1]Gas ducts are not pressurised and are treated as having rectangular cross-section.

P. O'Kelly, *Computer Simulation of Thermal Plant Operations*,
DOI 10.1007/978-1-4614-4256-1_10,

and k is the coefficient of thermal conductivity of the material, ρ is its density and c_p the specific heat capacity. For a cylindrical vessel, and ignoring variations in the azimuthal direction, this equation is written in cylindrical coordinates as

$$\frac{1}{r}\frac{\partial}{\partial r}\left(r\frac{\partial T}{\partial r}\right)+\frac{1}{r^2}\frac{\partial^2 T}{\partial \phi^2}+\frac{\partial^2 T}{\partial z^2}=\frac{1}{D}\frac{\partial T}{\partial t}. \tag{10.2}$$

If we initially neglect axial conduction ($\partial T/\partial z = 0$) and assume that the distribution of T is independent of ϕ (no azimuthal variation), Eq. 10.2 simplifies to

$$\frac{1}{r}\frac{\partial}{\partial r}\left(r\frac{\partial T}{\partial r}\right)=\frac{1}{D}\frac{\partial T}{\partial t}. \tag{10.3}$$

This equation describes the circular-symmetric radial distribution of temperature throughout a cross-section of a cylinder or cylindrical pipe or vessel. For a solid cylinder the radial ordinate r is defined for the range $0 \leq r \leq r_o$ and for a hollow cylinder $r_i \leq r \leq r_o$ where r_i is the inner radius of the hollow cylinder and r_o is the outer radius of both. The wall thickness of the hollow cylinder is $r_w = r_o - r_i$.

The steady-state profile of temperature $T(r)$ is obtained when all transients following a change in boundary conditions have decayed away. This profile is provided as the solution of Eq. 10.3 with the time derivative set to zero.

$$\frac{1}{r}\frac{\partial}{\partial r}\left(r\frac{\partial T}{\partial r}\right)=0 \tag{10.4}$$

which may be integrated twice to yield the general solution

$$T(r)=C_1\ln(r)+C_2 \quad \text{for } 0 < r \leq r_0. \tag{10.5}$$

C_1 and C_2 are constants selected to match the given boundary conditions. For the solid cylinder C_1 must be zero to ensure the solution remains finite at the centre of the cylinder ($r = 0$). Then

$$T(r)=C_2=T_\infty,$$

that is, the cylinder will reach a uniform temperature T_∞ throughout. For a hollow cylinder whose steady-state inner and outer surface temperatures are T_i and T_o we can write

$$C_1 = \frac{T_0 - T_i}{\ln(\frac{r_0}{r_i})}$$

$$C_2 = \frac{T_i\ln(r_0) - T_o\ln(r_i)}{\ln(\frac{r_0}{r_i})}.$$

If both inner and outer temperatures are equal to T_∞ the wall will also reach a uniform temperature T_∞ throughout.

Boundary Conditions

In order to obtain specific solutions to Eq. 10.3 it will be necessary to define the temperature at each surface in contact with the surrounding media with which the body exchanges heat. This can be done by defining either the respective surface temperature or a heat transfer mechanism. As in what follows we will be neglecting the axial temperature profile, we need to define boundary conditions at only the radial surface or surfaces. This definition can take any of the following forms [44]:

Dirichlet Very high surface heat transfer allows the setting of the body surface temperature $T(r_o, t)$ equal to that of the surrounding medium T_{ref}:

$$T(r_o, t) = T_{ref} \qquad t > 0.$$

Neumann Heat flux across the surface is specified to be $\dot{q}_o$:

$$\left.\frac{\partial T}{\partial r}\right|_{r=r_o} = -\frac{\dot{q}_o}{k} \qquad t > 0.$$

Robin Convective heat transfer to/from the surrounding medium at temperature T_{ref}; the surface radial gradient is defined by the surface heat transfer conditions:

$$\left.\frac{\partial T}{\partial r}\right|_{r=r_o} = -\frac{\alpha}{k}\left(T_{ref} - T(r_o, t)\right) \qquad t > 0.$$

10.1.1 Transient Conduction in a Solid Cylinder

In order to treat only homogeneous solutions, introduce the new variable ψ defined as

$$\psi(r, t) = T(r, t) - T_\infty(r), \tag{10.6}$$

where $T_\infty(r)$ is a steady-state solution of Eq. 10.7. Then

$$\frac{\partial \psi}{\partial t} = D\left(\frac{\partial^2 \psi}{\partial r^2} + \frac{1}{r}\frac{\partial \psi}{\partial r}\right). \tag{10.7}$$

Proceeding via the usual method of separation of variables we may write

$$\psi = \mathcal{R}\mathcal{T},$$

where $\mathcal{R}$ is a function of r only and $\mathcal{T}$ is a function of time only. After some further manipulations, we arrive at the solution [40]

$$\psi(r, t) = A\exp(-\lambda^2 Dt)\left[BJ_0(\lambda r) + CY_0(\lambda r)\right], \tag{10.8}$$

where $J_0(\lambda r)$ and $Y_0(\lambda r)$ are Bessel functions of zero order and of the first (J_0) and second (Y_0) kind. The separation parameter λ is to be defined by initial and/or boundary conditions.

$Y_0(\lambda r)$ behaves like $\ln(r)$ for small r. In order to ensure a finite solution in the vicinity of $r = 0$ it is necessary to set $C = 0$. Merging A and B into a new constant A we have

$$\psi(r,t) = A \exp(-\lambda^2 Dt) J_0(\lambda r). \tag{10.9}$$

Both A and λ are to be determined from the initial profile $\psi(r,0)$ and one boundary condition. Assuming the Dirichlet condition at the surface of the cylinder in contact with a medium at temperature T_{ref} we may write

$$\psi(r_o,t) = T(r_o,t) - T_{ref} = A \exp(-\lambda^2 Dt) J_0(\lambda r_o). \tag{10.10}$$

This will be true for all $t > 0$ only if λ is a root of

$$J_0(\lambda r_o) = 0. \tag{10.11}$$

There being an infinite number of roots of this equation and since Eq. 10.7 is linear, the individual solutions corresponding to each root λ_n may be summed to give the complete solution

$$\psi(r,t) = \sum_{n=1}^{\infty} A_n \exp(-\lambda_n^2 Dt) J_0(\lambda_n r), \tag{10.12}$$

where the summation is taken over all positive roots λ_n. The first six roots of $J_0(\lambda) = 0$ are 2.4048, 5.5201, 8.6537, 11.7915, 14.9309 and 18.0710.

The coefficients A_n are determined from the initial profile $\psi(r,0)$. For then, with $t = 0$

$$\psi(r,0) = \sum_{n=1}^{\infty} A_n J_0(\lambda_n r).$$

Utilising the orthogonality of the Bessel functions with respect to the weight function r, the coefficients A_n are given as the coefficients of the Fourier-Bessel expansion of $\psi(r,0)$.

$$A_n = \frac{2}{r_o^2 J_1^2(\lambda_n r_o)} \int_0^{r_o} r\psi(r,0) J_0(\lambda_n r) \mathrm{d}r. \tag{10.13}$$

At this stage it is expedient to introduce a further normalisation in the form of $\bar{r} = r/r_o$. Then

$$\psi(\bar{r},t) = \sum_{n=1}^{\infty} A_n J_0(\lambda_n \bar{r}) \exp(-\lambda_n^2 Dt), \tag{10.14}$$

and

$$A_n = \frac{2}{J_1^2(\lambda_n)} \int_0^1 \bar{r}\psi(\bar{r},0) J_0(\lambda_n \bar{r}) d\bar{r}. \tag{10.15}$$

Substituting in Eq. 10.14 for $\psi(\bar{r},t)$ from its definition, Eq. 10.6, we obtain

$$T(\bar{r},t) = T_\infty(\bar{r}) + \sum_{n=1}^{\infty} A_n J_0(\lambda_n \bar{r}) \exp(-\lambda_n^2 Dt). \tag{10.16}$$

To obtain the particular solution, that is, the relevant set of A_n, we must invoke a specific initial temperature profile and boundary condition.

Dirichlet Boundary Condition

$$T(\bar{r},0) = T_0 \quad T_\infty(\bar{r}) = T_\infty \qquad \text{where } T_0, T_\infty \text{ are both constants.}$$

Then $\psi(\bar{r},0) = T_0 - T_\infty$ and

$$\begin{aligned} A_n &= (T_0 - T_\infty)\frac{2}{J_1^2(\lambda_n)} \int_0^1 \bar{r} J_0(\lambda_n \bar{r}) d\bar{r} \\ &= (T_0 - T_\infty)\frac{2}{J_1^2(\lambda_n)} \frac{1}{\lambda_n} J_1(\lambda_n) \\ &= (T_0 - T_\infty)\frac{2}{\lambda_n J_1(\lambda_n)} \end{aligned}$$

and

$$T(\bar{r},t) = T_\infty + 2\exp(-\lambda_n^2 Dt)(T_0 - T_\infty) \sum_{n=1}^{\infty} \frac{J_0(\lambda_n \bar{r})}{\lambda_n J_1(\lambda_n)}. \tag{10.17}$$

Robin Boundary Condition This case is of greater practical interest than the Dirichlet condition but adds greater complexity to the solution. It defines the surface temperature gradient in terms of the local heat transfer coefficient and surface temperature differential. The earlier definition of ψ cannot be used here as the final temperature profile $T_\infty(r)$ is not known in advance of the solution. We must therefore revert to Eq. 10.3 but will retain the normalised $\bar{r}$. Also, to retain a homogeneous form of the equation we define

$$\vartheta(\bar{r},t) = T(\bar{r},t) - T_{ref}$$

and Eq. 10.3 becomes

$$\frac{1}{r}\frac{\partial}{\partial r}\left(r\frac{\partial \vartheta}{\partial r}\right) = \frac{1}{D}\frac{\partial \vartheta}{\partial t}, \tag{10.18}$$

with the general solution (from Eq. 10.9)

$$\vartheta(\bar{r},t) = A\ \exp(-\lambda^2 Dt) J_0(\lambda\bar{r}). \tag{10.19}$$

The boundary condition requires (noting that T_{ref} is constant)

$$\frac{\partial\vartheta(1,t)}{\partial\bar{r}} = -\frac{\alpha}{k}\vartheta(1,t).$$

From Eq. 10.19

$$\frac{\partial\vartheta(\bar{r},t)}{\partial\bar{r}} = A\exp(-\lambda^2 Dt)\frac{\partial J_0(\lambda\bar{r})}{\partial\bar{r}}$$

and noting that

$$\frac{\partial J_0(\lambda\bar{r})}{\partial\bar{r}} = -\lambda J_1(\lambda\bar{r}),$$

this yields

$$-\frac{\alpha}{k}AJ_0(\lambda\bar{r})\exp(-\lambda^2 Dt) = -\lambda AJ_1(\lambda\bar{r})\exp(-\lambda^2 Dt),$$

or

$$\frac{\alpha}{k}J_0(\lambda\bar{r}) = \lambda J_1(\lambda\bar{r}),$$

which, on the outer surface ($\bar{r} = 1$), becomes

$$\frac{\alpha}{k}J_0(\lambda) = \lambda J_1(\lambda). \tag{10.20}$$

This equation has an infinite number of roots λ_n, each giving a solution of Eq. 10.18. The summation of these solutions over the positive roots of Eq. 10.20 gives the general solution

$$\vartheta(\bar{r},t) = \sum_{n=1}^{\infty} A_n J_0(\lambda\bar{r})\exp(-\lambda_n^2 Dt), \tag{10.21}$$

where the coefficients A_n are obtained from the Fourier-Bessel expansion of the initial profile $\vartheta(\bar{r},0)$ [41,46]

$$A_n = \frac{2}{J_0^2(\lambda_n) + J_1^2(\lambda_n)}\int_0^1 \bar{r}\vartheta(\bar{r},0)J_0(\lambda_n\bar{r})\mathrm{d}\bar{r}. \tag{10.22}$$

Solutions of Eq. 10.20 have been tabulated by Carslaw and Jaeger [42] (Appendix 4, Table III) for various values of α and k.

In terms of the temperature $T(\bar{r},t)$ the solution is

$$T(\bar{r},t) = T_{ref} + 2\exp(-\lambda_n^2 Dt)\sum_{n=1}^{\infty}\frac{J_0(\lambda_n\bar{r})}{J_0^2(\lambda_n)+J_1^2(\lambda_n)}\int_0^1 \bar{r}\vartheta(\bar{r},0)J_0(\lambda_n\bar{r})\mathrm{d}\bar{r}. \tag{10.23}$$

10.1.2 Transient Conduction Through a Hollow Cylinder

Dirichlet conditions are assumed at both inner and outer surfaces and we can use ψ as defined earlier. This case requires solution of Eq. 10.3 within the region $r_i \le r \le r_o$. The behaviour of $Y_0(\lambda r)$ for small r is no longer a limiting influence, and we retain both coefficients B and $C \neq 0$. The solution is again of the form of Eq. 10.8, but the coefficients A_n are given from the finite Hankel transform of $\psi(r,0)$ [41]

$$H[\psi(r,t)] = \bar{\psi}_H(\lambda_n) = \int_{r_i}^{r_o} r\psi(r,0)\,[J_0(\lambda_n r)Y_0(\lambda_n r_i) - J_0(\lambda_n r_i)Y_0(\lambda_n r)]\,\mathrm{d}r, \tag{10.24}$$

where the λ_n parameters are the roots of the transcendental equation ([42], Appendix 4, Table IV)

$$J_0(\lambda_n r_o)Y_0(\lambda_n r_i) = J_0(\lambda_n r_i)Y_0(\lambda_n r_o). \tag{10.25}$$

The solution $T(r,t)$ then follows from $\psi(r,t)$

$$T(r,t) = T_\infty(r) + 2\exp(-\lambda_n^2 Dt)\sum_{n=1}^{\infty}\lambda_n^2\frac{J_0^2(\lambda_n r_o)\bar{\psi}_H(\lambda_n)}{J_0^2(\lambda_n r_i) - J_0^2(\lambda_n r_o)}$$
$$x\,[J_0(\lambda_n r)Y_0(\lambda_n r_i) - J_0(\lambda_n r_i)Y_0(\lambda_n r)]\,. \tag{10.26}$$

A Useful Approximate Solution

Assuming the initial temperature profile through the wall $T(r,0)$ is known and can be used, once Eq. 10.25 has been solved for the separation constants λ_n, to calculate $\bar{\psi}_H(\lambda_n)$ from Eq. 10.24, Eq. 10.26 can be used to calculate the time evolving solution $T(r,t)$. This is not a straight-forward process and for many cases of interest here can be replaced by a simple approximation. One such is based on the assumption that, for Dirichlet conditions at both the inner and outer surfaces which permit the definition of both initial and final temperature profiles through the wall, given the inner and outer surface temperatures, the transient profile may be defined by one which moves from the initial to the final profile in a manner and at a rate defined by Eq. 10.17. This equation is restated in terms of these two profiles as

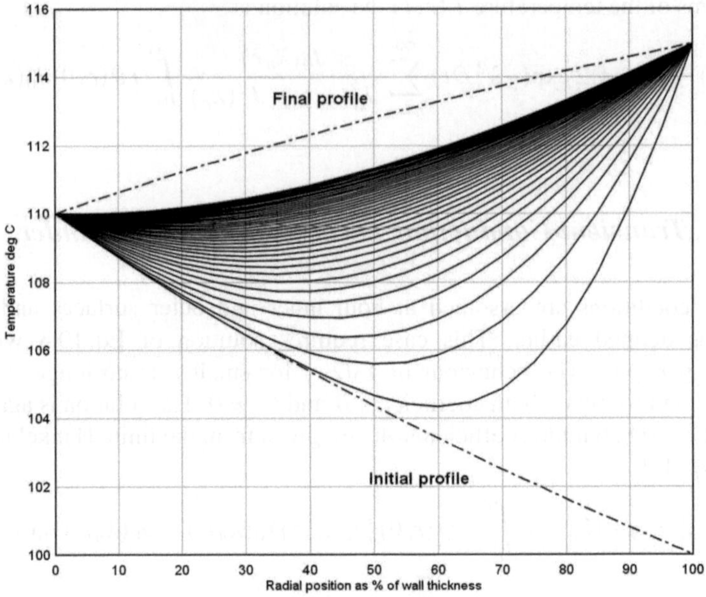

Fig. 10.1 Transient conduction through a cylinder wall—fixed surface temperatures

$$T(\bar{r},t) = T_\infty + 2\exp(-\lambda_n^2 Dt)\,(T_0(\bar{r}) - T_\infty(\bar{r}))\sum_{n=1}^{\infty}\frac{J_0(\lambda_n\bar{r})}{\lambda_n J_1(\lambda_n)}. \qquad (10.27)$$

This is clearly a generous approximation but, as the following computational results suggest, can yield useful results in many cases relevant to our simulation objectives. The behaviour of solutions to Eq. 10.27 is illustrated by Figs. 10.1, 10.2 and 10.3 which show the time evolution of the temperature profile through a steel cylindrical wall, using four terms of the summation. The wall dimensions used for Figs. 10.1 and 10.2 are $r_i = 0.3$ m and $r_o = 0.5$ m, giving a wall thickness of 0.2 m. Immediately prior to the disturbance, the inner wall temperature was 110 °C, the outer temperature was 100 °C and all transients had died out. At $t = 0+$ the outer temperature stepped up to 115 °C. The various temperature profiles are separated by 5 minutes.

Step changes in temperatures are not the norm in industrial equipment and this transient is not representative of temperature changes to be expected. The presumption in the preceding analysis that the temperature profile at $t = 0$ is a steady-state solution of the heat conduction equation precludes formal treatment of any *continuous* changes in boundary temperatures. However, all is not lost, and good results can be achieved with continuously varying boundary temperatures, as shown by the next example. This shows the temperature transients for the same conditions used for the preceding example but with the outer temperature approaching its final value exponentially with a time constant of 30 minutes.

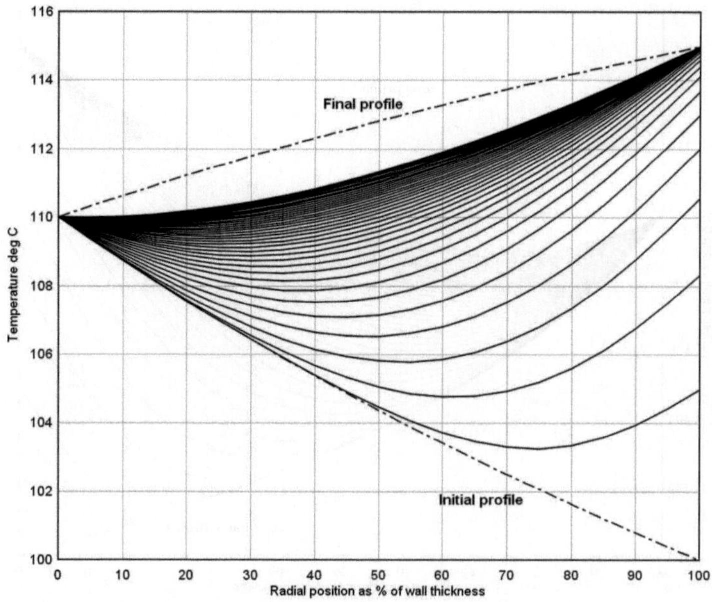

Fig. 10.2 Transient conduction—time-varying outer surface temperature

Figure 10.3 illustrates the influence of wall thickness on the transient behaviour. It shows the same behaviour as Fig. 10.2 but with a wall thickness of 0.4 m (r_i = 0.1 m, r_o = 0.5 m).

10.1.3 Calculation of Bessel Functions

Any ordinary Bessel function of order n can be represented by the infinite series

$$J_n(x) = \sum_{j=0}^{\infty} \frac{(-1)^j}{j!(n+j)!} \left(\frac{x}{2}\right)^{n+2j},$$

from which we have for Bessel functions of order zero

$$J_0(x) = 1 - \frac{x^2}{1!2^2} + \frac{x^4}{2!4^2} - \frac{x^6}{3!2^6} - \cdots,$$

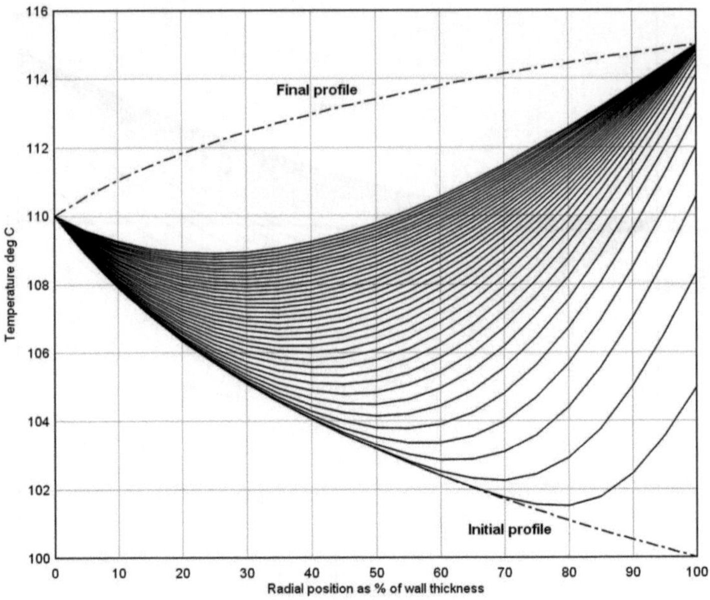

Fig. 10.3 Transient conduction through a thick cylinder wall

where successive terms a^j and a^{j+1} are related by

$$\frac{a^{j+1}}{a^j} = -\frac{x^2}{(2j)^2}$$

and for Bessel functions of order one

$$J_1(x) = \frac{x}{2} - \frac{x^3}{2^3 1!2!} + \frac{x^5}{2^5 2!3!} - \frac{x^7}{2^7 3!4!} + \cdots,$$

where successive terms a^j and a^{j+1} are related by

$$\frac{a^{j+1}}{a^j} = -\left(\frac{x}{2}\right)^2 \frac{1}{(j+1)(j+2)}.$$

Bessel functions cannot be calculated from these series for any value of the argument much greater than the first zero. Although formally convergent, the series quickly diverge in computation. However, Bessel functions may be successfully computed via rational polynomial functions which fit the solutions of the Bessel equation. Such functions are documented for Bessel functions of zero and first order in [1].

10.2 Spatial Discretisation of the Heat Conduction Equation

In the previous section, we have derived an analytical solution to the heat conduction equation in a flat cross-section through a pipe or cylindrical vessel, normal to the direction in flow. We have seen that, while revealing significant insights into the behaviour of the solution, and in particular suggesting an appropriate order of an approximation to the solution, the analytical solution is not very useful as a computational procedure for the real-time calculation of metal temperatures in a dynamic simulation. For this we must turn to a spatial discretisation method.

In the preceding section, we restricted attention to the solution of the heat diffusion (conduction) equation in only the radial ordinate. We will now add the axial (z) ordinate, and Eq. 10.2 becomes

$$\frac{1}{r}\frac{\partial T}{\partial r} + \frac{\partial^2 T}{\partial r^2} + \frac{\partial^2 T}{\partial z^2} = \frac{1}{D}\frac{\partial T}{\partial t}. \tag{10.28}$$

If we replace the continuous derivatives by their discrete time and central difference spatial approximations this equation may be replaced by its forward-time centered space (FTCS) equivalent

$$\frac{1}{D}\left(\frac{T_{j,k}^{n+1} - T_{j,k}^{n}}{\Delta t}\right) = \frac{1}{r_j}\left(\frac{T_{j-1,k} - T_{j+1,k}}{2\Delta r}\right) + \frac{T_{j-1,k} - 2T_{j,k} + T_{j+1,k}}{\Delta r^2} + \frac{T_{j,k-1} - 2T_{j,k} + T_{j,k+1}}{\Delta z^2}.$$

With $N_j = N_{rad}-1$, $N_k = N_{axl}-1$ and N_{rad} and N_{axl} the number of radial and axial zones, respectively, the index j spans the radial (r) coordinate cells $0 \leq j \leq N_j$ and k the axial (z) cells $0 \leq k \leq N_k$.

This equation can be rearranged into the form

$$\begin{aligned} T_{j,k}^{n+1} - T_{j,k}^{n} &= D\frac{\Delta t}{2r_j\Delta r}(T_{j-1,k} - T_{j+1,k}) \\ &\quad + D\frac{\Delta t}{\Delta r^2}(T_{j-1,k} - 2T_{jk} + T_{j+1,k}) \\ &\quad + D\frac{\Delta t}{\Delta z^2}(T_{j,k-1} - 2T_{jk} + T_{j,k+1}) \end{aligned}$$

and further manipulated into the implicit form

$$\begin{aligned} &-D\frac{\Delta t}{\Delta r^2}\left(1 - \frac{\Delta r}{2r_j}\right)T_{j-1,k}^{n+1} + \left[1 + 2D\frac{\Delta t}{\Delta r^2}\left(1 + \frac{\Delta r^2}{\Delta z^2}\right)\right]T_{j,k}^{n+1} \\ &-D\frac{\Delta t}{\Delta r^2}\left(1 + \frac{\Delta r}{2r_j}\right)T_{j+1,k}^{n+1} - D\frac{\Delta t}{\Delta z^2}\left(T_{j,k-1}^{n+1} + T_{j,k+1}^{n+1}\right) = T_{j,k}^{n}. \end{aligned} \tag{10.29}$$

Expanded for all j and k this equation forms a matrix equation which can be solved for the two-dimensional temperature field T_{jk}. However, its coefficient matrix will not be tridiagonal, although it will be sparse. A more efficient computational sequence can be used in which only tridiagonal matrices need be handled. As we have noted previously in Sect. 3.7, this offers significant advantages. Known as the *alternating direction implicit* or ADI method, it proceeds as follows.

We first notice that at a selected axial position k (axial cross-section), Eq. 10.29 can be used to construct a tridiagonal matrix equation in terms of the radial temperatures and that at a selected radial position j (cylinder of a given radius) that same equation can be used to construct a tridiagonal matrix equation in terms of the axial temperatures. The ADI method employs a two-step process by which the solution is advanced, first in the radial and then in the axial direction. Each step advances the solution by a time-step $\Delta t/2$ and uses the solution known from the previous half-step. The two equations to be solved are as follow:

Step 1: Radial profile for a given axial segment (given k)

$$-D\frac{\Delta t/2}{\Delta r_j^2}\left(1+\frac{\Delta r_j}{2r_j}\right)T_{j-1,k}^{n+1/2}+\left(1+2D\frac{\Delta t/2}{\Delta r_j^2}\right)T_{j,k}^{n+1/2}$$

$$-D\frac{\Delta t/2}{\Delta r_j^2}\left(1-\frac{\Delta r_j}{2r_j}\right)T_{j+1,k}^{n+1/2}=T_{j,k}^n+D\frac{\Delta t/2}{\Delta z}\left(\delta z^2\right)^n, \tag{10.30}$$

where

$$\begin{aligned}\delta z^2 &= T_{j,k-1}-2T_{j,k}+T_{j,k+1} && 1<k<N_k\\ &= -T_{j,0}+T_{j,1} && k=0\\ &= T_{j,(N_k-1)}-T_{j,N_k} && k=N_k\end{aligned}$$

and is known from the previous time step. Boundary conditions are introduced as follows.

$j=0$:

$$\left(1+2D\frac{\Delta t/2}{\Delta r_0^2}\right)T_{0,k}^{n+1/2}-D\frac{\Delta t/2}{\Delta r_0^2}\left(1-\frac{\Delta r_0}{2r_0}\right)T_{1,k}^{n+1/2}$$

$$=T_{j,0}^n+D\frac{\Delta t/2}{\Delta r_0^2}\left(1+\frac{\Delta r_0}{2r_0}\right)T_{-1,k}^{n+1/2}+D\frac{\Delta t/2}{\Delta z}\left(\delta z^2\right)^n.$$

$j=N_j$:

$$-D\frac{\Delta t/2}{\Delta r_{N_j}^2}\left(1+\frac{\Delta r_{N_j}}{2r_{N_j}}\right)T_{(N_j-1),k}^{n+1/2}+\left(1+2D\frac{\Delta t/2}{\Delta r_{N_j}^2}\right)T_{N_j,k}^{n+1/2}$$

$$=T_{N_j,k}^n+D\frac{\Delta t/2}{\Delta r_{N_j}^2}\left(1-\frac{\Delta r_{N_j}}{2r_{N_j}}\right)T_{+1,k}^{n+1/2}+D\frac{\Delta t/2}{\Delta z}\left(\delta z^2\right)^n.$$

Step 2: Axial profile for a given radial cylinder (given j)

$$-D\frac{\Delta t/2}{\Delta z^2}T^{n+1}_{j,k-1} + \left(1 + 2D\frac{\Delta t/2}{\Delta z^2}\right)T^{n+1}_{j,k} - D\frac{\Delta t/2}{\Delta z^2}T^{n+1}_{j,k+1}$$

$$= T^{n+1/2}_{j,k} + D\Delta t/2\left(\frac{\delta r}{\Delta r} + \frac{(\delta r^2)}{\Delta r^2}\right)^{n+1/2}, \qquad (10.31)$$

where

$$\begin{aligned}
\delta r &= \frac{T_{j,k-1} - T_{j,k+1}}{2} && 1 < j < N_j \\
&= T_{0,k} - T_{1,k} && j = 0 \\
&= T_{(N_j-1),k} - T_{(N_j),k} && j = N_j \\
\delta r^2 &= T_{j-1,k} - 2T_{j,k} + T_{j+1,k} && 1 < j < N_j \\
&= T_{0,k} - T_{1,k} && j = 0 \\
&= T_{(N_j),k} - T_{(N_j),k} && j = N_j
\end{aligned}$$

and are known from the previous full time step. Boundary conditions are introduced as follows.
$k = 0$:

$$\left(1 + 2D\frac{\Delta t/2}{\Delta z^2}\right)T^{n+1}_{j,0} - D\frac{\Delta t/2}{\Delta z^2}T^{n+1}_{j,1}$$

$$= T^{n+1/2}_{j,0} + D\frac{\Delta t/2}{\Delta z^2}T^{n+1}_{j,-1} + D\Delta t/2\left(\frac{\delta r}{\Delta r} + \frac{(\delta r^2)}{\Delta r^2}\right)^{n+1/2}. \qquad (10.32)$$

$k = N_k$:

$$-D\frac{\Delta t/2}{\Delta z^2}T^{n+1}_{j,(N_k-1)} + \left(1 + 2D\frac{\Delta t/2}{\Delta z^2}\right)T^{n+1}_{j,N_k}$$

$$= T^{n+1/2}_{j,N_k} + D\frac{\Delta t/2}{\Delta z^2}T^{n+1/2}_{j,+1} + D\Delta t/2\left(\frac{\delta r}{\Delta r} + \frac{(\delta r^2)}{\Delta r^2}\right)^{n+1/2}. \qquad (10.33)$$

and are known from the results of the half-time step calculation.

10.2.1 Treatment of Boundary Conditions

Boundary conditions are defined at the inner and outer radial surfaces $r = r_i$ and $r = r_o$ and at each end of the axial span $z = 0$ and $z = L$ where L is the length of the axial section. Across each of these surfaces the metal exchanges heat with its interfacing medium. At each radial surface the metal is in contact with a gas or fluid while at each axial boundary the metal is in contact with some interfacing metal body to which it is attached, usually rigidly but sometimes via a flexible attachment—a valve, pump, heat exchanger, tank, boiler component, turbine casing or flange, etc.

Denote by T_{bci} and T_{bco} the temperatures at the left (inner) and right (outer) radial boundary surfaces and by T_{bcl} and T_{bcr} the temperatures at the left and right axial boundary surfaces. We will assume Robin boundary conditions and convective heat transfer at the radial boundary surfaces, with heat transfer coefficients α_i and α_o, across a surface normal to the r direction. With λ the thermal conduction coefficient for the wall material, heat transfer across the inner and outer radial surfaces gives, for the kth axial segment,

$$\alpha_i(T_{0,k} - T_{bci}) = -\lambda \left.\frac{\partial T}{\partial r}\right|_{r=r_i} = -\lambda \frac{T_{-1,k} - T_{1,k}}{2\Delta r} \tag{10.34}$$

and

$$\alpha_o(T_{bco} - T_{N_j,k}) = -\lambda \left.\frac{\partial T}{\partial r}\right|_{r=r_o} = -\lambda \frac{T_{(N_j-1),k} - T_{+1,k}}{2\Delta r}. \tag{10.35}$$

At the axial boundaries heat may be transferred by convection to and from interfacing components with heat transfer coefficient α_e. If transfer is by conduction the corresponding factor is the material conductivity λ.

$$\lambda(T_{j,0} - T_{bcl}) = -\lambda \left.\frac{\partial T}{\partial z}\right|_{z=0} = -\lambda \frac{T_{j,1} - T_{j,-1}}{2\Delta z} \tag{10.36}$$

and

$$\lambda(T_{j,N_k} - T_{bcr}) = -\lambda \left.\frac{\partial T}{\partial z}\right|_{z=L} = -\lambda \frac{T_{j,+1} - T_{j,(N_k-1)}}{2\Delta z}. \tag{10.37}$$

The temperatures denoted by T_{-1} and T_{+1} are notional temperatures since their indices lie outside the span of the corresponding cell counts. From Eqs. 10.34–10.37 they can be replaced by the real gas/fluid temperatures and the left and right end metal temperatures in a form that includes the interfacial heat transfer coefficient α.

$$T_{-1,k} = T_{1,k} - \frac{2\alpha\Delta r}{\lambda}(T_{0,k} - T_{bci}) \tag{10.38}$$

and

$$T_{+1,k} = T_{(N_j-1),k} - \frac{2\alpha\Delta r}{\lambda}(T_{N_j,k} - T_{bco}) \tag{10.39}$$

at the radial boundaries and

$$T_{j,-1} = T_{j,1} - 2\frac{\Delta z}{\lambda}(T_{j,0} - T_{bcl}) \tag{10.40}$$

and

$$T_{j,+1} = T_{j,(N_k-1)} - 2\frac{\Delta z}{\lambda}(T_{j,N_k} - T_{bcr}) \qquad (10.41)$$

at the axial boundaries.

Equations 10.30 and 10.31 expanded for all j and k together with the boundary conditions of Eqs. 10.38–10.41 construct two tridiagonal matrices which can be solved for the two-dimensional temperature field T_{jk} using the ADI method, given the boundary condition temperatures T_{bci}, T_{bco}, T_{bcl} and T_{bcr}.

The application of this method to the calculation of the temperature field in a turbine shaft is described in Chap. 14.7. In anticipation, it is to be noted that the preceding discussion treated a cylindrical wall with heat transfer across both the inner and outer surfaces. The turbine shaft to be discussed later is a solid cylinder to which this analysis cannot be applied directly as the radius r is measured from the centre of the cylinder and can be zero. The simplest approach, which incurs negligible error, is to exclude the region about $r = 0$ by treating the shaft as a small bore hollow cylinder. This will often be true as many turbine shafts are manufactured with a small hollow core through which steam may be admitted for warming of the shaft.

10.2.2 Applicability of the Procedure

This procedure can be applied to the calculation of time-varying temperatures throughout thick components having regular geometries. It is not suitable for the *exact* treatment of temperature fields within components having complex boundaries such as turbine casings and complex valve chests. For these, the method of *finite element analysis* (FEA) is more suitable and is widely used. However FEA is not at all well suited to the calculations needed for or the objectives of real-time plant simulation. The preceding method is quite suitable to this purpose. Given that most components such as pipes and tubes, simple valve casings, simple vessels etc do not need finely resolved temperature profiling, a simpler procedure deriving directly from a consideration of a simple solid control volume can be used to develop an efficient calculation procedure. The following section considers such a procedure for the specific case of two interior radial temperatures within a conducting wall in contact with a heat transferring fluid or gas on both sides.

10.3 A Simplified Procedure for Two Radial Zones

The following figure depicts a cross-section through a cylindrical vessel or pipe and identifies the division of the radial cross-section into two zones centered about the radial distances r_{mi} and r_{mo} (Fig. 10.4).

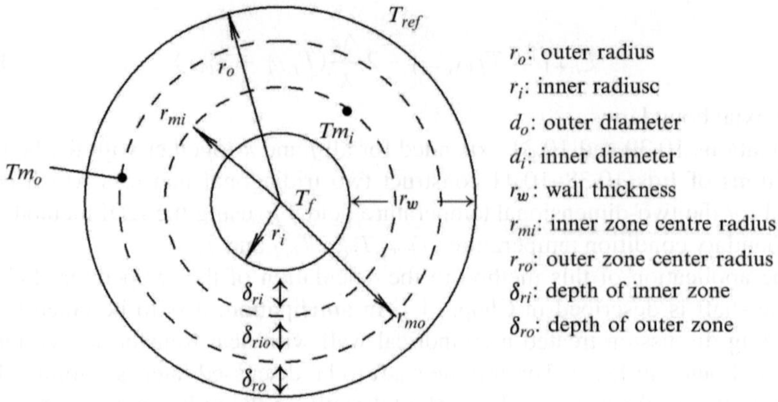

Fig. 10.4 Pipe cross-section and temperatures

Within each zone the temperature is represented by a single temperature defined at the radial centre of the zone, T_{mi} for the inner zone and T_{mo} for the outer.

The determination of the radial locations of the inner and outer zones is imprecise. Defining a parameter θ where $0 < \theta < 1$ and r_b is the location of the boundary between the inner and outer zones, we may write

$$r_b = r_i + \theta r_w = \theta r_o + (1 - \theta) r_i.$$

We then have

$$r_{mi} = r_i + \frac{\theta}{2} r_w \quad \text{and} \quad r_{mo} = r_o - \frac{\theta}{2} r_w.$$

The distance from each surface is then $\delta_{ri} = \frac{\theta}{2} r_w$ for the inner zone and $\delta_{ro} = \frac{1-\theta}{2} r_w$ for the outer zone. The distance between the two zonal centre radii is $\delta_{rio} = 0.5 r_w$.

The mass of tube metal in each zone is the mass of a cylindrical section

$$M_{mi} = \rho_m \pi (r_b^2 - r_i^2) l$$

for the inner zone and

$$M_{mo} = \rho_m \pi (r_o^2 - r_b^2) l$$

for the outer zone, where l is the length of the cell.

10.3.1 Heat Transfer

Heat transfer $\dot{q}_{hxm}$ from a fluid at temperature T_{fl} to a metal wall surface at temperature T_m is given by

$$\dot{q}_{hxm} = \alpha_{hxm} A_{hxm} (T_{fl} - T_m), \tag{10.42}$$

where α_{hxm} is the surface heat transfer coefficient and A_{hxm} is the surface heat transfer area.

This equation applies if T_m is the actual surface metal temperature in contact with the fluid. The surface temperature is not normally available, for calculation or for measurement, and we will therefore consider heat transfer as convection across the surface plus conduction through that section of the wall between the surface and the centre of the interface zone. We will use the actual surface area across which heat is being transferred and define a combined heat transfer coefficient $\bar{\alpha}_{hxm}$ which includes both of these heat transfer mechanisms.

For the inner surface we have

$$\frac{1}{\bar{\alpha}_{hxm_i}} = \frac{1}{\alpha_{hxm_i}} + \frac{\delta_{ri}}{\lambda}$$

and for the outer

$$\frac{1}{\bar{\alpha}_{hxm_o}} = \frac{1}{\alpha_{hxm_o}} + \frac{\delta_{ro}}{\lambda}.$$

λ is the coefficient of thermal conductivity of the wall material.

The inner metal temperature T_{m_i} for the jth cell is calculated from

$$c_m M_{m_i} \frac{\mathrm{d}T_{m_i}}{\mathrm{d}t} = (\dot{q}_{hxm_i} - \dot{q}_{mm}) + \dot{q}_{i,(j-1,j)} - \dot{q}_{i,(j,j+1)} \tag{10.43}$$

and a similar equation for the outer metal temperature.

$$c_m M_{m_o} \frac{\mathrm{d}T_{m_o}}{\mathrm{d}t} = -(\dot{q}_{hxm_o} - \dot{q}_{mm}) + \dot{q}_{o,(j-1,j)} - \dot{q}_{o,(j,j+1)} + \dot{q}_{rad}. \tag{10.44}$$

$\dot{q}_{rad}$ is heat transferred to the outer surface of the pipe or tube by external radiation such as from surrounding luminous gases, or lost as ambient radiation.

Observing the convention that positive heat flow is from the inner to the outer zone, heat conduction per unit length across the tube wall between T_{m_i} and T_{m_o} is

$$\dot{q}_{mm} = \lambda \frac{a_{\perp}}{\delta_{io}} (T_{m_i} - T_{m_o}). \tag{10.45}$$

$a_{\perp}$ is the cross-sectional area per unit length, normal to the radial heat flow. For a single pipe it is approximately $2\pi(r_i + 0.58\, r_w)$.

The longitudinal heat conduction between adjacent cells of the inner zone $\dot{q}_{i,(j-1,j)}$ and $\dot{q}_{i,(j,j+1)}$ is given from

$$\dot{q}_{i,(j-1,j)} = \lambda \frac{a_z}{l} (T_{m_i,(j-1)} - T_{m_i,j}) \tag{10.46}$$

$$\dot{q}_{i,(j,j+1)} = \lambda \frac{a_z}{l} (T_{m_i,j} - T_{m_i,(j+1)}) \tag{10.47}$$

and similar expressions for the outer zone. a_z is the wall cross-sectional area of the zone, normal to the longitudinal (in the axial direction) heat flow.

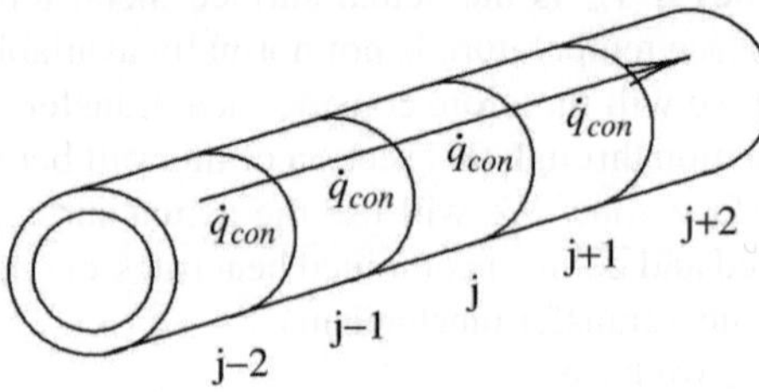

Equation 10.43 for the inner metal temperature may be expanded to

$$\begin{aligned} c_m M_{m_i} \frac{\mathrm{d}T_{m_i}}{\mathrm{d}t} &= f_s\,\bar{\alpha}_{hxm_i}\,A_{hxm_i}(T_{fl} - T_{m_i}) - \lambda \frac{a_\perp l}{\delta_{rio}}(T_{mi} - T_{mo}) \\ &\quad + \dot{q}_{i(j-1,j)} - \dot{q}_{i(j,j+1)} \\ &= -\left[f_s\,\bar{\alpha}_{hxm_i}\,A_{hxm_i} + \lambda \frac{a_\perp l}{\delta_{rio}}\right] T_{m_i} \\ &\quad + f_s\,\bar{\alpha}_{hxm_i}\,A_{hxm_i} T_{fl} + \lambda \frac{a_\perp l}{\delta_{rio}} T_{m_o} \\ &\quad + \lambda \frac{a_z}{l} T_{m_i,(j-1)} - 2\lambda \frac{a_z}{l} T_{m_i,j} + \lambda \frac{a_z}{l} T_{m_i,(j+1)}. \end{aligned} \tag{10.48}$$

f_s is a coefficient introduced to account for deterioration in heat transfer caused by boiler scale or other fouling on the inside surfaces of the tubes. Practical values tend to lie around 0.8–0.9.

Heat transfer from the secondary fluid to the pipe wall outer surface is given by

$$\dot{q}_{hxm_o} = f_f\,\bar{\alpha}_{hxm_o} A_{hxm_o}(T_{ref} - T_{m_o}). \tag{10.49}$$

f_f is a fouling factor which accounts for the deterioration of heat transfer caused by the fouling of the outside surface of the tube or pipe.

Equation 10.44 for the outer metal temperature T_{mo} may be expanded to

$$\begin{aligned} c_m M_{m_o} \frac{\mathrm{d}T_{m_o}}{\mathrm{d}t} &= -f_f\,\bar{\alpha}_{hmx_o}\,A_{hxm_o}(T_{m_o} - T_{ref}) + \lambda \frac{a_\perp l}{\delta_{rio}}(T_{m_i} - T_{m_o})) \\ &\quad + \dot{q}_{rad} + \cdots \\ &= -\left[f_f\,\bar{\alpha}_{hmx_o}\,A_{hxm_o} + \lambda \frac{a_\perp l}{\delta_{rio}}\right] T_{m_o} \\ &\quad + f_f\,\bar{\alpha}_{hmx_o}\,A_{hxm_o} T_{ref} + \lambda \frac{a_\perp l}{\delta_{rio}} T_{m_i} + \dot{q}_{rad} \\ &\quad + \lambda \frac{a_z}{l} T_{mo,(j-1)} - 2\lambda \frac{a_z}{l} T_{mo,j} + \lambda \frac{a_z}{l} T_{mo,(j+1)}. \end{aligned} \tag{10.50}$$

Fig. 10.5 Some pipes can be very thick! (Courtesy of Millmerran Operating Company, Australia)

10.3.2 Organisation of the Equations for Computation

Define for the jth cell

$$\begin{aligned}
cm_i &= c_m\, M_{mi},\\
cm_o &= c_m\, M_{mo},\\
cn &= \lambda\, a_\perp\, l/\delta_{rio},\\
cn_i &= \lambda\, a_z/l,\\
cn_o &= \lambda\, a_z/l,\\
\alpha_{hxm_i} &= f_s\, \bar{\alpha}_{hxm_i}\, A_{hxm_i},\\
\alpha_{hxm_o} &= f_f\, \bar{\alpha}_{hxm_o}\, A_{hxm_o}.
\end{aligned}$$

The defining differential equations for the inner and outer metal temperatures are then,

$$\begin{aligned}
cm_i \frac{\mathrm{d}T_{mi}}{\mathrm{d}t} &= -(\alpha_{hxm_i} + cn)T_{mi} + \alpha_{hxm_i}\, T_{ref} + cn\, T_{mo}\\
&\quad + cn_i\, T_{mi,(j-1)} - 2\, cn_i\, T_{mi} + cn_i\, T_{mi,(j+1)} \qquad (10.51)
\end{aligned}$$

and

$$\begin{aligned}
cm_i \frac{\mathrm{d}T_{mo}}{\mathrm{d}t} &= -(\alpha_{hxm_o} + cn)T_{mo} + \alpha_{hxm_o}\, T_{ref} + cn\, T_{mi}\\
&\quad + cn_o\, T_{mo,(j-1)} - 2\, cn_o\, T_{mo} + cn_o\, T_{mo,(j+1)} + \dot{q}_{rad}. \qquad (10.52)
\end{aligned}$$

Equations 10.51 and 10.52 may be restructured in discrete time and space terms as follows.

$$-\Delta t\, cn_i T_{mi,(j-1)} + [cm_i + \Delta t(\alpha_{hxm_i} + cn + 2\, cn_i)]\, T^{n+1}_{mi,j} - \Delta t\, cn_i T^{n+1}_{mi,(j+1)}$$
$$= cm_i T^n_{mi,j} + \Delta t\, \alpha_{hxm_i}\, T^n_{ref_i,j} + \Delta t\, cn_i\, T^n_{mo,j}.$$

$$-\Delta t\, cn_o T_{mo,(j-1)} + [cm_o + \Delta t(\alpha_{hxm_o} + cn + 2\, cn_o)]\, T^{n+1}_{mo,j} - \Delta t\, cn_o T^{n+1}_{mo,(j+1)}$$
$$= cm_o T^n_{mo,j} + \Delta t\, \alpha_{hxm_o}\, T^n_{ref_o,j} + \Delta t\, cn_i\, T^n_{mi,j} + \Delta t\, \dot{q}_{rad}.$$

Each of these equations may be written more compactly as

$$B_{j,j-1} T^{n+1}_{mi,(j-1)} + B_{j,j} T^{n+1}_{mi,j} + B_{j,j+1} T^{n+1}_{mi,(j+1)} = T^n_{mi,j} + Y^n_j,$$

$$C_{j,j-1} T^{n+1}_{mo,(j-1)} + C_{j,j} T^{n+1}_{mo,j} + C_{j,j+1} T^{n+1}_{mo,(j+1)} = T^n_{mo,j} + Z^n_j,$$

where

$$B_{j,j-1} = -\Delta t \frac{cn_i}{cm_i},$$

$$B_{j,j+1} = -\Delta t \frac{cn_i}{cm_i},$$

$$B_{j,j} = 1 + \frac{\Delta t}{cm_i}[\alpha_{hxm_i} + cn + 2\, cn_i],$$

$$Y_j = \frac{\Delta t}{cm_i}[\alpha_{hxm_i} T^n_{ref_i j} + cn\, T^n_{mo,j}],$$

$$C_{j,j-1} = -\Delta t \frac{cn_o}{cm_o},$$

$$C_{j,j+1} = -\Delta t \frac{cn_o}{cm_o},$$

$$C_{j,j} = 1 + \frac{\Delta t}{cm_o}[\alpha_{hxm_o} + cn + 2\, cn_o],$$

$$Z_j = \frac{\Delta t}{cm_o}[\alpha_{hxm_o} T^n_{ref_o j} + cn\, T^n_{mi,j} + \dot{q}_{rad}].$$

As j is taken over all cells these equations form two tridiagonal matrix equations in all inner and outer metal temperatures. The requisite boundary conditions at both ends of the component (T^n_{BL} at the left and T^n_{BR} at the right hand end) are provided by the appropriate metal temperatures of the interfacing components.

10.3.3 Treatment of Boundary Conditions

Using the inner temperature as an example (the principle applies to both), we have at the left hand end of a branch ($j = 1$)

$$B_{1,1}T_{mi,1}^{n+1} + B_{1,2}T_{mi,2}^{n+1} = T_{mi,1}^{n} + Y_1^n - B_{1,0}T_{BL}^{n}$$

and at the right hand end of a branch ($j = \mathrm{N}$)

$$B_{N,N-1}T_{mi,N-1}^{n+1} + B_{N,N}T_{mi,N}^{n+1} = T_{mi,N}^{n} + Y_n^N - B_{N,N+1}T_{BR}^{n}.$$

The boundary coefficients are

$$B_{1,0} = B_{N,N+1} = \Delta t\, cn_i = \Delta t\, \lambda\, a_{xi}/l.$$

10.3.4 Numerical Example

A Pendant Superheater Tube Bank

The final stage superheater in a large utility boiler is assumed to have the following specification.

Tube OD	54 mm
Tube wall thickness	7 mm
Average tube length	20 m
Number of tube panels	7
Number of tube bundles per panel	10
Number of tubes per bundle	3
Total number of tubes	210

Assuming a full-load steam flow of 350 kg/s through two parallel paths, or 175 kg/s per path, the flow per tube is 0.833 kg/s giving a specific mass flow ϕ of 663.14 kg/s/m^2.

The typical temperature rise across a single tube bank stage is around 25–35 K. Assume the following steam conditions at inlet and outlet of the stage.

Inlet:	240 bar, 520 °C	Enthalpy 3246.4 kJ/kg
Outlet:	238 bar, 560 °C	Enthalpy 3382.6 kJ/kg

The total energy picked up across the SH is then 136.2 kJ/kg × 175 kg/s = 23,835 kJ/s or 4,767 kW per cell if it is assumed that the heat pickup is evenly distributed across, say, five cells. This is not actually the case but will be adequate for illustration purposes.

The heat transfer calculations of the preceding sections combined with the heat transfer coefficient calculations of Sect. 8.2.4 produce the following temperature profiles across the (average) tube wall within the cell, from gas to steam. Assuming a target steam temperature T_s of 560 °C and an appropriate gas flow rate and bank factors we arrive at the following calculated gas (T_g), outer (T_{mo}) and inner (T_{mi}) tube metal temperatures for the transfer of 4,767 kW across the cell:

T_g	T_{mo}	T_{mi}	T_s
803 °C	583.6 °C	579.8 °C	560 °C

A Reheater Tube Bank

Reheater tube banks are located further down the fluegas path than the platen or pendant superheater banks and are therefore exposed to gas at lower temperatures. They are shielded from the furnace, and while receiving non-luminous radiation from the hot gas, they receive little if any direct radiation from the furnace. They must therefore be constructed from a larger number of tubes to increase the available heat transfer area. Being located after the HP turbine, they operate at a lower pressure, and their tubes will have thinner walls of possibly larger diameter. The larger number of larger tubes reduces the steam flow velocity through any individual tube, resulting in a reduced specific mass flow and a correspondingly reduced heat transfer coefficient. However, the increased surface area provided by the larger number of tubes compensates for this, and substantial heat transfer rates are achievable in regions of relatively lower gas temperatures.

The final reheat steam temperature will usually equal or exceed the final main steam temperature. For a drum boiler, the final main and reheat temperatures will often be 540/540 °C. For a supercritical boiler, these will be higher, say 565/587 °C. The final reheater stage must be placed as far up the fluegas path as possible in order to be in contact with the highest gas temperature. This usually results in its placement immediately after the final superheater. It will often be configured in counterflow arrangement to achieve the highest steam temperature.

The target temperature rise across the reheater is the difference between the HP turbine exhaust (cold reheat) and the final (hot) reheat temperatures. For the unit used as the basis for this example, the HP turbine exhaust temperature at full load is around 330 °C, and the final reheat temperature is 587 °C, giving a differential of 257 K. Spreading this equally over three reheat stages requires each stage to pickup around 80–90 K. The final stage reheater is assumed to have the following specification:

Tube OD	67 mm
Tube wall thickness	4 mm
Average tube length	20 m
Number of tube panels	30
Number of tube bundles per panel	12
Number of tubes per bundle	1
Total number of tubes	360 per side

With a target final steam temperature of 587 °C, conditions at inlet and outlet of the final reheater stage are

Inlet:	48 bar, 490 °C	Enthalpy 3412.6 kJ/kg
Outlet:	46 bar, 587 °C	Enthalpy 3638.7 kJ/kg

and flow rate (150 kg/s through each side of the reheat steam path), a pickup of 90 K requires transfer of 6,783 kW from gas to steam across the cell and gives rise to the following temperature profile.

T_g	T_{mo}	T_{mi}	T_s
756 °C	616.2 °C	615.3 °C	587 °C

These numerical results are indicative only as they have been calculated by isolating a single cell in the heat exchanger. As a result, they ignore (a) the influence of temperature profiles along the directions of gas and steam flows and (b) conduction along the tubes. They also ignore any radiant heat transfer that may be present to some degree, either as luminous from the furnace or non-luminous from the hot gas. However, they do indicate the relative temperature differentials across the steam and gas interfaces and the likely differentials through the tube walls.

Chapter 11
Furnaces, Fuels and Heat Release

Furnaces firing a variety of fuels have been used for steam production since the beginning of the industrial era (around 1823 or so). Over the years since, they have developed in size and sophistication to the highly efficient equipment now in operation for power generation, district heating and process steam production.

Boilers may be distinguished one from another by certain characteristics of their design and steaming rates, and by the applications in which they are used. We will now briefly summarise the major types.

11.1 Water-Tube Furnaces

These are steam generators of smaller steaming capacities up to, say, 80 t/h. For steaming rates up to 50 t/h, they can be supplied pre-fabricated as packaged boilers. For sizes greater than this they are usually erected on site (field erected boilers) though some degree of prefabrication may be used.

Water-tube boilers are arranged as two drums, the larger at the highest and the smaller at the lowest elevation of the boiler, connected together by vertical banks of tubes. One group of tubes encloses and forms the walls of the furnace chamber. These tubes, welded together to form a gas-tight enclosure, form so-called *membrane walls* and allow the furnace to be operated at a pressure greater than atmospheric. This simplifies operation of the furnace which then requires only a single forced draft fan for the supply of combustion air. Two vertical tube banks, referred to as the steam generating banks, are connected between the drums. They are arranged across the flue gas leaving the furnace, one upstream of the other in the gas flow. In smaller boilers the majority of steam is formed within these banks. These boilers have only natural circulation.

The floor of the furnace may be flat or sloping at a slight angle. The lower drum is always full of water and the floor tubes are exposed to the furnace. The upper drum is used for steam/water separation. Air and fuel—usually gas or light fuel

P. O'Kelly, *Computer Simulation of Thermal Plant Operations*,
DOI 10.1007/978-1-4614-4256-1_11,

Fig. 11.1 A package boiler before shipment (Courtesy of Foster-Wheeler Corp.)

Fig. 11.2 Sectional view of a D-type boiler

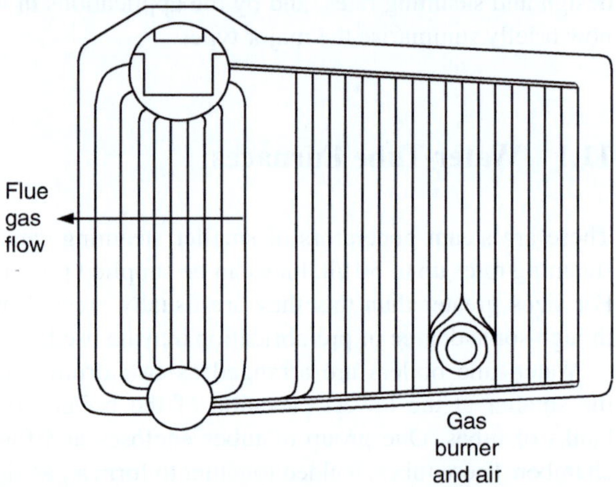

oil—are introduced into the furnace chamber through a single burner and flue gas exits across the steam generating tube banks. The downstream tubes of these banks, exposed to a lower gas temperature, act as a convectively heated downcomer bank and receive insufficient heat to form steam. The upstream tubes however, being in a higher temperature gas stream, do develop steam. The water-tubes forming the furnace walls act as risers, absorbing radiant and convective heat from the furnace and developing a part of the boiler's steam production (Fig. 11.1).

Because of their characteristic geometry, this style of packaged boiler is often referred to as a D-type boiler (Fig. 11.2). They are compact and quick to install. They are frequently found in marine applications due to their simplicity, compact form factor and reliability. In order to save space, marine boilers often operate at relatively high heat loadings and naval vessels' boilers will be designed for rapid load change rates.

Thanks to the small metal and water-steam masses they contain, simulation of packaged boilers presents few problems and from an operational point of view they can usually be adequately simulated using simple lumped parameter models.

11.2 Radiant Furnaces

Water-tube and D-type boilers rely principally on convective heating of the steam generating tubes for steam production. The radiative heat transfer to the furnace wall tubes is the lesser part of the heat transfer and steam production. Their larger cousins, the industrial bi-drum and utility boilers, relay principally on radiative heat transfer in the furnace chamber for steam production. Modern boilers with steam production capacities in excess of 250 t/h have dispensed with the steam-generating convective bank entirely and obtain in excess of 90 % of their steam production from radiative transfer alone. These boilers are generically lumped under the label "radiant furnace boilers". We will now look at them in more detail as their simulation will usually require a more detailed treatment.

11.2.1 Bi-drum Boilers

The industrial bi-drum boiler preserves the dual drum and steam generating tube arrangement of the D-type boiler but adds a substantial furnace, downcomer and waterwall tube banks to increase the steaming capacities to cover the range 120 t/h to around 450 t/h. Bi-drum boilers are available for electricity generation producing upwards of 1,200 t/h of steam though more often boilers of this capacity will be either single drum or once-through configurations (Fig. 11.3).

At the smaller end of the scale these boilers will operate at up to 5 MPa and around 45 % of the steam generated by the boiler will be produced in the steam generating tubes. This reduces to around 10 % for the higher steaming capacities and pressures, up to 15 MPa. The benefit of the bi-drum arrangement is the reduction in furnace size contributed by the convective steam-generating tube bank. Beyond a certain capacity this advantage is eroded by the increasing costs of two drums and the heavy tube bank and the single drum or once-through radiant furnace type becomes more economical.

Some industrial boilers of this class feature a spray condenser. This is a recuperative feedwater heater arranged as the final element in the feedwater path, upstream of the drum or, if provided, the economiser. It has an open steam line to the drum and provides a continuous supply of condensate to be used for steam desuperheating sprays. The system is self-regulating as variations in condensate

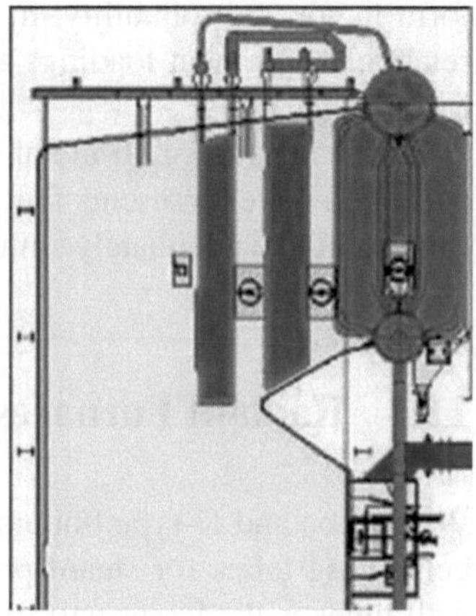

Fig. 11.3 Bi-drum boiler: superheaters and steam generating bank

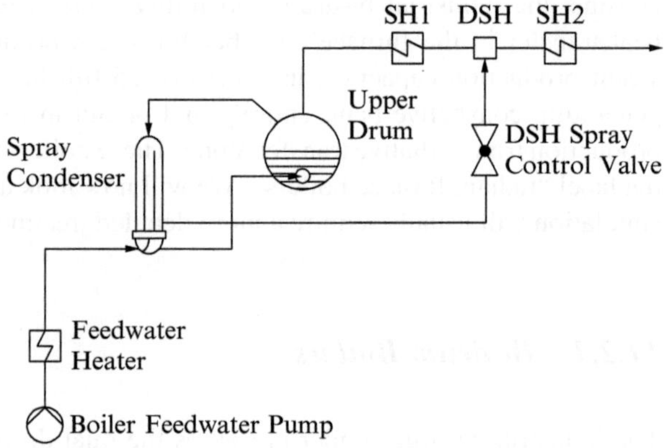

Fig. 11.4 General arrangement of a spray condenser

level vary the tube area available for condensation. The pressure differential between the spray condenser, effectively drum pressure, and the desuperheater provides the motive force for the spray flow. This is an efficient system as all of the energy of the steam bled from the drum is recovered, in part in the feedwater with the balance returned to the steam flow at the desuperheater (Fig. 11.4).

11.2.2 Furnaces Used for Power Generation

These are the largest steam generating units, with steam generation rates in excess of 3,500 t/h and are built in any of several standard configurations. The most common arrangements are shown by the elevation drawings on the following pages which show, in sequence, the tower arrangement and two variants of the folded arrangement, one with and one without a split flue gas backpass. In each case the furnace walls are totally enclosed by vertical riser tubes into which burners are set in each corner of the furnace or in the front and back walls. The corner-fired

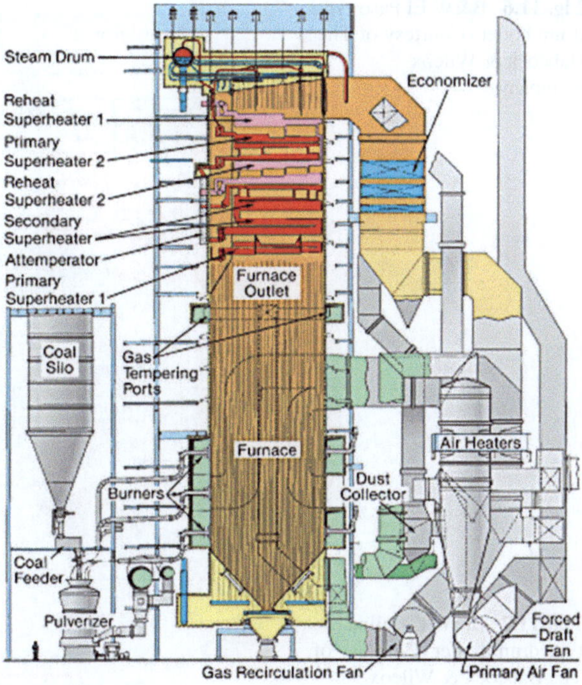

Fig. 11.5 B&W tower type drum boiler (Courtesy of The Babcock & Wilcox Company)

furnace usually provides means for mechanically tilting the burner ranges from the horizontal and a maximum deflection angle of 30° in either direction is typical.

Figure 11.5 shows a tower-type single drum boiler. In this design, the fluegas exits the furnace and continues vertically through a series of cross-flow superheating and reheating tube banks before being turned downwards to the economiser(s) and airheater(s). Being horizontal, the superheaters and reheater tubes are easily drained and are terminated at inlet and outlet by headers. Not having to negotiate the characteristic deflector or "nose" of a furnace with a backpass, the fluegas flow in a tower furnace can maintain a straight line into the superheater zone.

Figure 11.6 shows a drum boiler of the Babcock & Wilcox El Paso type with a single fluegas pass. In this design the fluegas exits the furnace through a 90° turn which includes the secondary and possibly tertiary superheaters arranged as pendant tube banks. There follows a short duct containing more superheater and reheater banks. The roof of the furnace and of this short duct section is enclosed by a small number of tubes but receive significant radiant heat from the furnace chamber. The fluegas is again turned downwards through 90° towards the remaining reheaters and primary superheater bank(s) after which it moves on to the economiser(s) and airheater(s). Being vertical, the pendant superheater tubes are not easily drained and precautions must be taken during startup as individual tubes may be blocked by condensation accumulated during the preceding shutdown.

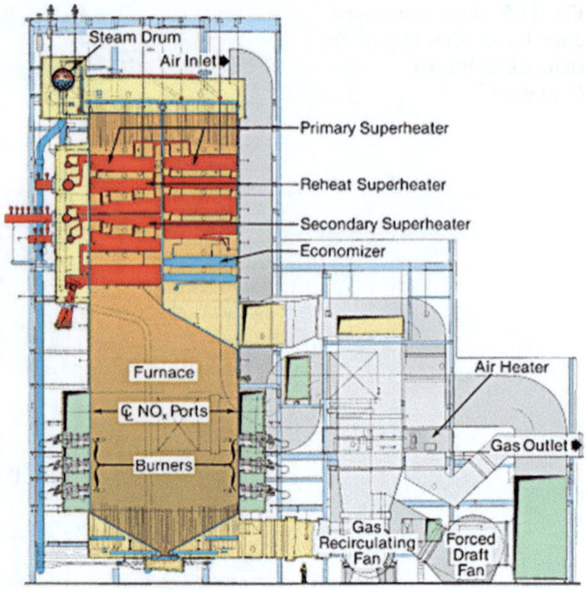

Fig. 11.6 B&W El Paso type drum boiler (Courtesy of The Babcock & Wilcox Company)

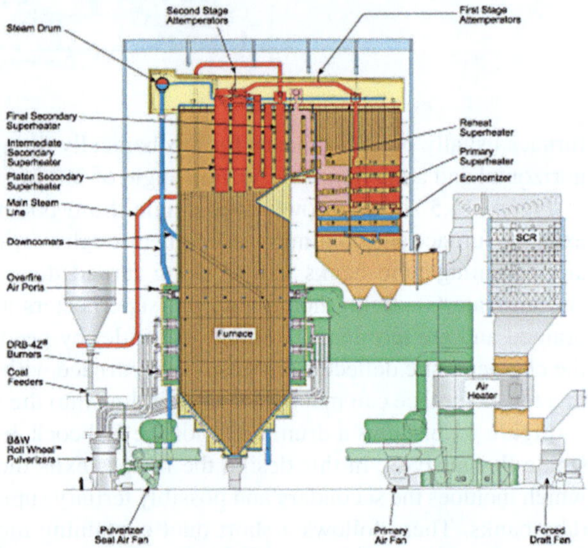

Fig. 11.7 B&W Carolina type drum boiler (Courtesy of The Babcock & Wilcox Company)

Figure 11.7 shows a drum boiler of the Babcock & Wilcox Carolina type with a split fluegas pass. As in the previous design, the fluegas exits the furnace through a 90° turn which includes the secondary and possibly tertiary superheaters, again arranged as pendant tube banks. The general arrangement is the same as the preceding type, but the fluegas is split into front and back ducts occupied on one side by the remaining banks of the reheater and on the other by the primary superheater.

In each case the furnace is defined to terminate at the lower surface of the first radiant superheating stage which, being immediately above and visible to the flame, is exposed to the most intensive radiant heat.

Steam flows from the drum (or water separator in a once-through configuration) to the primary superheater which is located at the point in the fluegas path, ahead of the economiser, where the gas-steam temperature differential is minimum. Thereafter, the steam returns to the secondary superheater, where the gas-steam temperature differential is maximum, before passing through the roof tubes to the tertiary and final superheater tube banks. Along the way some steam will pass through hanger tubes, used as part of the boiler support structure. Other tubes may be used to screen lower-temperature sections of the boiler from direct furnace radiation and are known, logically, as screen tubes. In summary, the arrangement of boiler steam tubing is designed to capture the heat released in the furnace to the greatest extent possible.

The heat released by the combustion of the fuel is distributed as

(a) Radiant heat to the waterwalls, superheaters, roof tubes and possibly reheaters.
(b) Convective heat transfer to the waterwall tubes.
(c) Heat used to raise the incoming air to flame temperature.
(d) Heat leaving the furnace in the flue gas.
(e) Heat used to evaporate the moisture in the fuel and combustion air.
(f) Waste heat leaving the furnace as hot ash.
(g) Heat leaving the furnace as hot water vapour.
(h) Heat losses to the environment by radiation and convection from the boiler structure.

The various radiant components are proportional to the fourth power of the absolute flame or gas temperature. Convective and other components are proportional to the local gas temperatures.

11.3 Fuels and Their Heating Values

The combustion process is a rapid and highly exothermic oxidation of the fuel material. Most industrial fuels exploit the oxidation of carbon, hydrogen and sulphur as almost all fuels are available as elemental carbon (such as coal) and various forms of hydrocarbons. Hydrocarbon fuels are available in both liquid and gaseous forms. Sulphur is often present, as an undesirable impurity. Biomass fuels add a range of complex combustible organic compounds, again made up primarily of carbon, nitrogen, hydrogen and oxygen atoms with a range of inorganic metal and non-metal compounds, including sulphur.

From a utilisation point of view the most important characteristics of fuels are their heating or calorific value and ash and moisture content. The composition and behaviour of post-combustion waste products are of mainly maintenance and environmental importance and of themselves have little bearing on plant operational simulation. However, they become important when plant operations are influenced by the need to minimise or avoid their effects.

The heating or calorific value of a fuel (CV) is usually determined experimentally using a calorimeter. The typical heating values of several common fuels are widely published and can be taken as appropriate values for simulation. In the case of liquid fuels, the given CV applies to the gaseous form of the fuel and does not include allowance for evaporation prior to combustion. For solid fuels it does not allow for the heat needed to bring the fuel to ignition temperature.

A distinction is made between gross and net heating value of the fuel. These are also referred to as the higher (or upper) and lower heating values. The higher heating value includes the latent heat of the moisture in the fuel as tested or which may be produced from combustion. For example, the combustion of methane proceeds according to

$$CH_4 + 2\,O_2 \longrightarrow CO_2 + 2\,H_2O$$

and the heat content of this moisture is included in the higher CV figure quoted for methane but not in the lower CV. Most solid fuels contain both ash and moisture as delivered. Liquid and gaseous fuels contain only small to vanishingly small amounts of both, if any at all. Gaseous fuels may be treated as ash and water free.

The CV of a fuel can be calculated from its given composition by adding the weighted CVs of the individual components. The following table [96] summarises the heating values of some common fuels.

	Chemical symbol	Net CV kJ/m^3	Net CV kJ/kg
Sulphur	S	–	9,276
Carbon monoxide	CO	12,633	10,103
Hydrogen	H_2	10,784	119,986
Methane	CH_4	35,885	50,015
Iso-butane	C_4H_{10}	123,053	45,624
Hydrogen sulphide	H_2S	23,413	15,248

11.3.1 Solid Fuels

The most widely used solid fuel is coal, commonly available as either lignite, otherwise known as brown coal, or anthracite, also known as black coal. The fuel is introduced into the furnace either by being blown in as a finely ground powder (pulverised fuel, widely known simply as "pf") or in non-pulverised form on a slowing moving metal chain grate travelling across the bottom of the furnace. Non-pulverised coal is used in the nugget form as delivered from the mine or supplier.

Combustion air is blown upwards from underneath the grate. Chain grate stokers are also used for several biomass fuels, such as sugar cane bagasse.

In modern high-efficiency furnaces, coal is always introduced to the furnace in pulverised form. Raw coal is reduced in size from the nuggets (average size 1–2 cm diameter) as delivered from the mine to a fine powder by grinding in pulverising mills. The particle size of pulverised coal is described statistically in terms of the fraction of particles which pass through a sieve of defined mesh or screen size. Individual countries use differing mesh standards and Steinmuller [9] quotes a good summary of these. The maximum particle size (max size) passing through the screen is quoted in millimeteres.

German DIN		UK Standards		USA ASTM	
Screen no	Max size	Screen no	Max size	Screen no	Max size
6	1.000	16	1.000	18	1.000
16	0.375	52	0.295	60	0.250
30	0.200	85	0.198	100	0.149
50	0.104	150	0.104	170	0.088
100	0.060	300	0.053	325	0.044

A distinction is made between the as-delivered and water-and-ash-free (*waf*) conditions. Both the as-delivered and *waf* CVs are determined by laboratory analysis, according to appropriate Standards. Physically, the calorific value of the coal is determined from its constituent components, also determined from laboratory analysis. Coal consists of volatiles (combustible hydrocarbons which rapidly evaporate and escape from the coal particle at relative low temperatures), ash, water and other combustible components, predominantly fixed carbon, sulphur and hydrogen. Using the quadratic form quoted by Scholz [96] the calorific value of *waf* coal is given as a function of the volatile fraction x (by weight) by the formula

$$CV_{waf} = 34430 + 123x - 3.7x^2.$$

There is usually no pre-combustion preparation of coal which is burnt in its as-delivered composition. Some drying of the coal is achieved in the pulverisers or mills. This is especially the case for brown coal which is delivered with up to 70 % moisture content.[1] The as-delivered CV is computed from

$$CV = CV_{waf}\left(1 - \frac{ash\,\%}{100}\right)\left(1 - \frac{H_2O\,\%}{100}\right),$$

[1]It should be noted that the properties of lignite can vary over wide ranges; for example, moisture content can be less than some forms of anthracite.

where *ash* % and H_2O % are the percentage (by weight) fractions of *ash* and moisture. The mass fraction of moisture in the fuel is relevant only to coal and will be known from the coal proximate analysis.

The analysis of the composition of coal is available in either of two forms. An *ultimate* analysis resolves the coal into its constituent chemical elements, the principal of which are carbon, hydrogen, sulphur, oxygen and nitrogen. The fractions of each component are based on the equivalent water-and-ash-free coal. This is a time-consuming procedure and is often replaced by a *proximate* analysis which treats the coal as composed of four constituents, fixed carbon, volatiles, ash and moisture. This is the form most commonly used for operational purposes.

11.3.2 Liquid Fuels

Liquid fuels fall into two major categories, natural occurring and synthetic. Naturally occurring fuels are derived almost exclusively from crude oil by a process of selective distillation. Principally on the basis of kinematic viscosity, they are subdivided into the categories extra light (EL), light (L), medium (M) and heavy (H). Modern power plants use only EL—in the form of diesel—and heavy. Heavy oils were frequently used in marine boilers and, being kept in ships' bunkers previously used for coal, were and still are classified as bunker oils and sub-classified as "A", "B" and "C". These classifications correspond to the ASTM[2] classifications Number 5 and Number 6 oils.

The CV of any oil may be determined from its density and sulphur content by the formula [96]

$$CV = 54{,}040 - 13.29\rho_{15} - 29.31\gamma_S \quad [\text{kJ/kg}],$$

where ρ_{15} is the oil density at 15 °C and γ_S is the mass fraction (by weight) of sulphur. The dependence on density can be explained in terms of the relative representation in the oil of hydrogen (contributing to a higher CV but lower density) and carbon (contributing to a lower CV but higher density) components.

Applying this formula to diesel (EL) and bunker oils (H) gives

Oil	Density kg/m3	Sulphur %	CV kJ/kg	Guideline CV kJ/kg per DIN 51603
Diesel	840.0	0.20	42,870	42,000
Bunker oils	970.0	3.5	41,046	39,800

[2] American Society for Testing and Materials.

Actual CVs of fuel oils show significant variability and, in the absence of specific figures for specific oils, these are guideline values only. However they will be found adequate for most practical purposes.

Synthetic liquid fuels are usually produced as combustible byproducts of chemical processes and tend to be consumed as fuels within the same process or at least on the same site. A well-known example is black liquor, produced as a byproduct of the wood pulping process.

Although they contain no or negligible ash, heavy fuel oils produce harmful residues as a product of combustion. These include metal oxides and other compounds present in the fuels as delivered. They are of little consequence to dynamic simulation analysis but, having strong fouling and corrosion consequences, they are of interest to plant maintenance and to the environment.

11.3.3 Gaseous Fuels

Natural gas is mostly methane and has a CV of around 42,000 kJ/kg (working range 32–47,000 kJ/m^3). Industrial boilers frequently use combustible byproducts as fuel. For example, a steelworks will generate blast furnace gas (bfg) and coke oven gas (cog) as process byproducts and consume them on site as fuels in boilers producing electric power and process steam. Typical compositions of bfg and cog are, together with their CVs,

Component	CV kJ/m^3	bfg %	cog %
H_2	10,784	4.8	59.7
CO	12,633	22.3	4.5
CH_4	35,885	0.0	25.4
Other hydrocarbons	50,000	0.0	3.2
N_2		49.5	3.2
CO_2		20.4	1.6
Water vapour		3.0	3.0
Gas CV kJ/m^3		3,335	17,270

The CV of bfg is lower than the lower limit of self-sustaining ignition of a fuel, usually taken at around 5,000 kJ/kg. For this reason, bfg must be fired together with another fuel to ensure sustained combustion and flame stability. This can be either mixed in the same burner or introduced as a pilot or secondary fuel.

The following table summarises the most commonly used gaseous fuels and their principal combustion properties. CO_2 is not a fuel but is included as a convenient place to list its physical properties.

	Chemical symbol	Molar mass kg/kmol	Molar norm vol m^3/kmol	Norm density kg/m^3	Gas constant kJ/kg K
Carbon monoxide	CO	28.0104	22.400	1.2505	0.29665
Hydrogen	H_2	2.01578	22.428	0.0899	4.12723
Methane	CH_4	16.043	22.360	0.7175	0.51701
Iso-butane	C_4H_{10}	58.123	21.550	2.6970	0.13754
Hydrogen sulphide	H_2S	34.076	22.192	1.5355	0.24158
Oxygen	O_2	31.9988	22.403	1.2504	0.25958
Nitrogen	N_2	28.0134	22.403	1.2504	0.29666
Carbon dioxide	CO_2	44.0098	22.261	1.9770	0.18763

11.3.4 Combustion Calculations

The basic combustion equations are

$$C + O_2 \longrightarrow CO_2,$$

$$H_2 + \frac{1}{2}O_2 \longrightarrow H_2O,$$

$$CO + \frac{1}{2}O_2 \longrightarrow CO_2,$$

$$S + O_2 \longrightarrow SO_2,$$

$$H_2S + \frac{3}{2}O_2 \longrightarrow S_2 + H_2O.$$

The following table summarises the exothermic heat released by oxidation of various substances.

	Gross heat release kJ/kg	Net heat release kJ/kg
C	32,823	32,823
CH_4	55,614	50,120
H_2	142,302	120,230
CO	10,124	10,124
C_4H_{10}	49,508	45,716
H_2S	16,536	15,243

In each of these reactions 1 kmol of each reagent reacts together to produce 1 kmol of the product. Thus

$$(1\texttt{kmol})\,C + (1\texttt{kmol})\,O_2 \longrightarrow (1\texttt{kmol})\,CO_2.$$

Inserting molar masses into the combustion equations allows us to write for the combustion of carbon,

$$12.011\,C + 31.9988\,O_2 = 44.0098\,CO_2$$

or, for 1 kg of carbon,

$$C + 2.6641\,O_2 = 3.6641\,CO_2.$$

The same exercise produces, for methane:

$$CH_4 + 3.9891\,O_2 = 2.7432\,CO_2 + 2.2459\,H_2O$$

for carbon monoxide:

$$CO + 0.5712\,O_2 = 1.5712\,CO_2$$

for sulphur:

$$S + 0.9981\,O_2 = 1.9981\,SO_2$$

for hydrogen:

$$H_2 + 7.9370\,O = 8.9370\,H_2O.$$

In each case the sums of masses on both sides of the equation balance, as required by conservation of mass.

The oxygen required for combustion is provided by air, whose composition, on a dry percentage basis, is set out by the next table.

Component %	O_2	N_2	Ar	Ne	CO_2
By volume	20.948	78.084	0.934	0.002	0.032
By weight	23.151	75.510	1.289	0.001	0.049

The complete combustion of carbon in air requires 2.6641/0.23151 or 11.507 kg of air. Similarly the complete combustion of methane requires 3.9891/0.23151 or 17.23 kg of air. The stoichiometric mass ratio R_S is the minimum mass of air needed for complete combustion of 1 kg fuel. The stoichiometric mass ratio for carbon is therefore 11.507 and for methane 17.23. Stoichiometric ratios may be expressed in terms of air volumes (m^3) by using the volumetric fractions of air composition.

These ratios apply to dry air. Air always contains moisture to a degree defined by its relative humidity. The presence of moisture reduces air density, calculated as follows.

$$\frac{1}{\rho_{moist}} = \frac{1 - x_{H_2O,air}}{1.2930} + \frac{x_{H_2O,air}}{0.8038}.$$

The density of dry air under standard conditions[3] is 1.2930 kg/m^3 and of H_2O vapour under the same conditions 0.8038 kg/m^3. The mass fraction of moisture in the air $x_{H_2O,air}$ is given from

$$x_{H_2O,air} = 0.622\frac{\psi p_s}{p - \psi p_s},$$

where ψ is the fractional relative air humidity and p_s is the saturation pressure at ambient temperature, calculated from the Tetens approximation [52],

$$p_s = 0.611 \exp\left(17.2694\frac{(T_{amb} - 273.16)}{(T_{amb} - 35.86)}\right)$$

with T_{amb} in degrees K.

The moist air stoichiometric ratio is obtained from the dry air ratio by multiplying by the ratio of dry to moist air densities. A higher mass flow of moist air is required because the lower density of the moist air reduces the number of oxygen molecules per kilogram of air.

11.3.5 Stoichiometric Ratios for Solid and Liquid Fuels

Gaseous fuels tend to be either simple components (H_2, CH_4, CO, etc.) or mixtures of simple components. Solid and liquid fuels tend to be complex blends of organic and inorganic substances, both volatile and non-volatile. The stoichiometric ratio for these fuels cannot be readily derived from a simple chemical decomposition though some useful approximations can be and often are made in this direction. Add to this the known geographical variability of common fuels and the prediction of combustion properties becomes more statistical in nature. In practice, it is frequently advisable to use typical or mean values based on a series of measurements in place of calculated values based on fuel composition.

It turns out that for each type of commonly used fuel (solid, liquid and gas) the stoichiometric ratio can be related linearly to the fuel's heating value, upper or lower. This matter is discussed comprehensively by Brandt [96], and the following formulae are quoted from that reference. In each case the lower heating value CV_L is to be used, expressed as MJ/kg. It should be noted that these relationships have been derived by fitting a straight line to a large number of experimental data, based on a selection of fuels from a given geographical area. They may need some adjustment to fit fuels from other areas.

Solid Fuels: Coal

$$R_S = 0.5678 + 0.3170\, CV_L.$$

Liquid Fuels: Oils (All Grades)

$$R_S = 0.4397 + 0.3241\, CV_L.$$

---

[3]Standard atmospheric pressure 101.3 kPa and 15 °C.

The stoichiometric ratios for gaseous fuels can be computed readily from composition. Natural gas has a typical composition of 82.3 % CH_4, 0.6 % C_2H_4, 10.0 % CO_2 and 7.1 % N_2 from which its R_S can be computed to be 15.067, close to 15.507 for pure methane.

Mixtures of Fuels

The stoichiometric ratio of a mixture is calculated as the weighted sum of the air required for each component of the mixture. For coal, the calculation proceeds from the coal composition which for a typical anthracite coal might be taken as (by weight and *waf*): 91 % C, 9 % volatiles.[4] One kilogram of the fuel contains 0.91 kg C which requires $2.6641 \times 0.91 = 2.4243$ kg of oxygen and 0.08 kg of methane which requires 0.3191 kg of oxygen, giving a total oxygen requirement of 2.7434 kg or 11.85 kg of air. The stoichiometric ratio for this coal is then 11.85. The combustion of the coal produces $3.3343 + 0.2195 = 3.5538$ kg of CO_2 and 0.1797 kg of water vapour. Adding the 9.11 kg of N_2 gives a fluegas mass of 12.85 kg per kg of coal.

The following table sets out the composition (by weight) of bfg and cog and the stoichiometric ratios calculated therefrom.

%	N_2	CO	CO_2	H_2	O_2	CH_4	R_{SA}	R_{SG}
bfg	50.0	23.0	22.0	2.0	3.0	0.0	1.22	2.22
cog	0.0	0.0	5.6	62.0	0.0	26.5	25.82	26.82

11.3.6 *Flue Gas Composition*

The fluegas is made up from

1. The incoming air minus the oxygen removed by combustion.
2. The gaseous products of combustion CO_2, SO_2, NOx.
3. The incoming air moisture plus any moisture created by combustion.
4. Moisture contributed by steam atomising of the fuel (liquid fuels only).
5. The nitrogen contributed by the incoming air minus any nitrogen converted into nitrous oxides.
6. Leakage air.

The theoretical fluegas mass or volume flows can be calculated from the known stoichiometric air flows and from the combustion products calculated from the

[4]The volatiles are taken as methane for the calculation. In fact methane is only a small component of the volatiles, the bulk coming from the thermal decomposition of the fuel as it enters the furnace.

known fuel composition. Using methane as an example we saw previously that the combustion of 1 kg of methane consumed 3.9891 kg of oxygen and therefore 17.23 kg of air. This reaction produced 2.7432 kg of CO_2 and 2.2459 kg of water vapour. Of the original air, assuming stoichiometric conditions, all of the oxygen was consumed and, neglecting the formation of nitrogen oxides, the original 13.241 kg of N_2 passed through unchanged. The original 17.23 kg of air produced 18.2309 kg of fluegas, including reaction moisture. Had the original air included moisture, this too would have passed through unchanged and is to be added to the fluegas mass. For methane, the dry air stoichiometric ratio is 17.23 and the stoichiometric fluegas ratio, including moisture produced by the reaction, is 18.23. Without the moisture it is 15.985. A similar calculation for some standard fuels produces the values in the following table for both the dry air R_{SA} and fluegas (including moisture) R_{SG} stoichiometric mass ratios.

Fuel	C	H_2	CH_4	CO	S	C_3H_8	C_4H_{10}
R_{SA}	11.507	34.2836	17.231	2.4673	4.3112	14.6724	15.4571
R_{SG}	12.507	35.2836	18.231	3.4673	5.3112	15.6724	16.4571

The following table presents the same information but on a volumetric basis of cubic meters of air and fluegas per kilogram of fuel.

Fuel	C	H_2	CH_4	CO	S	C_3H_8	C_4H_{10}
R_{SA}	18.8996	26.5139	13.3258	1.9081	3.3342	12.1205	11.9541
R_{SG}	8.8887	32.0769	14.7157	2.3031	3.3174	13.1288	12.9097

Excess air increases the fluegas mass flow by the excess air factor plus any moisture contained in that air.

11.3.7 Excess Air

Under ideal combustion conditions, were only stoichiometric air provided, the flue gas would contain no residual oxygen and all of the fuel would be burnt. However conditions are not ideal (e.g. mixing of the fuel and air is imperfect) and as far as technically possible a furnace is always fired with more than stoichiometric air. The ratio of actual air to stoichiometric air is called the excess air factor or coefficient.

The presence of excess air is indicated by the presence of oxygen in the fluegas and the amount of oxygen as a fraction (or percentage) of fluegas volume can be used to calculate the excess air factor. Denoting the excess air factor by κ we have

$$\kappa = \frac{\dot{m}_a}{\dot{m}_{RA}} = \frac{\dot{m}_{RA} + \Delta\dot{m}_a}{\dot{m}_{RA}} = 1 + \frac{\Delta\dot{m}}{\dot{m}_{RA}},$$

where $\dot{m}_{RA}$ is the stoichiometric air flow, $\dot{m}_a$ is the actual air flow and $\Delta\dot{m}$ is the excess air flow $= (\kappa - 1)\,\dot{m}_{RA}$. If we denote the stoichiometric fluegas mass flow by $\dot{m}_{RG}$ and by φ_{Rm} the ratio of stoichiometric fluegas to air mass flows we can write,

$$\frac{x_{O_2}}{0.2312} = \frac{(\kappa - 1)}{\varphi_{Rm} + (\kappa - 1)}, \tag{11.1}$$

where x_{O_2} is the mass fraction of oxygen in the fluegas. On a volumetric basis we have

$$\frac{y_{O_2}}{0.2095} = \frac{(\kappa - 1)}{\varphi_{Rv} + (\kappa - 1)}, \tag{11.2}$$

where y_{O_2} is the volume fraction of oxygen in the fluegas and φ_{Rv} is the ratio of stoichiometric fluegas to air flows on a volumetric basis. The latter method is to be preferred as the oxygen measurement is usually based on volume. The particular relationship is a function of the fuel as φ depends on the stoichiometric ratios which are fuel-dependent. For methane $\varphi_{Rm} = 1.038$ and for fuel oil $\varphi_{Rm} = 1.080$.

The following diagram depicts the relationship between fluegas oxygen concentration and excess air fraction for a variety of fuels.

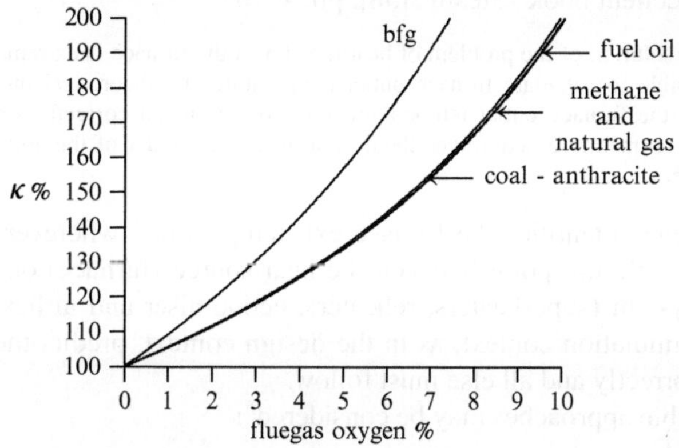

For many fuels φ_{Rm} is sufficiently close to 1.0 to allow us to write

$$\kappa = \frac{0.21}{0.21 - y_{O_2}}. \tag{11.3}$$

The use of excess air has disadvantages. The higher air flow increases draft fan power requirements. It reduces the flame temperature and thus the radiant heat release from the flame and necessitates a larger furnace. The disadvantages are offset by the advantages of improved flame stability, reduced unburnt fuel loss, reduced NOx formation and reduced ash slagging. The compromise between advantage and disadvantage produces an optimum excess air factor and fluegas oxygen concentration. For coal firing this is around 3 % oxygen and for oils and gas somewhat lower.

In practice the oxygen concentration is measured by taking a gas sample, in the process of which the vapour component of the gas will be removed by condensation to the temperature of the sampling apparatus. The measured oxygen concentration will therefore be indicated on a dry basis.

11.4 Furnace Heat Balance and Exit Temperature

The average temperature of the combustion products at the furnace exit is the furnace or flue gas exit temperature. The furnace exit is located notionally at a level at which the fuel can be considered completely consumed. For a tower boiler this is taken as slightly ahead of the first row of the first superheater. For a boiler with a backpass gas duct it is taken at the elevation of the furnace "nose", in line with or below the lowest point of the lowest pendant superheater.

There is no simple direct method for calculating the furnace exit temperature from a simplified lumped (i.e. not distributed) model. Indeed, to quote from Babcock & Wilcox's excellent book "Steam" [58], pp. 4–16

> An analytical solution of the problem of heat transfer in the furnace is extremely complex. It is not possible to calculate furnace outlet temperatures by theoretical methods alone. Nevertheless, the furnace outlet temperature must be predicted correctly since, in large measure, this temperature determines the design of the remainder of the unit, particularly the superheaters.

This last sentence identifies the furnace exit temperature, wherever and however defined, as the fulcrum point between the heat source (furnace) on one side and the flue gas system (superheaters, reheaters, economiser and airheater(s)) on the other. In the simulation context, as in the design context, predict the furnace exit temperature correctly and all else must follow.

Three possible approaches may be considered[5]:

1. Assume that the flame fills the furnace which is at a uniform temperature throughout; Eq. 11.4 can then be expressed in terms of the single flame temperature.
2. Define what we will call the "practical" flame temperature defined by a heat balance taken over the whole furnace, assuming that radiation from the flame is independent of the temperature of the receiving surface; use this temperature to calculate each of the radiant components of Eq. 11.5 and obtain the exit temperature from the heat in the flue gas leaving the furnace.

[5]These methods do not include the CFD approach which seeks to determine flow and thermodynamic conditions within the furnace zone using a fine multiple dimensional mesh covering the entire furnace. While of great value in identifying micro behaviour within the furnace, such as gas velocities, heat transfer and particle trajectories, these methods are too complex and computationally intensive for real-time simulation.

3. Use a distributed model of the furnace in which mass and energy balances are developed for a series of zones along the furnace vertical axis from the bottom of the furnace (ash hopper or travelling grate) to the furnace exit; integrate the energy balance along the flow path and obtain the furnace exit conditions at the end of that path.

The first procedure yields a flame temperature which is too low for realistic calculation of the radiant components and an exit gas temperature which is too high due to the dominance of the fourth power term in the heat balance equation.

The second is the basis of the lumped approach described in the next section. It preserves the correct balance between the fourth power and linear temperature terms in Eq. 11.5. It is the simplest computationally.

The third alternative is the most complex but does yield a good result and is to be preferred if the computation power is available. The method suffers from the disadvantage of needing an expanded set of plant parameters, with a corresponding increase in the complexity of its setup to match a specific plant. An example of this approach is described by Kwatny and Bauerle [95].

Section 11.4.2 develops a more complex model somewhat along the lines of the third alternative by treating a single flame and summing the outputs of all flames to arrive at the whole-of-furnace results. This method yields a good compromise between physical realism and computational complexity and provides a reduced set of parameters for adjustment to match a given plant's known or anticipated data.

11.4.1 Furnace Heat Balance: A Lumped Approach

The furnace heat balance may be described for any fired furnace by the following equation

$$\dot{Q}_i + \dot{Q}_{rel} = \dot{Q}_{rad} + \dot{Q}_{conv} + \dot{Q}_{fg} + \dot{Q}_{loss}, \tag{11.4}$$

$$\text{where}\begin{cases} \dot{Q}_i & \text{is the sensible heat of the incoming fuel and air flows,}\\ \dot{Q}_{rel} & \text{is the heat released by the combustion of the fuel,}\\ \dot{Q}_{rad} & \text{is the radiant heat transfer to the surrounding surfaces,}\\ \dot{Q}_{conv} & \text{is the heat transferred to the waterwalls by convection,}\\ \dot{Q}_{fg} & \text{is the heat leaving the furnace in the flue gas,}\\ \dot{Q}_{loss} & \text{is the allowance for losses.}\end{cases}$$

$\dot{Q}_{rel}$ is calculated as the sum of the heats released by the combustion of each individual fuel

$$\dot{Q}_{rel} = \sum_i \eta_{cb}\, \dot{m}_{f,i}\, CV_i,$$

where $\dot{m}_{f,i}$ is the mass flow of the ith fuel with heating value CV_i and η_{cb} is its combustion efficiency. This accounts for incomplete combustion of the fuel and for the fact that not all of the heat released by combustion is available for distribution within the furnace. Some is used to evaporate the moisture in the air (but not in the fuel if the lower heating value is used) and to heat it and any incombustible components of the fuel, such as ash, to flame temperature. It should be noted that the heat released by combustion is not released instantaneously. The rate of heat release is controlled by the dynamics of the combustion reaction, as will be discussed later.

The convective heat transfer

$$\dot{Q}_{conv} = c_{ww}\alpha_{ww}A_{ww}(T_{frn} - T_{ww})$$

treats the waterwalls as a flat plate of area A_{ww} equal to the total projected surface area of the wall tubes. The heat transfer coefficient α_{ww} is the standard Brandt coefficient. The parameter c_{ww} is introduced to account for the furnace geometry. In this equation T_{frn} and T_{ww} can be either local if matched to a distributed model of the waterwall temperatures or single representative values if a lumped model is used.

The radiant component is transferred to the waterwalls and other radiant receiving surfaces, including superheaters and reheaters.

$$\dot{Q}_{rad} = \dot{Q}_{rww} + \Sigma_j \dot{Q}_{rad,j},$$

where the index j denotes the jth receiving surface.

Fuels enter the furnace via burners. It is common furnace design practice to provide concentric fuel and primary air flow through a burner and these flows are therefore summed over the number of burners. Secondary and tertiary air flows are usually introduced into the furnace through separate ports, less in number than the burners.

The sensible heat of the incoming flows is the sum of the heat contents of the individual burner flows.

$$\dot{Q}_{in} = \sum_{i=1}^{n_{brn}} \dot{m}_{f,i} T_{f,i}\, cp_f + \sum(\dot{m}h)_{air},$$

where the total enthalpy of air entering the furnace is

$$\sum(\dot{m}h)_{air} = \left(\sum_{i=1}^{n_{brn}} \dot{m}_{pa,i} T_{pa,i} + \sum_{i=1}^{n_{sa}} \dot{m}_{sa} T_{sa,i} + \sum_{i=1}^{n_{ta}} \dot{m}_{ta} T_{ta,i}\right) cp_{air}$$

and

$$\begin{cases} \dot{m}_{f,i} & \text{is the fuel mass flow from the } i\text{th burner,} \\ \dot{m}_{pa,i} & \text{is the primary air ("pa") mass flow from the } i\text{th burner,} \\ \dot{m}_{sa} & \text{is the total secondary air ("sa") mass flow,} \\ \dot{m}_{ta} & \text{is the total tertiary air ("ta") mass flow,} \\ T_{f,i} & \text{is the temperature of the } i\text{th fuel flow,} \\ T_{pa,i} & \text{is the temperature of the } i\text{th primary air flow,} \\ T_{sa} & \text{is the temperature of the secondary air to the furnace,} \\ T_{ta} & \text{is the temperature of the tertiary air to the furnace,} \\ cp_{air} & \text{is the specific heat capacity of air.} \end{cases}$$

and all temperatures are degrees Celsius. Using the radiant heat calculation from Eq. 8.23, the heat balance equation may be written, with $\alpha_{fl} = \epsilon_{fl} c_R f_v$ and A_{fl} the surface area of a notional flame,

$$\begin{aligned} \dot{Q}_i + \dot{Q}_{rel} &= \alpha_{fl} A_{fl} \left(\frac{T_{fl} + 273}{100} \right)^4 + \alpha_{hx} A_{hx} (T_{fl} - T_{ww}) \\ &\quad + \alpha_{loss} A_{loss} (T_{ww} - T_{amb}) + \dot{m}_{fg} cp_{fg} T_{fg} \\ &\quad + r_m \left[\sum_{i=1}^{n_{brn}} x_{h2of,i} \dot{m}_{f,i} + x_{H_2O,air} \sum (\dot{m}_{air}) \right], \end{aligned} \tag{11.5}$$

where

$$\sum(\dot{m}_{air}) = \sum_{i=1}^{n_{brn}} \dot{m}_{pa,i} + \sum_{i=1}^{n_{sa}} \dot{m}_{sa} + \sum_{i=1}^{n_{ta}} \dot{m}_{ta}$$

is the total air flow into the furnace. n_{brn}, n_{sa} and n_{ta} are the number of burners, secondary and tertiary air ports respectively. r_m is the latent heat of evaporation of the moisture in the air and fuel, say 2,257 kJ/kg. The formulation of Eq. 11.5 allows differing moisture content for each burner, which can happen when raw coal in individual hoppers has been drawn from differing sources or has differing surface moisture.

The Adiabatic Flame Temperature

The temperature of a flame produced by the combustion of a fuel with no exchange of heat with the surroundings is denoted the adiabatic temperature. The maximum adiabatic temperature is calculated using stoichiometric air.

$$T_{ad} = \frac{\dot{Q}_{rel}}{\sum_{i=1}^{n_{brn}} \dot{m}_{f,i}\, cp_{f,i} + \sum(\dot{m}_{air})\, cp_{air}}.$$

A real flame will approximate to the adiabatic temperature at its core but its average value will be diminished by the non-adiabatic conditions in the furnace.

The Practical Flame Temperature

The practical flame temperature is calculated as the solution of a modified form of the heat balance Equation 11.5.

$$\dot{Q}_i + \dot{Q}_{rel} = \alpha_{fl} A_{fl} \left(\frac{T_{fl} + 273}{100} \right)^4 + \alpha_{hx} A_{hx} T_{fl} + \dot{m}_{fg} cp_{fg} T_{fg}. \quad (11.6)$$

In the lumped approach, α_{fl} is treated as if the entire heat is released from a single flame which is considered to fill the entire furnace volume but whose radiative heat is treated as if released from a single point, the "centre of flame".

Equation 11.6 may be solved by Newton-Raphson iteration.

$$T_{fl}^{n+1} = T_{fl}^{n} - \kappa \frac{f(T_{fl})}{f'(T_{fl})},$$

where

$$f(T_{fl}^n) = \dot{Q}_i + \dot{Q}_{rel} - \alpha_{fl} A_{fl} \left(\frac{T_{fl} + 273}{100} \right)^4$$
$$-(\alpha_{hx} A_{hx} T_{fl} + \dot{m}_{fg} cp_{fg} T_{fg}),$$
$$f'(T_{fl}^n) = 0.04 \alpha_{fl} A_{fl} \left(\frac{T_{fl} + 273}{100} \right)^3 - (\alpha_{hx} A_{hx} + \dot{m}_{fg} cp_{fg}).$$

The parameter κ is chosen to give good convergence. Experience suggests a value in the range 0.2–0.5 gives satisfactory convergence within 4–10 iterations.

This approach requires a second equation to determine the furnace exit temperature T_{fg}. The simplest method is to assign a value empirically based on the known parameters of a given furnace. For example, the furnace exit temperature can be made to follow a given characteristic as a function of air flow, heat release and configuration of the in-service burners.

When firing multiple fuels simultaneously separate flame temperatures should be calculated for each fuel type. Each flame is then treated as if it alone occupied the furnace and the individual radiant heat components from each flame superimposed.

11.4.2 Treatment of Individual Flames

A single flame from any fuel can be treated in two parts, one luminous and one non-luminous. The luminous part is the combustion zone and derives its luminosity from radiation emitted by the oxidation of the fuel. The physical appearance of a luminous flame is determined largely by its colour and shape. Colour is determined by the combusting material. Solid fuels produce bright yellow flames caused by the combustion of carbon particles and radiation from hot ash and char particles. With normal atomisation, liquid fuels burn with a pale yellow flame with a brightness depending on the degree of formation of carbon particles and the fuel's tendency to form suspended deposits.

High-intensity burners use very fine atomisation, and the flame is more bluish. They allow lower excess air which produces higher flame temperatures, and special arrangements must be used to avoid the formation of NOx, such as the use of air flows less than stoichiometric with supplementary air introduced around the burner to ensure complete combustion of the fuel. Gaseous fuels tend to produce blue flames from the combustion of hydrogen and hydrocarbons. The shape of the flame is defined largely by the burner which defines the effectiveness of mixing of fuel and air, the incoming air velocity and the distribution (spread) of the flame beyond the end of the burner. The luminous part usually has a recognisable (cigar) shape and is directed horizontally across the furnace chamber. The non-luminous part of the flame is the plume of hot combustion products. It is composed mainly of triatomic gases including CO_2, SO_2 and water vapour and fills the furnace. It may include some luminosity due to the presence of residual combusting particles and gases. The design of the furnace and its burners should ensure that the bulk of combustion is completed within the luminous combustion zone and all combustion should be complete before the gases reach the furnace exit. Poor milling of coal, poor atomisation of fuel oil or poor arrangement of secondary air flows can impede achievement of this objective.

Temperatures within the combustion flame zone range from a maximum of around 1,900 K within the initial ignition zone to a minimum of around 1,400 K on the surface [49, 50]. It is reasonable to assume that the maximum temperature, or something very close to it, is to be found at the geometrical centre of the luminous zone. The cited references examine the micro-structure within the flame to determine the spatial distribution of temperature and gas velocities. This lies beyond the scope of this treatment whose purpose is to describe useful macro-scale models.

Energy is radiated from every point within the flame. Radiation emitted within the flame must pass through the body of the flame and is partially scattered and absorbed along the way. Within the flame zone heat is released by the combustion reactions at a rate k whose temperature dependence is defined by the Arrhenius equation

$$k = A\mathrm{e}^{-E/(RT)},$$

where A is the frequency factor, with dimension second^{-1}, and E the activation energy, being the minimum required to initiate the reaction. A and E are known for each reaction. Regions of higher temperature will show higher reaction rates. The creation of high temperatures by efficient fuel/air mixing is important to the achievement of complete combustion of the fuel within the short residence time within the flame zone.

Fig. 11.8 A PF tangential burner register (Courtesy of Foster-Wheeler Corp.)

Essentially, $e^{-E/(RT)}$ is the probability that molecules having the appropriate energies will collide and react. The higher the temperature, the greater the probability of a collision. The following reaction equation defines the time variation of the specific mass of the reactant(s) as the reaction proceeds. Then, simply put, if c is the specific mass (density or concentration) of a reactant at time t after a reaction has commenced and the reaction involves only a single reactant with no intermediate reactant products,

$$\frac{\mathrm{d}c}{\mathrm{d}t} = -k.$$

Many reactions produce one or more intermediate reactants. For example, the combustion of carbon (reaction of carbon with oxygen) produces CO as an intermediate reactant whose subsequent oxidation (combustion) to CO_2 contributes significantly to the total exothermic energy release of the complete combustion of carbon.

Applied specifically to the combustion of pulverised coal (pf) in a furnace, Smith [53] quotes McKenzie et al. [54] for the following equation which relates the fraction of a pf particle remaining as a function of time as combustion proceeds.[6] It is derived from a combination of physical analysis and extensive test rig measurements.

$$\frac{\mathrm{d}y}{\mathrm{d}t} = -\frac{6}{d_0\sigma_0} y^{1-\alpha-\beta} \rho,$$

[6]The original equation is expressed in terms of u, the fractional burn-off.

where d_0 is the initial particle size, σ_0 is its initial density and ρ is the actual rate of combustion per unit external surface area of the particle. α and β are parameters relating to the changes in d and σ as combustion proceeds. Smith concludes that for particles that burn with steadily decreasing size and constant density, $\alpha = 1/3$ and $\beta = 0$. If size remains constant while density decreases, $\alpha = 0$ and $\beta = 1$. In the first case the reaction equation is,

$$\frac{\mathrm{d}y}{\mathrm{d}t} = -\frac{6}{d_0\sigma_0} y^{2/3}\rho$$

which approximates an exponential decrease, while in the second,

$$\frac{\mathrm{d}y}{\mathrm{d}t} = -\frac{6}{d_0\sigma_0}\rho$$

is a varying rate of decrease dependent on the density of the reactant present.

The real situation in an actual furnace is very complex, producing various intermediate reactant products. From an operational modelling point of view the exponential version provides the more stable framework for calculation of pf combustion rates and has been adopted for this purpose. This is discussed further in Sect. 11.4.10.

Flame Geometry

The luminous part of a single flame can be approximated in a simple fashion by an ellipsoid whose major axis is aligned with the fuel/air flow direction.[7] The volume V of a symmetrical[8] prolate (cigar-shaped) ellipsoid having equal minor semi-axes r_1 and major semi-axis r_2 ($r_1 < r_2$) is

$$V = \frac{4}{3}\pi r_1^2 r_2 = \frac{4}{3}\pi r_1^3\,\mathbf{r}$$

and $\mathbf{r}$ is the aspect ratio r_2/r_1. The surface area A is

$$A = 2\pi r_1^2\left(1 + \mathbf{r}\frac{\arcsin(\vartheta)}{\vartheta}\right),$$

[7] An alternative shape which has been used elsewhere is a frustum of a cone.

[8] An ellipsoid with a circular equator.

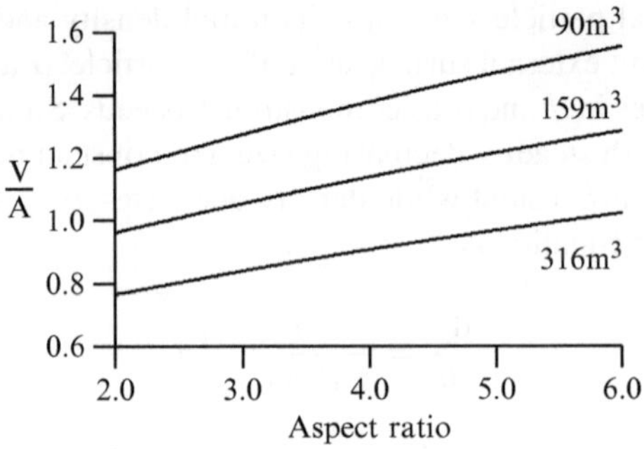

where

$$\vartheta = \sqrt{1 - (1/\mathbf{r}^2)}.$$

For a given volume V, the minor semi-axis is given by

$$r_1 = \sqrt[3]{\frac{3\,V}{4\pi\,\mathbf{r}}}.$$

The preceding diagram plots the volume/area ratio of a prolate ellipsoid as a function of aspect ratio for three values of volume, 90 m^3, 159 m^3 and 316 m^3, representative of the flames in large utility and industrial boilers.

Heat Transfer from the Flame

We will assume that the volume of the flame is proportional to the heat being released within the flame [92],

$$V_{fl} = k_{vol}\,\dot{Q}_{rel}$$

and that the radiant heat $\dot{Q}_{rad}$ is proportional to the flame surface area A_{fl}. Given that the heat released by the flame will be known from fuel flow and its properties, the relationship between heat release and radiative component can be related to the aspect ratio of the approximating ellipsoid which in turn can be related to fuel type. Oil and gas flames tend to be longer and thinner (larger aspect ratio) than those of pulverised coal. Given the flame volume and aspect ratio, the flame semi-axes r_1 and r_2 are calculated and from them, the flame area.

Assuming spherically symmetric radiation from the flame, the energy falling on the waterwalls will be inversely proportional to the square of the distance from the flame (the "slant distance" in the accompanying diagram), taken from the geometric centre of the approximating ellipsoid at the burner elevation. Given an assumed symmetrical firing, this can be taken to coincide with the furnace vertical centreline. It might be expected that individual flames will be offset from the centreline but, when the contributions of all flames are added, the individual offsets can be assumed to mutually cancel.

11.4.3 Heat Absorption and Transmission Within the Furnace

Part of the luminous radiation will be absorbed by the fluegas along the transmission path. The absorption of energy not only diminishes the energy arriving at the receiver but also heats the absorbing gas which then retransmits part of that energy as non-luminous radiation to the walls. The radiative transmission Eq. 8.28 in Sect. 8.3.1 defined the rate of absorption of radiant energy over a path length s. Introducing the pressure dependency of the absorption coefficient the solution of this equation can be written

$$I(r) = I_e(1 - \mathrm{e}^{-\kappa\, k_g\, p_g\, s}), \tag{11.7}$$

where p_g is the partial pressure of the absorbing gas and k_g is its absorption coefficient. As mentioned in Sect. 8.3.1, k_g is a complex function of pressure and gas composition. κ is an absorber-specific weighting constant. For CO_2 $\kappa = 0.15$ and for H_2O $\kappa = 0.18$.

The partial pressure of an absorber gas in a mixture is

$$p_g = P\, x_g\, \frac{R_g}{R},$$

where P is the total pressure of the gas, x_g is the mass fraction of the absorber gas, R_g is the gas constant of the absorber gas and R is the gas constant of the gas mixture. The major absorbers in the furnace are CO_2 at around 20 % ($R_g = 0.18763$) and H_2O at around 7 % ($R_g = 0.4615$), depending on the moisture content of the fuel and moisture produced as a combustion product.[9] The partial pressure of CO_2 in the furnace will then be 13.5 kPa (0.1333 bar) and that of moisture 13.32 kPa (0.1315 bar).

Lin in Chap. 8 of [47] (p. 403) gives a formula for the calculation of k_g applying to absorption by tri-atomic gases over a transmission distance s [m] (beam length).

$$k_g = k_y\, x_g,$$
$$k_y = \left(\frac{7.8 + 16\, y_{H_2O}}{3.16\sqrt{x_g\, P\, s} - 1}\right)\left(1 - 0.37\frac{\mathcal{T}_{fx}}{1000}\right), \tag{11.8}$$

where x_g is the mass fraction of the absorber gas, y_{H_2O} is the volumetric fraction of water vapour, P is the total gas pressure [MPa] and $\mathcal{T}_{fx}$ is the furnace exit temperature [K].

The following diagram shows values of the absorption factor $(1 - \mathrm{e}^{-kps})$ for CO_2 and H_2O calculated using Eqs. 11.7 (full lines) and 11.8 (dotted lines) with the furnace parameters quoted above. Absorption is indicated as a fraction of the source energy after passing through a distance s m.

---

[9] This is significant when firing methane or natural gas as significant moisture is produced.

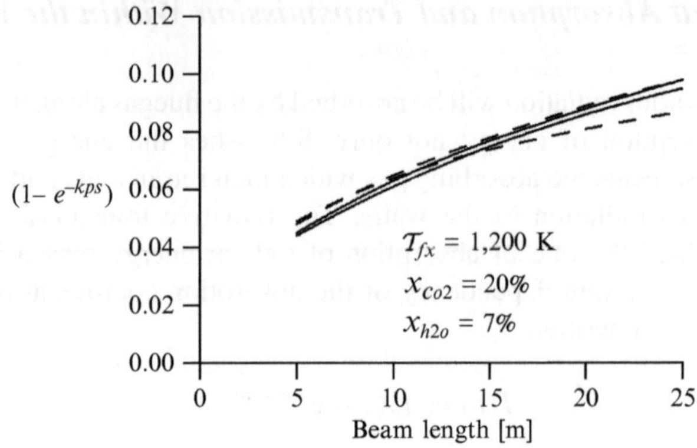

Radiant Heat from the Flame to the Waterwalls

$\dot{q}_{ab}$ is that part of the radiant heat from the flame absorbed by CO_2 and H_2O along the path from the flame to the receiving surface at the nth zone, calculated using the absorption coefficients of Eq. 11.8. Each zone, both above and below the flame, is traversed by each ray passing from the flame zone to each zone beyond it. For a spatially discretised path, intensity absorption ΔI_r^{n+1} in the $(n+1)$th zone is calculated from Eq. 8.30 for each ray traversing the zone

$$I_r^{n+1} = I_r^n \, e^{-k_{ab}\Delta s} \tag{11.9}$$

from which we have

$$\Delta I_r^{n+1} = I_r^{n+1} - I_r^n = -k_{ab}\Delta s \, I_r^n. \tag{11.10}$$

The total energy absorbed in any zone is the sum of the individual energies absorbed from each transiting ray.

Energy from each flame arrives at each wall zone along a single ray path. The intensity of the arriving radiation is the original flame intensity $I_{fl} = \dot{Q}_{rad}/A_{fl}$ minus the energy absorbed along the path. The total energy for the wall zone is obtained by multiplying the received intensity (kJ/s/m^2) by the projected wall tube area.

With multiple flames, the aggregate energy received at any waterwall zone can be calculated as the sum of the individual flame contributions. This incurs some error as a flame is largely opaque to radiation from other flames and will block some individual rays.

Non-luminous Radiation from the Furnace Gases

$\dot{q}_{nlr}$ is the non-luminous radiation energy emitted by the gas in the jth cell and received by the section of waterwall tubing bordering the cell. It is calculated as

$$\dot{q}_{nlr} = c_R \epsilon_w \left[(\epsilon_{CO_2} + \epsilon_{H_2O}) \left(\frac{T_{g,j}}{100} \right)^4 - A \left(\frac{T_{ww,j}}{100} \right)^4 \right],$$

where the surface absorptivity A is calculated using the wall surface temperature. For simulation of most phases of furnace operation, it is sufficient to treat absorptivity and emissivities as equal.

The emissivities ϵ_{CO_2} and ϵ_{H_2O} can be calculated from $E_{CO_2,H_2O}/(c_R T_g^4)$ where E_{CO_2} and E_{H_2O} are calculated from Eqs. 8.25 and 8.26 in Sect. 8.3.3. However, given that the furnace temperatures are predictable for most furnaces at around 1,500–1,900 K, a value in the range 0.15–0.2 may be assumed for each emissivity coefficient.

11.4.4 Spatial Profile of Furnace Gas Temperature

A simplified two-dimensional model can be used to calculate the profiles of flame temperature and heat transfer to the waterwalls from a single flame source to the furnace exit. When firing multiple flames (which can include differing fuels), the aggregate heat fluxes are obtained by summing the individual flame contributions. Aggregate flame and gas temperatures are obtained by averaging the individual zonal temperatures of all flames. Clearly, this is a major simplification and cannot compare in detail or accuracy with a full three-dimensional mixed flow model typical of CFD analyses. Experience suggests that, in the macro sense, adequate results can be obtained for small furnaces with as few as 20 zones and less than 50 for large furnaces, computationally more realisable for real-time simulation that the 50,000 to more than 120,000 cells used in some CFD analyses [97]. It should be appreciated however that the two approaches are directed at very different objectives.

Figure 11.9 shows the furnace divided vertically into a number of contiguous cells or zones, for each of which a simplified energy balance is used to calculate the local gas temperature on the basis of radiation energy into and from the cell and the transport of fluegas enthalpy from the flame to the furnace exit.

Each zone represents a horizontal slice across the furnace. A zone contains M_g kg of gas at temperature T_g°C , pressure p_{frn} kPa and specific volume v_g m³/kg, calculated from

$$v_g = R_g(T_g + 273)/p_{frn},$$

where R_g is the gas constant for fluegas, calculated from

$$R_g = 0.2869 + 0.0746\, x_{CO_2} + 0.1746\, x_{H_2O} \quad [\text{kJ/kg K}].$$

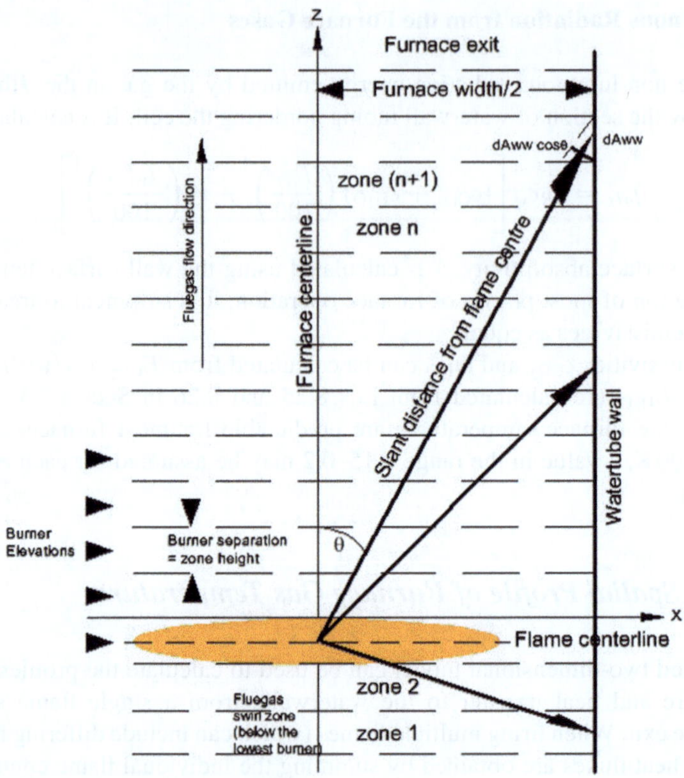

Fig. 11.9 Division of the furnace into axial zones

The gas temperature is calculated for the jth zone from the dynamic heat balance

$$c_p M_g \frac{\mathrm{d}T_{g,j}}{\mathrm{d}t} = \dot{m}_{fg} c_{p_{fg}} (T_{g,j-1} - T_{g,j}) + \sum (\dot{m}_{air} T_a c_{p_a}) + \dot{q}_{ab} - \dot{q}_{nlr} - \dot{q}_{conv} + \dot{q}_{src}, \qquad (11.11)$$

where $T_{g,j-1}$ is the gas temperature from the preceding cell; $\dot{m}_{fg}$ is the fluegas mass flow rate [kg/s] (assumed constant along the vertical axis); $\sum(\dot{m}_a T_a c_{p_a})$ is the rate of sensible energy addition [kJ/s] from primary, secondary or tertiary air flows entering the zone from burners or air ports, each at its own temperature T_a; $\dot{q}_{ab}$ is the energy absorbed in the cell from the flame radiation; $\dot{q}_{nlr}$ is the energy transmitted as non-luminous radiation from the cell and $\dot{q}_{conv}$ is the convective heat transfer from the cell to the riser tubes. $\dot{q}_{src}$ accounts for any heat generated by combustion which may be continuing within the cell.

With

$$D = \Delta t \frac{\dot{m}_{fg}}{M_g}$$

this equation can be reorganised as a fully implicit discrete time solution procedure for all cells, indexed $j \in (1, N_z)$, as

$$T_{g,j}^{n+1} = \left(\frac{1}{1+D}\right) T_{g,j}^{n} + \left(\frac{D}{1+D}\right) T_{g,j-1}^{n+1} + \frac{\Delta t}{c_p M_g} \left(\frac{1}{1+D}\right) \left(\sum(\dot{q}) + \sum(\dot{m}_a T_a c_{p_a})\right), \quad (11.12)$$

where the sum of heat exchanges $\sum(\dot{q}) = \dot{q}_{ab} - \dot{q}_{nlr} - \dot{q}_{conv} + \dot{q}_{src}$.

Each term on the right hand side of this equation is known at the nth time step and for the current and preceding $(j-1)$ spatial zones. The procedure advances the solution for each zone to the $(n+1)$th time step, treating separately movement from the flame elevation (a) towards the furnace exit and (b) towards the bottom of the furnace.

The fluegas flow rate $\dot{m}_{fg}$ is the gas flow leaving the flame towards the furnace exit. With a single flame, it is equal to the net combustion air (primary, secondary and tertiary) entering the flame zone from the burner or associated air ports. Not all combustion air leaves the flame directly towards the furnace exit. A small part circulates below the flame elevation, and the lower furnace is a region of swirl of no clear local direction. This random flow pattern should be considered when treating the heat balance in the zones below the lowest active burner.

With multiple burners in service the total fluegas leaving the furnace is the sum of all individual burner flows plus all supplementary air flows entering the furnace, plus/minus any leakage into or from the furnace.

11.4.5 Burners and Air Flows into the Furnace

There is a wide variety of burner designs, each optimised for the type of fuel (gas, liquid or solid) and for the specific characteristics of the fuel. Burner designs are matched to the furnace configuration; for example, coal burners will differ in design if used for single wall, opposing wall or corner-firing configuration. Special features may be incorporated to meet certain performance requirements such as reduced formation of oxides of nitrogen (low NOx).

Whatever their particular form, burners are designed to ensure optimised mixing of fuel and air at the point of combustion and to create a stable flame front ahead of the burner, sufficiently removed from the burner to avoid heat-induced damage. In oil burners, the oil flow is broken into a dense spray of fine particles by the air or steam atomiser and evaporated by radiant heat from the furnace. Combustion air is

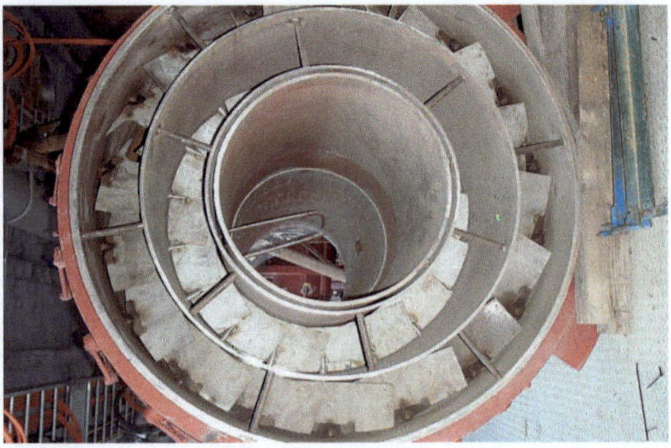

Fig. 11.10 A pf burner nozzle seen from the furnace (Courtesy of Millmerran Operating Company, Australia)

supplied from air registers which direct an air flow into the atomised spray. Some whirl is usually added to this flow to assist mixing. Oil burners include a flame holder mechanism which locates the flame and prevents entry of the flame into the burner. Some burners use a movable impeller for this purpose. The impeller must be inserted prior to lighting the burner. The impeller imparts a controlled swirl and can vary the ratio of directly mixed air and combustion air directed around the flame.

Burners for pulverised coal are usually constructed as round or rectangular nozzles. The powdered coal is carried from the mill to the burner by a stream of air (primary or "carrying" air) at a velocity greater than that at which the fuel would drop out of suspension. Minimum air flow velocities are usually around 16–22 m/s. The fuel-air stream enters the burner along a direct path, usually without swirl. The mass flow ratio of air to fuel, at 1.2–1.6, is significantly less than the required stoichiometric ratio of around 6–11 and additional air must be supplied from the secondary air registers. In some burners this is added in a premix zone within the burner. The mixing ratio is usually set to achieve 3–7 % excess air in the burner but any ratio may be used, including stoichiometric or even deficient air. The burner ratio is set during plant commissioning using adjustable vanes. Figure 11.10 shows the furnace end of a pf burner nozzle. The individual deflector vanes used for optimisation of fuel/air mixing are clearly visible.

The burner design aims to produce the highest flame temperature consistent with other performance objectives, such as low NOx production. A high flame temperature ensures maximum radiant heat transfer from the flame, maximum combustion reaction rates and lowest coal particle dwell time in the furnace, and maximum flame stability as the incoming coal is ignited by radiant heat from the flame. The balance of the requisite combustion air is added to the flame zone from the surrounding air ports, either surrounding the burner in a coaxial burner or located above and below the burner in a corner-fired register. This additional air cools the flame outer zone and establishes the temperature of the flame plume that moves towards the furnace exit.

Burners designed for low NOx are intended to do one of two things, or both: reduce the temperature of the flame and reduce the oxygen available for combustion.

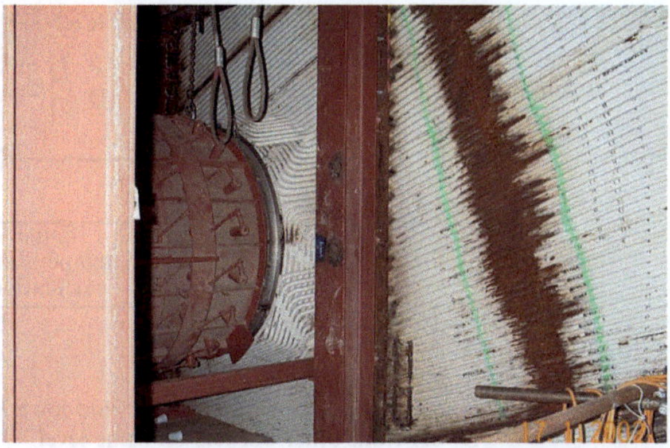

Fig. 11.11 A wall-mounted pf burner with surrounding wall tubes (Courtesy of Millmerran Operating Company, Australia)

Initial combustion in a sub-stoichiometric mixture achieves the first of these but produces a high flame temperature. This is subsequently offset by the later mixing of secondary and tertiary air to reduce the flame temperature in regions of higher oxygen concentration, that is, in regions of higher excess air.

Secondary air is supplied to the furnace from the forced draft (FD) fans. Different manufacturers use different plant configurations and differing control strategies for this task. In a wall-fired furnace a common arrangement allows individual control of combustion air to each burner level from a level-specific windbox with airflow to each level regulated by its own regulating damper following instructions from the plant's combustion control system. Some burners are provided with a further level of local control which maintains a given windbox–furnace pressure differential. The total FD fan flow is controlled to maintain a given secondary air duct pressure. Some designs use a single windbox common to all burner levels to which combustion air is supplied direct from the FD fans via the recuperative airheater. Airflow to individual burners is regulated by the local secondary air dampers, usually one per burner. The total airflow can be controlled by direct regulation of the FD fan control elements, either an inlet damper or inlet vane position.[10] For a corner-fired furnace, combustion air is supplied to each corner register and distributed above, below and directly to the flame by fixed dampers adjusted to produce the desired air flow distribution. These dampers are usually set in position during initial plant commissioning and may be adjusted from time to time to maintain optimal furnace conditions.

This burner geometry can be used to configure a routine for the distribution of both primary and secondary air flows to the furnace. We will assume that the flow of air direct to the flame is always close to the stoichiometric air required by the given fuel flow. Greater or lesser air is accommodated by an excess air adjustment parameter. The order of calculation is depicted by the following schema for a furnace with four burners per burner level.

[10]In many instances variable speed drives are now replacing these control methods.

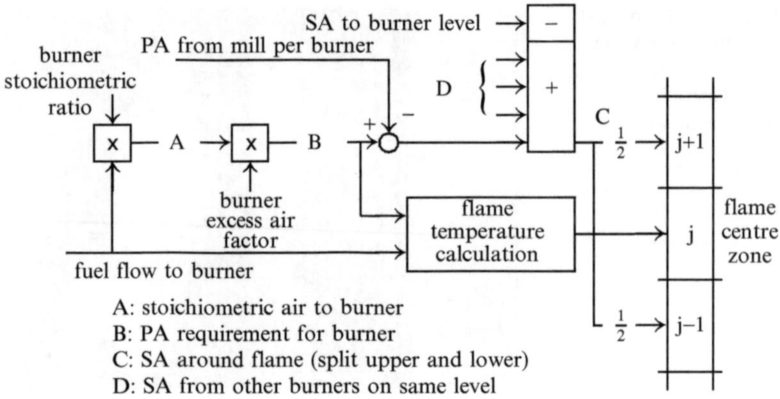

11.4.6 Evaluation of Reference Furnace Parameters

The following data is typical of full load operation of a 500 MW coal-fired boiler. The plant has five coal mills available but needs only four for full load operation. The furnace is tangentially fired from four corners with five levels of burners per corner, each fed by one mill. The furnace is 40 m high with a cross-section of 10 × 12.5 m. The burner vertical separation is 2.5 m, with secondary air ports between each pair of burners.

Main steam flow	400 kg/s
Feedwater inlet temp	275 °C
Feedwater inlet enthalpy	1,430 kJ/kg
Steam pressure in drum	18,340 kPa
Saturation steam enthalpy at drum pressure	2,498 kJ/kg
Design full-load heat release	1,475,000 kJ/s

The heat needed for evaporation of 400 kg/s at this pressure is (2,498–1,430) kJ/kg × 400 kg/s or 427,200 kJ/s. Assuming 8 % convective heat transfer the radiant heat to the waterwalls will be $0.92 \times 427,200$ or 393,024 kJ/s.

With a fuel CV of 28,000 kJ/kg the stoichiometric ratio is 11.85 and the full load fuel flow for the design heat release is 52.68 kg/s. Assuming four equally loaded coal mills in operation, the coal flow per mill is 13.2 kg/s or 3.3 kg/s for each of four burners. Total stoichiometric airflow is 624 kg/s giving, with 20 % excess air (3 % fluegas oxygen), an actual airflow of 749 kg/s.

The plant design data quotes total heat pickup in the secondary superheater to be 105,125 kJ/s and 146,000 kJ/s in the reheater. Assuming convective heat pickup in the secondary superheater is 67 % of the total heat pickup, the radiant heat transfer to this superheater will be 34,691 kJ/s. If we assume 8 % radiant heat transfer to the reheater its radiant heat absorption will be 11,680 kJ/s. Adding these to the waterwall radiant heat component gives total radiant heat transfer from the furnace as $\dot{Q}_{rad} = 439{,}395$ kJ/s or some 30 % of the total furnace heat release.

The heat leaving the fluegas will be the 70 % of released heat not present as radiant heat, or 1,032,500 kJ/s. Ignoring leakage flows, the actual fluegas flow rate can be calculated using the fluegas stoichiometric ratio of 12.85 and the excess air factor of 20 %, giving a fluegas flow of 812.3 kg/s. At this temperature the specific heat capacity of the fluegas is 1.2179 and the furnace exit temperature is calculated to be $1{,}032{,}500/(812.3 \times 1.2179) = 1{,}043°$ C.

This particular plant is equipped with four igniters and four oil guns for supplementary heat input. Each igniter is rated at a heat release of 21,060 kJ/s corresponding to an oil flow rate of 0.585 kg/s and an oil CV of 38,000 kJ/s. Each oil gun is rated at a heat release of 52,920 kJ/s corresponding to an oil flow rate of 1.47 kg/s.

11.4.7 Furnace Heat Flux and Temperature Computations

Application of this method is illustrated by a series of calculated flux and furnace gas temperature profiles, together with the breakdown of totalised heats as produced and used within the furnace. The totalised heats are compared with data from the furnace just described. This furnace was designed for pulverised coal (pf) and heavy fuel oil (hf). Firing with natural gas (ng) was explored to give a comparison with a different and increasingly common fuel. The natural gas-fired furnace is assumed to have front and rear wall firing, with four levels each with four burners. The various firing arrangements and rates are summarised by the following table.

Case	Fuel	Rating %CMR	Rows i/s
1	Pulverised coal	100	4
2	Pulverised coal	75	4
3	Pulverised coal	75	3
4	Pulverised coal	50	4
5	Natural gas	100	4
6	Natural gas	75	4
7	Natural gas	75	3
8	Natural gas	50	4

The parameter settings used for the determination of flame geometry and furnace heat distribution are as follows.

Fuel	Flame ϵ	AR const.	Stoich. ratio	Fuel CV	Excess air	Flame const.
pf	0.45	0.125	11.85	28,000	1.20	0.0035
hf	0.40	0.130	14.03	39,800	1.15	0.003
ng	0.35	0.145	17.20	42,000	1.12	0.003

The following table summarises the numerical results, calculated for 40 vertical positions. Case 0 is the reference coal-fired plant described earlier, and all heat values are quoted in kJ/s.

Case	Qrel	Qlum	Qnlum	Qfg	FEGT	Qww	QSH/RH
0	1,475,000	393,024	n/a	1,032,500	1,043	427,200	46,200
1	1,489,000	384,245	24,420	1,007,812	1,077	435,580	55,416
2	1,109,190	292,684	23,414	741,236	1,057	342,540	42,212
3	1,110,210	288,062	18,874	755,157	1,076	326,548	41,684
4	746,190	210,150	21,603	476,543	1,019	248,350	29,011
5	1,448,600	381,808	34,016	1,015,663	1,185	446,492	55,665
6	1,115,337	286,528	26,296	760,659	1,183	334,952	41,419
7	1,114,000	290,550	32,452	737,063	1,159	353,008	41,904
8	740,940	199,004	29,692	470,334	1,110	257,728	28,701

The table headings have the following meanings:

- Qrel: Total heat released in the furnace by combustion of the fuel
- Qlum: Heat to the waterwalls originating as luminous radiation
- Qnlum: Heat to the waterwalls originating as non-luminous radiation
- Qfg: Heat leaving the furnace as sensible heat in the fluegas
- FEGT: Furnace exit gas temperatures °C
- Qww: Total heat received by the waterwalls
- QSH/RH: Total radiative heat to the radiant superheaters and reheaters

The total radiative heat available may differ from the heat actually received as this will depend on the boiler geometry and view factors.

The lower emissivity of the gas flame means lower levels of luminous radiation and consequently higher flame and gas temperatures. Higher gas temperatures produce the higher non-luminous radiative heat transfers recorded for natural gas firing. Natural gas firing usually involves lower excess air factors than with, say, coal-firing and this also contributes to the higher flame temperature.

The following Fig. 11.12 display a series of heat flux profiles obtained by applying the methods described here to the computation of conditions within the furnace. In place of total heats the diagram shows heat fluxes [kW/m^2] at the waterwalls. Total heat received is the product of heat flux and receiving area. Because of the flame symmetry assumed by the calculation the results do not

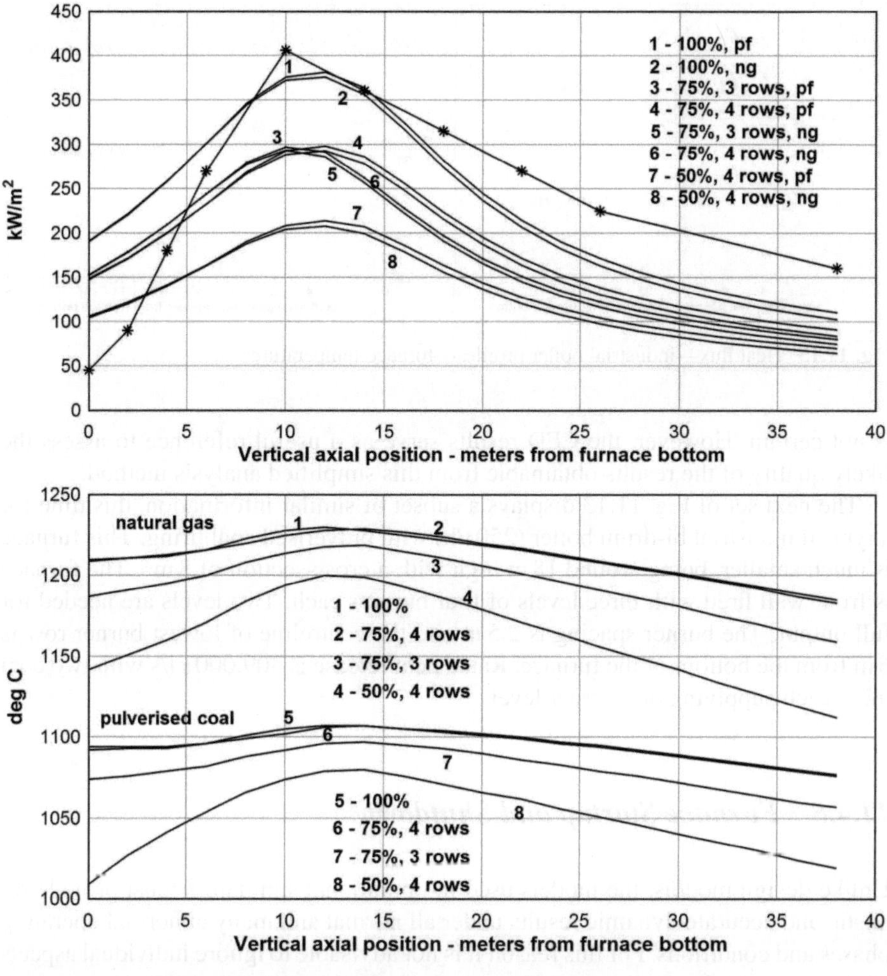

Fig. 11.12 Heat flux—utility boiler profiles—furnace gas temperature

distinguish between wall and corner-firing. While not an inherent limitation of the method, treatment of asymmetry adds significantly to the numerical computation load.

The heat flux profile marked $*$ has been created using results from the CFD analysis reported by Cho [97]. The same study quotes a measured furnace exit temperature of 1,221 °C with a fuel flow rate per mill of 15.2 kg/s and coal CV (higher) of 30,012 kJ/kg. This comparison is to be used with some caution. It is difficult to establish an exact correspondence between this macro-modelling approach and the CFD results since the latter are local heat fluxes within an extensive three-dimensional map and identification of the equivalent axial location

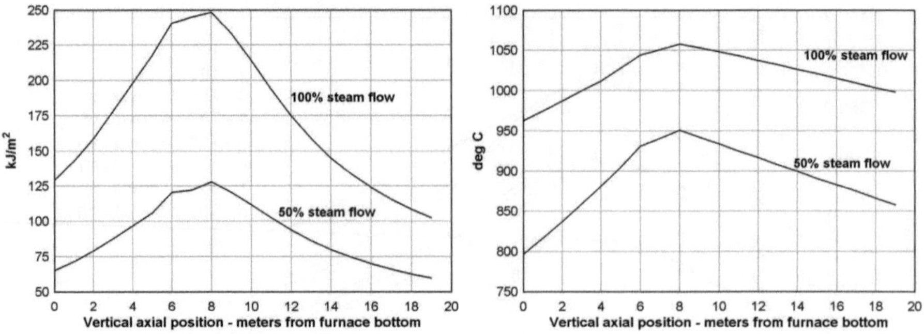

Fig. 11.13 Heat flux—industrial boiler profiles—furnace temperature

is not certain. However, the CFD results serve as a useful reference to assess the likely quality of the results obtainable from this simplified analysis method.

The next set of Fig. 11.13 displays a subset of similar information, this time for a typical industrial bi-drum boiler (250 t/h) with pulverised coal firing. This furnace is much smaller, being around 18 m high with a cross-section of 8 m^2. The furnace is front-wall fired with three levels of four burners each. Two levels are needed for full output. The burner spacing is 2.5 m and the centreline of lowest burner row is 6 m from the bottom of the furnace. Rated heat release is 309,000 kJ/s with two coal mills each supplying one burner level.

11.4.8 Furnace Startup and Shutdown

Unlike design models, the models used in operational simulators must provide realistic and accurate dynamic results under all normal and many abnormal operating phases and conditions. For this reason it is not advisable to ignore individual aspects of the process physics because they seem negligible compared to others at, say, full load because those same aspects can become significant under other operating conditions.

Normal operations include startups and shutdowns. A cold start requires modelling of the furnace with no airflow, establishment of minimum airflow or furnace purge prior to initial ignition, igniter sequence start, then progressive addition of furnace energy release up to the 100 % continuous maximum rating (CMR) condition. A startup can be conducted with the furnace and its associated piping and fittings in any thermodynamic state, reachable from an orderly or emergency shutdown. Shutdowns can be orderly reductions of load to some minimum level from which a furnace trip can be initiated, or can be initiated as an emergency or unintentional master fuel trip (MFT) from any load.

Initial Start of Igniters

Each primary and secondary fuel burner is associated with an igniter. As its name suggests, an igniter is intended to provide the minimum energy needed to initiate the combustion reaction of a main fuel. An igniter is a relatively low energy source and usually burns light fuel oil (diesel) or a fuel gas such as natural gas or butane. The flow of fuel to an igniter is fixed, that is, the igniter is either on or off, but some adjustment of igniter flows might be introduced by varying the fuel supply pressure, in which case all igniters will respond equally.

Igniter heat release is small compared to the primary fuels. In the absence of any other fuel source, igniters operate in an air-rich environment. The minimum airflow for ignition is usually many times the stoichiometric requirement and during operation with igniters only, fluegas oxygen levels will exceed 15 %.

Furnace Trip

A furnace trip is usually triggered by the furnace protection or burner management system (BMS). It may be initiated manually by the plant operator. In every case it leads to an immediate shutoff of all fuel to the furnace and may initiate a runback of furnace airflow to that required for a furnace purge. This flow is usually maintained until re-ignition of burners. Immediately following the trip the furnace exit temperature, at 1,100–1,300 °C prior to the trip, drops rapidly to around 300–400 °C before increasing slowly as the furnace airflow absorbs heat from the hot waterwalls and furnace fittings (convective heat transfer). The heat pickup continues as the airflow passes through the superheaters, reheater and economiser before arriving at the airheater across which it transfers part of its heat to the incoming combustion air. As the gas temperature to the airheater increases so too does that of the hot air to the furnace. This recuperative cycle continues until the boiler metal cools sufficiently to allow the air temperatures to begin reducing. This may take up to 20–30 min after the trip, after which the complete system begins to decrease in temperature. Cool-down rates of around 20 °C per hour might be expected.

Since a furnace trip implies no ignition in the furnace the calculation method described in Sect. 11.4.4 must be modified. The flame temperature defined at the reference burner level must be replaced by some other temperature reference, from which the furnace gas profile can be calculated. An appropriate reference is the temperature of the incoming combustion air flow.

11.4.9 Effect of Burner Tilt

Tangentially fired furnaces usually have provision for changing the inclination of the complete burner register relative to the horizontal and thereby the "pointing" direction of the flame. All burners at a single corner are mechanically linked together and move as a single unit. The registers at the individual corners are electrically

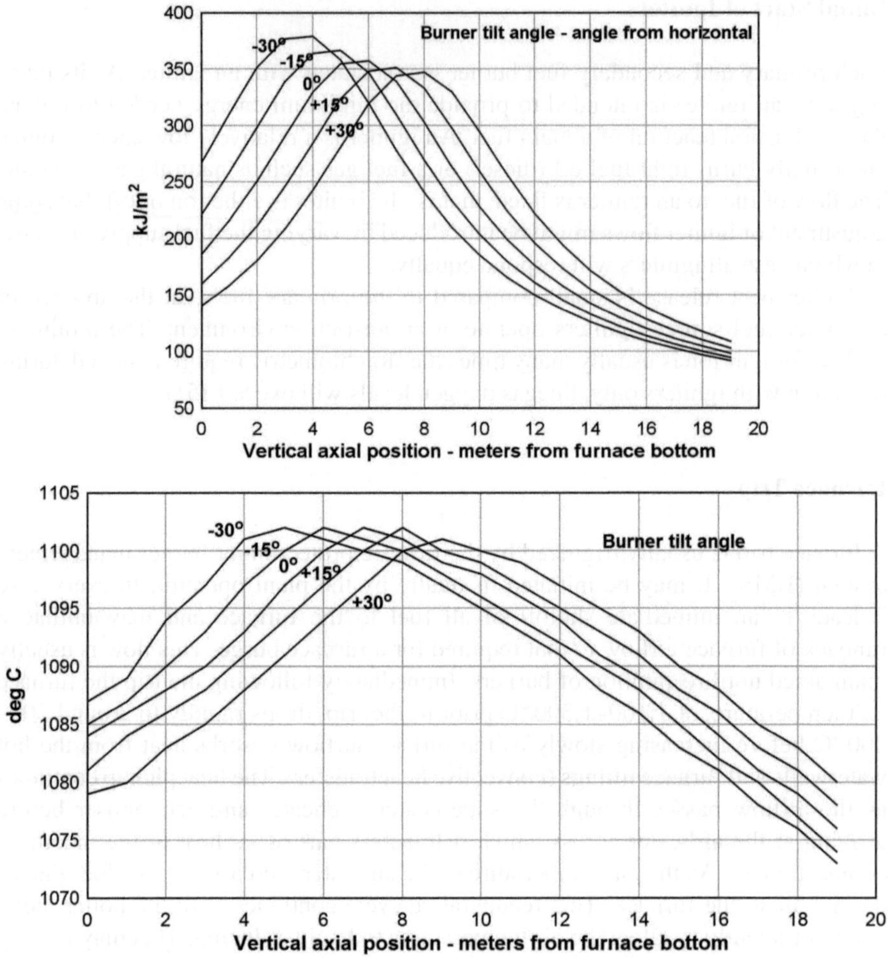

Fig. 11.14 Effect of burner tilts on furnace heat and temperature profiles

driven by a common control signal and move in synchronism. As the complete set of burners moves up or down, the centres of the individual flames move with them within the furnace. This has the effect of displacing the heat flux and temperature profiles by a similar distance. The flame displacement can be calculated simply from the projection of the tilt angle on the furnace centre line.

$$\Delta z_{fl,bt} = \frac{WOF}{2}\tan(\theta_{bt}),$$

where WOF is the width of the furnace at the burner register level. θ_{bt} is the angle through which the burners tilt and is usually limited to $\pm 30°$. The effect is illustrated by Fig. 11.14 which applies to the utility furnace with coal firing.

The overall effects can be summarised as:

- Tilt up: less heat to the waterwalls and an increase in furnace exit temperature
- Tilt down: more heat to the waterwalls and a decrease in furnace exit temperature

Tilting burners are usually used for the control of reheater steam temperatures. The change in heat to the waterwalls means that some change to firing rate must be anticipated to maintain steaming rate and load.

11.4.10 The Dynamics of Furnace Heat Release

Combustion rates of fuels as they are introduced into the furnace zone in which sufficient energy is available to initiate combustion can be determined from the Arrhenius equation. For gaseous and volatile liquid fuels the combustion rate is very high and can safely be treated as instantaneous, certainly when compared with the dominant furnace dynamics. Coal is different and the calculation of the combustion dynamics is based on the following assumed sequence of events.

1. The incoming coal flow is raised to ignition temperature by the heat already available in the furnace. This heat is provided either by the igniters or by the combustion of fuel from other burners. A minimum heat release of, say, 5,000 kJ/s might be assumed as a pre-requisite for combustion.
2. The pulverised particles begin combusting after a negligible delay and continue to combust and release heat for some seconds. This time is characterised by the burnout time constant for the coal which may be likened to the mean lifetime of a pulverised coal particle.

With
- M_j — the mass of combusting pf particles from the jth burner
- $\dot{m}_{pf,j}$ — the mass flow of pf leaving the jth burner
- k — the reaction rate constant for a pf particle [s^{-1}]

and assuming an exponential form of the reaction equation for the pf particles,

$$\frac{\mathrm{d}c}{\mathrm{d}t} = -k\,c$$

the mass balance for M_j for the jth burner is simply,

$$\frac{\mathrm{d}M_j}{\mathrm{d}t} = \dot{m}_{pf,j} - k\,M_j$$

since kM_j is the rate of consumption of the incoming pulverised coal stream. In order to transform the reaction rate constant k into a more tangible form, introduce the new variable $S_j = k\ M_j$. Then

$$\tau \frac{\mathrm{d}S_j}{\mathrm{d}t} = \dot{m}_{pf,j} - S_j. \tag{11.13}$$

The burner combustion time constant (seconds) $\tau = 1/k$ can be identified with the mean pf particle life, typically in the range of 2–3 s. It is assumed here that all of the fixed carbon is consumed. In practice, a small fraction may remain (the "unburnt carbon in ash") as the ash layer formed on the outside of the coal particle during combustion can insulate a small residual mass of carbon and prevent its complete combustion.

The heat made available by combustion of the fuel from the jth burner is the product of the S_j, the CV of the fuel and the combustion efficiency. The total heat release is $\sum(S_j)$, summed over all burners. The combustion efficiency accounts for the heat lost in raising the fuel to combustion temperature, the evaporation of the volatile and moisture contents and the heat lost with the ash.

Influence of Coal and Pulveriser Parameters

The pulverised coal parameters can be related to the performance of the milling plant by some assumptions which reflect operational experience. For example, worn mills produce larger pulverised fuel particles which burn longer (perhaps up to 8 seconds) at a lower temperature and propagate combusting particles further along the furnace gas path. These operational issues can best be treated empirically, without the need for complex particle combustion characterisation which might include the influence of, for example, ash content and particle size on particle burning rates and dwell times in the furnace.

Chapter 12
Boiler Circuits and Steam Generation

As described in the previous chapter, furnaces and boilers are manufactured in a wide range of types and sizes. Regardless of steaming capacity, the steam generation process in any boiler is the same. Heat released in the furnace by the combustion of a fuel is transferred to water in tubes in which steam is formed. Boilers are distinguished one from another by their physical size, the balance between radiative and convective heat transfer from the furnace to the water tubes, the rate of heat release and steam production, the method of separation of the steam and water phases and the operating pressure and temperature of the steam produced.

The simulation objective will vary, depending on the purpose to which the simulation will be put. Smaller boilers in process plants are generally part of the site's energy recovery and generation, and the operational interest in the boilers is frequently their integration with the site's overall steam and fuel supply systems. The detailed behaviour of the boiler plant itself is of secondary interest, and a lower level of simulation detail will suffice. The operational interest for a power utility plant will be operator training and the development and pre-commissioning tuning of a plant's automation and control system. Increasingly this interest is being shown by the process industries and therefore even for industrial boilers, the detailed behaviour of the boiler plant, during both normal and abnormal operations, will be of interest and may require a more precise treatment of the boiler and furnace dynamics.

12.1 Power Generation Utility Boilers

Large steam power-generating plants fired with fossil fuels are built in any of two basic configurations. In regions with relatively low fuel costs, boilers below about 600 MW have usually been of the drum type, with one or two horizontally mounted steam/water separating drums mounted at the top elevation of the waterwall. Drum boilers are limited in rating to less than 700 MW or so because of the high costs of the drums and of the structure needed to support them. Some reduction in structural

P. O'Kelly, *Computer Simulation of Thermal Plant Operations*, DOI 10.1007/978-1-4614-4256-1_12,

costs in very large boilers can be gained by using twin but smaller drums. These boilers operate with continuous circulation of water from the drum to the bottom of the evaporator (the riser) and thence back to the drum. They are limited to operating pressures less than 19 MPa, and the fluid emerging from the top of the riser contains up to 33 % steam by mass.

If fuel costs are relatively high or if advantage is to be taken of the improved efficiencies achievable at higher steam temperatures, once-through boilers may be preferred for units with rated outputs above 400 MW. These plants do not maintain a circulation flow during normal loaded operation but instead introduce feedwater directly into the riser where it is raised first to saturation temperature and then converted progressively to 100 % steam with some superheating while still in the riser. There being no water phase left in the riser exit flow, there is no need for water/steam separation, and a drum is not required. However, when starting up, these boilers operate in recirculating mode and during this phase water/steam separation is required.

Once-through plants are built as *subcritical*, with rated steam conditions below the critical pressure (22.4 MPa), or *supercritical.* Operating conditions of supercritical plant vary throughout the world, being typically around 25 MPa (3,700 psi) and 537°C (1,000°F) in the USA and 28.5 MPa and 620°C (1,100°F) in Europe. Future designs, termed ultra-supercritical (USC), are planned with operating pressures up to 34 MPa and temperatures up to 700°C [93]. The two most common types are Benson (licence held by Siemens AG, Germany) and Sulzer (Sulzer Bros., Switzerland). The Benson design is intended for once-through operation at supercritical pressures. The Sulzer design was originally developed as a means of circumventing the high-water-purity requirement of the Benson design and was intended to operate at subcritical pressures with a small continuous recirculation flow. Continuous blowdown from the water separator assists water quality control. This arrangement is still in use and maintains a riser exit mass fraction of around 97%. This type of plant is frequently used with lignite (brown coal) firing in which furnace exit temperatures must be held below the ash fusion temperature to avoid serious slag formation problems in the upper furnace and superheating regions. The following table summarises some salient features and merits of each type.

The choice of a particular configuration will always be influenced by construction costs but may include operational considerations, for example, the ash slagging issue.

12.1.1 Boiler Design Parameters

The vertical profile of heat flux to the waterwalls is not uniform and exhibits a peak in the region of the burners. In most furnaces these are located in the lower half with the elevation of the uppermost burner row about 1/3 to 1/2 the height of the furnace. The waterwall tubes are welded together to form a membrane wall, with a tube centre line spacing typically around 1.5–2 times the tube outer diameter.

	Drum boiler	Once-through
Capital cost	Ultimately limited by cost of drum and support structure	High cost of special steel thick-walled tubing less important at higher ratings Sensitivity to water purity penalises cost by requiring expensive demineralising plant
Construction features	Riser formed from large number of smooth-bore tubes	Small number of spiral-wound tubes in lower evaporator section with a larger number of smooth vertical tubes in the upper section May use rifled tubes to enhance heat transfer in the spiral-wound section
Efficiency	Maximum efficiency limited by attainable pressure and temperature ($< p_{crit}$)	High efficiencies achievable at supercritical pressures and temperatures
Operations	Start-up and load-change rates limited by drum stress limits (<50°C/h)	Fast start-up and shutdown due to less restrictive stress limits
	Simple operation and control	Complex start-up procedure and on-load control
	Normally operated with fixed pressure and temperature	Well suited to sliding pressure operation

The design of the heat flux distribution attempts to minimise large temperature differentials which could cause uneven expansion and thermal stresses within the walls. Because of their higher operating temperatures, once-through boilers use spiral-wound tubing in the lower furnace area—the region of highest heat flux—to assist this objective. Spiral-wound tubes are longer than vertically arranged tubes for the same heat transfer area. The tubes being longer, the residence time of the riser fluid in the tubes is greater. The gravitational head, expressed in terms of the resolved vertical component of the flow, is $g \sin \beta$ where β is the angle of inclination of the tubes to the horizontal. Typical inclination angles are 10–20° giving tube lengths 3–5 times the vertical height. In other boiler designs, the riser tubes are grouped into separate banks in order to optimise heat flux and temperature distribution.

Basu et al. [15] quote recommended values of flow velocities and specific mass flows (kg/s/m^2) for various boiler configurations. The following table presents selected extracts from that reference. All figures are quoted on a per tube basis.

Natural- and assisted-circulation boilers—water velocities

Drum pressure MPa	4–6	10–12	14–16	17–19
Water velocity at riser inlet m/s	0.5–1.0	1.0–1.5	1.0–1.5	1.5–2.0
Downcomer inlet velocity m/s	≤3	≤3.5	≤3.5	≤4

Once-through boilers—specific mass flow rates [kg/s/m^2] in the lower radiative region

	Vertical tube panel		
	Single pass	Multi pass	Spiral-wound
Subcritical	1600–2700	1200–2000	1500–2500
Supercritical	2100–2700	1600–2000	2000–3000

These guidelines can be used to determine the principal design parameters of an evaporator, given its type and rating. This can be useful if detailed plant data is not available.

Example 1: Supercritical Benson boiler

Rated MCR	435 MW
Rated steam flow	350 kg/s
Rated operating pressure	24.5 MPa (242 bar)
Rated steam temperature at boiler exit	589°C
Water spec. vol. at riser inlet v'	0.001556 m^3/kg
Riser inlet temperature (from economiser)	290°C
Target mass flow density ϕ	2,000 kg/s/m^2
Riser inlet flow velocity $= \phi\, v'$	3.1 m/s
Assumed tube ID	0.036 m
Inlet flow rate per tube $= \phi\, A$	2.04 kg/s
Number of riser tubes	172

Example 2: Natural-circulation drum boiler

Rated MCR	500 MW
Rated steam flow	400 kg/s
Rated operating pressure	18.24 MPa (180 bar)
Rated steam temperature at boiler exit	540°C
Water spec. vol. at riser inlet v'	0.00104 m^3/kg
Riser inlet temperature (from economiser)	320°C
Target circulation ratio	7:1
Target circulation flow rate	2,940 kg/s
Target mass flow density ϕ	1,500 kg/s/m^2
Riser inlet flow velocity $= \phi\, v'$	1.56 m/s
Assumed tube ID	(i) 0.054 (ii) 0.064 m
Inlet flow rate per tube	(i) 3.45 (ii) 4.83 kg/s
Number of riser tubes	(i) 852 (ii) 609

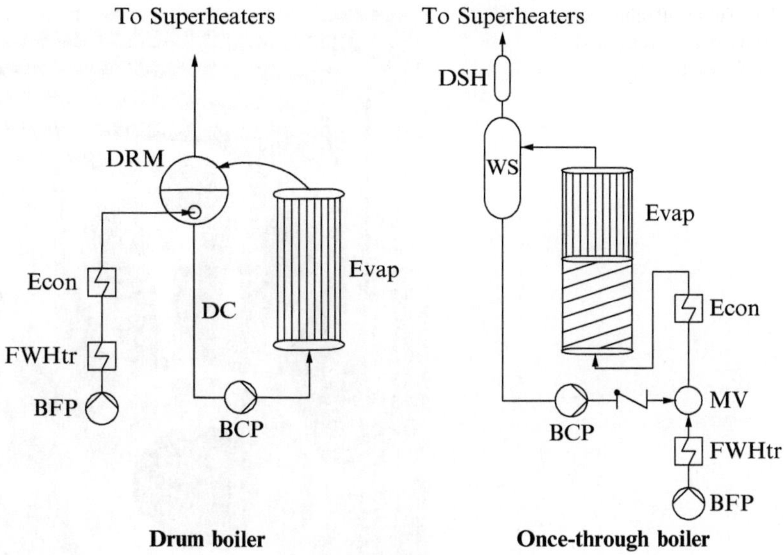

DRM Steam drum
WS Water separator
BCP Boiler circulationpump
BFP Boiler feedwaterpump
Evap Evaporator
DC Downcomers
Econ Economiser
FWH Feedweaterheater
MV Mixing vessel
DSH Desuperheater

Fig. 12.1 Water circuit arrangements of a power plant boiler

12.1.2 Evaporator Water Circuits

The general arrangement of the evaporator circuit is shown by Fig. 12.1 for a drum-type and a once-through boiler.

Drum Boilers

Wet steam emerging from the riser is separated into its steam and water components in the drum(s). The steam phase leaves the drum as saturated steam and flows to the superheaters while the water phase is mixed with incoming feedwater and recirculated. Water travels from the drum to the bottom of the boiler via the downcomer tubes, a set of typically 3–9 large pipes set into the bottom of the drum. The entry orifice to each downcomer tube is fitted with a coarse grid of crossed plates to prevent the formation of vortices (vortex suppressors). The minimum water

Fig. 12.2 Tubes attached to a lower boiler header (courtesy of Foster-Wheeler corp.)

volume permitted in the drum during operation is chosen to ensure the absence of vortices which could draw steam into the downcomers. The presence of a vapour phase in the downcomers can adversely affect the circulation flow and initiate hydrodynamic flow instability.

The downcomer pipes receive no heat from the furnace but lose heat to ambient. They terminate at the bottom of the boiler in the bottom header(s) (Fig. 12.2) from which the flow divides into a large number of parallel streams (individual riser tubes) which ascend back to the boiler drum. These are usually divided into sets of vertical banks of tubes, distributed around and forming the walls of the furnace. A number of boiler circulation pumps (BCPs) (usually less than four) may be installed between the bottom header(s) and the riser banks. A boiler with circulation pumps is termed an *assisted-circulation* boiler, and the pumps run continuously. One without pumps is termed a *natural-circulation* boiler, the circulation flow being induced by the buoyancy of the steam phase in the risers. Circulation is maintained around the loop during all phases of operation, except of course when the plant is out of service. While this may seem a trite observation, it highlights the need, from a modelling perspective, to ensure validity of the model when flow is zero. In a drum boiler with assisted circulation, water will leave the riser with a steam mass fraction of up to around 33 % at rated output. The riser exit mass fraction in a natural-circulation boiler is less, at around 12–18 % at full load. In these boilers, to reduce the pressure losses around the circuit, the riser and downcomer tubes are of larger diameter than those in assisted-circulation boilers.

The ratio of the full-load steady-state steam flow leaving the evaporator—equal to the steam production rate—to the total circulation flow is the *circulation ratio*. It is the inverse of the mean steam mass fraction at the riser exit and is typically around 3:1 for an assisted-circulation boiler, 7:1 for natural circulation and 10:1 for an industrial boiler.

The vertical height differential of the riser tubes (the difference between the top and bottom elevations) will typically be of the order of 20–50 m for a drum boiler and up to 90 m for large supercritical boilers, which tend to be built in tower configuration. Tube numbers vary in the range of 150–1,200 with an outer diameter of between 35 and 85 mm. The wall thickness depends on the tube internal diameter and operating pressure and usually lies in the range 3–6 mm. The heat transfer area is large, and the tubes are exposed on the furnace side to radiant and convective heat transfer from the furnace chamber. In consequence of the vertical head, a static pressure differential exists between top and bottom of the riser (of the order of 300–800 kPa prior to boiling when full of water, reducing with the formation of steam), creating local variations of saturation conditions which will influence the local formation of steam, particularly under the low-pressure conditions during start-up.

Feedwater is introduced into the drum where it mixes with the drum water contents. The sub-cooling effect of the incoming feedwater keeps the water temperature in the drum below the saturation temperature. This and the higher static pressure at the riser inlet ensure that the riser inlet flow is sub-cooled. It should be noted that prior to first boil-off, there will be no feedwater flow to the drum, unless needed to make up water lost through the drum blowdown which, in some plants, is a normal part of the start-up procedure in order to maintain the necessary boiler water quality.

Once-Through Evaporators

The flow up the riser is progressively heated by heat absorbed from the tube walls. In a drum boiler under load, the riser flow reaches the top of the riser with up to around 2/3 water content. In a once-through boiler operating in once-through mode, the riser converts completely to steam somewhere before the riser exit. The point along the riser flow path at which the water phase disappears (steam mass fraction becomes 1) is termed the dry-out point. During start-up, shutting down and at low load, dry out will not be achieved, and the riser exit flow will need separation of the phases. A vertical water/steam separator vessel is provided between riser exit and inlet, with a forced recirculation path back to the riser inlet. During operation with recirculation, feedwater is introduced into and mixed with water leaving the separator vessel in a mixing chamber below the bottom of the water separator. In some designs, the mixing chamber may be located upstream of the economiser. The mixture is passed to the bottom of the riser and ascends in the usual way but now forced by the feedwater pumps, an arrangement called *forced circulation* operation. The total flow to the riser is the sum of the feedwater flow from the feed pump(s) and the recirculation flow from the boiler recirculation pump. At least a minimum flow to the evaporator tubes is guaranteed at all times to ensure adequate cooling of the economiser and boiler tubes. Circulation around the separator loop is maintained up to around 40–50 % load (start-up and low-load operation) or at some predetermined steam pressure (typically around 12–14 MPa) at which time the riser exit flow to the water separator is close to or has passed the dry-out point. At this stage, the recirculation flow is stopped, the recirculation line isolated and once-through operation commenced. Without the recirculation mixing, the riser inlet temperature is the final feedwater temperature (economiser outlet) which, to avoid steam formation in the economiser, will be less than the local saturation temperature and again, the incoming water is sub-cooled.

In boilers subject to frequent starts and shutdowns, an alternative recirculation arrangement may be used whereby the liquid phase from the water separator is passed through a recuperative heat exchanger before being passed to the deaerator. This arrangement can improve overall efficiencies by recovering blowdown losses. The return of the drain to the deaerator raises the deaerator water temperature and reduces bled steam demand, both by the deaerator and HP feedwater heaters [94].

Steam Formation

As the water flows upwards along the riser tubes, it absorbs heat from the tubes and its enthalpy increases. Since it is moving upwards, static pressure will decrease with height and with it the saturation temperature. If sufficient heat is absorbed from the tubes, the water enthalpy will be increased to the local saturation enthalpy. At this point, the flow is no longer sub-cooled and the addition of any further heat will cause boiling to commence. The point of transition between saturated water and saturated two-phase fluid is denoted the "boiling boundary". This is not a sharp demarcation point and is preceded by a sub-cooled zone in which small steam bubbles form.[1] If insufficient heat is absorbed by the water flow, because either the inlet enthalpy or the riser tube temperature is too low, the riser flow will not reach saturation enthalpy and will remain sub-cooled over its complete length. This is the case during start-up prior to the first boil-off and can also be established some time after a unit trip.

Experimental data indicates that local evaporation commences in the riser before the bulk flow has reached saturation enthalpy. Due to the high heat flux at the tube/fluid interface, the fluid temperature at the interface will be higher than the bulk fluid temperature, and steam bubbles will form at this interface before bulk boiling commences. The bubbles detach from the tube wall and are entrained by the upward riser flow. The fluid velocity is highest and hence the static pressure is lowest along the centre line of the riser flow. The steam bubbles from the tube interface will move towards this region of local low pressure. Thus, although steam is produced predominantly at the tube wall interface, the riser flow in cross section

[1] A more detailed discussion of this and other boiling processes is given in Chap. 13.

shows a higher concentration of steam bubbles towards the centre line than at the interface. Bubbles coalesce into clumps and large voids, passing through the various flow regimes described in Chap. 13.

Above the boiling boundary the riser flow develops into two phases as an increasing fraction of the water flow converts to steam. This gives rise to increases in,

(a) The two-phase pressure loss along the riser attributable to the effective reduction of the water-phase flow area by the steam phase.
(b) The volume occupied by the steam phase (increases in the ratio of $v'' : v'$) with a corresponding displacement of water towards the upper reaches of the riser.
(c) The upward buoyancy force due to the reduction in fluid density.
(d) The two-phase heat transfer due to the effects of local turbulence caused by the production and movement of the steam bubbles.

In the once-through evaporator, as the dry-out point is approached, the wall heat transfer coefficient decreases markedly, causing an attendant sharp rise in tube wall temperature. Beyond the dry-out point, the steam is single phase and the addition of further heat increases its temperature.

Various valves attached to the evaporator circuit include

- Safety valves
- Blowdown valves
- Waterwall upper and lower header drains
- Water separator drains
- Boiler circulation drains

12.2 Modelling Treatment

The primary interest of the evaporator model is the calculation of the profiles of riser fluid and metal temperatures, circulation flow rate, steam production rate and the relative fractions of water and steam phases in the riser tubes.

Many of the operational characteristics of the evaporator loop result from its essentially distributed nature, that is, the influence of processes arising from variation of state and property variables in and across the direction of flow. By treating only single-dimensional variations in the axial flow direction, we will tacitly assume uniform fluid properties normal to the direction of flow.

The differential equations determining the dynamic behaviour of the evaporator system are derived by developing the non-stationary mass, momentum and energy balances for the drum or water separator, the downcomer and riser and the overall circulation flow. The following table summarises the salient differences of drum and once-through boilers from the modelling perspective.

	Drum	Once-through mode Recirculation	Once-through
Pressure defined by	Drum	Water separator	FW pumps
Circulation	Natural or assisted	Forced	Forced
Feedwater to	Drum	Riser	Riser
Riser flow =	Downcomer flow	Recirc + feedwater	Feedwater

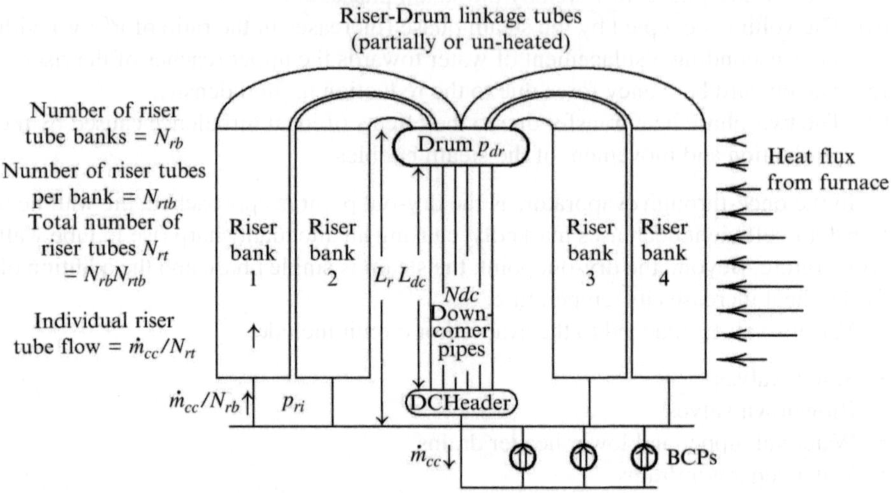

Fig. 12.3 Schematic representation of drum boiler circulation paths

12.2.1 Drum Boilers

The total circulation mass flow $\dot{m}_{cc}$ is obtained from the balance of individual pressure rises and drops around the evaporator circuit.

drum static head − riser static head + circulation pumps developed head
= downcomer head loss + riser head loss (one/two phase).

With reference to Fig. 12.3, the first term is simply $g\,L_{dc}/v_{dc}$. The riser static head is identical, with the downcomer water specific volume v_{dc} replaced by the mean riser specific volume v_{rm}. L_r denotes the vertical length of the riser and L_{dc} that of the downcomer. The net buoyancy head is then $g\,(L_{dc}/v_{dc} - L_r/v_{rm})$. The effective geodetic heights of both riser and downcomer may be taken as equal by referring each to the same upper and lower datums. The tubing linking the riser exit to the drum may be taken as contributing zero net height.

The circulation flow is assumed evenly distributed across the downcomer tubes and the individual downcomer tube flow $\dot{m}_{dc} = \dot{m}_{cc}/N_{dc}$. The downcomer head loss is the friction loss per tube $k_{dc}\dot{m}_{dc}^2$. The friction loss coefficient is given from Eq. 6.6 as

$$k_{dc} = \frac{\xi L_{dc} v_{dc}}{2\, d_{dc} A_{xdc}^2}$$

A_{xdc} is the flow cross-sectional area of a downcomer pipe and ξ is the Fanning friction factor. Alternatively, Zlin [47] Chap. 8, quotes

$$k_{dc} = \left(\xi \frac{L_{dc}}{d_{dc}} + \sum \xi_M \right) \frac{\nu^2}{2} \rho_l,$$

where ν is the flow velocity = $\dot{m}v/A_{xdc}$. The friction factor ξ is calculated from

$$\xi = 0.25\,[\log(3.7\, d_i/k)]^{-2}$$

and k is the hydraulic roughness [mm].

The downcomer bank pressure loss coefficient is

$$K_{dc} = \frac{1}{N_{dc}^2} k_{dc}.$$

Figure 12.3 depicts the riser as divided into four parallel and, for the present purpose, identical banks of N_{rtb} tubes per bank. In general, it may be divided into any number N_{rb} of unequal banks. This is consistent with the usual construction of boilers in which the riser may be divided into front, rear, left and right sidewalls and upper, lower, screen and sundry other tube bank designations. If N_{rt} is the total number of riser tubes (= $N_{rb} N_{rtb}$), the mass flow per tube is $\dot{m}_{rt} = \dot{m}_{cc}/N_{rt}$. The single-phase pressure gradient along the riser is $k_{1\phi} \dot{m}_{rt}^2$ where the pressure loss coefficient is

$$k_{1\phi} = \frac{\xi\, v_r}{2\, d_{i,r} A_{x,r}^2}.$$

The riser head loss is composed of the single-phase friction drop in the sub-cooled zone (water) plus the two-phase pressure drop in the saturated or boiling zone. Two-phase effects are accommodated by multiplying the single-phase pressure loss coefficient $k_{1\phi}$ by an appropriate two-phase multiplier, calculated as a function of local flow conditions.[2] For example, using the Doležal multiplier (Eq. 13.3),

$$\phi_{lo}^2 = 1 + x\left(\frac{v''}{v'} - 1\right) + 8\frac{\dot{q}v''}{k_{r,1ph}\mathbf{r}v_0}.$$

Assuming identical thermal and flow conditions in each tube, the per tube pressure loss coefficient $k_{1\phi}$ in the riser converts to a bank-based coefficient $K_{1\phi}$ as

$$K_{1\phi} = k_{1\phi}/N_{rt}^2.$$

[2] Refer to Sect. 13.1.5.

As discussed in Sect. 7.2.1, the BCP head-flow characteristic can be approximated by a parabolic curve passing through selected points on the test characteristic.

$$\Delta p_{bcp}(n_{bcp}) = dp_0(\omega^2) - k_{bcp}\dot{m}_{cc}^2$$

dp_0 is the zero-flow head of the pump and is proportional to the square of the pump rotational speed ω. n_{bcp} denotes the number of BCPs in service.

Combining these terms gives

$$g\left[\frac{L_{dc}}{v_{dc}} - \frac{L_r}{v_{rm}}\right] + dp_0(\omega_{bcp}^2) - k_{bcp}(n_{bcp})\dot{m}_{cc}^2 = K_{dc}\dot{m}_{cc}^2 + K_{1\phi}\phi_{lo}^2\dot{m}_{cc}^2. \quad (12.1)$$

This quadratic equation can be solved for the total circulation flow $\dot{m}_{cc}$, given the combined head/flow characteristic of the in-service BCPs.

12.2.2 The Boiler Steam Drum

A boiler may be equipped with one or more steam drums, depending on rating and construction type. The drum serves several purposes:

(a) Separates the water and vapour phases of the fluid emerging from the riser.
(b) Mixes the incoming feedwater with water in the drum to provide the water source for the evaporator.
(c) Establishes a reference point for control of the water mass within the evaporator system.
(d) Provides a droplet-free supply of steam to the main steam system.

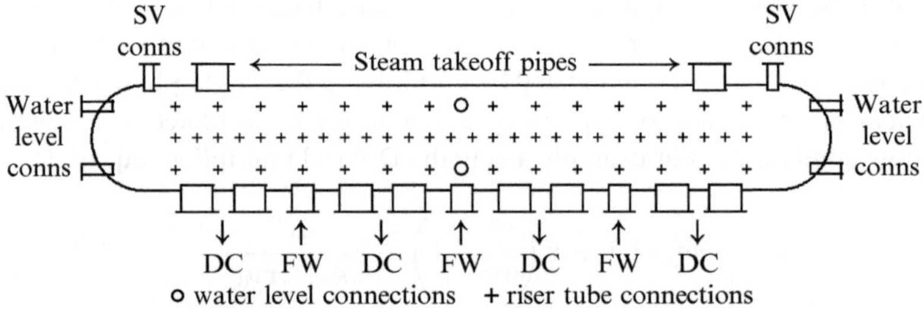

From the simulation point of view, the drum provides the thermodynamic boundary conditions for both the evaporator and the downstream steam system and, most importantly, steam pressure and its rate of change. In the following discussion, the drum is treated as an isolated component, connected to its interfacing systems by conditions defined at its boundaries. Although the drum is relatively long in comparison with its diameter, the distributed effects of the complex flow patterns within the drum will be ignored. For example, it will be assumed that the incoming feedwater mixes immediately and perfectly with the existing water phase to produce a single representative drum water temperature. In the real drum, this is not the case, and temperature variations exist throughout the water, both lengthwise along the drum and vertically.

Uneven patterns of flow to the downcomers and emerging from the risers produce local variations of water level along the length of the drum. To detect these it is usual practice to provide sets of tapping points for water-level measurements at a minimum of three locations along the drum—one at each end and one in the middle. It has been observed during operation that the level profile varies with operating condition such as load, but it is difficult to identify generalised repeatable and predictable patterns as they depend strongly on plant construction details. An important influence parameter is the method of connection of feedwater piping to the drum. The preceding diagram shows feedwater connection stubs interleaved between the downcomers. An alternative method is to connect the feedwater inflow to one end of the drum, in larger plants via two stubs, and distribute it along the drum through perforations in the pipe(s), under the normal water level.

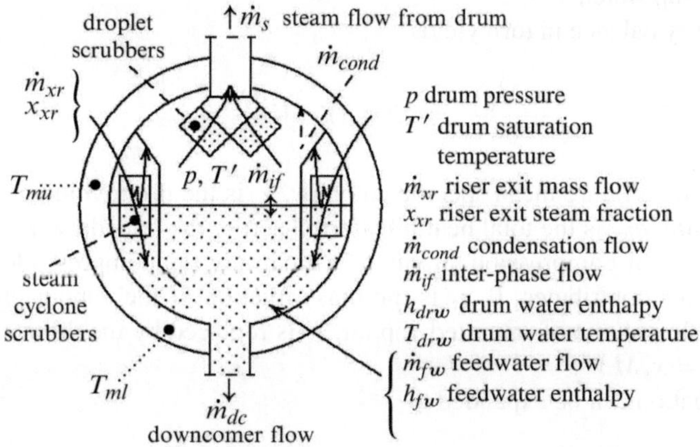

The water and steam phases occupy well-delineated spaces which meet at the surface of the water space. Given the large mass and energy flows within the drum and their turbulent conditions, this interface surface is unlikely to be either smooth or steady. However, for this level of modelling, it will be expedient to treat it as smooth and sufficiently well defined to serve as a measurable line of separation.

Differing approaches can be taken to the calculation of the thermodynamic state of the drum contents. The simplest is to treat them as a saturated homogeneous two-phase mixture of liquid and vapour phases coexisting in thermodynamic equilibrium. The temperature of the mixture is the saturation temperature corresponding to the drum pressure. Mass and energy balances for the contents can then be written in terms of the net mass $\Sigma(\dot{m})$ and enthalpy $\Sigma(\dot{m}\,h)$ flows crossing the boundary.

This approach of necessity ignores any inter-phase mass or energy transfer processes which might be taking place within the drum.

The Homogeneous Contents Approach

The mass balance taken over the drum contents, assumed homogeneous, yields

$$\frac{\mathrm{d}M}{\mathrm{d}t} = \Sigma(\dot{m}), \tag{12.2}$$

where M is the total mass of the drum contents. The net mass balance of all flows crossing the drum's boundary is

$$\Sigma(\dot{m}) = \dot{m}_{fw} + \dot{m}_{rx} - \dot{m}_{dc} - \dot{m}_s - \dot{m}_{bl}$$

$\dot{m}_{bl}$ is the blowdown flow from the drum. This is taken from the drum as part of the chemical water treatment system and is used to reduce dissolved solids in the boiler water. It may be an intermittent or continuous flow, depending on the quality of the boiler make-up water.

The energy balance in turn yields

$$\frac{\mathrm{d}(M\,h)}{\mathrm{d}t} = \Sigma(\dot{m}\,h) - V\frac{\mathrm{d}p}{\mathrm{d}t} + \dot{q}_{hx},$$

where h is the mixture mean specific enthalpy, V is the drum volume and p is the drum pressure. $\dot{q}_{hx}$ is the total heat transferred across the boundary. $V\,\mathrm{d}p/\mathrm{d}t$ is the rate of change of compression energy to which only the compressible part of the drum contents contributes. If x_s is the mass fraction of the vapour phase and v'' is the specific volume of saturated vapour, V is replaced by the drum steam-phase volume $V_s = x_s M\,v''$.

This equation can be expanded to

$$h\frac{\mathrm{d}M}{\mathrm{d}t} + M\frac{\mathrm{d}h}{\mathrm{d}t} = \Sigma(\dot{m}\,h) - V_s\frac{\mathrm{d}p}{\mathrm{d}t} + \dot{q}_{hx}$$

from which

$$M\frac{\mathrm{d}h}{\mathrm{d}t} + x_s M\,v''\frac{\mathrm{d}p}{\mathrm{d}t} = \Sigma(\dot{m}\,h) - h\Sigma(\dot{m}) + \dot{q}_{hx}. \tag{12.3}$$

Equations 12.2 and 12.3 contain four unknowns, M, h, p and x_s, leaving two further equations to find. The first of these is provided by the assumption that the drum contents exist in a state of thermodynamic equilibrium. This allows the replacement of h by its representation in terms of x_s and the saturation specific enthalpies of the liquid h' and vapour h'' phases.

$$h = x_s h'' + (1 - x_s)h'.$$

We could then write

$$
\begin{aligned}
\frac{\mathrm{d}h}{\mathrm{d}t} &= \frac{\mathrm{d}}{\mathrm{d}t}(x_s h'' + (1-x_s)h') \\
&= x_s\frac{\mathrm{d}h''}{\mathrm{d}t} + h''\frac{\mathrm{d}x_s}{\mathrm{d}t} + (1-x_s)\frac{\mathrm{d}h'}{\mathrm{d}t} - h'\frac{\mathrm{d}x_s}{\mathrm{d}t} \\
&= x_s\frac{\mathrm{d}h''}{\mathrm{d}t} + (h''-h')\frac{\mathrm{d}x_s}{\mathrm{d}t} + (1-x_s)\frac{\mathrm{d}h'}{\mathrm{d}t} \\
&= x_s\frac{\mathrm{d}h''}{\mathrm{d}t} + \mathbf{r}\frac{\mathrm{d}x_s}{\mathrm{d}t} + (1-x_s)\frac{\mathrm{d}h'}{\mathrm{d}t},
\end{aligned}
$$

where $\mathbf{r} = h'' - h'$. Since h' and h'' are functions of the single variable p, the derivatives of h' and h'' may be written in terms of $\mathrm{d}p/\mathrm{d}t$ using their partial derivatives.

$$
\frac{\mathrm{d}h}{\mathrm{d}t} = \left(x_s\frac{\partial h''}{\partial p} + (1-x_s)\frac{\partial h'}{\partial p}\right)\frac{\mathrm{d}p}{\mathrm{d}t} + \mathbf{r}\frac{\mathrm{d}x_s}{\mathrm{d}t}.
$$

This equation would allow the elimination of $\mathrm{d}h/\mathrm{d}t$ from Eq. 12.3 but introduces a new time derivative $\mathrm{d}x_s/\mathrm{d}t$ for which an explicit expression must be found. While this can be found from the definition of x_s the resulting equations become quite cumbersome, and an alternative approach should be sought.

The Separated Phases Approach

We first observe that the assumed homogeneity of the drum contents is a poor representation of reality. There are clear divisions between the liquid and vapour phases which occupy different spaces at different temperatures. The flows from the drum to the downcomers and to the steam system are of different phases and specific enthalpies, and the individual terms making up $\Sigma(\dot{m}\,h)$ cannot be expressed simply in terms of the mixture enthalpy h. Further, the assumption of saturation conditions in the drum obviates the need for a separate energy balance, and therefore, the energy equation 12.3 is unnecessary. Were we to relax the saturation assumption, we would need a separate energy balance for each phase to define the conditions for return to saturation and would need to describe a mechanism for mass and energy exchange between the phases.

These issues can be avoided if we proceed as follows. The vapour-phase mass fraction x_s is defined to be

$$
x_s = 1 - \frac{M_w}{M}. \tag{12.4}
$$

The total mass of the drum contents M is given from Eq. 12.2. The mass of the liquid (water) phase M_w is obtained from a separate water-phase mass balance.

$$
\frac{\mathrm{d}M_w}{\mathrm{d}t} = \Sigma(\dot{m})_w, \tag{12.5}
$$

where

$$\Sigma(\dot{m})_w = \dot{m}_{fw} - \dot{m}_{dc} + (1 - x_{rx})\dot{m}_{rx} - \dot{m}_{bl} + \dot{m}_{if}$$

and $(1 - x_{rx})\dot{m}_{rx}$ is the water fraction of the flow emerging from the riser(s). The total mass of the drum contents is the sum of the two individual phase masses $M_s + M_w$. The mass of steam M_s is therefore

$$M_s = M - M_w.$$

Differentiating this equation with respect to time gives

$$\frac{\mathrm{d}M_s}{\mathrm{d}t} = \frac{\mathrm{d}M}{\mathrm{d}t} - \frac{\mathrm{d}M_w}{\mathrm{d}t}$$

from which we obtain

$$\frac{\mathrm{d}M_s}{\mathrm{d}t} = x_{rx}\dot{m}_{rx} - \dot{m}_s - \dot{m}_{if} - \dot{m}_{cond} = \Sigma(\dot{m})_s$$

$\dot{m}_{cond}$ is the rate of condensation on the inner surface of the upper drum in contact with steam. Heat transfer from the steam to the wall is $\dot{q}_{hxs}$ where

$$\dot{q}_{hxs} = \alpha_{hxs} A_{hxs}(T' - T_{dru}) = \dot{m}_{cond}\,\mathbf{r},$$

where α_{hxs} and A_{hxs} are the heat transfer coefficient and contact surface area between the steam and drum wall, respectively. T' is the saturation temperature, and T_{dru} is the temperature of the upper wall in contact with the steam. Therefore,

$$\dot{m}_{cond} = \dot{q}_{hxs}/\mathbf{r}$$

$\dot{m}_{if}$ is the interface flow (condensation or evaporation) passing between the two phases across the steam-water interface. This is a complex mechanism of relatively little operational consequence. Here it is heuristically represented as being proportional to the difference between saturation and drum water temperatures.

$$\dot{m}_{if} = k_{if}(T' - T_w).$$

Since the drum water mean temperature will usually be less than the saturation temperature due to sub-cooling by the feedwater, this representation produces a predominantly condensation interface flow. However, it will almost always be small and of no consequence to normal operations. Its largest contribution will appear during large transient variations of drum pressure and after shutdown of the plant when, together with condensation on the drum wall surface, it will contribute to the slow decrease of drum pressure.

An alternative expression for $\mathrm{d}M_s/\mathrm{d}t$ is, with $M_s = V_s/v''$,

$$\frac{\mathrm{d}M_s}{\mathrm{d}t} = \frac{\mathrm{d}}{\mathrm{d}t}\left(\frac{V_s}{v''}\right)$$

$$= \frac{1}{v''}\frac{\mathrm{d}V_s}{\mathrm{d}t} - \frac{V_s}{v''^2}\frac{\mathrm{d}v''}{\mathrm{d}t}$$

$$= \frac{1}{v''}\frac{\mathrm{d}V_s}{\mathrm{d}t} - \frac{V_s}{v''^2}\frac{\partial v''}{\partial p}\frac{\mathrm{d}p}{\mathrm{d}t}$$

from which we have

$$-\frac{V_s}{v''^2}\frac{\partial v''}{\partial p}\frac{\mathrm{d}p}{\mathrm{d}t} = -\frac{1}{v''}\frac{\mathrm{d}V_s}{\mathrm{d}t} + \Sigma(\dot{m})_s.$$

From Eq. 4.34,

$$\frac{\partial v''}{\partial p} = -\frac{v''}{\gamma\, p},$$

which leads to

$$\frac{V_s}{\gamma\, p}\frac{\mathrm{d}p}{\mathrm{d}t} = v''\Sigma(\dot{m})_s - \frac{\mathrm{d}V_s}{\mathrm{d}t}. \tag{12.6}$$

The volume of the steam phase equals the total drum volume V_{drm} less the volume of the water phase. Then with $V_w = v' M_w$,

$$V_s = V_{drm} - V_w = V_{drm} - v' M_w.$$

Differentiating this equation with respect to time gives

$$\frac{\mathrm{d}V_s}{\mathrm{d}t} = -v'\frac{\mathrm{d}M_w}{\mathrm{d}t}$$

assuming v' is constant in the short term. After substitution for $\mathrm{d}V_s/\mathrm{d}t$, Eq. 12.6 may be written.

$$\left(\frac{V_s}{\gamma\, p}\right)\frac{\mathrm{d}p}{\mathrm{d}t} = v''\Sigma(\dot{m})_s + v'\Sigma(\dot{m})_w.$$

Noting that

$$\frac{\mathrm{d}(\ln p)}{\mathrm{d}t} = \frac{1}{p}\frac{\mathrm{d}p}{\mathrm{d}t}$$

this last equation may be written

$$\frac{\mathrm{d}(\ln p)}{\mathrm{d}t} = \frac{\gamma}{V_s}\left[v''\Sigma(\dot{m})_s + v'\Sigma(\dot{m})_w\right] \tag{12.7}$$

and $p = \exp(\ln p)$.

The right-hand side of Eq. 12.7 depends implicitly on p as the flows to and from the drum depend on drum pressure. As this analysis treats the drum as an isolated system, it assumes that these interfacing variables are defined by external systems which are aware of the drum pressure.

Drum Water Energy Balance The temperature of the drum water phase is determined from the mixing of all water flows into and out of the drum. Then,

$$\frac{\mathrm{d}(M_w h_w)}{\mathrm{d}t} = \Sigma(\dot{m}h)_w - \dot{q}_{hxw}, \tag{12.8}$$

where net energy flow into and from the drum water is

$$\Sigma(\dot{m}h)_w = \dot{m}_{fw}h_{fw} + (1 - x_{rx})\dot{m}_{xr}h' - \dot{m}_{dc}h_w - \dot{m}_{bl}h_w + \dot{m}_{if}(h'' - h')$$

for which we assume that the water phase of the riser flow is saturated. During start-up prior to first boil-off, this will not be the case, and the actual riser exit enthalpy must be used in place of h'.

During start-up and initial pressure raising, feedwater flow to the drum is minimal, comprising the mass flow of steam filling and pressurising the steam system piping, including superheaters and headers, and blowdown to control water chemistry. The normal sub-cooling effects of the feedwater are therefore minimal during this phase. The static head in the riser will allow the water phase of the flow exiting the riser to reach a higher temperature than the drum saturation temperature, and part of the riser flow will evaporate in the drum ($\dot{m}_{if}$). The remainder will mix with the drum water and increase its temperature.

The heat transfer to the lower wall of the drum in contact with the water phase is

$$\dot{q}_{hxw} = \alpha_{drw}A_{drw}(T_{drw} - T_w),$$

where α_{drw} and A_{drw} are the heat transfer coefficient and contact surface area between the drum water and wall, respectively.

Equation 12.8 expands to

$$\frac{\mathrm{d}(M_w h_w)}{\mathrm{d}t} = h_w\frac{\mathrm{d}M_w}{\mathrm{d}t} + M_w\frac{\mathrm{d}h_w}{\mathrm{d}t}$$

from which we have

$$\begin{aligned} M_w\frac{\mathrm{d}h_w}{\mathrm{d}t} &= h_w\frac{\mathrm{d}M_w}{\mathrm{d}t} + \Sigma(\dot{m}h)_w - \dot{q}_{hxw} \\ &= -h_w\Sigma(\dot{m})_w + \Sigma(\dot{m}h)_w - \dot{q}_{hxw} \end{aligned}$$

or

$$M_w\frac{\mathrm{d}h_w}{\mathrm{d}t} + h_w\Sigma(\dot{m})_w = \Sigma(\dot{m}h)_w - \dot{q}_{hxw}. \tag{12.9}$$

After expansion of the two Σ terms, Eq. 12.9 can be written

$$M_w \frac{dh_w}{dt} + h_w(\dot{m}_{fw} + (1 - x_{rx})\dot{m}_{rx} + \dot{m}_{if})$$
$$= \dot{m}_{fw}h_{fw} + (1 - x_{rx})\dot{m}_{rx}h' + \dot{m}_{if}\mathbf{r} - \dot{q}_{hxw} \tag{12.10}$$

from which the drum water temperature $T_w = h_w/c_w$. c_w is the specific heat capacity of water.

Drum Wall Metal Temperatures The wall of the drum in a high-pressure boiler is extremely thick, around 200 mm for a 500 MW plant. Like all thick-walled components subject to high temperatures, the maximum temperature differentials through the wall must be limited in order to avoid excessive stresses. This imposes maximum rates of warming during start-up, and the rate of change of drum saturation temperature is usually limited to a maximum of 50–70 °C per hour. This can be the most significant factor in determining the minimum cold start duration. Operating staff and supervisory systems are provided with measurements of drum metal temperatures and their upper/lower differentials for monitoring purposes.

The drum is also subject to bending stress induced by the differential between the temperatures of the upper and lower hemispheres of the drum. The upper part of the drum is in contact with steam with poor heat transfer, and the lower is in contact with lower temperature water but with good heat transfer. Temperature differentials can be calculated independently for the upper and lower sections of the drum. For simulation purposes, a simplified representation of these temperatures is usually sufficient. The derivation of a set of simple equations to calculate two representative temperatures T_{dru} and T_{drl} is outlined, using the heat transfer terms $\dot{q}_{hxw}$ and $\dot{q}_{hxs}$ mentioned in the preceding derivation. The calculation includes a simplified provision for conduction from the upper to the lower sections as well as ambient heat loss through the drum insulation.

$$c_m M_{mu} \frac{dT_{dru}}{dt} = -\lambda A_{xm}(T_{dru} - T_{drl}) + \dot{q}_{hxs} - \dot{q}_{amb},$$
$$c_m M_{ml} \frac{dT_{drl}}{dt} = \lambda A_{xm}(T_{dru} - T_{drl}) + \dot{q}_{hxw} - \dot{q}_{amb},$$

where M_{mu} and M_{ml} are the masses of metal in the upper and lower hemi-cylindrical halves and c_m is the specific heat capacity of the metal (0.465 kJ/(kg C) for steel).

After substitution of the various heat transfer terms, we obtain

$$c_m M_{mu} \frac{dT_{dru}}{dt} = -\left(\lambda A_{xm} + \alpha_{hxs}A_{hxs} + \lambda_{insul}\frac{A_{insul}}{d_{insul}}\right)T_{dru}$$
$$+ \lambda A_{xm}T_{drl} + \alpha_{hxs}A_{hxs}T' - \lambda_{insul}\frac{A_{insul}}{d_{insul}}T_{amb} \tag{12.11}$$

for the drum upper metal temperature and

$$c_m M_{ml} \frac{\mathrm{d}T_{drl}}{\mathrm{d}t} = -\left(\lambda A_{xm} + \alpha_{hxw} A_{hxw} + \lambda_{insul} \frac{A_{insul}}{d_{insul}}\right) T_{drl} + \lambda A_{xm} T_{dru} + \alpha_{hxw} A_{hxw} T_w - \lambda_{insul} \frac{A_{insul}}{d_{insul}} T_{amb} \quad (12.12)$$

for the drum lower metal temperature. These equations are linear in the two temperatures and can be solved by any of the usual methods.

The indicative stresses in both the upper and lower sections can be estimated from the instantaneous difference between the representative wall temperature and the appropriate reference source temperature, steam saturation or drum water temperature.

Water Temperature at the Downcomer Outlet

Water enters the downcomer at a temperature close to the drum or water separator water temperature. Assuming that no steam has been entrained from the drum or separator vessel, the fluid entering the downcomer is always water.

In Sect. 5.2.3, the general one-dimensional energy equation was resolved into

$$c_p \frac{\partial T}{\partial t} + c_p v \frac{\partial T}{\partial z} = v\left(\frac{\partial p}{\partial t} + \frac{\partial p}{\partial z}\right) - \frac{v}{A}\alpha_{hx} A_{hx}(T - T_{ref}) + v v \left.\frac{\partial p}{\partial z}\right|_{loss},$$

which, with water as the working fluid, may be written for the downcomer as

$$c_w \frac{A_{dc}}{v_{dc}} \frac{\partial T_{dc}}{\partial t} + c_w \dot{m} \frac{\partial T_{dc}}{\partial z} + \alpha \pi d_{dc} T_{dc} = \alpha \pi d_{dc} T_{amb}. \quad (12.13)$$

The boundary conditions are defined by the initial temperature profile $T_{dc}(z, 0)$ for all z at t = 0 and the downcomer inlet temperature $T_{dc}(0, 0) = T_{drw}$.

In Sect. 5.2.3, the analytic solution $T_{dc}(z, t)$ of Eq. 12.13 for all (z, t) with the given boundary conditions was shown to be

$$T_{dc}(z, t) = T_{dc}(z, 0)\mathrm{e}^{-\tau_t t} + T_{amb}(1 - \mathrm{e}^{-\tau_t t}) + (T_{dc}(0, 0) - T_{amb})(\mathrm{e}^{-\tau_z z} - \mathrm{e}^{-\tau_t t})H(t - z/\nu), \quad (12.14)$$

where

$$\tau_t = \frac{\alpha \pi d_{dc}}{c_w \frac{A_{dc}}{v_{dc}}}, \qquad \tau_z = \nu\, \tau_t$$

and the flow velocity is

$$\nu = \frac{v_{dc} \dot{m}_{dc}}{A_{dc}}.$$

The downcomer outlet temperature $T_{dcx}(t) = T_{dc}(L_{dc}, t)$ is obtained from Eq. 12.14 by setting $z = L_{dc}$.

While this gives a true reproduction of the transport or residence time of the water in the downcomer, this time (of the order of a few seconds) is insignificant compared to the other thermal delays within the flow loop. For example, for a downcomer 30 m long and a typical flow velocity of 10 m/s, the transit time is 3 s. Not much heat will be lost to the pipe wall, but any changes of drum water temperature will be delayed by 3 s before entering the riser. The downcomer outlet temperature therefore can be approximated effectively and more simply as follows. Given that the volume V_{dc} in the downcomer is constant, the mass of water in the downcomer is

$$M_{dc} = \frac{V_{dc}}{v_{dc}}.$$

Ignoring water compressibility, mass flow in = mass flow out. From the energy balance, we then have for the water in the downcomer

$$M_{dc}\frac{\mathrm{d}}{\mathrm{d}t}h_{dc} = \dot{m}_{dc}(h_{drw} - h_{dcx}) - \dot{Q}_{amb}, \tag{12.15}$$

where the ambient heat loss from the downcomer is

$$\dot{Q}_{amb} = \alpha_{dc,amb} A_{dc}(T_{dc} - T_{amb}).$$

In Eq. 12.15, h_{dc} is the mean or representative water enthalpy in each downcomer tube, while h_{dcx} is the enthalpy at the downcomer exit. A second relationship between these two variables is required in order to complete the calculation. This may be established in various ways but most conveniently by setting the outlet enthalpy equal to the mean enthalpy.

$$h_{dc} = h_{dcx}.$$

With the substitution $T_{dcx} = c_w h_{dcx}$ and $T_{drw} = c_w h_{drw}$, assuming the water specific heat c_w is the same for both, we can write

$$M_{dc}c_w\frac{\mathrm{d}T_{dcx}}{\mathrm{d}t} + (\dot{m}_{dc}c_w + \alpha_{dc,amb}A_{dc})T_{dcx} = \dot{m}_{dc}c_w T_{drw} + \alpha_{dc,amb}A_{dc}T_{amb}.$$

The solution of this equation can be expressed as that returned by the library function `fnlag`, introduced in Sect. 3.3.2. With $T_{dcx,targ}$ the target (steady-state) value of the solution and τ_{dcx} the associated time constant, the downcomer exit temperature T_{dcx} is returned as

$$T_{dcx} = \mathtt{fnlag}(T_{dcx}, T_{dcx,targ}, \tau_{dcx}), \tag{12.16}$$

where

$$T_{dcx,targ} = \frac{\dot{m}_{dc} c_w T_{drw} + \alpha_{dc,amb} A_{dc} T_{amb}}{\dot{m}_{dc} c_w + \alpha_{dc,amb} A_{dc}},$$

$$\tau_{dcx} = \frac{M_{dc} c_w}{\dot{m}_{dc} c_w + \alpha_{dc,amb} A_{dc}}.$$

The inclusion of ambient losses ensures this formulation remains valid for zero downcomer flow.

12.3 Steam Generation in the Riser

Starting from the basic conservation equations, we will now derive of a set of equations which describes the dynamic behaviour of the key state and subsidiary variables of the riser. This applies equally to the risers in both drum and once-through configurations. It will be assumed that the direction of flow is upwards, inclined to the horizontal at an angle β.

The development is presented for a single tube. The multiple tubes of a real riser are handled by either assuming that all tubes are identical and simply summing the single tube result over all tubes or by grouping similar tubes into banks, within each of which all tubes can be regarded as identical. Individual banks can differ, one from another.

Two different methods are employed. The first uses spatial discretisation and explicit integration of each of the conservation equations. The second applies the serial method, described in Sect. 5.4.1, which leads to almost the same result but with an important and significant difference.

In each derivation all variables are functions of the axial ordinate z and time t. As this is the evaporator we can assume that the flow can be single or two phase (liquid and vapour) and, if two phase, the fluid is in the saturated state.

12.3.1 The Profile of Specific Mass Flow from the Mass Balance

The mass balance equation is

$$\frac{\partial \phi}{\partial z} = \frac{1}{v^2}\frac{\partial v}{\partial t} = \frac{1}{v^2}\left(\frac{\partial v}{\partial p}\frac{\partial p}{\partial t} + \frac{\partial v}{\partial x}\frac{\partial x}{\partial t}\right).$$

Since

$$v = x\,v'' + (1-x)v'$$

we have

$$\frac{\partial v}{\partial x} = v'' - v'$$

and, with $\partial v/\partial p$ approximated by $x\, \partial v''/\partial p$,

$$\frac{\partial \phi}{\partial z} = \frac{1}{v^2}\left(x\frac{\partial v''}{\partial p}\frac{\partial p}{\partial t} + (v'' - v')\frac{\partial x}{\partial t}\right). \tag{12.17}$$

This equation quantifies the influence of both pressure and steam fraction changes on the local flow rate. It shows, for example, that since $\partial v''/\partial p$ is negative, an increase in local pressure induces a decrease in local mass flow to a degree proportional to the local steam content and inversely proportional to the local pressure. It also shows that, for pressures for which $v'' \gg v'$ (less than, say, 20 MPa), a local increase in steam content, caused, for example, by increased local steam generation, will cause an increase in local mass flow inversely proportional to the local fluid specific volume. This effect is particularly pronounced at the lower pressures and steam mass fractions encountered during start-up. At the onset of boiling, the small local change in steam mass fraction will cause a large increase in mass flow which can destabilise the calculation by causing an excessively large change in ϕ within one time step.

The term $\partial x/\partial t$ can be eliminated as follows.

The riser fluid mass fraction is defined in terms of local fluid enthalpy as

$$h = xh'' + (1 - x)h' = h' + x\mathbf{r},$$

where the latent heat of evaporation $\mathbf{r} = h'' - h'$ is a function of local pressure. The partial differentials of h with respect to time t and space z are

$$\frac{\partial h}{\partial t} = \frac{\partial h'}{\partial t} + x\frac{\partial \mathbf{r}}{\partial t} + \mathbf{r}\frac{\partial x}{\partial t} \tag{12.18}$$

and

$$\frac{\partial h}{\partial z} = \frac{\partial h'}{\partial z} + x\frac{\partial \mathbf{r}}{\partial z} + \mathbf{r}\frac{\partial x}{\partial z}. \tag{12.19}$$

The time derivative can written

$$\begin{aligned}\frac{\partial h}{\partial t} &= \frac{\partial h'}{\partial t} + x\frac{\partial h''}{\partial t} - x\frac{\partial h'}{\partial t} + \mathbf{r}\frac{\partial x}{\partial t}\\ &= (1 - x)\frac{\partial h'}{\partial t} + x\frac{\partial h''}{\partial t} + \mathbf{r}\frac{\partial x}{\partial t}\\ &= \left((1 - x)\frac{\partial h'}{\partial p} + x\frac{\partial h''}{\partial p}\right)\frac{\partial p}{\partial t} + \mathbf{r}\frac{\partial x}{\partial t}\end{aligned}$$

giving

$$\frac{\partial h}{\partial t} = \frac{\partial h}{\partial p}\frac{\partial p}{\partial t} + \mathbf{r}\frac{\partial x}{\partial t}$$

or

$$\frac{\partial x}{\partial t} = \frac{1}{\mathbf{r}}\frac{\partial h}{\partial t} - \frac{1}{\mathbf{r}}\frac{\partial h}{\partial p}\frac{\partial p}{\partial t}.$$

Therefore,

$$\begin{aligned}(v'' - v')\frac{\partial x}{\partial t} &= \left(\frac{v'' - v'}{h'' - h'}\right)\frac{\partial h}{\partial t} - \left(\frac{v'' - v'}{h'' - h'}\right)\frac{\partial h}{\partial p}\frac{\partial p}{\partial t}\\ &= \frac{\partial v}{\partial h}\frac{\partial h}{\partial t} - \frac{\partial v}{\partial h}\frac{\partial h}{\partial p}\frac{\partial p}{\partial t}.\end{aligned}$$

Equation 12.17 can now be written

$$v^2\frac{\partial \phi}{\partial z} = \left(x\frac{\partial v''}{\partial p} - \frac{\partial v}{\partial p}\right)\frac{\partial p}{\partial t} + \frac{\partial v}{\partial h}\frac{\partial h}{\partial t}.$$

But

$$x\frac{\partial v''}{\partial p} = \frac{\partial v}{\partial p}$$

leaving

$$v^2\frac{\partial \phi}{\partial z} = \frac{\partial v}{\partial h}\frac{\partial h}{\partial t}. \tag{12.20}$$

12.3.2 The Profile of Pressure from the Momentum Balance

The discussion presented in Sect. 5.2.2 revealed that wave effects are introduced through the momentum equation via local compression of the steam phase. While wave effects create the conditions for choked flow and shock effects and are important *per se*, they have little relevance to the dynamics of steam generation in the evaporator under any but extreme operational conditions. From a computational perspective, we should remove local compression effects and reduce the momentum equation to

$$\frac{\partial p}{\partial z} = -\left(\frac{\partial \phi}{\partial t} - \phi^2\frac{\partial v}{\partial h}\frac{\partial h}{\partial z} - \left.\frac{\partial p}{\partial z}\right|_L - \rho\, g\, \sin\beta\right) \tag{12.21}$$

that is, the profile of pressure along the riser is statically defined by friction losses, gravity and local thermal effects. The term $\partial\phi/\partial t$ accounts for flow inertia.

In steady state $\partial\phi/\partial t = 0$, and the distribution of pressure is defined by geodetic head differences and flow losses. The volume expansion term $\phi^2\frac{\partial v}{\partial h}\frac{\partial h}{\partial z}$, although a non-zero contribution if $\partial h/\partial z \neq 0$, is generally small and can be neglected for most purposes.

The following is to be noted regarding units. If a consistent set of units has been used, each term in Eq. 12.21 is expressed in units of Pascals. Terms which include constants may be calculated implicitly in kiloPascals (kPa) if the constants are expressed in non-consistent units. For example, use of $g = 9.80665\,\text{m/s}^2$ in this equation is consistent and renders the geodetic height contribution in Pascals. For water ($\rho \approx 1{,}000\,\text{kg/m}^3$), this yields $\Delta p = \rho g = 9{,}806.65\,\text{Pa/m} = 9.80665\,\text{kPa/m}$. By contrast, the volume expansion term is typically of the order of only a few hundred Pa per metre of riser length.

12.3.3 The Profile of Fluid Enthalpy from the Energy Balance

Ignoring kinetic and gravitational potential energy, the energy conservation equation 5.7

$$\frac{v}{A}\frac{\mathrm{d}\dot{q}}{\mathrm{d}z} = \left(\frac{\partial h}{\partial t} + v\frac{\partial h}{\partial z}\right) - v\frac{\partial p}{\partial t} + v\left(\frac{\partial v}{\partial t} + v\frac{\partial v}{\partial z} + g\,\sin\beta\right)$$

may be reduced to

$$v\frac{\partial h}{\partial z} = \frac{v}{A}\frac{\mathrm{d}\dot{q}}{\mathrm{d}z} - \frac{\partial h}{\partial t} + v\frac{\partial p}{\partial t}. \qquad (12.22)$$

The term $\partial h/\partial t$ accounts for thermal inertia. The energy balance in the evaporator is driven by the heat transfer term $\mathrm{d}\dot{q}/\mathrm{d}z$. In an open-flow system such as this, the compression energy term $v\,\partial p/\partial t$ is relatively small and can be neglected.

In steady state, $\partial p/\partial t = 0$ and $\partial h/\partial t = 0$ and mass flow ϕ will be uniform along the flow path ($\partial\phi/\partial z = 0$). Equation 12.22 then indicates that the spatial gradient of enthalpy $\mathrm{d}h/\mathrm{d}z$ is defined by the heat transferred to the flow $\mathrm{d}\dot{q}/\mathrm{d}z$, as expected.

$$\dot{m}\frac{\mathrm{d}h}{\mathrm{d}z} = \frac{\mathrm{d}\dot{q}}{\mathrm{d}z}.$$

12.3.4 Solution of the Conservation Equations

The conservation equations applied to the riser are summarised as follows.

Mass:

$$v\frac{\partial\phi}{\partial z} - \frac{1}{v}\frac{\partial v}{\partial h}\frac{\partial h}{\partial t} = 0. \qquad (12.23)$$

Momentum:

$$\frac{\partial p}{\partial z} + \frac{\partial \phi}{\partial t} - \phi^2 \frac{\partial v}{\partial h} = \rho g \sin \beta + \left. \frac{\partial p}{\partial z} \right|_L . \tag{12.24}$$

Energy:

$$-v \frac{\partial p}{\partial t} + \frac{\partial h}{\partial t} + \phi v \frac{\partial h}{\partial z} = \frac{v}{A} \frac{\mathrm{d}\dot{q}}{\mathrm{d}z} . \tag{12.25}$$

Heat transfer is defined by the supplementary equation

$$\frac{\mathrm{d}\dot{q}}{\mathrm{d}z} = -\alpha_{hx} a_{hx} (T_{fl} - T_w), \tag{12.26}$$

where a_{hx} is the heat transfer area per unit length. The flow specific volume v is given from the saturation properties relationship

$$v = v(p, h) \tag{12.27}$$

and the fluid temperature from

$$T_{fl} = T(p, h). \tag{12.28}$$

These equations can be solved numerically via an appropriate time and space discretisation scheme. Since the local compression term has been removed from the momentum equation, the solution will not include high-speed acoustic effects and will allow the use of longer time steps.

The solution of these equations is no small task for a large number of riser cells. We will now look at the serial solution method as an alternative approach.

12.3.5 Application of the Serial Method of Solution

The serial method of solution of the three conservation equations was described in Sect. 5.4.1 where the following three equations were derived.

$$p_j^{n+1} = p_{j-1}^{n+1} + \left[f_{1,j} + \phi \frac{\partial v}{\partial h} f_{2,j} \right] \Delta z, \tag{12.29}$$

$$h_j^{n+1} = h_{j-1}^{n+1} + \left[v f_{1,j} + f_{2,j} \right] \Delta z, \tag{12.30}$$

$$\phi_j^{n+1} = \phi_{j-1}^{n+1} + \frac{1}{\theta \Delta t} \left[S_\rho^n - \rho_j^{n+1} \right] \Delta z. \tag{12.31}$$

The functions f_1 and f_2 were given as

$$f_1(z) = \frac{1}{\theta \Delta t}\left[S_\phi^n - \phi^{n+1}\right] + (\Delta p_{loss} - g\,\rho \sin\beta)^n\,, \tag{12.32}$$

$$f_2(z) = \frac{1}{v\theta\Delta t}\left[S_h^n - h^{n+1} + v^{n+1}p^{n+1}\right] + \frac{1}{\phi}\frac{1}{A}\frac{\mathrm{d}\dot{q}}{\mathrm{d}z}$$

$$+v\Delta p_{loss} \tag{12.33}$$

and the recursive functions are

$$S_\rho^{n+1} = \frac{1}{\theta}\rho^{n+1} - \frac{1-\theta}{\theta}S_\rho^n \tag{12.34}$$

with the boundary condition

$$S_\rho^0 = \rho^0(z),$$

$$S_\phi^{n+1} = \frac{1}{\theta}\phi^{n+1} - \frac{1-\theta}{\theta}S_\phi^n \tag{12.35}$$

with the boundary condition

$$S_\phi^0(z) = \phi^0(z),$$

and

$$S_h^{n+1} = \frac{1}{\theta}\left[h^{n+1} - v^{n+1}p^{n+1}\right] - \frac{1-\theta}{\theta}S_h^n \tag{12.36}$$

with the boundary condition

$$S_h^0(z) = h^0(z) - v^0 p^0(z).$$

Together with the properties relationship

$$v^{n+1} = v(p^{n+1}, h^{n+1})$$

these equations may be applied to the steam generation process directly in this form. In keeping with the discrete-space formulation, it will be necessary to divide the flow path into spatial cells for which the tube wall temperature and fluid properties (including steam mass fraction) will be calculated. Boundary conditions at each end of the flow path will be defined by the configuration of the steam generation system, configured either as a circuit (drum boiler or once-through unit operating in circulation mode with a water separator) or for once-through flow.

The term $(S_u - u)/\Delta t$, with $u = \phi$, p or h, introduces an important difference to the earlier formulation represented by Eqs. 12.23–12.25. It can be interpreted as a numerical approximation to the derivative of u. Differentiation can act as a destabilising influence as it amplifies small numerical differences. If Eqs. 12.31–12.33 are used as they stand, the implied differentiation may cause numerical instability. Where signal differentiation is used in digital process control, it is always combined with a low-pass filter. A similar technique can be applied here by averaging the derivatives over consecutive time steps. If k steps are used, then $1/(\theta \Delta t)$ can be replaced by $1/(k\, \theta \Delta t)$ at the cost of some loss of dynamic accuracy.

Computation experience suggests that a riser cell length of around 1 m is a good compromise between complexity and accuracy and ensures stable and reliable results. A riser 20 m long would then be represented by 20 cells, one 40 m long by 30–40 cells, and so on. The cell length defines the minimum resolution of the location of the boiling boundary and the volume of the fluid converting *instantaneously* from the liquid to the vapour state. The larger this volume, the greater the effect of phase change on fluid flow and volume displacement from the cell. Both of these can have an adverse effect on the stability and accuracy of the solution.

Treatment of Phase Change Within the Flow

Equation 12.31 is suitable for single-phase flow without phase change. The following approach is more appropriate to the treatment of two-phase flows.

With $a = x\partial v''/\partial p$ and $\bar{v} = v'' - v'$ Eq. 12.17 can be written

$$a\frac{\partial p}{\partial t} + \bar{v}\frac{\partial x}{\partial t} = v^2\frac{\partial \phi}{\partial z}. \tag{12.37}$$

With the time derivatives replaced by their first-order discrete equivalents this becomes

$$a\frac{p^{n+1} - p^n}{\Delta t} + \bar{v}\frac{x^{n+1} - x^n}{\Delta t} = v^2\frac{\mathrm{d}\phi}{\mathrm{d}z}$$

from which

$$\begin{aligned} a\, p^{n+1} + \bar{v}\, x^{n+1} &= a\, p^n + \bar{v}\, x^n + \Delta t v^2\frac{\mathrm{d}\phi}{\mathrm{d}z} \\ &= a\, p^n + \bar{v}\, x^n + \Delta t v^2\frac{\mathrm{d}}{\mathrm{d}z}[\theta\, \phi^{n+1} + (1-\theta)\phi^n] \\ &= a\, p^n + \bar{v}\, x^n + (1-\theta)\Delta t v^2\frac{\mathrm{d}\phi^n}{\mathrm{d}z} + \theta \Delta t\, v^2\frac{\mathrm{d}\phi^n}{\mathrm{d}z}. \end{aligned}$$

Grouping terms denominated at the $(n+1)$th and n-th steps gives

$$\begin{aligned} \theta\, \Delta t\, v^2\frac{\mathrm{d}\phi^{n+1}}{\mathrm{d}z} - a\, p^{n+1} - \bar{v}\, x^{n+1} &= -(1-\theta)\Delta t\, v^2\frac{\mathrm{d}\phi^n}{\mathrm{d}z} - a\, p^n - \bar{v}\, x^n \\ &= -S_\rho^n \end{aligned}$$

from which

$$\theta\, \Delta t\, v^2 \frac{\mathrm{d}\phi^{n+1}}{\mathrm{d}z} = a\, p^{n+1} + \bar{v}\, x^{n+1} - S_\rho^n \tag{12.38}$$

with

$$S_\rho^n = (1-\theta)\Delta t\, v^2 \frac{\mathrm{d}\phi^n}{\mathrm{d}z} + a\, p^n + \bar{v}\, x^n.$$

We can then write

$$\begin{aligned} S_\rho^{n+1} &= (1-\theta)\Delta t\, v^2 \frac{\mathrm{d}\phi}{\mathrm{d}z}^{n+1} + a\, p^{n+1} + \bar{v}\, x^{n+1} \\ &= -\theta\, \Delta t\, v^2 \frac{\mathrm{d}\phi}{\mathrm{d}z}^{n+1} + a\, p^{n+1} + \bar{v}\, x^{n+1} + \Delta t\, v^2 \frac{\mathrm{d}\phi}{\mathrm{d}z}^{n+1} \\ &= S_\rho^n + \Delta t\, v^2 \frac{\mathrm{d}\phi}{\mathrm{d}z}^{n+1} . \end{aligned}$$

Equation 12.38 can now be written

$$\Delta t\, v^2 \frac{\mathrm{d}\phi^{n+1}}{\mathrm{d}z} = -\frac{1}{\theta} S_\rho^n + \frac{1}{\theta}\left(a\, p^{n+1} + \bar{v}\, x^{n+1}\right) \tag{12.39}$$

with

$$\begin{aligned} S_\rho^{n+1} &= \frac{1}{\theta} S_\rho^n + \frac{1}{\theta}\left(a\, p^{n+1} + \bar{v}\, x^{n+1} - S_p^n\right) \\ &= -\frac{(1-\theta)}{\theta} S_\rho^n + \frac{1}{\theta}\left(a\, p^{n+1} + \bar{v}\, x^{n+1}\right) \\ &= -\frac{(1-\theta)}{\theta} S_\rho^n - \frac{1}{\theta}\left(v''(x^{n+1} - x^n) - v'\, x^{n+1}\right). \end{aligned}$$

Again replace $\mathrm{d}\phi/\mathrm{d}z$ in Eq. 12.39 by its discrete-time equivalent to obtain

$$\frac{\theta\, \Delta t\, v^2}{\Delta z}(\phi_j^{n+1} - \phi_{j-1}^{n+1}) = a\, p^{n+1} + \bar{v}\, x^{n+1} - S_\rho^n$$

from which

$$\phi_j^{n+1} = \phi_{j-1}^{n+1} + \frac{\Delta z}{\theta\, \Delta t\, v^2}\left(a\, p^{n+1} + \bar{v}\, x^{n+1} - S_\rho^n\right).$$

Finally, replace a by $x\, \partial v''/\partial p = -x\, v''/p$ to give

$$\phi_j^{n+1} = \phi_{j-1}^{n+1} + \frac{1}{v^2}\frac{\Delta z}{\theta\, \Delta t}\left[v''(x^{n+1} - x^n) - S_\rho^n\right]. \tag{12.40}$$

Boundary Conditions at the Riser Inlet

The inlet boundary conditions p_0, h_0 and ϕ_0 are defined by interfacing systems. With these known and using the profile of riser metal temperature known from the previous time step, solution of this system of equations will yield the complete picture of riser fluid conditions at the (n+1)th time step, up to and including the last riser cell which defines the riser discharge conditions.

The following table summarises the identification of riser inlet conditions for each of the two possible plant configurations, circulation loop and once-through.

Circulation loop

Inlet enthalpy $h_0 = h_{ri}$	The downcomer outlet enthalpy or circulation pump discharge enthalpy
Inlet flow ϕ_0	The per riser tube specific flow
Inlet pressure $p_0 = p_{ri}$	The drum or water separator pressure plus the downcomer gravity term minus the downcomer friction loss

Once-through

Inlet enthalpy $h_0 = h_{ri}$	The enthalpy of the feedwater to the boiler
Inlet flow ϕ_0	The per tube specific flow, calculated as total feedwater ϕ divided by the number of riser tubes
Inlet pressure $p_0 = p_{ri}$	The feedwater pressure at the boiler inlet

12.3.6 *Calculation of the Riser Tube Wall Temperatures*

We will assume that each riser tube wall cell can be adequately represented by a single metal temperature T_m. A simple heat balance on the j-th cell gives

$$\begin{aligned} c_m M_m \frac{\mathrm{d}T_{m,j}}{\mathrm{d}t} &= \dot{q}_{frn,j} - \dot{q}_{tf,j} + \lambda A_x (T_{m,j} - T_{m,(j+1)})/dz \\ &\quad + \lambda A_x (T_{m,(j-1)} - T_{m,j})/dz - \alpha_{hxf} A_{tf} (Tm_j - T_{f,j}) \\ &= \dot{q}_{frnR,j} + \lambda_{frn} A_{frn} (T_{frn} - T_{m,j}) \\ &\quad + \lambda A_x (T_{m,j} + T_{m,(j+1)} - 2\,T_{m,j})/dz \\ &\quad - \alpha_{hxf} A_{tf} (Tm_j - T_{f,j}) \end{aligned}$$

where
- $\dot{q}_{frnR,j}$ radiant heat to the j-th riser tube cell
- $\dot{q}_{tf,j}$ convective heat from furnace to the j-th riser tube cell
- A_x longitudinal heat flow cross-sectional area
- A_{frn} outer tube area receiving convective heat from the furnace
- A_{tf} inner tube area between tube and riser fluid
- T_{frn} furnace temperature used for convective heat transfer
- $T_{f,j}$ temperature of the riser fluid in the j-th cell
- α_{frn} convective heat transfer coefficient furnace and riser tube
- α_{hxf} convective heat transfer coefficient riser tube/fluid

λ is the coefficient of thermal conductivity of the riser tube and c_m the specific heat capacity of the tube metal. M_m is the mass of the tube metal per unit length. δz is centre line spacing of adjacent cells, assumed the same for all cells.

As usual we replace the time derivative by its discrete-time equivalent. Using a fully implicit formulation and with

$$D = \frac{\Delta t}{c_m M_m} \quad \text{and} \quad X = \alpha_{frn} A_{frn} + \alpha_{hxf} A_{hxf}$$

we obtain

$$-D\frac{\lambda A_x}{\delta z} T^{n+1}_{m,(j-1)} + T^{n+1}_{m,j}\left[1 + D\left(\frac{\lambda A_x}{\delta z} + X\right)\right] + D\frac{\lambda A_x}{\delta z} T^{n+1}_{m,(j+1)}$$

$$= T^n_{m,j} + D(\dot{q}_{frnR,j} + \alpha_{hxf} A_{hxf} T_{f,j} + \alpha_{frn} A_{frn} T_{frn,j}). \tag{12.41}$$

For a set of N_{rsz} riser cells, this may be expanded into a tridiagonal matrix equation with boundary conditions defined by the temperatures of the components to which the riser tubes connect at each end. At the inlet end this will be the metal temperature of the boiler lower header. At the outlet it will be the drum metal (upper or lower) or water separator.

The structure of the coefficient matrix is defined by its main and off-diagonal elements.

Main diagonal $\quad 1 + D\left(\dfrac{\lambda A_x}{\delta z} + X\right)$

Lower off-diagonal $\quad -D\dfrac{\lambda A_x}{\delta z}$

Upper off-diagonal $\quad D\dfrac{\lambda A_x}{\delta z}$

If longitudinal conduction is ignored, the calculation of the cell metal temperatures reduces to a single equation for each, decoupled from its neighbours. However, inclusion of conduction is certainly a better representation of reality, and significant spatial variations of riser metal temperatures should be expected. This is particularly the case in the once-through configuration, with notably higher temperatures in the vicinity of the dry-out zone and beyond. An additional benefit brought by the longitudinal conduction terms is improved numerical stability of the calculation, particularly in the vicinity of dry out.

12.3.7 Shrink and Swell of Drum Level

In a drum boiler the water level in the drum is known to exhibit predictable but large variations in response to changes in furnace firing rates or steam flows. When they occur during normal operations, these variations tend to occur in a direction opposite to the drum water-level control system expectations and can create level control difficulties. These drum-level variations are collectively referred to as "shrink and swell" and are caused principally by changes in the volume of the steam phase in the riser, with some less significant contribution from any steam phase which might be present within the drum water. Riser-based effects occur either as the result of changes in the rate of steam production or changes in pressure local to the point of steam production.

Two distinct versions of this swell behaviour can be distinguished. The first is the rapid and in some plants extreme rise in drum level which can follow the commencement of boiling. Water-level increases in excess of 50 % of drum diameter can occur which, if not correctly anticipated, can lead to a furnace trip on high drum level. Similar but less severe events can occur at higher pressures following changes in furnace firing configuration.

Immediately prior to initial boil-off, the riser is full of sub-cooled water, close to the saturated state. If a small increase in heat absorbed by an incremental volume of water increases its enthalpy beyond saturation, a bubble of steam is formed. At or near atmospheric pressure the volume occupied by a unit mass of steam is some 1,700 times that of the same mass of water. In the absence of any flow inertia this volume of water would be displaced immediately from the riser into the drum by the steam bubble, causing a rise in drum level. This potentially very large flow is restricted by the inertia and friction losses of the riser fluid and a much smaller displacement actually occurs. The constraint on the local displacement flow causes the local pressure to increase, with an attendant rise in local saturation enthalpy and temperature. At the inception of boiling, the good heat transfer between the riser tube wall and fluid will keep their temperature differential small and the initial boiling will cease, temporarily. As the riser tube temperature continues to increase this pattern will be repeated, with further displacements of water to the drum. Steam produced in the riser is partially reabsorbed by cooler riser fluid and partially transported to the drum where its accumulation raises the drum pressure. Increasing drum pressure raises the saturation enthalpy and temperature everywhere along the riser and inhibits further steam production until more heat is transferred from the furnace through the riser tubes by the increasing tube temperature. Eventually, a stable rate of steam production is achieved, with a raised but stable drum level.

If the riser consisted of a single tube, this process would cause the drum level to increase by a succession of small discrete increments. However, the riser consists of a large number of tubes, each of which exhibits this behaviour but to a slightly different extent and at a slightly different time. The net effect is to smear the individual effects together resulting in a smooth change of drum level.

As the pressure increases, the relative displacement of water from the riser decreases with the increasing saturated steam specific volume. The ratio of the specific volumes of steam and water decreases from around 1,700 at atmospheric pressure to 1,236 at 140 kPa, 841 at 200 kPa, 176 at 10,000 kPa and 88.9 at 18,000 kPa.

The second manifestation of drum swell is associated with momentary changes of drum pressure which can occur at any pressure and during any phase of plant operations. Any pressure change causes an immediate contraction (pressure increase) or expansion (pressure decrease) of the steam bubbles in the riser (and in the drum, if any) with a consequent change in flow rate to the drum. In both natural- and assisted-circulation systems the total circulation flow is also influenced by changes in the 2-phase pressure drop in the circuit with changes in local mixture quality. Thus a decrease in pressure causes an increase in riser steam-phase volume and an acceleration of flow to the drum, exacerbated by a decrease in flow from the drum to the downcomers due to the (now) higher 2-phase pressure loss in the circuit. The net mass gain in the drum increases the level. A pressure increase has the opposite effect on both of these influences, and the drum level decreases.

Equation 12.17 identifies and quantifies each of these influences as being dependent on the rate of change of pressure and steam mass fraction.

$$\frac{\partial \phi}{\partial z} = \frac{1}{v^2}\left(x\frac{\partial v''}{\partial p}\frac{\partial p}{\partial t} + (v'' - v')\frac{\partial x}{\partial t}\right).$$

12.4 Numeric Example

Figure 12.4 presents the results of computation of an idealised boiler start-up from first light-off to achievement of normal operating pressure. The computation uses the drum model described in Sect. 12.2.2 integrated with the serial solution of the riser as described in Sect. 12.3.5 with $\phi(z)$ calculated using Eq. 12.40. The magnitude of the riser swell contribution was adjusted to produce a drum-level increase of some 50 % of the drum diameter.

Figure 12.5 shows an enlarged section of the preceding figure around the moment of first boil-off. The sudden expulsion of fluid from the riser into the drum and the resulting rise in drum level are clearly evident, as are three distinct expulsion phases, of diminishing magnitude, as the formation of the vapour phase spreads through the riser.

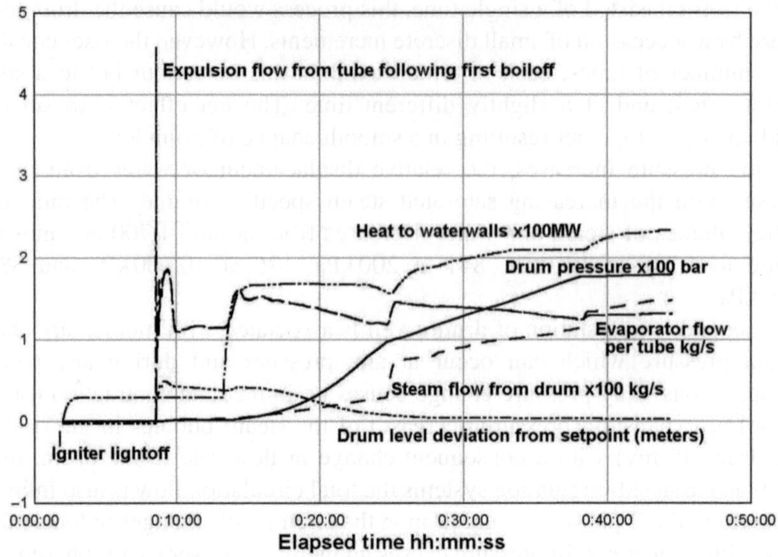

Fig. 12.4 Drum boiler start-up from first light-off

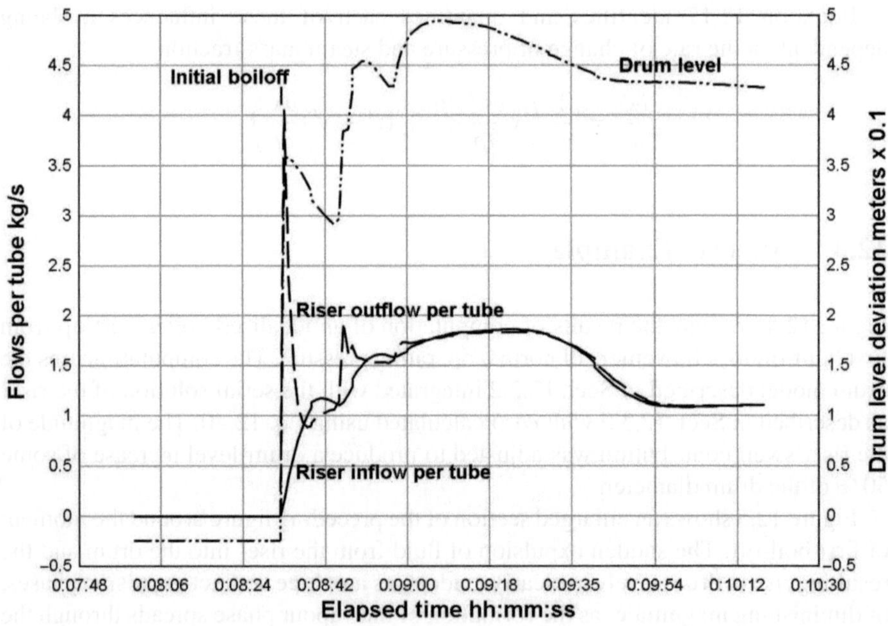

Fig. 12.5 Drum boiler start-up—Expanded view around first boil-off

12.5 Other Boiler Models

The analysis method outlined above is obviously not the only way in which the dynamic behaviour of these components can be modelled and alternatives can be found in the relevant literature, for example [59, 60]. Many of these models are of low order and are intended for use in control system investigations and development. They usually represent the entire boiler as an integration of furnace (heat release related directly to fuel flow), evaporator, often divided into two or three zones covering sub-cooled water, a two-phase saturated fluid zone and, for once-through evaporators, a superheated single-phase steam zone. The drum or water separator is treated as the location of water-steam separation and defines the system pressure and water mass. Depending on the purpose to which the model is to be put, the model may show more detail in one or other of these regions.

Of particular note is the model of a drum boiler developed by Åström and Bell. In a series of three papers [55–57], the authors have described the evolution of a compact low (4th)-order integrated model of the evaporator and drum of a typical drum boiler. Despite its small size, this model is capable of good reproduction of the salient features of a drum boiler over a reasonable range of its on-load operation. It is a practical and useful model for cases in which the detailed behaviour of the boiler itself is not the primary interest but rather its integration with its control systems and interaction with interconnected systems.

Chapter 13
Two-Phase Effects in Steam Flows

Although the title of this chapter refers to steam flows the material applies generally to any two-phase flow. The empirical correlations to be introduced in this chapter have been developed using data obtained using refrigerants and gas/water mixes as well as water/steam flows. Of particular relevance are those correlations which treat two-phase flows in which the ratio of water to steam changes along the flow path as a result of progressive evaporation or condensation. Not all two- or multi-phase flows experience this phenomenon. For example, crude oil flows from the wellhead to the processing platform maintain their oil–water–gas mix ratios. While these ratios might change along the flow path as a result of relative phase slippage there is no physical change to the components.

In evaporating and condensing environments, two-phase flow develops when a part of the liquid stream converts to the gas or vapour phase, or vice versa. In the case of evaporation, this can occur by the addition of heat to the liquid which raises the local fluid temperature above the local fluid saturation temperature. A reduction of local pressure below the fluid vapour pressure causes the same effect by lowering the local saturation temperature below the fluid temperature.

The existence of a vapour phase mixed with the liquid phase changes both the local frictional pressure gradient and the interface heat transfer coefficient. The manner and degree of these changes depends strongly on the vapour content of the fluid and the distribution of the vapour phase throughout the fluid.

Within the context of this treatment, we consider only flow systems, with particular emphasis on water–steam mixes, in which the fluid is moving along a defined path, induced by either natural convection or pumping, or a combination of both. We will not consider the formation of a vapour phase in a static fluid in, for example, a tank, by heating or sudden depressuration. This process is generally described as pool boiling.

The physical characteristics of two-phase flow vary widely depending on the amount of the vapour phase present and the intensity of the heat flux. The latter defines the gradient of the temperature profile across the wall/fluid interface and throughout the bulk of the fluid.

P. O'Kelly, *Computer Simulation of Thermal Plant Operations*,
DOI 10.1007/978-1-4614-4256-1_13,

The literature devoted to two-phase flow phenomena, their parameters and estimation methods, is rich and voluminous and no attempt will be made to survey it. It can be noted, however, that a small number of seminal publications have served to define the principal elements of the field which, in the absence of a comprehensive and tractable theory, remains strongly based on the fitting of empirical functions to experimental data.

The following discussion relates to water moving in inclined or vertical tubes, heated by an external source. In water–steam systems this most commonly means heating by radiation and convection in a fired furnace.

13.1 Two-Phase Flow Patterns

On the basis of density, size and distribution of bubbles, two-phase flow has been broadly divided into the following typical categories or flow regimes.

Bubble Flow This occurs at the low heat fluxes and low vapour phase densities which typify the early stages of boiling. The vapour phase consists of small bubbles more or less evenly distributed throughout the liquid phase and moving with nearly the same velocity. As the fluid bulk temperature increases so the size and density of bubbles increase. In upwards-inclined flow, their increasing buoyancy propels them faster than the liquid phase. The drag of the ascending bubbles pulls the liquid phase with them, and the bulk flow velocity increases. As bubble density increases individual bubbles begin to coalesce, and the flow pattern changes to slug flow.

Slug Flow The coalescence of bubbles gives rise to vapour voids within the liquid which separates into discrete "lumps" or slugs of liquid. This is an intermediate stage and is sustainable only at relatively low flow velocities.

Annular Flow During both bubble and slug flow the wall temperature gradient and density of bubbles are sufficiently low to maintain a continuous liquid layer in contact with the heating surface. As bubble density and flow velocity increase, the bubbles tend to move to the centre of the flow—the region of lowest pressure—and the flow changes to a continuous central vapour core surrounded by a continuous liquid cylinder. These two zones are not composed solely of vapour or liquid. The central vapour region contains entrained liquid particles which eventually convert to vapour. The interfacial liquid layer contains numerous vapour bubbles, some of which move through the liquid layer into the central vapour core. The number, size and motion of these bubbles contribute significantly to the heat transfer behaviour at the wall–liquid interface. Annular flow is usually associated with high vapour velocities and high heat fluxes.

Dispersed Flow The residual liquid film in contact with the wall becomes thinner as liquid is entrained into the vapour phase flow as droplets. These evaporate in the vapour phase leaving only a single liquid phase in contact with the wall.

This is also known as film boiling, occurs at very high vapour fractions and vapour flow velocities and is marked by a sharp decrease in water–fluid heat transfer coefficients.

Within each of these flow regimes both pressure gradients and heat transfer coefficients are strongly dependent on the concentration and flow velocity of each phase, and these in return depend on the flow regime. Computations handling phase change and two-phase effects must therefore be iterative, both locally and globally, if they wish to reproduce even theoretical flow behaviour with good accuracy.

Dry Out The disappearance of the liquid film in contact with the heat transfer surface marks the end of two-phase boiling and is denoted as the point of dry out. Heat transfer to the now complete vapour phase is characterised by a single-phase convective heat transfer process. This phase should never be reached in a drum boiler but is a necessary phase of once-through evaporator operation. It is marked by relatively high temperature differentials across the wall/vapour interface.

13.1.1 Bubble or Nucleate and Film Boiling

Heat transfer to the fluid at the interface between the heating element and the fluid is influenced by the manner and effectiveness with which the transferred heat is removed from the interface zone. This occurs via the simultaneous processes of conduction through the fluid and mass transport by the convective movement of the fluid away from the interface. These are simultaneously hydrodynamic and thermal processes.

The presence of bubbles within the steam formation zone greatly enhances both these processes. From the preceding discussion it is clear that the size, number and distribution of bubbles depend on the local boiling regime (flow pattern). For calculation purposes it is necessary to identify the flow regime using suitable criteria and apply the heat transfer and friction loss coefficients appropriate to that regime.

The phases of bubbly, slug and annular flow give rise to the presence of bubbles within the interfacial zone. Heat transfer to the fluid is enhanced, but friction losses increase in consequence of the reduced cross section available to fluid flow. These are also the phases of "nucleate" boiling. The improved heat transfer allows the transfer of heat to the fluid at small wall–fluid temperature differentials.

As the fluid interfacial layer thins towards the end of the annular flow phase, the process moves to the dispersed flow phase. The fluid layer in contact with the wall becomes too thin to support a bubble population and forms a continuous film. The heat transfer coefficient decreases abruptly as the boiling leaves the nucleate phase and enters the film boiling phase. This is also known as the point of "departure from nucleate boiling" or DNB. For transfer of the same heat flux the wall temperature must increase sharply to a temperature which will be maintained and exceeded after dry out.

We will now review some of the most commonly used methods for the estimation of two-phase effects on friction pressure gradients and heat transfer coefficients.

13.1.2 Modelling of Two-Phase Flow Processes

Two-phase flow systems have traditionally been treated in either of two forms. The homogeneous approach treats the two phases as uniformly mixed within the bulk of the fluid and therefore moving with identical velocities. The mixture properties are calculated as the weighted averages of the properties of each individual phase. The separated flow model treats each phase as if it were flowing independently of the other. Separate mass, momentum and energy balances are computed for each phase, with appropriate provision for exchanges of mass, energy and momentum between the phases.

The separated flow model is necessarily more complex than the homogeneous model and introduces some difficult concepts when treating interphase transfers. If the significantly higher computation load and theoretical uncertainties are accepted, the separated flow approach can yield better results in the microanalysis of evaporation and condensation processes. However, in terms of the simulation of plant operations (macro-analysis), the homogeneous model provides good results within the limits of other uncertainties. Some overlap between the two methods is provided by using a slip model (i.e. by treating the differing flow velocities of the two phases) to calculate the otherwise homogeneous mixture properties.

Experimental data compared to flow modelling suggests that separated flow models give better prediction of conditions at low pressures, flow rates and heat fluxes while homogeneous models give at least equal results at higher pressures and flows.

The parameters of interest in two-phase flow systems are the friction pressure loss and heat transfer coefficients. In the absence of tractable physical analyses for their calculation they must be empirically estimated. This can be done by correlating these parameters with other physically measurable or calculable quantities. Correlations take the form of empirically determined and sometimes quite complex functions which have been fitted to some set of experimentally measured data. The correlations are often identified by the name of the experimenter under whose name(s) they were first published. Before using a published correlation it is important to be aware of the test conditions under which it was determined in order to ensure its applicability to the problem at hand. Many correlations can be applied outside the test data range with satisfactory accuracy but with reduced certainty. Parameter predictions based on empirical correlations can have considerable error, sometimes exceeding 50 %.

We will start with some definitions. In what follows, the frequently used subscripts have the following meanings.

- g indicates the gas or vapour phase
- l indicates the liquid phase
- lo indicates that the total flow has liquid phase properties
- go indicates that the total flow has gas or vapour phase properties
- f the friction term
- a the acceleration or momentum term

gr The gravitation term (geodetic height)
m derived from a homogeneous mixture model—2 phase
s derived from a separated flow (slip) model—2 phase
2ph two-phase flow model, general

We will consider only one-dimensional flow along the flow axis, denoted z. The treatment of only a single coordinate axis does not imply ignorance of the crossflow velocity or mixture properties but implies simply that mixture properties are treated as averaged patterns across the plane, normal to the flow direction. Specific flow patterns in this plane can be associated to one or other of the flow types introduced in the earlier section of this chapter.

The ratio $G = \dot{m}/A$, where $\dot{m}$ is the mass flow [kg/s] and A is the area [m^2] across which the mass flows, is variously referred to as the specific mass flow, the mass flux or the mass velocity.[1] For a single tube of diameter d, $A = \pi/4\,d^2$. In single-phase flow the mass flow $\dot{m}$ occupies the entire flow cross-section area. In two-phase flow each phase is considered to flow through only part of the total flow cross section. The ratio of the mass of the gas (or vapour) phase to the total is known as the mixture mass fraction, commonly denoted as x.

$$x = \frac{\dot{m}_g}{\dot{m}_g + \dot{m}_l}.$$

The voidage Ω is expressed as the ratio of the gas phase flow cross-sectional area to the total.

$$\Omega = \frac{A_g}{A_g + A_l}.$$

Voidage can also be calculated as the ratio of the volumetric flow of the gas/vapour phase to the total volumetric flow.

$$\Omega = \frac{\dot{V}_g}{\dot{V}_g + \dot{V}_l} = \frac{\dot{m}_g v_g}{\dot{m}_g v_g + \dot{m}_l v_l},$$

where v_g and v_l are the specific volumes of the gas and liquid phases. If both the liquid and vapour phases move with the same velocity these two calculation methods produce the same result. However, if the phases are considered to move with differing velocities—v_l for the liquid phase and v_g for the gas/vapour phase—then we must write

$$\Omega = \frac{S\dot{V}_g}{S\dot{V}_g + \dot{V}_l},$$

[1]Elsewhere in this book $\dot{m}/A$ is designated ϕ. In this chapter it is designated G to maintain consistency with the two-phase literature and to avoid confusion with the various flow multipliers, also designated ϕ.

where S is the "slip" ratio defined as

$$S = \frac{v_g}{v_l}.$$

For the homogeneous model the mixture specific volume v_m is given from

$$v_m = x\, v_g + (1 - x)v_l.$$

For a non-homogeneous flow with slip S, v_m is given from

$$v_m = \frac{x\, v_g + S(1 - x)v_l}{x + S(1 - x)}.$$

This expression can also be written

$$v_m = v_l + \frac{x\,(v_g - v_l)}{S - x(S - 1)}$$

from which the following derivatives can be developed [92].

$$\frac{\partial v_m}{\partial x} = \frac{S(v_g - v_l)}{(S - x(S - 1))^2},$$

$$\frac{\partial v_m}{\partial p} = -\frac{\frac{x(1-x)(v_g - v_l)}{S - x(S-1)}\frac{\mathrm{d}S}{\mathrm{d}p} - S(1 - x)\frac{\mathrm{d}v_l}{\mathrm{d}p} - x\frac{\mathrm{d}v_g}{\mathrm{d}p}}{S - x(S - 1)}.$$

13.1.3 Two-Phase Slip

The slip ratio S is not easily determined from theoretical flow models and must be estimated from other parameters. Various correlations may be used to determine S. These are based on:

(a) Direct correlation with pressure-dependent fluid properties
(b) Correlation with the void fraction
(c) Correlation with fluid properties and flow regime

Butterworth and Hewitt [63] give correlations of the first and second types. For example, a simple correlation of the first type derived by Zivi [88] suggests, for a saturated steam flow,

$$S = \left(\frac{v''}{v'}\right)^{0.33},$$

where the double prime indicates the saturated vapour state and the single prime indicates the saturated water phase.

Doležal [91] cites Thom [65] to provide the following table relating the group $S(v'/v'')$ to pressure in bar.

Pressure (bar)	1.0	7	35	70	105	140	175	221.5
$S(v'/v'')$	0.153	0.3	0.5	0.6	0.67	0.725	0.79	1.0

For pressures < 17.73 MPa this tabulated relationship can be readily computed with good accuracy from a simple quadratic approximating function.

$$S(v'/v'') \approx 0.153 + 0.11\,\hat{p} + 0.075\,\hat{p}^2.$$

Input to the approximating function is $\hat{p} = \log_{10}(p/100)$ where p is expressed in kPa absolute. For pressures between 17.73 MPa and the critical pressure $S(v'/v'')$ can be approximated by a linear variation between its value at 17.73 MPa (0.79) and its value at the critical pressure (1.0). The estimate of slip S can then be extracted from $S(v'/v'')$, given v' and v'' from pressure.

Correlations of the third type are proposed by the transient reactor analysis code (TRAC-PD2) report [87]. These require prior knowledge of the voidage coefficient Ω and determine the relative slip velocity $v_r = v_g - v_l$. Quoting from the TRAC report,

for the bubbly regime ($0 \leq \Omega \leq 0.1$),

$$v_r = \frac{1.41}{(1-\Omega)}\left[\frac{\sigma\, g\,(\rho_l - \rho_g)}{\rho_l^2}\right]^{0.25}$$

for the slug flow regime ($0.2 \leq \Omega \leq 0.65$),

$$v_r = \frac{0.345}{(1-\Omega)}\left[\frac{g\, D_H(\rho_l - \rho_g)}{\rho_l^2}\right]^{0.5}$$

for the annular flow regime ($0.85 \leq \Omega \leq 0.9$),

$$v_r = \frac{v_m}{\left[\frac{\rho_g(76-75\Omega)}{\rho_g\,\Omega^{0.5}}\right]^{0.5} + \Omega\frac{\rho_g}{\rho_m}},$$

where ρ denotes the density of the liquid (l) or vapour (g) phase, g is the gravity constant [m s^{-2}], σ is the interfacial surface tension [N/m] and D_H is the tube hydraulic diameter [m]. A linear variation of v_r is used in each the transition regions from bubbly to slug flow ($0.1 \leq \Omega \leq 0.2$), slug to annular flow ($0.65 \leq \Omega \leq 0.85$) and annular flow to dry out ($0.9 \leq \Omega \leq 1.0$). The two-phase mixture density is

$$\rho_m = \Omega\,\rho_g + (1-\Omega)\rho_l.$$

Note that this is not equal to the homogeneous mixture density given from

$$\frac{1}{\rho_m} = \frac{1-x}{\rho_l} + \frac{x}{\rho_g}.$$

The slip ratio is determined from ν_r as

$$S = 1 + \frac{\nu_r}{\nu_l}, \tag{13.1}$$

where ν_l is the velocity of the liquid phase.

13.1.4 Voidage Correlations

The two-phase mass fraction x and the voidage Ω can be related. From

$$x = \frac{m_g}{m_g + m_l} = \frac{1}{1 + \frac{m_g}{m_l}},$$

we have

$$\frac{1-x}{x} = \frac{m_l}{m_g}.$$

For homogeneous flow, $\Omega = \Omega_H$ where

$$\Omega_H = \frac{V_g}{V_g + V_l} = \frac{m_g v_g}{m_g v_g + m_l v_l} = \frac{1}{1 + \frac{m_l}{m_g}\frac{v_l}{v_g}} = \frac{x\,v_g}{(1-x)v_l + x\,v_g}.$$

Substituting $\frac{1-x}{x}$ for $\frac{m_l}{m_g}$ and replacing specific volume v by density $\rho = 1/v$ to be consistent with the two-phase literature, we obtain

$$\frac{1-\Omega_H}{\Omega_H} = \left(\frac{1-x}{x}\right)\left(\frac{\rho_g}{\rho_l}\right)$$

or

$$\frac{1}{\Omega_H} = 1 + \left(\frac{1-x}{x}\right)\left(\frac{\rho_g}{\rho_l}\right).$$

For the non-homogeneous case the picture is somewhat more complex. Chisholm [69] suggests

$$\frac{1}{\Omega} = 1 + K\left(\frac{1-x}{x}\right)\left(\frac{\rho_g}{\rho_l}\right),$$

where

$$K = \left(\frac{\rho_l}{\rho_m}\right)^{0.5}$$

and

$$\frac{1}{\rho_m} = \frac{1-x}{\rho_l} + \frac{x}{\rho_g}.$$

By comparison Butterworth proposed a void correlation which combines several existing correlations.

$$\frac{1}{\Omega} = 1 + 0.28\left(\frac{1-x}{x}\right)^{0.64}\left(\frac{\rho_g}{\rho_l}\right)^{0.36}\left(\frac{\mu_l}{\mu_g}\right)^{0.07}.$$

Other voidage correlations are those of Zivi [88], Armand [89] and Collier [74]

$$\text{Zivi: } \frac{1}{\Omega} = 1 - \left(\frac{1-x}{x}\right)\left(\frac{\rho_g}{\rho_l}\right)^{2/3},$$

$$\text{Armand: } \Omega = (0.833 + 0.167\,x)\Omega_H,$$

$$\text{Collier: } \Omega = 1 - \frac{1-x}{\phi_{lo}},$$

where ϕ_{lo} is the friction pressure loss multiplier, to be introduced in the next section.

13.1.5 Pressure Gradients

The momentum equation 5.3 relates the total pressure gradient along a flow path to the individual contributions of frictional drop, acceleration or momentum changes, and gravitation.

$$\left.\frac{\partial p}{\partial z}\right|_T = \left.\frac{\partial p}{\partial z}\right|_f + \rho\left(\frac{\partial v}{\partial t} + v\frac{\partial v}{\partial z}\right) - g\rho\sin\beta$$

or

$$\left.\frac{\partial p}{\partial z}\right|_T = \left.\frac{\partial p}{\partial z}\right|_f + \left.\frac{\partial p}{\partial z}\right|_a + \left.\frac{\partial p}{\partial z}\right|_{gr}.$$

For steady single-phase flow in uniform pipes, the friction pressure gradient (pressure loss per unit length) can be written

$$\left.\frac{\partial p}{\partial z}\right|_f = 2\frac{f G^2}{\rho D_H}, \tag{13.2}$$

where f is the Fanning friction factor, G is the specific mass flow and D_H is the hydraulic diameter. ρ is the density of the fluid at the prevailing temperature and pressure.

Equation 6.6 is an alternative representation of this equation and relates the friction pressure drop over a flow length L to the mass flow rate, density of the working medium and the friction factor ξ. To express this equation as a pressure gradient, $\partial p/\partial z$, we replace L by the unit increment Δz and allow $\Delta z \rightarrow 0$. The partial differential is retained to emphasise that pressure is a function of other variables and to identify this term with the pressure loss term in the fundamental momentum and energy equations.

If only a single phase is present, these parameters are completely specified by the physical properties of the fluid or gas. If two phases are present the friction factor and density depend on the amount of each phase present.

The Blasius formula for single-phase flow can be used to calculate the friction factor f

$$f = C_f Re^{-n},$$

where $Re = \rho v/\mu D_H$ is the flow Reynolds Number. For smooth tubes $n = 0.25$ and for rough tubes $n = 0$, that is, independent of Re. $C_f = 0.079$ for $Re < 10^5$.

The classic work by Lockhart and Martinelli [66] and Martinelli and Nelson [67] set the ground for the systematic study of two-phase flow effects using the flow multiplier. The basic idea was to define a factor which multiplies the single-phase pressure gradient to yield the two-phase gradient. Lockhart and Martinelli defined two multipliers as

$$\phi_l^2 = \left.\frac{(\mathrm{d}p/\mathrm{d}z)_{2ph}}{(\mathrm{d}p/\mathrm{d}z)_{lo}}\right|_f$$

and

$$\phi_g^2 = \left.\frac{(\mathrm{d}p/\mathrm{d}z)_{2ph}}{(\mathrm{d}p/\mathrm{d}z)_{go}}\right|_f .$$

In practice the liquid multiplier ϕ_l^2 is most commonly used. The single-phase liquid-only pressure gradient is more accurately and stably predictable than the gas or vapour gradient, which is influenced by pressure and temperature variations to a greater degree. Then

$$\left.\frac{\partial p}{\partial z}\right|_{f,2ph} = \phi_{lo}^2 \left.\frac{\partial p}{\partial z}\right|_{f,1P} .$$

Lockhart and Martinelli (1948–1949) defined what is now known as the Martinelli factor X_{tt}. This defines the ratio between the pressure gradients along a flow path, where each phase flowing alone.

$$X_{tt}^2 = \left.\frac{\phi_l^2}{\phi_g^2}\right|_f .$$

The subscript tt on X indicates that the flow of each phase is considered to be turbulent. Then, for the phases,

$$\left.\frac{\partial p}{\partial z}\right|_{lo,f} = C_{f,l}\left[\frac{(1-x)^2 G^2}{\mu_l}\right]^{-n}$$

and

$$\left.\frac{\partial p}{\partial z}\right|_{go,f} = C_{f,g}\left[\frac{x^2 G^2}{\mu_g}\right]^{-m}.$$

The specific mass flow G is the same for both phases as the correlation assumes that each phase flows at the total mass flow rate. Lockhart and Martinelli assumed $n = m = 0.2$ and equal values for $C_{f,l}$ and $C_{f,g}$. Then

$$X_{tt} = \left(\frac{1-x}{x}\right)^{0.9}\left(\frac{\mu_l}{\mu_g}\right)^{0.1}\left(\frac{\rho_g}{\rho_l}\right)^{0.5}.$$

Alternatively, assuming $n = m = 0.25$ (the Blasius index for smooth tubes) gives

$$X_{tt} = \left(\frac{1-x}{x}\right)^{0.875}\left(\frac{\mu_l}{\mu_g}\right)^{0.125}\left(\frac{\rho_g}{\rho_l}\right)^{0.5}.$$

The two-phase multiplier is widely supported and well developed. We will now review several correlations used to estimate multipliers. The Martinelli factor appears explicitly in some of these correlations.

Wallis [68] (1969) This correlation assumes $n = 0.25$ (smooth tubes) in the Blasius equation and uses a homogeneous model to calculate mixture density and viscosity. Then,

$$\phi_{lo} = \left(1 + \frac{\rho_l - \rho_g}{\rho_g}\right)\left(1 + x\frac{\mu_l - \mu_g}{\mu_g}\right)^{-0.25}.$$

It is valid for Re > 2100 (turbulent flow) and $0 < x < 0.7$.

Chisholm (1967–1973) Given the Martinelli factor X_{tt}, Chisholm [69] proposed

$$\phi_l^2 = 1 + \frac{C}{X_{tt}} + \frac{1}{X_{tt}^2}$$

and

$$\phi_g^2 = 1 + C\,X_{tt} + X_{tt}^2.$$

The coefficient C takes a value defined by the flow regime, according to the following table:

Liquid	Gas	C
Turbulent	Turbulent	20
Laminar	Turbulent	12
Turbulent	Laminar	10
Laminar	Laminar	5

For most industrial cases the turbulent-turbulent combination applies giving $C = 20$.

In a later publication [70] Chisholm proposed a variant to this scheme which includes the influence of mass flow in the calculation of the coefficient C. Then,

$$C = \left[\lambda + (C_2 - \lambda)\left(\frac{\rho_g}{\rho_l}^{0.5}\right)\right]\left[\left(\frac{\rho_l}{\rho_g}\right)^{0.5} + \left(\frac{\rho_g}{\rho_l}\right)^{0.5}\right] \quad \text{with } \lambda = 0.5\,(2^{2-n} - 2)$$

and n is the Blasius index. For $G < 2000\,\text{kg/m}^2\text{s}$, $C_2 = 2000/\text{G}$ and $\lambda = 0.75$ for smooth tubes ($n = 0.25$). For rough tubes ($n = 0.0$), $C_2 = 1500/\text{G}$ and $\lambda = 1$.

These factors apply only to low flows and pressures. Chisholm [71] extended the validity range by matching to the extensive graphical results prepared by Baroczy [75]. He defined a property coefficient Γ as

$$\Gamma = \frac{\mathrm{d}p/\mathrm{d}z|_{f,go}}{\mathrm{d}p/\mathrm{d}z|_{f,lo}}$$

related to the Martinelli factor by

$$\Gamma = \frac{1}{X_{tt}}\left(\frac{1-x}{x}\right)^{(2-n)/2},$$

where again n is the coefficient in the Blasius equation. The friction multiplier is then given from

$$\phi_{lo}^2 = 1 + (\Gamma^2 - 1)\left[B\,x^{(2-n)/2}(1-x)^{(2-n)/2} + x^{(2-n)}\right].$$

The parameter B was chosen to match the Baroczy data for a series of flow regimes, identified by the property parameter Γ.

For $0 < \Gamma < 9.5$	$B = 55/G^{0.5}$	$\text{G} \geq 1900\,\text{kg/m}^2\text{s}$
	$B = 2400/G$	$500 < G < 1900\,\text{kg/m}^2\text{s}$
	$B = 4.8$	$\text{G} \leq 500\,\text{kg/m}^2\text{s}$
For $9.5 < \Gamma < 28$	$B = 520/(\Gamma G^{0.5})$	$\text{G} \leq 600\,\text{kg/m}^2\text{s}$
	$B = 21/\Gamma$	$\text{G} > 600\,\text{kg/m}^2\text{s}$
For $\Gamma > 28$	$B = 1500/(\Gamma^2 G^{0.5})$	

These correlations fitted smooth tube pressure gradients but underestimated pressure gradients in rough tubes. Chisholm suggested the following correction for rough tubes. The subscript "R" indicates rough tubes, and "S" indicates smooth tubes.

$$B_R = B_S \left\{ 0.5 \left[1 + \left(\frac{\mu_g}{\mu_l} \right)^2 + 10^{-600\Omega/D_H} \right] \right\}^{(0.25-n)/0.25}.$$

Friedel [77] (1979) This correlation appears to provide good agreement with experimental data over a wide range of operating conditions and is now widely used in industry. It is well suited to applications for which the viscosity ratio $\mu_l/\mu_g < 1{,}000$, a condition generally satisfied for most industrial applications.

$$\phi_{lo}^2 = E + \frac{3.24\, F\, H}{Fr^{0.045}\, We^{0.035}},$$

$$E = (1-x)^2 + x^2 \left(\frac{\rho_l f_{go}}{\rho_g f_{lo}} \right),$$

$$F = x^{0.78}(1-x)^{0.224},$$

$$H = \left(\frac{\rho_l}{\rho_g} \right)^{0.91} \left(\frac{\mu_g}{\mu_l} \right)^{0.19} \left(1 - \frac{\mu_g}{\mu_l} \right)^{0.7}.$$

The Froude number Fr is given from

$$Fr = \frac{G^2}{gD\rho_m^2}$$

and the Weber number We from

$$We = \frac{G^2 D}{\rho_m \sigma}.$$

Finally, we have

$$\frac{1}{\rho_m} = \frac{x}{\rho_g} + \frac{1-x}{\rho_l}.$$

Bankhoff [76] (1979) The friction multiplier is ϕ_{Bf}, used as follows.

$$\left. \frac{\mathrm{d}p}{\mathrm{d}z} \right|_f = \phi_{Bf}^{7/4} \left. \frac{\mathrm{d}p}{\mathrm{d}z} \right|_{lo}.$$

The liquid friction pressure gradient is given from

$$\left. \frac{\mathrm{d}p}{\mathrm{d}z} \right|_{lo} = f_{lo} \frac{2G^2}{D_H \rho_l}$$

and the liquid friction factor f_{lo} is given from the Blasius equation with n = 0.25.

$$f_{lo} = \frac{0.079}{Re_{lo}^{0.25}} \text{ where } Re_{lo} = \frac{G\,D_H}{\mu_l}.$$

The two-phase multiplier ϕ_{Bf} is then given by

$$\phi_{Bf} = \frac{1}{1-x}\left[1-\gamma\left(1-\frac{\rho_g}{\rho_l}\right)\right]^{3/7}\left[1+x\left(\frac{\rho_l}{\rho_g}-1\right)\right],$$

where

$$\gamma = \frac{0.71+2.35\left(\frac{\rho_g}{\rho_l}\right)}{1+\frac{1-x}{x}\frac{\rho_g}{\rho_l}}.$$

This method is applicable to two-phase flows with a mass fraction range of $0 < x < 1$.

Cicchitti [78] (1960) The attraction of this correlation is its simplicity. Despite this it appears to work reasonably well for all flow regimes and mass fractions. It quotes the two-phase pressure gradient in terms of specific mass flow G, mass fraction x and the local physical properties of both liquid and gas phases, without the need for the calculation of a set of intermediate factors.

$$\left.\frac{\mathrm{d}p}{\mathrm{d}z}\right|_f = \frac{0.092}{D^{1.2}}G^{1.8}\left[\mu_l - x(\mu_l-\mu_g)\right]^{0.2}\left[v_l + x(v_g - v_l)\right].$$

Müller–Steinhagen and Heck [85] (1986)

$$\left.\frac{\mathrm{d}p}{\mathrm{d}z}\right|_f = F(1-x)^{1/3} + Bx^3,$$

$$A = \left.\frac{\mathrm{d}p}{\mathrm{d}z}\right|_{lo} = f_{lo}\frac{2G^2}{D_H\rho_l},$$

$$B = \left.\frac{\mathrm{d}p}{\mathrm{d}z}\right|_{go} = f_{go}\frac{2G^2}{D_H\rho_g},$$

$$F = A + 2(B-A)x$$

and the friction factors f_{lo} and f_{go} are given from the standard Blasius equation, with n = 0 (rough tubes).

$$f_{lo} = \frac{0.079\,\mu_l}{G\,D_H},$$

$$f_{go} = \frac{0.079\,\mu_g}{G\,D_H}$$

Doležal [91] (1972) This has been developed specifically for water/steam applications and is the only method which explicitly uses the local heat flux $\dot{q}$. Using a physical model of the mass exchange between the surface boundary and the bulk fluid flow to create buoyancy and upwards flow pressure and flow resistance, Doležal shows that the 2-phase friction factor can be calculated from the single-phase factor by

$$f_{2ph} = f_{1P} + 8\left(\frac{\dot{q}}{\mathbf{r}}\frac{v''}{v'}\frac{1}{\rho_l v_0}\right)\left(\frac{1}{1 + x_s\,(v''/v' - 1)}\right),$$

where $\dot{q}$ is the heat flux $[W/m^2]$ normal to the mass flow, v_0 is the flow velocity at the entrance to the boiling cell, $\mathbf{r} = (h'' - h')$ is the latent heat of vaporisation, f_{1P} is the single-phase pressure drop coefficient and v', v'', h', h'' are the fluid water and steam saturation specific volumes and enthalpies. x_s is the steam mass fraction allowing for the effects of slip.

$$x_s = \frac{S\,(v'/v'')x}{1 - [1 - S(v'/v'')]x}.$$

The two-phase gradient is calculated from the single-phase gradient

$$\frac{\partial p_{2ph}}{\partial z} = \phi_{lo}^2 \frac{\partial p_{1P}}{\partial z}$$

using as multiplier

$$\phi_{lo}^2 = 1 + \frac{8}{f_{1p}}\frac{1}{r}\frac{v''}{v'}\frac{\dot{q}}{\rho_l v_o} - S\frac{(v'/v'')(v''/v' - 1)}{1 - S(v'/v'')}\left\{1 + \frac{\ln\left[1 - (1 - S(v'/v'')2\,x_s\right]}{(1 - S(v'/v'')2\,x_s}\right\}$$

and

$$\frac{\partial p_{1P}}{\partial z} = 2\frac{f_{1P}\,v' G^2}{D_H}.$$

In practice, the slip effect can be ignored and the pressure multiplier replaced by the much simplified form

$$\phi_{lo}^2 = 1 + x\left(\frac{v''}{v'} - 1\right) + 8\frac{\dot{q}v''}{f_{1p}\mathbf{r}v_0}. \qquad (13.3)$$

13.2 Heat Transfer Coefficients

Section 13.1 introduced several categories of two-phase flow patterns. The subsequent discussion of pressure gradients related these flow regimes to one or other pressure gradient multiplier correlation. On a micro-scale various authors have refined the definition of these flow patterns to give an improved match to selected heat transfer features. On the macro-scale adopted here we will continue to use these same pattern categories for the discussion of correlations to be used for the calculation of heat transfer coefficients.

13.2.1 Sub-cooled Boiling

This is the first stage of vapour phase development and occurs when the wall temperature is above the local saturation temperature, the fluid bulk saturation temperature is below the saturation temperature and the heat flux across the wall/fluid interface is sufficient to promote bubble formation but insufficient to promote nucleation. Small bubbles form at the interface. Some adhere to the wall and the remainder are entrained by the flow. Some of these re-condense in the cooler fluid while others are carried by the flow to warmer regions before they can re-condense. The bubble density (voidage) is insufficient to have a significant effect on the pressure gradient but the heat transfer coefficient is greatly enhanced. The onset of sub-cooled boiling can be determined from the differential between the local wall and bulk fluid temperatures. There is general agreement that this must exceed some 5 K for bubble formation. Butterworth and Hewitt [63] quote correlations by Jens and Lottes (1951) and Thom et al. (1965–1966) which predict the temperature differential between the wall and fluid saturation temperatures required for sub-cooled boiling.

$$\text{Jens and Lottes:} \quad \Delta T_{sat} = 25(Q/A)^{0.25}\mathrm{e}^{-p/62}$$
$$\text{Thom et al.:} \quad \Delta T_{sat} = 22.65(Q/A)^{0.5}\mathrm{e}^{-p/87}$$

with (Q/A) quoted in MW/m^2 and p in bar (absolute). They also note that these formulae are not in good agreement with each other and bracket the index 0.33 quoted by Rohsenow [43]. We might therefore feel free to modify one or other of these correlations for the present purpose and postulate a revised formula, now with $\dot{q} = Q/A$ given in the more usual W/m^2, and p in bar.

$$\Delta T_{sat} = 0.235\,\dot{q}^{0.33}\mathrm{e}^{-p/87}.$$

Using this formula with actual operating data from a real industrial boiler yields the following results.

For low furnace heat at start-up,

$$\dot{q} = 9{,}554\,\mathrm{W/m^2},\ p = 5\,\mathrm{bar}$$
$$\Delta T_{sat} = 4.7\,\mathrm{K}$$

For full-load operation,

$$\dot{q} = 39{,}900\,\mathrm{W/m^2},\ p = 50\,\mathrm{bar}$$
$$\Delta T_{sat} = 4.4\,\mathrm{K}$$

These are close to the 5 K suggested by Butterworth and Hewitt.

We can work backwards from this expression for ΔT_{sat} to establish a usable formula for a heat transfer coefficient for sub-cooled boiling. We can express the heat transferred to the fluid in terms of the local saturation temperature or in terms of the local bulk fluid temperature. Working from the previous formula for the temperature differential between wall and *saturation* temperature, we can write

$$\dot{q} = \alpha_{sc}\Delta T_{sat} = \alpha_{sc}(T_w - T_{sat}) = 0.235\,\dot{q}^{0.33}\mathrm{e}^{-p/87}$$

from which we have

$$\frac{1}{\alpha_{sc}} = 0.235\,\dot{q}^{-0.67}\mathrm{e}^{-p/87}$$

or

$$\alpha_{sc} = 4.255\,\dot{q}^{0.67}\mathrm{e}^{p/87}, \tag{13.4}$$

which is the desired expression for the sub-cooled boiling heat transfer coefficient α_{sc}. Using this formula to predict the heat transfer coefficient for the typical industrial boiler quoted earlier gives,

For low furnace heat at start-up,

$$\dot{q} = 9,554\ W/m^2,\ p = 5\,\mathrm{bar}$$

$$\alpha_{sc} = 1,540\ W/m^2/K$$

For full-load operation,

$$\dot{q} = 39,900\ W/m^2,\ p = 50\,\mathrm{bar}$$

$$\alpha_{sc} = 9,130\ W/m^2/K$$

The first of these would apply during boiler start-up when circulation flow is very low (G [kg/s/m^2]<10). The calculated heat transfer coefficient is some 6–10 times that which might be expected for single-phase water under the same low-flow conditions. The second, representing normal operation of the boiler, is some three times that of single-phase water. These are reasonable figures. The second result also suggests that, during normal on-load operations, the sub-cooled boiling zone will be quite short.

The sub-cooled boiling heat transfer coefficient calculated from this formula is independent of the mass flow rate and is similar in form to that appropriate to pool boiling. This is significant during start-up of a natural circulation boiler since the flow in the evaporator circuit will be zero prior to any formation of bubbles. Nevertheless, the heat transfer coefficient can be calculated with reference to the heat flux and pressure only. The calculation need not be iterative if it is assumed that, for ΔT_{sat} <5 K, heat transfer is convective to single-phase water only and for ΔT_{sat} >5 K, sub-cooled boiling applies.

The sub-cooled boiling regime can be considered finished once the vapour mass fraction x exceeds 5 % (0.05) or when the bulk fluid temperature has reached the local saturation temperature, whichever comes first.

13.2.2 Onset of Nucleate Boiling

This coincides with the end of sub-cooled boiling and may therefore be identified by the same condition ($x > 0.05$ and/or $T_{fl} = T_{sat}$). The physical picture is one in which the wall–fluid heat flux is large enough to ensure that the bubble nucleation sites are stable and allow bubble populations to grow. This minimum heat flux is denoted $\dot{q}_{onb}$, and simple physical models have been used to estimate it.

Once nucleate boiling has become established the boiling zone is described as a region of "fully developed boiling" or "nucleate boiling". This region continues until either the evaporative section of tubing ends, that is, no further heat transfer to the fluid, or the point of DNB has been reached. In terms of the flow regimes described earlier this region spans both the slug and annular flow regimes.

Throughout the nucleate boiling region the fluid will be at its saturation temperature[2] and heat transfer from the wall can be treated as a combination of nucleate boiling transfer across the interface and forced convection transfer across the liquid layer. This can be represented by a generic power law combination of two heat transfer coefficients, one applying to nucleate boiling (α_{nb}) and the second to convective heat transfer through the liquid layer (α_{cb}).

$$\alpha_{2ph} = \left[\alpha_{nb}^{m} + \alpha_{cb}^{m}\right]^{(1/m)}, \tag{13.5}$$

where m is an integer.

It is worth noting that the heat transfer coefficient in the nucleate boiling region is very large, and the consequences of errors in its prediction are less significant to the overall accuracy of the simulation than other errors which may arise from other approximations.

13.2.3 Heat Transfer Coefficients for Annular Flow

As noted previously, the slug flow region is short, and the nucleate boiling region is predominantly annular flow. Methods based on empirical correlations for the estimation of two-phase heat transfer coefficients can be divided into the following frequently cited categories:

- Those correlated with the Lockhart–Martinelli factor as X_{tt}^{-1} in the generic form

$$\alpha_{2ph} = \alpha_{1ph}\, a \left(\frac{1}{X_{tt}}\right)^{b},$$

[2]In inclined and vertical flows the local saturation temperature will decrease with height as the static geodetic head decreases.

where α_{1ph} may be calculated from any of the several single-phase correlations using either liquid-only (lo) or liquid-phase (l) properties

- Those using a modification of the Dittus–Boelter equation

$$Nu = 0.023\, Re^{0.8} Pr^{0.42}$$

or the Sieder–Tate formulation

$$Nu = 0.027\, Re^{0.8} Pr^{0.33} \left(\frac{\mu_B}{\mu_l}\right)^{0.14}.$$

The Nusselt number Nu is calculated from

$$Nu = \frac{\alpha D_H}{\lambda}.$$

- Other more recent correlations using the methods of Gungor and Winterton and Steiner and Taborek and the Gnielinski [82] correlation

$$Nu = \frac{(f/8)(Re_l - 1000)Pr}{1 + 12.7(f/8)^{0.5}(Pr^{2/3} - 1)}, \tag{13.6}$$

where the Fanning friction factor f is given by

$$f = [0.7904 \ln(Re) - 1.64]^{-2} = [1.82 \log_{10}(Re) - 1.64]^{-2}.$$

Assuming that for most practical purposes the ratio of viscosities of the liquid and vapour phases (μ_l and μ_g) in the sub-critical region remains constant at around 720, the viscosity term in the Sieder–Tate formulation may be set to 1.93.

Martinelli-Based Multipliers

Several correlations based on this formulation have been published, distinguished by their individual selection of the coefficients a and b and by the choice of the single-phase reference as either α_l or α_{lo}. For example [63],

- Dengler and Addoms (1956): $\alpha_{2ph} = \alpha_{lo}\, 3.5 \left(\frac{1}{X_{tt}}\right)^{0.5}$
- Pujol and Stenning (1968): $\alpha_{2ph} = \alpha_{lo}\, 4.0 \left(\frac{1}{X_{tt}}\right)^{0.37}$
- Bennett, Collier, Pratt and Thornton (1961): $\alpha_{2ph} = \alpha_l\, 0.64 \left(\frac{1}{X_{tt}}\right)^{0.74} \dot{q}^{0.11}$

Each of these correlations is quoted as valid for $x < 0.7$. Ghajar [90] tabulates in excess of 25 heat transfer correlations attributed to 20 authors, applying to various fluid/gas combinations, including water/steam and refrigerants, and ranges of validity of their reference variables. It is widely reported in the two-phase literature that various formulae can give widely differing predictions for α_{2ph}, which emphasises the difficulties in finding the "right" correlation to use for any specific application.

Power Law Combinations of Nucleate and Convective Boiling Transfer

Chen (1961) The most widely recommended form is that developed by Chen [84]. This and similar correlations assume that the heat transfer mechanism involves components due to nucleate boiling (α_{nb}) and convective heat transfer through the liquid layer (α_{cb}). The two-phase heat transfer coefficient is then given by the superposition of these two mechanisms

$$\alpha_{2ph} = \alpha_{nb} + \alpha_{cb}.$$

The Chen correlation is reported as delivering a good match to its reference experimental data but is complex to implement and is not recommended for macro-level simulation.

Gungor and Winterton (1986–1987) A recent variation of the Chen correlation is that proposed by Gungor and Winterton [79]. It replaces the convective boiling contribution by a single-phase liquid term α_l calculated from the Dittus–Boelter equation using the local liquid fraction of flow $(x - 1)\,\dot{m}$. Then

$$\alpha_{2ph} = E\,\alpha_l + S\,\alpha_{nb}$$

E and S are calculated from a series of intermediate calculations.

The original 1986 correlation was replaced by a revised and simplified version [80] which dropped the explicit nucleate boiling term in favour of a liquid-based term only. The new term included the influence of heat flux and mass flow.

$$\alpha_{2ph} = E\,\alpha_l, \tag{13.7}$$

where

$$E = 1 + 3000 B_o^{0.86} + 1.12 \left(\frac{x}{1-x}\right)^{0.75} \left(\frac{\rho_l}{\rho_g}\right)^{0.41}$$

and the Boiling number B_o is given from

$$B_o = \frac{\dot{q}}{\dot{m}\mathbf{r}},$$

where $\dot{m}$ is the total mass flow rate and $\mathbf{r}$ is the latent heat of vaporisation. This correlation is reported to match its database to $\approx 22\,\%$.

The Steiner–Taborek Asymptotic Correlation (1989) [81] This method sets $m = 3$ in Eq. 13.5, and the two-phase heat transfer coefficient is calculated from

$$\alpha_{2ph} = \left[(\alpha_{nb,0} F_{nb})^3 + \left(\alpha_{lo} F_{2ph}\right)^3 \right]^{1/3}.$$

If the vapour mass fraction $x = 0$ (boiling not yet initiated), only the convective contribution is considered ($F_{nb} = 0$) unless the heat flux $\dot{q} > \dot{q}_{onb}$ in which case both contributions are included. The heat flux necessary to sustain nucleate boiling $\dot{q}_{onb}$ is calculated as

$$\dot{q}_{onb} = 2 \frac{\sigma \, T_{sat} \, \alpha_{lo}}{r_0 \rho_g \mathbf{r}}, \tag{13.8}$$

where σ is the interphase surface tension and r_o is a standard bubble nucleation diameter in meters, taken to be 100 μm (0.0001 m).

F_{2ph} is the two-phase flow multiplier for convective heat transfer through the annular layer. If $\dot{q} > \dot{q}_{onb}$ and $x < 1$ at the end of the boiling zone (normally the case for drum boilers but not for once-through evaporators),

$$F_{2ph} = \left[(1-x)^{1.5} + 1.9 x^{0.6} \left(\frac{\rho_l}{\rho_g} \right)^{0.35} \right]^{1.1}.$$

This expression is valid for the range $3.75 < (\rho_l/\rho_g) < 5{,}000$.

If $\dot{q} < \dot{q}_{onb}$ only forced convection is considered and

$$F_{2ph} = \left\{ \left[(1-x)^{1.5} + 1.9 x^{0.6} (1-x)^{0.01} \left(\frac{\rho_l}{\rho_g} \right)^{0.35} \right]^{-2.2} + \left[\left(\frac{\alpha_{go}}{\alpha_{lo}} \right) x^{0.01} \left(1 + 8(1-x)^{0.7} \right) \left(\frac{\rho_l}{\rho_g} \right)^{0.67} \right]^{-0.2} \right\}^{-0.5}.$$

The nucleate boiling coefficient F_{nb} is given from

$$F_{nb} = F_{pf} \left(\frac{\dot{q}}{\dot{q}_0} \right)^{nf} \left(\frac{d_i}{d_{i0}} \right)^{-0.4} \left(\frac{R_p}{R_{p0}} \right)^{0.133} F(M)$$

F_{pf} is a pressure-dependent correction factor, given from

$$F_{pf} = 2.816 \, p_r^{0.45} + \left[3.4 + \left(\frac{1.7}{1 - p_r^7} \right) \right] p_r^{3.7}.$$

The nucleate boiling exponent nf is given from

$$nf = 0.8 - 0.1 \, \mathrm{e}^{1.75 \, p_r}.$$

The "reduced pressure" $p_r = p/p_{crit}$, with p and p_{crit} quoted in bar. The standard reference diameter d_{io}=0.01 m (10 mm) and the standard surface roughness R_{p0}=1 μm (0.001 mm).

The residual molecular weight correction factor $F(M)$ is given by

$$F(M) = 0.377 + 0.199\,ln(M) + 0.000028427\,M^2.$$

The maximum value $F(M)$ may take is 2.5. For water, $M = 18$ and $F(M) = 0.961$.

The single-phase heat transfer coefficient α_{lo} is given by the Gnielinski formula, Eq. 13.6.

Ghajar (2005) Finally, Ghajar [83] quotes a correlation based on voidage and a power law relating the two-phase heat transfer coefficient to mass fraction, voidage, Prandtl number and the usual viscosity ratio,

$$\alpha_{2ph} = (1 - \Omega)\alpha_l \left\{1 + C\left[\left(\frac{x}{1-x}\right)^m \left(\frac{\Omega}{1-\Omega}\right)^n \left(\frac{Pr_g}{Pr_l}\right)^p \left(\frac{\mu_g}{\mu_l}\right)^q\right]\right\}. \quad (13.9)$$

Quoting from test results obtained for slug, bubbly/slug and bubbly/slug/annular flow regimes Ghajar recommends the following values for the indices in this expression

$$C = 2.86, m = 0.42, n = 0.35, p = 0.66, q = -0.72,$$

and the following validity ranges for the process variables

Re_{sl}	$\frac{x}{1-x}$	$\frac{\Omega}{1-\Omega}$	Pr_g/Pr_l	μ_g/μ_l
2,468→ 35,503	6.9x10^{-4} → 0.03	0.36→ 3.45	0.102→ 0.137	0.015→ 0.028

The Reynolds number Re_l is evaluated using fluid properties according to the formula

$$Re_l = \frac{4\,\dot{m}_l}{\pi\sqrt{1-\Omega}\,\mu_l\,D_H}.$$

13.2.4 Critical Heat Flux and Dry Out

Towards the end of the annular flow regime, the high vapour velocities create high drag forces (interfacial shear). The liquid film becomes unstable, and droplets torn away contribute to the separated (or mist) flow phase. The point at which the annular phase changes to separated flow is determined by the "critical steam fraction" denoted x_{crit}. This is sometimes equated to the "critical heat flux" or $\dot{q}_{crit}$. Once this point has been reached, the heat transfer coefficient decreases abruptly and changes to single-phase convection. The value of x_{crit} can be estimated from the following correlations [86], with pressure p expressed in bar absolute.

Pressure range (bar)	x_{crit}
$4.9 \leq p \leq 29.4$	$25.6\,\dot{q}^{-0.125}\,\dot{m}^{-0.333}\,d_i^{-0.07}\,e^{-0.01715\,p}$
$29.4 \leq p \leq 98.0$	$46.0\,\dot{q}^{-0.125}\,\dot{m}^{-0.333}\,d_i^{-0.07}\,e^{-0.00255\,p}$
$98.0 \leq p \leq 196.0$	$76.6\,\dot{q}^{-0.125}\,\dot{m}^{-0.333}\,d_i^{-0.07}\,e^{-0.00795\,p}$

These figures apply within the validity ranges:

Mass flow $\dot{m}$	$200 \leq \dot{m} \leq 5{,}000$ [kg/s/m^2]
Inner diameter d_i	$0.04 \leq d_i \leq 0.032$ [m]

Post dry out, the heat transfer coefficient is calculated from an appropriate single-phase correlation. The standard Dittus–Boelter equation is widely used, both for the sub- and supercritical pressure range.

Chapter 14
Steam Turbines

Chapter 7 dealt with the class of rotating machines called "turbomachines". These devices are usually driven by some form of prime mover such as an electric motor or steam turbine. As well as driving pumps and compressors, and as the most widely used form of marine propulsion, the most common usage of large steam turbines is driving electric generators in central power stations.

An early steam turbine was demonstrated in 1896 by C. H. Parsons in a small but fast torpedo boat, the "Turbinia". The steam turbine's advantages of high speed, high power, compact dimensions and good reliability led to its widespread adoption for marine propulsion, and it quickly replaced the heavy and slow reciprocating engines of the day.

Early turbines had a single row of blades. Parsons' turbine used several rows of reaction blading to distribute the pressure drop over several stages. This was followed in 1896 by the invention by Curtis of the multi-row velocity-compound turbine. Modern turbines use a combination of stage types but still employ the physical principles of these pioneering designs.

14.1 Steam Expansion Through the Turbine

A single turbine expansion stage consists of

(a) A fixed convergent/divergent Laval nozzle which converts potential energy represented by pressure to kinetic energy represented by velocity.
(b) A rotating set of blades which convert momentum (kinetic energy) of the steam flow to mechanical work

Two types of turbine stage are used, impulse and reaction. In the impulse stage most pressure conversion takes place in the stationary nozzles with little change in pressure in the moving blades. In the reaction stage most pressure conversion takes place in the rotating blades. The following analysis does not distinguish between

P. O'Kelly, *Computer Simulation of Thermal Plant Operations*,
DOI 10.1007/978-1-4614-4256-1_14,

Fig. 14.1 A boiler feedpump turbine, open for inspection (courtesy of Millmerran Operating Company, Australia)

the two types as the overall effects of each are sufficiently similar. Most turbines use one impulse stage as the first HP stage for speed and power control with the others being reaction stages. It is usual to operate the impulse stage in the choked condition for speed control; otherwise, stable speed control, with the turbogenerator unsynchronised, is difficult.

Figure 14.1 shows a turbine used to drive a boiler feed pump. As the turbine is operated with variable speed, the first stage is an impulse row followed by (in this case) four reaction rows. The characteristic shapes of the individual blades—impulse or reaction—are clearly visible in the picture.

Following Traupel [61], the mass flow through a turbine expansion stage is calculated from the general nozzle equation

$$\dot{m}^2 = \left(C_T \sum_i a_i \varepsilon_i \right)^2 \frac{\gamma}{\gamma+1} \frac{p_1}{v_1} \left[1 - \left(\frac{p_1}{p_2} \right)^{\frac{\gamma+1}{\gamma}} \right], \tag{14.1}$$

where γ is the adiabatic index defined by the state equation for steam

$$pv^{\gamma} = \texttt{constant}.$$

The constant C_T is the stage flow coefficient and is substantially independent of operating conditions. The term $\sum_i a_i \varepsilon_i$ is the summation of the individual admission areas where a_i is the full area of the i-th admission stage and ε_i is the admission fraction of the i-th stage. This is relevant only to the first (control) stage and only to turbines provided with partial arc admission control. The subscript “1” denotes the upstream variable, “2” the downstream variable.

This equation can be written in terms of the stage design conditions, denoted by the subscript "0", assuming that C_T is constant for all operating conditions and that the stage operates with full arc admission under design conditions.

$$\begin{aligned}\left(\frac{\dot{m}}{\dot{m}_0}\right)^2 &= \left(\frac{p_1 v_{10}}{p_{10} v_1}\right)\left(\frac{1-(p_2/p_1)^{\frac{\gamma+1}{\gamma}}}{1-(p_{20}/p_{10})^{\frac{\gamma+1}{\gamma}}}\right) \\ &= \left(\frac{p_1}{p_{10}}\right)^2\left(\frac{p_{10} v_{10}}{p_1 v_1}\right)\left(\frac{1-(p_2/p_1)^{\frac{\gamma+1}{\gamma}}}{1-(p_{20}/p_{10})^{\frac{\gamma+1}{\gamma}}}\right) \\ &= \left(\frac{p_1}{p_{10}}\right)^2\left(\frac{\mathcal{T}_{10}}{\mathcal{T}_1}\right)\left(\frac{1-(p_2/p_1)^{\frac{\gamma+1}{\gamma}}}{1-(p_{20}/p_{10})^{\frac{\gamma+1}{\gamma}}}\right),\end{aligned} \tag{14.2}$$

where $\mathcal{T}_{10}$ and $\mathcal{T}_1$ are expressed in degrees K.

This equation is valid only for an infinite series of turbine stages. For a finite number of stages, and particularly for a single stage, the pressure ratio function must be modified to reflect the existence of a critical ratio beyond which the flow velocity reaches sonic velocity, and the mass flow rate becomes independent of downstream conditions (the choked flow condition mentioned earlier). An appropriate function valid for all pressure ratios ≤ 1 is, with $\Pi = p_2/p_1$,

$$e = \left[1 - \left(\frac{\Pi - \Pi_{cr}}{1 - \Pi_{cr}}\right)^{\frac{\gamma+1}{\gamma}}\right]^{1/2}$$

$$e = 1 \quad \text{for} \quad \Pi \leq \Pi_{cr}.$$

We may then write

$$\frac{\dot{m}}{\dot{m}_0} = \left(\frac{p_1}{p_{10}}\right)\sqrt{\frac{\mathcal{T}_{10}}{\mathcal{T}_1}}\left(\frac{e}{e_0}\right). \tag{14.3}$$

Since e is a function of the ratio of downstream to upstream pressures, this equation contains all the information necessary for the computation of the complete condition line on the basis of conservation of mass in each stage.

The normalised flow ratios calculated using Eqs. 14.2 and 14.3 are shown in Fig. 14.2 for a range of p_2 from zero to p_1. The turbine stage parameters are assumed to have their nominal or rated values.

As shown in Fig. 4.1, γ can be expected to lie in the range 1.25–1.3 for most operating conditions of interest. The ratio $(\gamma + 1)/\gamma$ is then sufficiently close to 2 to allow the writing of Eq. 14.3 as

$$e^2 = 1 - \left(\frac{\Pi - \Pi_{cr}}{1 - \Pi_{cr}}\right)^2. \tag{14.4}$$

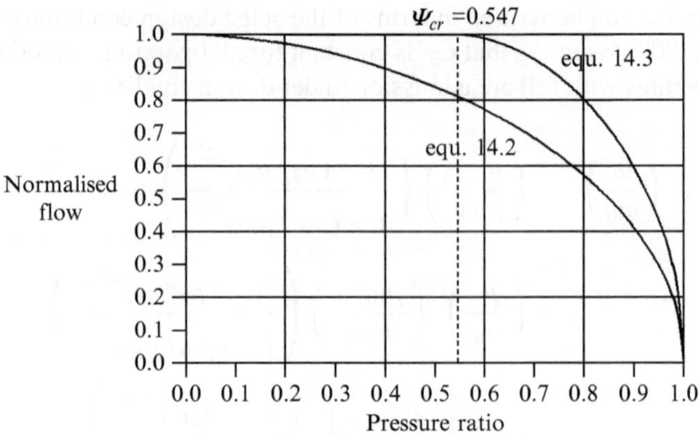

Fig. 14.2 Steam turbine flow characteristics

With this approximation, and noting that $p_2 = p_1 - \Delta p$, where Δp is the pressure differential across the stage, Eq. 14.4 may be resolved to

$$e^2 = \left(\frac{2}{1-\Pi_{cr}}\right)\frac{\Delta p}{p_1} - \left(\frac{2}{(1-\Pi_{cr})^2}\right)\left(\frac{\Delta p}{p_1}\right)^2$$

for $\Pi \geq \Pi_{cr}$. Since $\Delta p \ll p_1$, this may be written without the quadratic term as

$$e^2 = \left(\frac{2}{1-\Pi_{cr}}\right)\frac{\Delta p}{p_1} \tag{14.5}$$

allowing Eq. 14.3 to be written as

$$\frac{\dot{m}}{\dot{m}_0} = K_t\left(\frac{p_{in}}{p_{in0}}\right)\sqrt{\frac{T_{in0}}{T_{in}}}\left(\frac{1}{e_0}\right)\sqrt{\frac{\Delta p}{p}}, \tag{14.6}$$

where p_{in} and T_{in} are the steam pressure and temperature before the first blade row and as above, the subscript "0" denotes the rated value. Turbines are usually designed to operate with approximately equal pressure ratios across all stages, and the flow function e_0 can be evaluated using the design ratio Π_0.

$$e_0 = \left[1 - \left(\frac{\Pi_0 - \Pi_{cr}}{1-\Pi_{cr}}\right)^{\frac{\gamma+1}{\gamma}}\right]^{1/2}.$$

Ideally,

$$K_t = \sqrt{2/(1-\Pi_{cr})} = 2.1012 \tag{14.7}$$

but the approximations involved in the reduction of Eq. 14.3 to Eq. 14.6 introduce some error. The constant can be adjusted if necessary to match the known characteristic of a reference turbine.

If the dynamics of steam flow through the turbine are ignored, the following algebraic procedure may be used to calculate steam conditions after each blade row.

Inlet flow:

$$\dot{m}_{in} = \dot{m}_0 \left(\frac{p_{in}}{p_{in0}}\right) \sqrt{\left(\frac{\mathcal{T}_{in0}}{\mathcal{T}_{in}}\right)}. \tag{14.8}$$

For the downstream rows:

$$\begin{aligned} p_j &= \Pi_0 p_{j-1}, \\ \dot{m}_j &= \dot{m}_{j-1} - \dot{m}_{bl,j} - \dot{m}_{dr} \pm \dot{m}_{gss}, \\ T_j &= (T_{j-1} + 273)\left(\frac{p_j}{p_{j-1}}\right)^{\frac{\gamma-1}{\gamma}} - 273, \\ v_j &= f(p_j, T_j), \end{aligned}$$

where $\dot{m}_{bl}$, $\dot{m}_{dr}$ and $\dot{m}_{gss}$ are bled steam, drain and gland sealing flows.

The assumption of equal pressure ratios for all blade rows is satisfactory if no extraction flows are taken or if extraction flows are much less than the main steam flow, as is usually the case. A general solution allowing variable pressure ratios to accommodate extraction flows, drain and gland sealing flows can be developed for the complete cylinder from a statement of conservation of mass for each blade row. Then, with the introduction of a blade row "capacitance" C_j, we can write for the j-th row,

$$C_j \frac{\mathrm{d}p_j}{\mathrm{d}t} = \dot{m}_{j-1} - \dot{m}_j - \dot{m}_{bl,j} - \dot{m}_{dr} \pm \dot{m}_{gss}. \tag{14.9}$$

The mass flow $\dot{m}_j$ is obtained from Eq. 14.6. Then, for the j-th blade row,

$$\begin{aligned} \dot{m}_j &= \dot{m}_0 K_t \left(\frac{p_{in}}{p_{in0}}\right) \sqrt{\frac{\mathcal{T}_{in0}}{\mathcal{T}_{in}}} \left(\frac{1}{e_0}\right) \sqrt{\frac{\Delta p_j}{p_j}} \\ &= \mathcal{A}_t \sqrt{\frac{\Delta p_j}{p_j}} \\ &= \mathcal{A}_t \frac{1}{\sqrt{(p_j \Delta p_j)}} (p_{j-1} - p_j), \end{aligned} \tag{14.10}$$

where

$$\mathcal{A}_t = \dot{m}_0 K_t \left(\frac{p_{in}}{p_{in0}}\right) \sqrt{\frac{T_{in0}}{T_{in}}} \left(\frac{1}{e_0}\right)$$

whose component terms are known from design data and the current turbine inlet conditions. The linearisation

$$\sqrt{\frac{\Delta p_j}{p_j}} = \frac{\Delta p_j}{\surd(p_j \Delta p_j)}$$

assumes the denominator terms of the right-hand side are known from the preceding time step.

Equation 14.10 may be expanded for all blade rows and inlet and outlet boundary conditions. The resulting tridiagonal equation yields the profile of pressures through the turbine for the given boundary conditions and extraction, drain and gland sealing flows.

Conditions at the turbine inlet are defined by the nozzle box pressure p_{in}, steam enthalpy h_{in} and temperature T_{in} together with the resulting inlet flow $\dot{m}_{in}$ calculated from Eq. 14.8. The pressure after the last row is defined by conditions external to the turbine exhaust. Where the turbine exhausts to a steam line, as in a back-pressure industrial turbine drive or a turbogenerator exhausting to an air-cooled condenser, the pressure drop in the exhaust pipe or duct must be included in the calculation of the turbine exhaust pressure.

The following table presents design data for a large steam turbine used in a supercritical generating unit. The quoted pressure ratio assumes the same pressure ratio for all blade rows in the cylinder.

Cylinder	Rated flow kg/s	Inlet pressure bar	Inlet temp deg C	Exhaust pressure bar	Exhaust temp deg C	Blade rows	Pressure ratio
HP	354	242.2	566	52.62	329	27	0.945
IP	302	47.35	593	5.783	283	17	0.884
LP	259	5.782	283	0.184	58	7	0.73

The expansion line calculation can produce steam conditions at each blade row, for example, for each of the 27 HP cylinder rows. For most simulation purposes this detail is not required and several rows can be grouped to give a lesser number of equivalent rows. If the number of grouped rows were selected to yield a pressure ratio of 0.85–0.87, the 27 rows of the tabulated HP cylinder can be reduced to eight rows between inlet and the first bleed point and three rows to the exhaust. The design pressure ratio is selected to match the design data of the reference turbine. The following table sets out the calculated pressures and temperatures at the bleed and exhaust rows for four different loads, calculated using the expanded version of Eq. 14.10, together with the actual design figures.

% CMR	100	75	50	25
Design inlet pressure kPa	24,220	18,950	12,510	11,220
Design inlet temperature °C	566	566	544	515
Design inlet flow kg/s	354	254	170	91
Calculated inlet flow kg/s	350	261	175	91
Calculated first-stage pressure kPa	24,220	18,950	12,510	6,358
Design bleed point pressure kPa	7,655	5,603	3,760	1,913
Calculated bleed point pressure kPa	7,652	5,715	3,865	2,002
Design bleed point temperature °C	379	380	369	334
Calculated bleed point temperature °C	382	394	377	341
Design exhaust pressure kPa	5,262	3,878	2,613	1,354
Calculated exhaust pressure kPa	5,262	3,888	2,646	1,354
Design exhaust temperature °C	329	332	323	293
Calculated exhaust temperature °C	345	341	327	297

The inlet pressure and temperature values quoted are those before the turbine throttle valves. The table is based on sliding pressure operation down to around 50% CMR, with fully open throttle valves. Below 50% turbine power is controlled by the throttle valves while holding steam pressure constant. For the calculation of turbine conditions at 25% CMR for this table, the throttle valves were closed to match the known design flow.

14.2 Developed Work in the Energy Balance

An additional term must be added to the energy balance equation for a turbine stage to provide for the mechanical work done by the turbine stage which does not appear as heat transferred. This is denoted W_j [kJ/s] being the mechanical work done by steam expansion in the j-th expansion stage.

The mechanical work done is computed as the product of the steam flow through the stage and the real specific enthalpy drop across the stage.

$$W_j = \dot{m}_j \Delta h_j. \tag{14.11}$$

The total work done by all stages is the sum of the individual stage works.

$$W_\Sigma = \sum_j W_j. \tag{14.12}$$

It can be shown that, for adiabatic expansion from p_j to p_{j+1}, the ideal enthalpy difference across the stage is

$$\Delta h_{s,j} = \frac{\gamma}{\gamma - 1} p_j v_j \left[1 - \left(\frac{p_{j+1}}{p_j} \right)^{\frac{\gamma-1}{\gamma}} \right]. \tag{14.13}$$

The real specific enthalpy differential is given as $\Delta h_s \eta_p$, where η_p is the stage expansion efficiency.

Expansion in a turbine stage should be treated as a polytropic process as the stage incurs irreversible losses. The adiabatic index in Eq. 14.13 is replaced by the polytropic index, defined in Sect. 4.2 as

$$\frac{n}{n-1} = \frac{1}{\eta_p} \frac{\gamma}{\gamma - 1} \tag{14.14}$$

as we are considering expansion. After making this substitution Eq. 14.13 becomes, now for the real enthalpy differential across the stage,

$$\Delta \hat{h}_{s,j} = \frac{n}{n-1} p_j v_j \left[1 - \left(\frac{p_{j+1}}{p_j} \right)^{\frac{n-1}{n}} \right]. \tag{14.15}$$

14.3 Moisture Production Within the Turbine Stage

Moisture will be formed within the turbine under the following conditions:

- During start-up warming, steam will condense on cold parts; the condensate is removed through casing drains. Condensate mass flow rates are computed for all operating conditions (i.e. not just start-up) on the basis of heat transfer from the steam to the condensing surfaces. Under normal loaded operation, condensation is unlikely as the condensing surfaces will be close to steam temperature.
- During normal operation (well-developed steam flow through the blading) a water phase develops wherever the steam enthalpy reduces below the local saturation enthalpy. The mass flow rate is defined by the steam wetness factor and the total steam flow rate. The water phase is removed via the moisture separator for each stage and the total moisture passed to the drains cooler.

Modern steam turbines can achieve HP and IP thermal efficiencies of around 93–96%. LP turbines achieve lower values, but overall efficiencies of multi-cylinder sets will exceed 90% at full load.

The moisture flow rate for the j-th stage is calculated from

$$x = \frac{h_s - h'}{h'' - h'},$$

$$\dot{m}_w = (1 - x)\dot{m}_j, \tag{14.16}$$

where x is the steam mass fraction, h_s is the local steam enthalpy, h', h'' are the local saturation enthalpies, $\dot{m}_j$ is the steam flow into the j-th stage and $\dot{m}_w$ is the moisture flow from the stage.

14.4 Low Flow Operation

Steam turbines are optimised for full- or near-full-load operation. At low flows (typically $< 25\%$) and particularly during start-up, steam flows are very low and the flow patterns through blades can depart from the ideal flow regimes. This leads to reduced efficiencies in all cylinders but because of its long blades and particularly those of the last row, the LP turbine is particularly vulnerable to water droplet erosion by wet steam, heating due to friction losses and blade vibration due to turbulent and vortex flows between the blades. Instances of blade failure have been attributed to these flow pattern disturbances [62]. Specifically the last row of the LP turbine, the exhaust area of which strongly influences the power capability of the turbine, must be protected from these effects and so-called turbine exhaust hood sprays are provided to keep LP turbine exhaust temperatures down during start-up and periods of low steam flow.

14.5 Gland Sealing Steam Flows

The casing enclosing a rotating turbomachine (turbine or compressor) is supported at each end by a pedestal which also carries the shaft bearing and the shaft gland seal. Gland seals are intended both to prevent the entry of air into a casing through the gap between shaft and casing and to minimise loss of the working medium from the casing through the gap. In the absence of a seal, with subatmospheric pressure in the casing, air can enter the casing along the shaft. Similarly, with a high pressure in the casing, the working medium can flow outwards along the shaft. The entry of air is prevented by creating a positive (> atmospheric) pressure between the outside air and the casing. This is established by introducing steam[1] at an intermediate point along the seal and by creating a low admittance flow path to limit the flow of the sealing medium along the shaft. The same sealing arrangement minimises loss from the casing via the same low admittance flow path.

The most common form of gland seal is the labyrinth seal. The general arrangement is shown schematically by the following (Fig. 14.3).

[1] Water or oil are used in other applications.

Fig. 14.3 Turbine shaft gland seals (courtesy of Millmerran Operating Company, Australia)

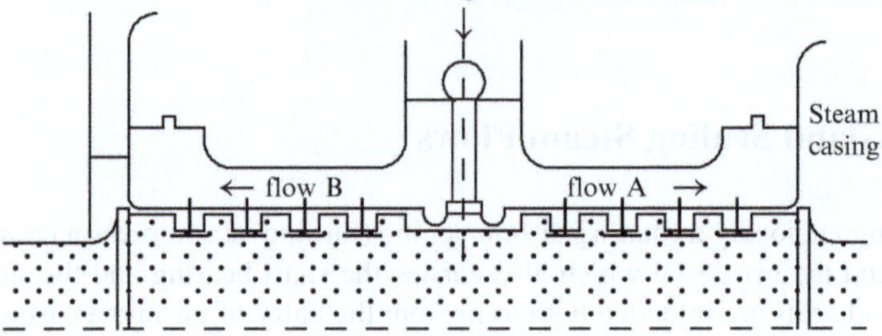

The seal consists of two circumferential rows of short, interlocking, narrowly spaced, parallel baffles arranged along the axis of the shaft. One row, mounted on the shaft, protrudes a few millimetres from the surface and meshes with a matching row inserted into the stationary casing. The complete arrangement creates a long, small cross-section path and presents high resistance to flow. We will consider the operation of a steam turbine gland seal but the principle applies to any sealing medium. Supplied from an external source to a point roughly central in the labyrinth, steam will flow both inwards into the turbine casing (flow A) and outwards towards a collection header or atmosphere (flow B). In small turbines this flow will be either collected and returned to the system or vented. In large turbines it will always be collected and returned to the system. As pressure is developed inside the turbine casing the sealing steam flow A will reverse when the casing pressure exceeds the pressure at the gland seal inlet point. A small but continuous flow of steam is lost from the turbine through the seal during normal operation.

Since the minimum labyrinth clearance c_l is much less than the shaft diameter d_{sh}, the minimum flow cross-sectional area a_{gs} is approximately $\pi\, d_{sh}\, c_l$. The flow through the seal can be calculated simply as

$$\dot{m} = a_{gs}\sqrt{\frac{p_1^2 - p_2^2}{l\; p_1 v_1}},$$

where l is the steam flow path length. Given that all stages of the seal are identical, we can replace the path length l by a factor $(1/k^2)\, n_{stg}$ where n_{stg} is the number of stages along the flow path. The preceding equation can then be written

$$\dot{m}^2 = k^2\, a_{gs}^2 \left(\frac{p_1^2 - p_2^2}{n_{stg}\; p_1 v_1}\right). \tag{14.17}$$

This can be reorganised into the form $\dot{m}^2 = \mathcal{A}\Delta p$, with $\Pi = p_2/p_1$. Then,

$$\dot{m}^2 = \frac{k^2\, a_{gs}^2}{n_{stg} v_1}(1 + \Pi)\Delta p. \tag{14.18}$$

It is the nature of gland seals that the pressure ratio across them is very large, allowing the assumption that $\Pi \ll 1$. Then

$$\dot{m}^2 = \frac{k^2\, a_{gs}^2}{n_{stg} v_1}\Delta p. \tag{14.19}$$

The small clearances between the rotating and stationary sections of the seal set tight limit on the allowable longitudinal movement of the turbine shaft during those operational phases during which the turbine shaft may be expanding or contracting. These in turn set limits on the allowable rates of change of turbine shaft temperatures during both fast and slow transients.

Traupel [61] suggests that the minimum labyrinth clearance be determined according to the following rule, with the shaft diameter d_{sh} expressed in millimetres:

$$c_l = A\,\frac{d_{sh}}{1000} + 0.25\ \text{mm}$$

$$\text{with}\ \begin{cases} \text{A} &= \text{0.6 for compressors} \\ \text{A} &= \text{0.85 for turbines fabricated from ferritic steel} \\ \text{A} &= \text{1.3 for turbines fabricated from austenitic steel} \end{cases}$$

The various values of A reflect the differing temperature environments in which compressors and turbines operate, with compressors being exposed to relatively small temperature variations and austenitic steel turbines exposed to the highest temperatures and largest temperature gradients, both static and dynamic.

The order of magnitude and dependence of gland flows on operating conditions are illustrated by the following numerical examples. These have been calculated for a gland seal on a high-pressure turbine casing with superheated steam at pressures of 1,000 kPa and 10,000 kPa in the casing (p_1) and at the point of connection of the gland sealing steam, a counter-pressure (p_2) of 200 kPa.

For a shaft diameter d_{sh} = 500 mm; then, using A = 1.3, the gland clearance is 0.9 mm giving a flow cross-section area of 1.4e−3 m^2. The gland steam flows from the HP turbine casing for the assumed steam conditions are tabulated.

Steam pressure	Steam temperature	Stages	Gland steam flow
1,000 kPa	300°C	5	0.039 kg/s
		10	0.028 kg/s
10,000 kPa	340°C	5	0.204 kg/s
		10	0.144 kg/s

Heat Transfer from Gland Sealing Steam

The steam flow through the glands will transfer heat to the shaft in the region of the glands. In view of the low velocity of the steam, heat transfer rates will be low and total heat will be low given the small flow rates. Heat transfer from the gland flows should be considered in view of the local shaft heating and expansion effects though low accuracy should suffice.

14.6 Heat Transfer from Steam to the Casings and Shaft

Heat transfer to the casing and shaft is used for the calculation of casing and shaft temperatures, casing and shaft thermal expansions and differential expansions, shaft and casing stresses, stress limits and shaft fatigue life estimation.

Heat is transferred by convection from the steam to the turbine blades, both moving and stationary, and to the turbine casing(s). Little heat is transferred directly to the shaft or internal casing, heat reaching these predominantly via conduction through the blading. A small quantity of heat is transferred directly to the shaft in the gland seals.

Figure 14.4 shows the arrangement of shaft, rotating and stationary blading, support structure and outer casing for a typical HP turbine cylinder.

Steam enters the turbine via the cavity ahead of the first-stage nozzles. Steam leaving the nozzles impinges on the first blade row (usually an impulse stage) and passes through successive rows of stationary and rotating blades (reaction stages) to the turbine exhaust. Steam leaving the last row of blades is directed back through the annular region (the annulus) between the outer casing and the inner structure

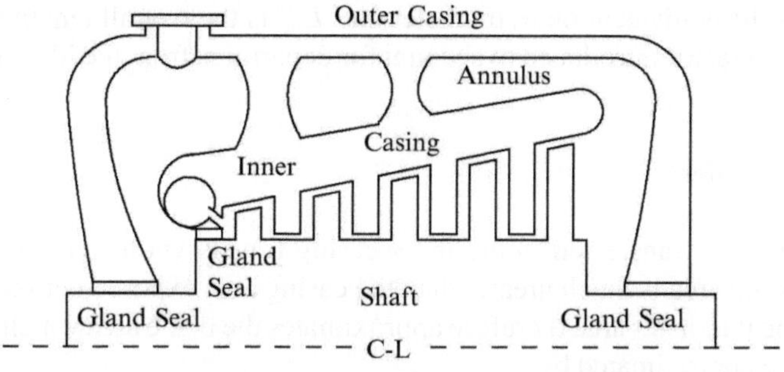

Fig. 14.4 Schematic arrangement of a return flow HP turbine cylinder

supporting the stationary blading, the inner casing. The inner casing is exposed to steam on both sides. On the blading side, the steam is at the stage temperature as given by the expansion line calculation. On the annulus side, it is at a temperature approximating the temperature of the steam leaving the stage, the turbine exhaust temperature. It is reasonable to assume this temperature since steam transits the annulus very quickly and the outer casing is well insulated. The annulus receives a small steam flow through the high-pressure shaft gland between the impulse row and the annulus. It also receives or sends a small flow through the outer casing shaft seals, depending on turbine operation.

14.6.1 Heat Transfer to the Casings

Inner Casing Annulus Side

The estimation of the heat transfer area is complicated by the geometry of the outer surface of the stationary blading support structure. This is not a smooth cylindrical surface, having a basically regular surface but including support rings connecting to the outer casing and various other departures from a regular surface. The construction details vary greatly among manufacturers. It is reasonable to define an outer diameter of an equivalent smooth cylinder which will be close to that of the identifiable inner casing cylindrical outline. Since in most HP and IP cylinder designs the stationary blading is installed along only some 2/3 of the total inner casing length, the area used per stage for the outer surface of the inner casing can be calculated as the total area divided by the number of configured stages. The total outer surface area of the inner casing is

$$A_{oc} = f_s \frac{\pi}{4} d_o^2 L_{oc}, \tag{14.20}$$

where d_0 is the equivalent outer diameter and L_{oc} is the overall length of the inner casing. f_s is a factor introduced to account for departures from the idealised surface.

Stationary Blades

The bulk of heat transferred to the inner casing travels via the stationary blading whose exposed area is much greater than the casing area exposed between the blade rows. The heat transfer area therefore approximates the blade area which for a single blade can be approximated by

$$A_{bl} = 2\, l_{bl}\, c_{bl}\, f_a, \tag{14.21}$$

where l_{bl} is the average length of blading (from root to tip) and c_{bl} is the average chord length of the blades (distance from leading to trailing edge). The factor 2 includes both sides of the blades. Because the blades have an airfoil profile, the factor f_a is applied to allow for surface curvature and might be expected to take a value in the range 1–1.2.

Heat transfer to the stationary blades can be treated on a per-blade-pair basis using the heat transfer coefficient from Eq. 8.13. The mass flow density ϕ is calculated from the turbine exhaust flow $\dot{m}_s$ as

$$\phi = \frac{\dot{m}_s}{n_{bls}} \frac{1}{A_{bls}},$$

where n_{bl} is the number of blades in the j-th row, $n_{bls} = n_{bl} - 1$ is the number of inter-blade spaces and, with δb the average blade-blade separation, $A_{bls} = l_{bl}\delta b$ is the flow area between adjacent blades. The equivalent diameter d_{eq} is calculated from

$$d_{eq} = \frac{4\, A_{bls}}{U_{fs}},$$

where $U_{fs} = 2(l_{bl} + \delta b)$.

Conduction along the blading is included by assuming that this can be represented by a conduction path per blade of length equal to 1/2 of the average blade length l_{bl}. The overall heat transfer coefficient for heat transfer from the steam to the inner casing is then given as

$$\frac{1}{\alpha_1} = \frac{1}{\alpha_0} + \frac{l_{bl}/2}{\lambda A_{x,bl} f_c}, \tag{14.22}$$

where $A_{x,bl}$ = (chord x thickness) is the mean blade cross-sectional area $[m^2]$. Again, f_c is a factor introduced to account for the non-rectangular cross-section of the blade and will take a value in the range 0.5–0.75.

Heat transferred to the inner casing at the j-th turbine stage is then

$$\dot{q}_1^j = \alpha_1^j \, a_{bl}^j \, n_{bl} (T_s^j - T_m^j), \tag{14.23}$$

where T_s^j is the local steam temperature and T_m^j is the local casing temperature.

Heat Transfer from the Inner Casing to the Annulus Steam

Heat is conducted through the inner casing and by convection to the steam in the annulus. The heat transfer coefficient on the annulus side is given by the standard Hausen correlation, Eq. 8.9, with flow and steam thermodynamic conditions defined by the turbine stage exhaust. The steam mass flow density ϕ used by this correlation is determined as

$$\phi = \dot{m}_{exh} / A_{ann},$$

where $\dot{m}_{exh}$ is the mass flow leaving the last blade row and A_{ann} is the representative annular area between the inner and outer casings. This area is of course not uniform, but uniformity is a workable assumption.

The total heat transferred to the steam in the annulus from the inner casing is the sum of the heat transferred from each inner casing cell, assuming a uniform steam temperature in the annulus. For most operational phases, this is a probably true.

$$\dot{q}_{ic} = \sum_{j=1}^{j=n_c} \dot{q}_{ic}^j, \tag{14.24}$$

where, with T_{sa} is the representative temperature of the steam within the annulus,

$$\dot{q}_{ic}^j = \alpha_0 a_{ic} \left(T_{ic}^j - T_{sa} \right) \tag{14.25}$$

$a_{ic} = \pi/4 \, d_{eq}^2 L$ is the heat transfer area of the inner casing cell on the annulus side. d_{eq} is the equivalent diameter of the outer surface of the inner casing and L is its length. α_0 is the heat transfer coefficient of Eq. 8.13.

14.6.2 Calculation of Inner Casing Temperatures

The temperatures through the inner casing are of no great consequence to any of the thermodynamic processes taking place in the turbine. However, two situations of interest can be identified for which a more precise calculation of inner casing temperatures will be needed. Both are related to turbine supervisory measurements. The first is the expansion of the inner casing and, derived from it, the differential between turbine shaft and inner casing expansions and the monitoring of gland seal clearances. The second is the estimation of rotor temperatures used for rotor stress calculations. Since rotor stresses cannot be directly measured, at least given the present state of technology, they must be estimated. Some turbine monitoring systems use the measured casing temperatures as proxy for the adjacent shaft surface temperature as the starting point for estimation of rotor temperatures and stress margins.

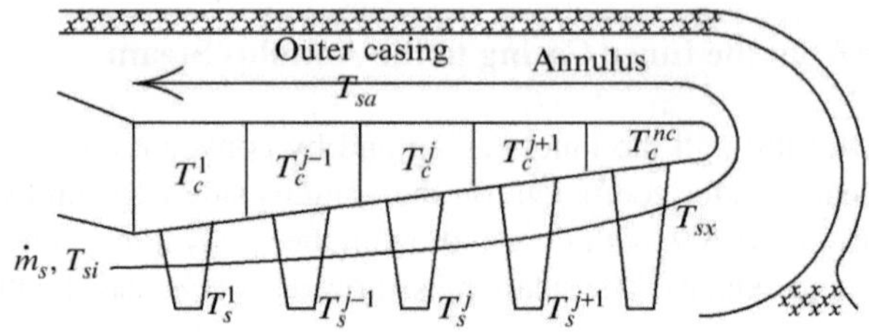

The notation can be made more compact by defining

$$\bar{\alpha}_{bl,j} = (\alpha_{bl}\, a_{bl}\, n_{bl})^j \quad \text{and} \quad \bar{\alpha}_{ic,j} = (\alpha_{ic}\, a_{cas})^j .$$

A simple heat balance for each inner casing cell, including heat transfers on the blading ($\dot{q}_{bl}$) and annulus ($\dot{q}_{ic}$) sides and spatial discretisation of axial conduction along the casing, yields the following ordinary differential equation for the temperature of the j-th cell:

$$\begin{aligned}(M_m c_m)\frac{\mathrm{d}T_{ic}^j}{\mathrm{d}t} &= \dot{q}_{bl}^j - \dot{q}_{ic}^{j-1} + \lambda\frac{A_x}{dl}\left(T_{ic}^{j-1} - 2T_{ic}^j + T_{ic}^{j+1}\right)\\ &= \bar{\alpha}_{bl,j}\left(T_s^j - T_{ic}^j\right) - \bar{\alpha}_{ic,j}\left(T_{ic}^j - T_{sa}\right)\\ &\quad + \lambda\frac{A_x}{dL}\left(T_{ic}^{j-1} - T_{ic}^j + T_{ic}^{j+1}\right)\\ &= -\left(\bar{\alpha}_{bl,j} + \bar{\alpha}_{ic,j} + 2\lambda\frac{A_x}{dL}\right)T_{ic}^j + \lambda\frac{A_x}{dL}\left(T_{ic}^{j-1} + T_{ic}^{j+1}\right)\\ &\quad + \bar{\alpha}_{bl,j}T_s^j + \bar{\alpha}_{ic,j}T_{sa},\end{aligned} \tag{14.26}$$

where M_m and c_m are the mass and material specific heat capacity of the casing metal in the cell. dL is the length of a casing cell, assumed equal for all cells.

Under normal on-load flow conditions, the steam transit time through the annulus is very short, and the steam temperature within the annulus T_{sa} will be indistinguishable from the turbine exhaust temperature T_{sx}. Even under conditions of very low or zero flow, such as during initial warming or during cooldown after a turbine trip, these temperatures are unlikely to be noticeably different. However, the calculation of the outer casing temperature is facilitated if we treat the annulus temperature as distinct from the temperature of the steam at the turbine blading exhaust.

Setting

$$\varpi = \bar{\alpha}_{bl,j} + \bar{\alpha}_{ic,j} + 2\lambda \frac{A_x}{dL} \quad \text{and} \quad \tau = \frac{M_m c_m}{\varpi}$$

this equation can be rewritten

$$\tau \frac{\mathrm{d}T_{ic}{}^{j}}{\mathrm{d}t} + T_{ic}^{j} = \frac{\lambda\, A_x}{dL} \left(T_{ic}^{j-1} + T_{ic}^{j+1} \right) / \varpi + \left(\bar{\alpha}_{bl,j} T_s^j + \bar{\alpha}_{ic,j} T_{sx} \right) / \varpi. \quad (14.27)$$

Taking j over all n_c inner casing cells, Eq. 14.27 creates a system of n_c differential equations for the inner casing cell temperatures, given the appropriate boundary temperatures and steam flow.

Given the presence of terms for temperature in the (j-1)th and (j+1)th cells in Eq. 14.27, an obvious approach to its solution would be the replacement of the time derivative by its discrete time equivalent and reorganisation of the equation into a tridiagonal form incorporating the boundary temperatures at each end of the casing. However, given that the casing temperatures vary more slowly than the surrounding steam temperatures, the individual equations may be solved sequentially. Using the boundary condition defined for $j = 0$, the procedure starts from $j = 1$ and moves through all cells to $j = n_c$, using the just calculated value of T_{ic}^{j-1} and the previously calculated value of T_{ic}^{j+1}. The calculation for the final cell will use the boundary temperature defined for $j = n_c+1$.

As Leyzerovich points out [62], there is little theoretical basis for choosing a reliable heat transfer coefficient formula for the casing and field data must be relied on, ultimately. However, in the absence of such data, reasonable estimates can be provided by the conventional formula and adjusted to fit a given turbine. Leyzerovich presents a selection of field data which can assist.

14.6.3 Outer Casing Temperature

The outer casing of an HP turbine is relatively thick and well insulated. It is heated by convective transfer from the steam and to a much lesser extent by conduction from connecting parts, such as gland seals and bearings. Since conduction is relatively small compared with convective heat transfer under most circumstances, it will be represented in a very much simplified form, chosen more to facilitate the representation of cooldown than warm-up, for which it is of less importance.

The interest in the outer casing stems primarily from the need to calculate the differential expansion between the casing and the turbine shaft. Turbine supervisory measurements use this to monitor outer casing gland seal clearances. Of secondary but still significant importance is the reproduction of the cooldown behaviour of the turbine.

As mentioned previously, a simplified approach is sufficient for calculation of the annulus conditions. With h_{sa} the representative enthalpy of the steam within the annulus and c_{sa} its nominal heat capacity, a simple heat balance on the steam within the annulus produces

$$c_{sa}\frac{h_{sa}}{dt} = \dot{m}_{sx}(h_{sx} - h_{sa}) + \sum_{j} \dot{q}_{ic,j} - \dot{q}_{oc}, \tag{14.28}$$

where the heat transfer $\dot{q}_{ic,j}$ from the steam within the j-th zone of the turbine blading is given by Eq. 14.25. If we assume that the outer casing can be represented sufficiently well by a single temperature T_{oc}, heat transfer to the outer cylinder $\dot{q}_{oc}$ is

$$\dot{q}_{oc} = \alpha_{oc} A_{oc}(T_{sa} - T_{oc}), \tag{14.29}$$

where A_{oc} is the area in contact with steam. α_{oc} is the heat transfer coefficient at the inner surface of the outer cylinder and will be adequately approximated by the standard Hausen form. The annulus steam temperature follows from

$$T_{sa} = T(h_{sa}, p_{sa}).$$

The casing temperature follows from

$$(cm)_{oc}\frac{\mathrm{d}T_{oc}}{\mathrm{d}t} = \dot{q}_{oc} - \dot{q}_{amb} - C_{cool}(T_{oc} - T_{ref}), \tag{14.30}$$

where the heat loss through the outer cylinder to ambient is

$$\dot{q}_{amb} = \bar{\alpha}_{insul}(T_{oc} - T_{amb})$$

T_{ref} is some temperature towards which the casing will cool in the long term. It may be a bearing temperature or ambient. Given the complex geometries of turbine components, the precise identification of the various possible conduction paths is extremely complex and need not be attempted. Instead, the coefficient C_{cool} can be chosen to match the known or expected long-term cooldown rate of the casing as in practice this will be the principal behaviour of interest.

A further comment of Leyzerovich [62] is of interest. As regards the long-term cooldown behaviour of the turbine, convective heat exchange by steam within the turbine chambers is considered to play no role in defining the cooldown transients, these being dominated by axial conduction from the shaft and casings to the end seals and bearings.

14.6.4 Heat Transfer to the Moving Blades

The main interest in heat transfer to the moving blades is the calculation of the turbine shaft temperatures and the derived thermal stresses since most heat transferred to the turbine shaft passes through the moving blades. The greater emphasis in the literature is given to gas turbines for which the problem of blade temperatures is much more acute and greater attention must be paid to maximum blade temperatures and cooling requirements. Where relevant it has been assumed that the conclusions developed for gas turbine blading will apply equally to steam turbine blading with the substitution of steam parameters and properties.

With the turbine rotating, steam will flow axially along the turbine shaft, with minimal radial component outwards along the blade face. At any point the velocity of the steam relative to the blade surface is the rotational velocity of the blade $w = r\omega$, where r is the radial distance [m] of the steam from the shaft centre line and ω is the rotational speed of the shaft in radians/s. The Stanton number St is defined as

$$St = \frac{\alpha_1 v}{c_p\, w},$$

where α_1 is the heat transfer coefficient at the blade surface [kW/(m^{2}K)], v is the steam specific volume and c_p is the steam specific heat [kJ/(kg K)]. We then have the heat transfer coefficient from

$$\alpha_1 = \frac{c_p\, w\, St}{v}. \tag{14.31}$$

The Stanton number can be obtained from the Prandtl number P_r and Fanning friction factor ξ via the following relationship:

$$St = \frac{\xi/2}{1 + f(P_r)\sqrt{\xi/2}}. \tag{14.32}$$

By fitting to Fig. 8.4.23, p. 426 of Traupel, Vol. I [61], the function $f(P_r)$ is approximated by

$$f(P_r) = -8.0194 + 8.3791 P_r - 0.3597 P_r^2. \tag{14.33}$$

The Prandtl number for steam moves in the range from 5.8 at 20°C to around 0.9 at 600°C. It has a second-order dependency on pressure which can be ignored with little error. The following table summarises the values of P_r for steam for the indicated range of pressure and temperature (Perry [17], Table 3-304, p. 3-242).

At a pressure of 20 bar, the Prandtl number may be approximated from this table by

$$P_r = 1.616 - 0.2506\left(\frac{T}{100}\right) + 0.02177\left(\frac{T}{100}\right)^2 \tag{14.34}$$

to within 1%.

We may assume that the turbine blades are hydraulically smooth and that a reasonable constant for ξ will be in the range 0.035–0.05 at normal steam flows, increasing to around 0.1 at low steam flows (low Reynolds numbers $< 60{,}000$,

Bar	Temperature (K)								
	300	350	400	450	500	600	700	800	900
1	5.81	2.32	0.980	0.967	0.955	0.936	0.920	0.906	0.891
5	5.82	2.32	1.340	0.983	0.973	0.947	0.925	0.907	0.892
10	5.82	2.32	1.340	0.983	1.028	0.947	0.925	0.907	0.892
20	5.80	2.32	1.340	0.979	1.190	0.999	0.946	0.912	0.893
40	5.78	2.31	1.340	0.977	0.862	1.080	0.975	0.924	0.895
60	5.74	2.31	1.340	0.976	0.859	1.190	1.008	0.934	0.899
80	5.72	2.31	1.340	0.975	0.856	1.330	1.046	0.946	0.902
100	5.69	2.31	1.340	0.975	0.853	1.740	1.088	0.960	0.905
150	5.64	2.30	1.340	0.974	0.842	–	1.220	0.994	0.916
200	5.59	2.29	1.340	0.974	0.833	–	1.380	1.014	0.925

typical of start-up warming). Given the unpredictable complexities of flows under start-up conditions, these values are to be treated with some caution. Reasonable deviations found when fitting to actual plant data should not cause any surprises.

14.7 Turbine Shaft Metal Temperatures

The allowable rate of warming of a turbine shaft is the most significant parameter in defining the maximum start-up rate of the steam turbine. The primary objective of the start procedure is to minimise the start-up time while remaining within the allowable shaft temperature stress limits. These are related directly to the gradient of temperature in the radial direction. Most modern turbines, and all large turbines, are equipped with some form of shaft stress monitoring and limitation system. Since shaft temperatures are not measured, these systems require estimation of the profile of temperatures at selected points through the shaft. High-pressure valve chests and turbine casings and flanges may also be subject to stress monitoring and limitation. For these, actual temperature measurements are provided by thermocouples embedded at key locations within the component.

14.7.1 Calculation of the Temperature Field

For simulation purposes, a spatial field of time-varying metal temperatures in shafts and casings will be required:

1. As estimates of temperatures to be used for stress calculations by monitoring systems
2. For the calculation of shaft and casing expansion and differential expansion measurements

The calculation of metal temperature fields is greatly complicated by the irregular shapes and boundaries of the components. For design purposes, this usually

Fig. 14.5 A view of an LP turbine showing the steam flow path (courtesy of Millmerran Operating Company, Australia)

necessitates the use of finite element analysis with a large number of elements generating a fine mesh field. For operational simulation purposes, the accuracy requirements can be relaxed, permitting the assumption of more regular shapes and a smaller number of individual elements or cells.

Temperatures within a cylindrical shaft are defined by the unsteady heat conduction equation, Eq. 10.2 of Chap. 10.

$$\frac{1}{r}\frac{\partial T}{\partial r} + \frac{\partial^2 T}{\partial r^2} + \frac{\partial^2 T}{\partial z^2} = \frac{1}{D}\frac{\partial T}{\partial t}.$$

Boundary condition temperatures are related to the interfacing systems by specific instances of the generic Eqs. 10.38 and 10.39 for the radial (r) ordinate and Eqs. 10.40 and 10.41 for the axial (z) ordinate. The solution follows from application of the *alternating direction implicit* or ADI method, discussed in Sect. 10.2.

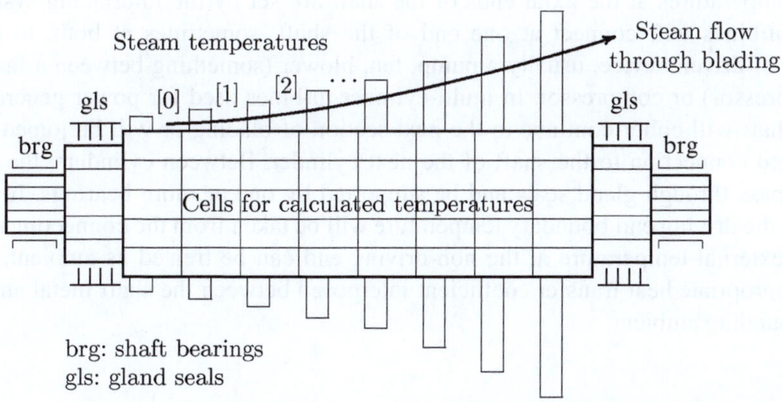

Consider a cylindrical shaft of the following geometry.

The central section of the shaft carries the blade rows. Each row consists of a number n_{bl} of individual identical blades set into the shaft and defines a cylindrical disk normal to the axis of rotation. This disk is treated as defining a single axial zone of length dL where $dL = L_{sb}/n_{rows}$ and n_{rows} is the number of blade rows distributed over the length of the blade zone L_{sb}. The reference or boundary temperature for the j-th axial zone is the local steam temperature Ts_j in contact with the blades. Of the generic boundary condition equations at the surface of an axial zone, Eqs. 10.38 and 10.39, only one need be applied for heat transfer at the outer surface of the shaft.

$$T_{-1} = T_1 - \frac{2\alpha_j \Delta r}{\lambda}(T_0 - Ts_j). \tag{14.35}$$

Recall that the earlier discussion of the ADI method applied to a hollow cylinder with heat transfer across both inner and outer surfaces. For the turbine shaft, we use symmetry to calculate the temperature distribution through half the shaft ($r_i < r < r_o$) and assume that the radial gradient of temperature along the centre line is zero, that is, $\partial T/\partial r = 0$ for $r = 0$. As mentioned in Chap. 10, we artificially assume a small $r_i > 0$, as if the shaft has a small bore hollow centre, which in reality it may indeed have, and the inner surface boundary condition is then more correctly $\partial T/\partial r = 0$ for $r = r_i$.

The heat transfer coefficient at the blade interface is given by Eq. 14.31. Two zones are provided at each end of the shaft linking the shaft to the gland sealing steam and to the shaft bearings. These provide the radial boundary conditions for the respective zones. It is apparent that the bearing lubricating oil must absorb heat conducted along the hot shaft. This will add to and usually exceed by a large margin the heat generated within the oil film in the journal bearings. These bearings usually operate at a higher temperature than others, in excess of 80°C, and require a suitable high temperature lubricating oil.

Most turbines are equipped with a thrust bearing to prevent lateral displacement of the shaft. This is commonly a Babbitt-type tilting pad bearing located between the HP and IP cylinders.

Temperatures at the axial ends of the shaft are set by the interfacing systems. All turbines will connect at one end of the shaft, sometimes at both, to some form of driven device, usually a pump, fan, blower (something between a fan and compressor) or compressor. In multi-cylinder turbines used for power generation, the shaft will either continue to the next section of blading or will be joined by a flanged connection to the shaft of the next cylinder. Between cylinders, the shaft will pass through gland seals and be supported by one or more bearings. In each case, the driving end boundary temperature will be taken from the connecting shaft. The external temperature at the non-driving end can be treated as ambient, with an appropriate heat transfer coefficient interposed between the shaft metal and the surrounding ambient.

In the radial direction the shaft is treated as a series of concentric cylindrical zones centred about the shaft centre line. Since symmetrical heating can usually be assumed[2], the spacing of the individual zones can be equal for all. The number of zones is selected to give a balance between computation load and accuracy.

14.8 Turbine Supervisory Measurements

Turbine supervisory instrumentation or TSI is widely used for the supervision, control and condition monitoring of all sizes of turbines, both steam and gas. In many applications, TSI is a mandatory requirement. It encompasses a range of measurements including vibrations, thermal expansion of shaft and casings, shaft eccentricity, rotor axial position, thrust bearing movement and rotor speed. Its purpose is to enhance the reliability of the turbine by supervising the operating parameters of the machine, identifying departures from normal which may indicate incipient damage and provide the quantitative basis for sound operational decisions, both automated responses and manual intervention by operating personnel.

14.8.1 Vibrations

Vibrations have been discussed in Sect. 7.5.3 within the context of bearings. It is true that most vibration measurements on rotating machines are taken from bearing housings as these provide the closest access to the shaft and the earlier material applies here to turbines without change. Because of their size and complexity, being composite assemblies of multiple rotating elements, and in consideration of their significant economic asset and revenue values, vibration issues of large steam turbines deserve individual attention.

Vibrations are measured individually for each bearing. In modern turbines, three vibrations are measured or resolved—the equivalent or total vibration and its x- and y-components. Movement of the shaft within the bearing may also be measured, using differential proximity measurements. During operation, the shaft will precess within the bearing, and its path can be displayed on a polar plot of shaft position. This information can be valuable to the diagnosis of potential operating problems but is rarely used for on line operational guidance.

The representation of the vibration/shaft speed relationship for any bearing has been discussed in Sect. 7.5.3 and can be applied here without change.

---

[2]Some turbines may be accelerated during run-up using only a single throttle valve. This can induce asymmetric flow through the blading and uneven heating.

14.8.2 Shaft Eccentricity

This is a measure of the degree of bending of the shaft. After an extended outage a turbine shaft, being of considerable length and mass suspended between two points (bearings), will bend downwards under its own weight, the point of maximum deflection being located about the mid-point of the axial separation of the suspension points. This will cause the ends of the shaft in the bearings to deflect upwards (while the shaft is stationary) by a much smaller amount, from which deflection the central deflection can be calculated. The end-of-shaft deflection is measured as the *eccentricity*. Before the turbine can be restarted the eccentricity must be reduced to within an acceptable range. This is done by rotating the shaft slowly using the turning gear motor. Significant time may be needed to remove the shaft bend or bow before the turbine may be accelerated. A similar situation can arise following a trip of the turbine, in which case the bowing of the shaft is a result of sudden and uneven cooling. The situation mentioned earlier using a single valve for start-up can also produce increased eccentricity because of uneven heating of the shaft. This can be a problem for turbines using partial arc admission.

Eccentricity is measured by a proximity device at the hot end of the HP turbine shaft, as far from the casing as possible to minimise thermal influences. The measurement is effective only at low shaft speeds, such as turning gear speed, and in some designs is disabled above some preset speed. With the shaft turning at this low speed, eccentricity will slowly reduce and, once below its permissible threshold, this condition for speed raising will be satisfied. Failure of the eccentricity to reduce may indicate a permanent bow of the shaft, with expensive consequences.

Acceleration of a turbine with excessive eccentricity will result in high levels of vibration induced by the eccentric rotating force of the shaft. It is probable that a turbine attempting to start under this condition will be runback or shutdown by its protective systems.

14.8.3 Expansion and Differential Expansion

During start-up of the turbine, the shafts and casings of the individual cylinders are exposed to wide variations of steam temperatures and heat transfer conditions. These vary from cylinder to cylinder, with the HP and IP cylinders exposed to the largest variations and the LP cylinder(s) exposed least. In each cylinder, the rotor and casing(s) will warm towards the local steam temperature and expand. The reverse process—cooling—will occur during shutdown. Being of differing materials and exposed to differing steam conditions, rotors and casings will expand by differing amounts. In addition, while the shaft inside the casing may expand and contract without restriction, movement of the casing will be constrained by connected pipework. As the moving and stationary parts are separated by small clearances in places such as shaft seals and blade tips, care must be taken to avoid contact between surfaces to avoid mechanical damage. The shaft is permitted little axial movement relative to its bearings, and the construction must ensure that bearings and shafts move together.

Thermal expansion of rotors and casings occurs in all three dimensions and to avoid excess stresses and distortions they must be free to move to accommodate these dimensional changes. The turbine is therefore not constructed as a solid unit but rather as a series of rigid components which can move freely with respect to each other, albeit in a carefully constrained fashion designed to preserve exact geometric alignments. Since rotation is not allowed, thermal movements are constrained to the axial and transverse directions. During initial commissioning some slight rotation is allowed to align the various sections of rotor axially but, once aligned, further rotation is prevented.

The term "differential expansion" is applied to the difference between shaft and casing expansions, as measured at a specific reference point associated with each casing, usually at the hot end of the casing. The rotating shaft is free to expand along its entire length and may therefore move differentially with respect to the casing at any point. The axial movement of a shaft is referred to some fixed or reference point, usually taken as the LP cylinder which moves relatively little. The LP cylinder may be fixed in relation to its associated condenser if water cooled or to a fixed foundation point if the condenser is air cooled and therefore not attached directly to the turbine cylinder. If there are multiple cylinders, each is referred to its own datum.

It is normal practice to measure the actual position of the thrust bearing. As previously mentioned, this is usually located on its own pedestal, together with a bearing, between the HP and IP cylinders. A fixed reference is established, called the *key phasor*, which serves as a reference position for the thrust bearing. This bearing's position is taken as an indication of the balance of axial forces imposed by the axial steam flow on the shaft.

Expansion is expressed as a positive length, being the expansion of the component from its state at the reference temperature for which expansion is zero. Differential expansion can be positive or negative.

Each bearing is supported on a concrete pedestal which supports both bearings and casings. In some designs, HP and IP and/or IP and LP shafts may share a common bearing in which case both casings will be supported on the same pedestal, together with the bearing. Bearings and casings are free to slide axially, and casings laterally, along sliding keys embedded in the pedestal support structure. Saying that the casings move freely is optimistic. A casing or bearing imposes a considerable weight on its sliding support. With time, the sliding interface can become contaminated, and significant stiction can develop. Thermally induced movement of the casing may then not be smooth but can exhibit sudden steps as the expansion force increases beyond the stiction force and is relaxed by a sharp movement.

As the range of temperatures experienced by the LP cylinder, perhaps 160°C or so, is much less than that seen by either the HP or IP cylinders—in the range of 540–600°C—thermal expansion of the LP cylinder(s) is of less operational concern.

However, the expansion and differential expansion of all cylinders will normally be monitored.

The expansion of any component can be calculated simply from

$$e_{xp} = \beta \Delta T\, L, \tag{14.36}$$

where e_{xp} is the expansion, β is the material's coefficient of linear thermal expansion (K^{-1}) and L is the length of the component. e_{xp} will take the units of L. β is typically around 11.5e–6 K^{-1}. For a HP/IP shaft with a combined length of 15 m, a change of volumetric mean temperature of 350 K would produce an elongation of 63 mm over this length.

This is greatly in excess of observed data, a deviation due to the assumption that the entire shaft reaches this volumetric mean temperature. In reality only the heated part of the shaft, being that part of the shaft carrying the blades, might achieve this temperature, the remainder receiving no external heat and being comparable in length to the heated section. A more realistic calculation of shaft temperatures reduces this to around 45 mm. From a simulation perspective, these variations indicate the importance of reasonably accurate calculation of shaft and casing temperatures if parameters derived from them are to be realistic.

14.9 Shaft Stresses and Stress Evaluators

The simulation interest in turbine stress calculations stems from the need to drive the turbine's shaft and valve chest monitoring and limitations systems. These are commonly known as wall (WSE), rotor (RSE) or turbine (TSE) stress evaluators. Individual turbine manufacturers have proprietary systems which use black-box models of thick-walled and shaft components to estimate temperature profiles within the component and from these to estimate thermal stresses. By comparing estimated stresses to material limits, the system can determine allowable operating margins. Should any margin reduce below a preset limit, a warning is issued to operating staff and some form of active intervention may be initiated by the automation system. This will usually take the form of the blocking of further shaft speed or turbine load increase but may initiate a reduction in either of these, as appropriate to the plant's operating phase. The material limit values are selected to protect the component against plastic deformation or surface embrittlement (which can lead to surface cracking) or, as is now often the case, to minimise loss of material fatigue life.

The fields of elasticity, stress analysis and their influence on mechanical design are large and mature. It is well beyond the scope of this book to cover any of these fields to any depth. We will therefore confine discussion to a summary of the most fundamental principles and of the operational considerations which motivate this interest.

The turbine shaft is exposed at all times to deforming stresses. These are induced by centrifugal forces imposed by the shaft's rotation and by thermal expansion and contraction induced by heat transfer from and to the steam or gas. Bending forces on the blades are imposed at the root by centrifugal force and as reaction forces to the pressure of steam on the blades. The shaft itself may be subject to strong compressive forces imposed by blade rings which are heat-shrunk onto the shaft. Moreover, the complex shape of the shaft and blade connection arrangements create local regions of stress concentration.

During the design stage detailed stress analyses are conducted, usually using FEA techniques and exact shaft and blade profiles. These methods serve to define expected and minimum stresses in the rotor but are much too complex and time-consuming for online use. They are replaced for operational stress monitoring by very much simplified stress estimators based on simplified geometries, measurements of actual steam and casing metal temperatures and estimates of shaft metal temperatures.

Somewhere between these two sits simulation, where everything is calculated. Computer performance limitations dictate the use of simplified models for the calculation of shaft stresses and for the calculation of the casing, shaft and blade metal temperature fields upon which the stress calculations are based.

We will consider stresses arising from centrifugal and thermal effects only. We will assume that the turbine shaft (and blades where appropriate) can be represented by simple shapes. The turbine shaft is treated as a uniform extended cylinder, solid or bored, of radius r_o and the blades as a point mass located at a rotational radius r_e from the axis of rotation. The stress on the blade root is notionally located at the outer surface of the rotor cylinder.

For the shaft, thermal and centrifugal shaft stresses are calculated for each of the three cylindrical coordinates, tangential, radial and axial. Only the radial centrifugal stress is considered for the blades.

By way of introduction we will briefly summarise the fundamental concepts of elasticity.

14.9.1 Fundamental Concepts

When subject to a deforming force (tensile or compressive), any solid material will change its shape in each dimension by an amount related to the value of the deforming force. If the deformation returns to zero after removal of the deforming force, that is, the material regains its original shape, the deformation is said to have been elastic. If some deformation remains after removal of the deforming force, the deformation is said to be plastic and the material has exceeded its elastic limit. The stress to which the material is subjected is the component of the deforming force in the selected coordinate direction divided by the area of the material normal to the force and has the dimension of pressure. For a force which remains below the material's elastic limit, the ratio of the applied stress, denoted σ, to the resulting *relative* deformation[3] called strain and denoted ϵ, is a constant. This is summarised by Hooke's law.

$$\frac{\sigma}{\epsilon} = E, \tag{14.37}$$

where E is the Young's Modulus of the material.

Application of a force to a slab of material in, say, the z direction (Cartesian coordinates) will cause a deformation in the z direction. This will be an elongation for a tensile force and a reduction for a compressive force. At the same time and in response to the same force, most materials will deform in each of the other directions but in the opposite sense. An elongation in the z direction will be accompanied by a reduction in both the x and y directions. The measure of the degree to which this occurs is the Poisson's ratio, denoted ν. It is the negative ratio of the transverse strain ϵ_x to the axial (in the direction of the force) strain ϵ_z.

$$\nu = -\frac{\epsilon_x}{\epsilon_z} \tag{14.38}$$

and is constrained to the range (0, 0.5).

In the general case it is necessary to consider the stresses developed in each of the three coordinate directions. These are the principal stresses and, in Cartesian coordinates, are denoted σ_x, σ_y and σ_z. In cylindrical coordinates, they are the radial stress σ_r, the tangential or azimuthal stress σ_θ and the axial stress σ_z. Generalised to three coordinates, Hooke's law becomes

Cartesian Coordinates

$$\begin{aligned}
\epsilon_x &= \frac{1}{E}\left(\sigma_x - \nu(\sigma_y + \sigma_z)\right), \\
\epsilon_y &= \frac{1}{E}\left(\sigma_y - \nu(\sigma_x + \sigma_z)\right), \\
\epsilon_z &= \frac{1}{E}\left(\sigma_z - \nu(\sigma_x + \sigma_y)\right).
\end{aligned}$$

Cylindrical Coordinates

$$\begin{aligned}
\epsilon_r &= \frac{1}{E}(\sigma_r - \nu(\sigma_\theta + \sigma_z)), \\
\epsilon_\theta &= \frac{1}{E}(\sigma_\theta - \nu(\sigma_r + \sigma_z)), \\
\epsilon_z &= \frac{1}{E}(\sigma_z - \nu(\sigma_r + \sigma_\theta)).
\end{aligned}$$

[3] If an element of length L changes length by ΔL, $\epsilon = \Delta L/L$.

Assuming the material to be isotropic, both E and ν are the same in all directions.

Thermal expansion can be included in each of these strain expressions by the addition of the term $\beta \Delta T$, where β is the coefficient of thermal expansion of the material and ΔT is the differential between the local temperature of the material and some reference temperature for which the thermal expansion can be taken to be zero. In many cases, this will be taken as the volumetric mean temperature of the material. Then, in cylindrical coordinates, the individual strain terms, including thermal expansion, become

$$\epsilon_r = \frac{1}{E}(\sigma_r - \nu(\sigma_\theta + \sigma_z)) + \beta\,\Delta T,$$
$$\epsilon_\theta = \frac{1}{E}(\sigma_\theta - \nu(\sigma_r + \sigma_z)) + \beta\,\Delta T,$$
$$\epsilon_z = \frac{1}{E}(\sigma_z - \nu(\sigma_r + \sigma_\theta)) + \beta\,\Delta T.$$

For a body subjected to multidimensional deforming forces, the approach towards a stress limit can be difficult to predict on the basis of principal stresses, taken individually. It is therefore useful to define a composite stress factor which includes contributions from the individual stresses. The equivalent volumetric stress σ_{eq} due to von Mises (1923) can be used for this purpose. This is calculated as

$$\sigma_{eq} = \frac{1}{\sqrt{2}}\sqrt{(\sigma_1 - \sigma_2)^2 + (\sigma_2 - \sigma_3)^2 + (\sigma_3 - \sigma_1)^2}. \tag{14.39}$$

Being a scalar σ_{eq} can be easily compared to a reference value and used to indicate an approaching limit. As formulated by von Mises, it was hypothesised that a material would reach its yield point should the stress σ_{eq} reach some critical value, known as the von Mises yield strength. In what follows here σ_{eq} will be calculated from operational conditions and compared to a known limit value, which may itself depend on operating conditions, to initiate an appropriate automated intervention.

Within a significant band centred on the normal operating range of a turbine shaft, the shaft and casing materials remain within their elastic limits. We may therefore assume that the materials remain isotropic and that stresses arising from differing causes may be summated linearly to give composite results.

14.9.2 Centrifugal and Thermal Stresses in a Solid Cylindrical Shaft

This subject is covered comprehensively in Traupel [61] Vol. 2. With some reorganisation of the results quoted there, we may write the following equations for the centrifugal radial $\sigma_{r\omega}$, tangential $\sigma_{\theta\omega}$ and axial $\sigma_{z\omega}$ stresses at any radius r measured from the axis of rotation.

$$\sigma_{r\omega} = \frac{3-2\nu}{8(1-\nu)}\rho\,\omega^2(r_o^2 - r^2), \tag{14.40}$$

$$\sigma_{\theta\omega} = \frac{3-2\nu}{8(1-\nu)}\rho\,\omega^2 r_o^2 - \frac{1+2\nu}{8(1-\nu)}\rho\,\omega^2 r^2, \tag{14.41}$$

$$\sigma_{z\omega} = \frac{\nu(3-2\nu)}{4(1-\nu)}\rho\,\omega^2 r_o^2 - \frac{\nu}{2(1-\nu)}\rho\,\omega^2 r^2 \tag{14.42}$$

ω is the speed of angular rotation of the shaft in radians/second. ρ is the density of the shaft material[4], and β is the coefficient of linear expansion.[5] The Poisson's ratio ν can be taken as 0.3 for steels used in turbine construction.

Traupel derives expressions for the thermal stresses in each of the cylindrical coordinates. For the radial and tangential stresses in a solid cylinder, we have

$$\sigma_{rT} = \frac{E\beta}{1-\nu}\int_r^{r_o}\frac{1}{r^3}\left(\int_0^r r^2\frac{dT}{dr}dr\right)dr, \tag{14.43}$$

$$\sigma'_{rT} = \frac{E\beta}{1-\nu}\frac{1}{r^3}\int_0^r r^2\frac{dT}{dr}dr, \tag{14.44}$$

$$\sigma_{\theta T} = r\sigma'_{rT} + \sigma_{rT}. \tag{14.45}$$

The equations for σ_{rT} and σ'_{rT} can be expanded as follows, using integration by parts. Recall that integration by parts follows the rule

$$\int v\frac{du}{dr}dr = uv - \int u\frac{dv}{dr}dr.$$

To expand the integral

$$I = \int_r^{r_o}\frac{1}{r^3}\left(\int_o^r r^2\frac{dT}{dr}dr\right)dr$$

set

$$I = \int_r^{r_o}\frac{1}{r^3}I_1 dr \text{ with } I_1 = \int_o^r r^2\frac{dT}{dr}dr.$$

[4]usually steel, around 7,800 kg/m^3.

[5]typically around 11.5e−6 per K.

To integrate by parts, set

$$u = T; \quad v = r^2 \longrightarrow \frac{\mathrm{d}v}{\mathrm{d}r} = 2r; \quad u\frac{\mathrm{d}v}{\mathrm{d}r} = 2rT; \quad uv = Tr^2.$$

Then

$$I_1 = \left[Tr^2 - \int 2Tr\,\mathrm{d}r\right]_0^r = Tr^2 - 2\int_0^r Tr\,\mathrm{d}r.$$

Therefore,

$$I = \int_r^{r_o} T\frac{1}{r}\mathrm{d}r - 2\int_r^{r_o} \frac{1}{r^3}\left(\int_0^r Tr\,\mathrm{d}r\right)\mathrm{d}r.$$

Consider next the integral

$$I_2 = \int_r^{r_o} \frac{1}{r^3}\left(\int_0^r Tr\,\mathrm{d}r\right)\mathrm{d}r.$$

Again, integrating by parts, set

$$\frac{\mathrm{d}u}{\mathrm{d}r} = \frac{1}{r^3} \longrightarrow u = -\frac{1}{2r^2}; \quad v = \int_0^r Tr\,\mathrm{d}r \longrightarrow \frac{\mathrm{d}v}{\mathrm{d}r} = Tr.$$

Then

$$\begin{aligned} I_2 &= \left[-\frac{1}{2}\frac{1}{r^2}\int_0^r Tr\,\mathrm{d}r + \frac{1}{2}\int T\frac{1}{r}\mathrm{d}r\right]_r^{r_o} \\ &= -\frac{1}{2}\frac{1}{r_o^2}\int_0^{r_o} Tr\,\mathrm{d}r + \frac{1}{2}\frac{1}{r^2}\int_0^r Tr\,\mathrm{d}r + \frac{1}{2}\int_r^{r_o} T\frac{1}{r}\mathrm{d}r. \end{aligned}$$

Therefore,

$$\begin{aligned} I &= \int_r^{r_o} T\frac{1}{r}\mathrm{d}r + \frac{1}{r_o^2}\int_0^{r_o} Tr\,\mathrm{d}r - \frac{1}{r^2}\int_0^r Tr\,\mathrm{d}r - \int_r^{r_o} T\frac{1}{r}\mathrm{d}r \\ &= \frac{1}{r_o^2}\int_0^{r_o} Tr\,\mathrm{d}r - \frac{1}{r^2}\int_0^r Tr\,\mathrm{d}r. \end{aligned}$$

Define

$$\vartheta_m = \frac{2}{r_o^2}\int_0^{r_o} Tr\,\mathrm{d}r$$

to be the volumetric mean shaft temperature. The radial stress then becomes

$$\sigma_{rT} = \frac{E\beta}{1-\nu}\left[\frac{1}{2}\vartheta_m - \frac{1}{r^2}\int_0^r Tr\,\mathrm{d}r\right]. \tag{14.46}$$

The second bracketed term can be interpreted as half the volumetric mean temperature of that volume of the shaft between the centre line and the radius r.

The equation for σ'_{rT} can be expanded into

$$\sigma'_{rT} = -\frac{E\beta}{1-\nu}\frac{1}{r^2}\left[Tr^2 - 2\int_0^r Tr\,\mathrm{d}r\right]$$

giving for $\sigma_{\theta T} = r\sigma'_{rT} + \sigma_{rT}$

$$\sigma_{\theta T} = \frac{E\beta}{1-\nu}\left[\frac{1}{2}\vartheta_m + \frac{1}{r^2}\int_0^r Tr\,\mathrm{d}r - T\right]. \quad (14.47)$$

Finally the axial stress σ_{zT} may be written, from

$$\begin{aligned}\sigma_{zT} &= \nu(\sigma_{rT} - \sigma_{\theta T}) - E\beta\,\Delta T \\ &= \frac{\nu}{1-\nu}(\vartheta_m - T(r)) - E\beta\,\Delta T.\end{aligned}$$

The temperature differential ΔT is the difference between the local temperature $T(r)$ and a reference temperature. Setting the reference temperature to the volumetric mean temperature ϑ_m allows us to reduce this equation to

$$\sigma_{zT} = \frac{E\beta}{1-\nu}(\vartheta_m - T(r)). \quad (14.48)$$

Once a temperature field within the shaft has been determined using, for example, the analysis method of Sect. 14.7, the respective mean volumetric temperatures can be determined from numerical integration and the stresses calculated for any radial position.

During warming of the shaft the outer zone is subject to thermal compressive forces and the inner zone to thermal tensile forces. During cooling this situation is reversed. The centrifugal force is always tensile and is greatest at the shaft centre, reducing to zero at the surface. However, the blades add a radial centrifugal tensile stress normal to the rotor surface plane.

The accuracy of the stresses calculated using Eqs. 14.46–14.48 for a cylindrical shaft will ultimately depend on the accuracy of the metal temperatures used for their calculation. Even the proprietary TSE systems must be "calibrated" against design calculations, and a similar approach can be taken here by the provision of calibration constants to allow for both matching to expected stress values and to compensate for the effects of possibly coarse temperature cell quantization. And of course, the fairly gross simplifications introduced by the assumed geometry of shaft and blading will have a significant and adverse influence on overall accuracy.

Fig. 14.6 A view of LP turbine blading and its attachment (courtesy of Millmerran Operating Company, Australia)

14.9.3 Rotor Construction

The construction of the turbine rotor has considerable influence on the development of stresses in the shaft and blade disks. Individual manufacturers prefer their own construction methods; HP, IP and LP rotors show differing characteristic methods and steam turbines differ significantly from gas turbines and axial flow compressors. It is not possible to provide a general stress calculation method which applies equally to all cases, but if attention is confined only to the actual shaft, that is, the rotating cylinder to which the blades are attached, the methods outlined in this section can be applied with reasonable expectations of success.

Construction of the turbine rotor, which includes the shaft and its attached blade rows, generally takes any of three principal forms. HP and IP steam turbines are frequently constructed as a solid cylindrical or near-cylindrical shaft into which the individual blades are set in grooves let into the cylinder surface. The blades are retained in the grooves by a variety of methods including lock-ways, bolts and welds. HP turbine blades are usually small and impose relatively small centrifugal forces. IP turbine blades are larger than those of HP turbines but still impose relatively small centrifugal forces. The principal HP and IP turbine stresses are thermal, and the shaft will sometimes be manufactured with a hollow bore to reduce thermal warming times. The centre bore might also be used for shaft warming.

The situation for LP turbines is very different, and other construction methods are used. While rotating at the same speed as the HP and IP shafts, unless in a cross-compound arrangement they rotate at half-speed, the LP blades are much longer (in excess of 1,100 mm in large machines), are more robust than those of the HP and IP turbines and impose much greater centrifugal and bending forces on their points of attachment. The LP blades are usually mounted into disks either slid onto the shaft and located using a key-way and some form of locking device or are heat-shrunk onto the shaft. In some designs, the disks are integral with the shaft and the blades slide into key-ways set into the periphery. In all such cases, the shaft is of much smaller radius than the disks and the most significant stresses are centrifugal and bending at the foot of the blade rather than thermal. For this reason, operational stress monitoring is applied to the HP and IP turbines but not usually to the LP turbine, although some LP turbines are provided with thermal stress monitoring at the steam inlet.

The use of heat shrinking attachment imposes a compression force to the shaft in the plane of the disk. This exceeds the centrifugal force of the blades by a considerable margin. The compression force reduces with increasing shaft speed as the blade centrifugal force increases. However, the net force on the shaft remains compressive up to shaft speeds well in excess of the rated speed.

Gas turbines and axial flow compressors are frequently assembled as rows of disks retained by bolts, either a series of bolts at around half radius or a single central bolt. By reducing the mass of the rotor, this construction method reduces its thermal mass and is better suited to rapid speed and load changes. Some designs use welded disks in preference to a bolted assembly. The reader is referred to specialist texts such as [61, 62] for further details and drawings.

14.9.4 Rotor Stress Variations During a Cold Start

The variables of principal interest are the equivalent shaft stress σ_{eq}, the shaft volumetric mean temperature ϑ_m, the rotor surface temperature, being here the mean temperature of the outermost rotor radial zone, and the differential between this and the shaft mean volumetric temperature.

The maximum zonal temperature differential is that seen by the outermost radial zone which therefore experiences the greatest radial thermal stress and, as it happens, the greatest tangential thermal stress, both of these stresses being compressive. The outermost zone is subject to the blade centrifugal stress which, being tensile, counteracts the compressive thermal stress. As the rotor speed increases, the centrifugal stresses increase and heat transfer to the blades increases with the increasing heat transfer coefficient and steam temperature. The shaft (here, blade) surface temperature increases towards the steam temperature and at 3,000 rpm approaches it closely even at a relatively low steam flow rate.

Figures 14.7–14.10 show trends of various calculated variables for three different idealised cold start schemes. Each figure uses the following curve assignments and identifiers.

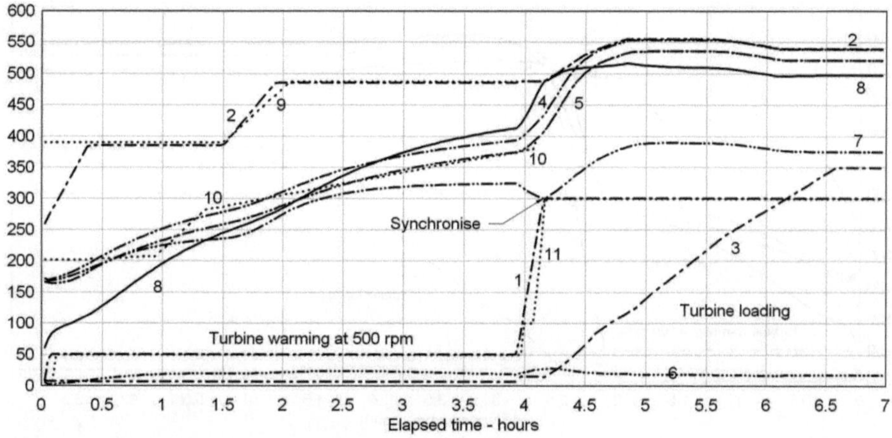

Fig. 14.7 Turbine cold start—warming at 500 rpm

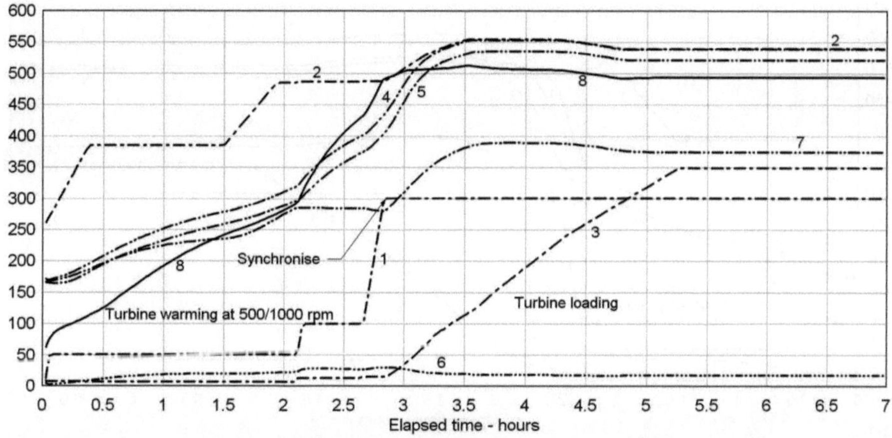

Fig. 14.8 Turbine cold start—warming at 500/1,000 rpm

1 — · — Turbine shaft speed [rpm/10]; 2 — · · — Inlet steam temperature [°C]; 3 — · — Steam flow to turbine [kg/s]; 4 — · ·— Shaft surface temperature [°C] 5 — · ·— Shaft average temperature [°C]; 6 — · · · — Shaft differential (sfc - avg) temperature [°C]; 7 — · ·— Turbine inner casing temperature [°C]; 8 ——— Shaft equivalent stress σ_{eq} [MPa]

The first of these, Fig. 14.7, which we shall call the base case, shows results of a simulation matched to data available from a cold start of a 450 MW generating unit. The turbine model was set up with the design data of the HP cylinder. Steam and shaft temperatures were initialised to match as nearly as possible the plant's measured data, with the exception of main steam temperature which was allowed

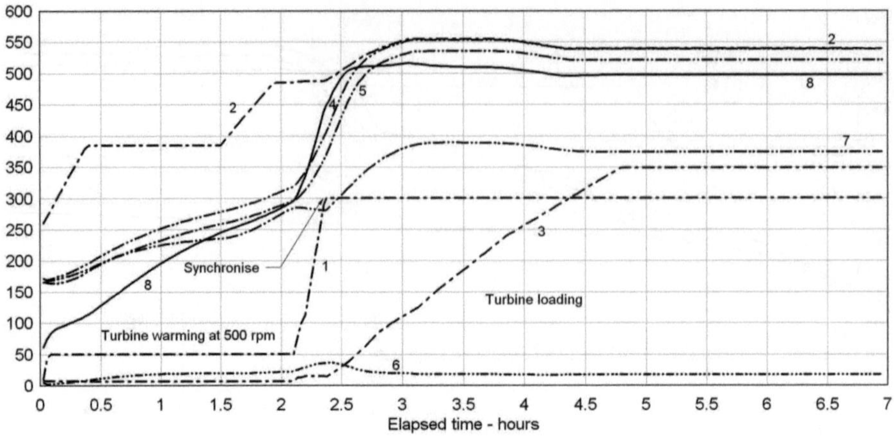

Fig. 14.9 Turbine cold start—warming at 500/1,000 rpm with early acceleration

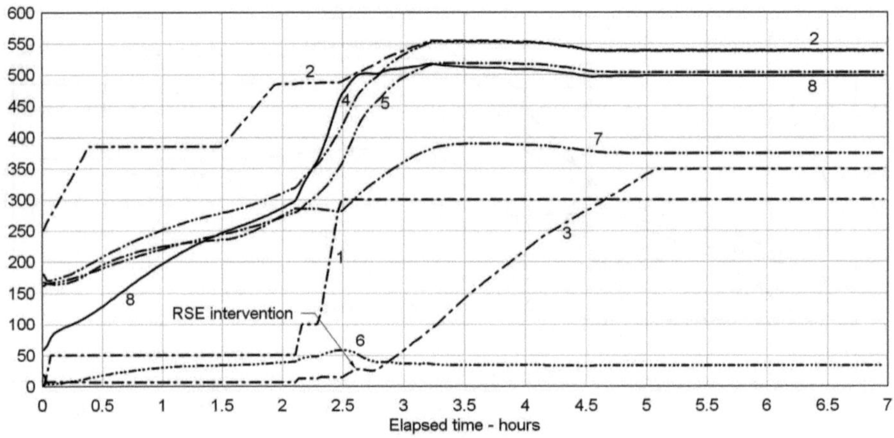

Fig. 14.10 Turbine cold start—early acceleration with RSE intervention

to increase at an operationally realistic rate from an initial 250°C to the measured 380°C. A lengthy period of pre-warming had established shaft surface, core and average temperatures at around 200°C, with the turbine rotated by the turning gear. The subsequent real plant time variations of turbine speed and steam temperatures were reproduced in the simulation by drivers.

The model behaviour was matched to that of the plant by adjustment of the turbine blade coefficient of heat transfer (Eq. 14.32) with the turbine rotating at 500 rpm. A value of 0.145 gave a good fit to the time taken to reach the real plant's shaft average temperature of 380°C, a value used during the cold start as release for shaft acceleration to 3,000 rpm for synchronising. Figure 14.7 includes selected plant data—shaft rotation speed (9, rpm/10), steam inlet temperature (10, °C) and shaft average temperature (11, °C)—superimposed on the model calculated trends to allow comparison of the modelled and real plant values. No measured data has been included for the post-synchronisation phase as the real and simulated procedures differed in some detail. However, the general behaviour patterns were similar.

The "measured" shaft average temperature is of course not a physical measurement but a value calculated by a proprietary routine supplied by the turbine manufacturer as part of the rotor stress evaluator (the RSE, TSE or WSE mentioned earlier). The details of these calculations are not known to the simulation, and the various turbine "measurements"—shaft surface, core and average metal temperatures, as well as the stresses and margins calculated from them—must be generated by simulation-specific calculations such as those described here. Matching a simulation to a specific installation will require some knowledge of the RSE calculation method or, if this is unavailable, at least access to recordings of the RSE-calculated variables during a variety of unit starts.

Figures 14.8 and 14.9 show the same calculated trends for the same turbine but with differing speed profiles and acceleration criteria.

For Fig. 14.8, the turbine was held at 500 rpm for 2 h 10 min, then accelerated to 1,000 rpm and held at this speed until the shaft average temperature[6] exceeded the same threshold value of 380°C. This allowed the unit to be synchronised some 80 min earlier than the base case at no significant cost in the form of increased shaft stress or differential temperature increase. Figure 14.9 followed the same speed profile of Fig. 14.8 but allowed acceleration of the turbine once the shaft average temperature had exceeded the lower value of 310°C. The unit was synchronised some 100 min earlier than the base case but at the cost of increases in maximum stress and shaft differential temperature from 28°C for the base case to around 36°C.

To illustrate intervention by the RSE, most likely in the form of a hold on loading until the differential temperature returned below some lower threshold value, the method of calculation of the surface/shaft average temperature differential was changed slightly to cause it to increase above an expected limit of some 50°C. This situation is shown by Fig. 14.10, which repeats Fig. 14.9 with intervention triggered at $>$48°C and removed at $<$42°C. The action of the RSE delayed full load by around 20 min.

Although only a few of an infinite selection of possible start-up schedules, these examples serve to illustrate the process and suggest opportunities for start-up optimization.

---

[6] All of these temperatures and stresses are calculated for the inlet section, being the most highly stressed.

Chapter 15
Steam Condensation

This chapter considers the condensation process taking place in large-scale condensing equipment of the type commonly found in power generation and process utility plant. In power generation, condensers are used almost exclusively for the condensation of steam at the exhaust end of steam turbines. In the process industries, condensers are more widespread, being used both for this duty and with distillation columns, heat recovery and reboilers and refrigeration plant.

Condensers are physically large components with banks of thin-walled tubes within or onto which steam and condensible vapours condense and through which heat is transferred to the cooling medium. In process plant applications condensing single-component vapours, condensers may be horizontal or vertical with condensation on the outside of the tubes and on the inside of vertical tubes if multiple component.

At the present time the majority of condensers in utility and industrial service use water as the cooling medium. To increase the flexibility of siting choice and in response to increasingly acute water consumption limitations, air-cooled condensers (ACC) are becoming more widely used. These have the great advantage of adding nothing to a site's water consumption and are appearing increasingly frequently in dry or arid areas.

In a power generation plant the main turbine condenser establishes the turbine exhaust conditions of pressure and exhaust steam temperature by condensing all of the steam leaving the low-pressure turbine(s).[1] The rate of condensation is matched to the rate of heat removal by movement of the steam temperature. A lower condensation rate causes an increase in pressure which, given saturation conditions within the condenser steam space, raises the saturation temperature. The pressure

[1]The condenser receives steam not only from the turbine but also from the turbine bypass, sometimes from other turbine drives and from the several drain connections to the condenser flash tank. These can contribute a small flow of air as well as steam.

P. O'Kelly, *Computer Simulation of Thermal Plant Operations*,
DOI 10.1007/978-1-4614-4256-1_15,

stabilises at a higher value corresponding to a temperature at which the increased heat transfer to the condenser cooling medium matches the heat load. If no stable condition can be reached, the steam source flow must be reduced.

Condenser pressure is one of the major performance indicators in power generation practice and off-standard back pressure can give rise to significant annual operating losses. For this reason, deterioration of condenser pressure, via its various causes, is a popular element of simulator training exercises and malfunction analyses.

Common to both water-cooled condenser (WCC) and ACC is the objective of maximum heat transfer on both sides of the tubes to maximise the condensation rate and with it to achieve the greatest practical reduction of the partial pressure of steam in the condenser. However, each goes about the task by following a somewhat different principle. The WCC condenses steam on the outside of water-cooled tubes to which it transfers heat through the tube wall. The tubes are arranged into large multi-row banks of horizontal smooth tubes upon which condensate forms and falls through the inter-tube space to the collecting hotwell beneath. In the air-cooled condenser, the tubes are arranged into single or double rows of heavily finned tubes inclined at up to 60 °C to the horizontal to promote draining of the condensate inside. Groups of tubes are associated with a single fan which provides the cooling airflow.

15.1 Water-Cooled Condensers

Condensers are assembled from banks of tubes manufactured from high thermal conductivity and enhanced corrosion resistant materials such as Admiralty brass. Tubes are arranged physically to allow the condensate to fall under gravity from the higher to the lower condensation regions and into the condensate storage well without accumulating on the lower tubes. The tubes can also be arranged to allow the cooling water to make several sequential passes through the condensing region before leaving the condenser (Fig. 15.1).

It is beyond our scope to consider all of these variants here so we will restrict ourselves to single-component (steam) condensation on the outside surfaces of horizontal tubes arranged in multiple banks in a single pass from inlet to outlet. The tubes are supported at each end by a tube sheet into which the ends of each tube are expanded and welded to ensure an air-tight fit. The length of the cooling water path is the horizontal distance between the tube sheets.

Condenser performance, as measured by the pressure of steam in the condenser, is a vital parameter in the optimisation of the economic performance of the plant. Condenser pressure should be as low as possible, consistent with the requirement to

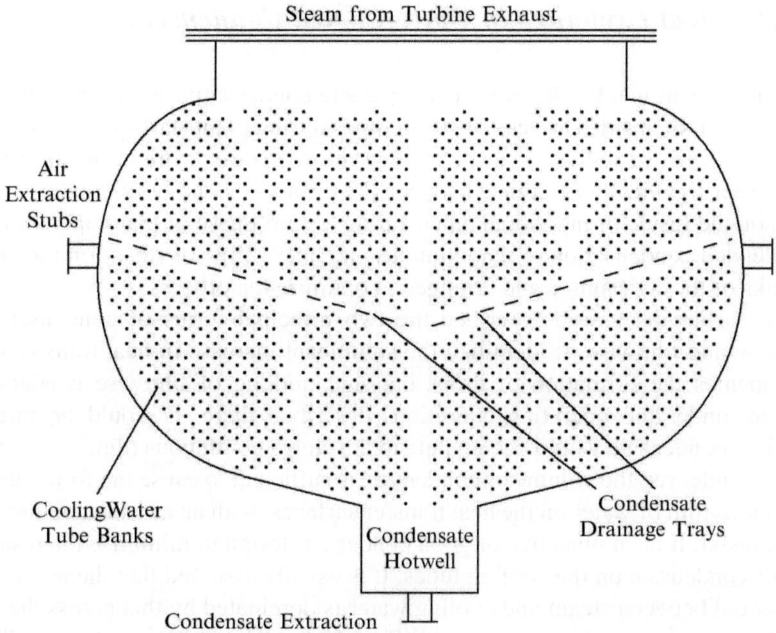

Fig. 15.1 Physical arrangement of a single-pass steam turbine condenser

transfer heat to the cooling water tubes.[2] The following terms are commonly found in condenser design and applications literature and specifications. The definitions given here are adapted from the Heat Exchange Institute Standard for Steam Surface Condensers [101].

Inlet Temperature Difference The difference between the condensing steam temperature and the cooling water inlet temperature.

Terminal Temperature Difference The difference between the condensing steam temperature and the cooling water outlet temperature.

Condensate Depression The difference between the condensing steam temperature and the temperature of the condensate in the hotwell.

[2]Because a condenser operates at less than atmospheric pressure, that is, under vacuum, the pressure is often referred to as "vacuum". A high "vacuum" signifies a low condenser pressure and *vice versa*. Here, to be consistent with the use of SI units, we will refer to condenser "pressure", in absolute units of kPa.

15.1.1 Heat Exchange in Water-Cooled Condensers

This subject was met briefly in Sect. 8.2.5 where condensation enhancement of heat exchange on the inside surfaces of steam pipes during plant start-up was discussed. In this chapter, we will look at the condensation process in more detail. There is of course a significant difference between the occasional and brief condensation of steam on the inside of individual pipes during a short phase of plant operations and the intended condensation of a continuous and heavy flow of steam on the outside of banks of tubes designed and arranged to optimise the process.

The earlier discussion described the two principal forms of condensation as dropwise and filmwise. It identified the significant increase in heat transfer which accompanies the formation of drops but said nothing of filmwise condensation because, under the operating conditions described there, it would be rare that sufficient condensate would accumulate to develop a continuous film.

In a condenser, the volume of condensate is sufficient to cause the formation of a continuous film of water on the heat transfer surfaces, with an attendant reduction in heat transfer. It is an objective of good condenser design to minimise the residence time of condensate on the cooling tubes. It is usually assumed that the temperature differential between steam and cooling water is dominated by that across the water film. By comparison, the temperature differential across the tube wall is negligible, which obviates the need for calculation of the tube temperature.

In condenser parlance, it is usual to speak of an *overall* heat transfer coefficient, either for a single tube or a tube bank. This is to be distinguished from the tube surface coefficients (inner and outer) which are available from the usual Colburn or Nusselt correlations. The overall heat transfer coefficient, conventionally denoted by U, is defined by the equation for heat transfer

$$\dot{Q} = U \cdot A \cdot \Delta T_{LM}, \tag{15.1}$$

where ΔT_{LM} is the log mean temperature differential

$$\Delta T_{LM} = \frac{\Delta T_2 - \Delta T_1}{\ln \frac{\Delta T_2}{\Delta T_1}}$$

Tw_1, Tw_2 are the inlet and outlet cooling water temperatures and Ts_1 and Ts_2 are the steam inlet and outlet temperatures. Then,

$$\Delta T_1 = Ts_1 - Tw_2,$$
$$\Delta T_2 = Ts_2 - Tw_1.$$

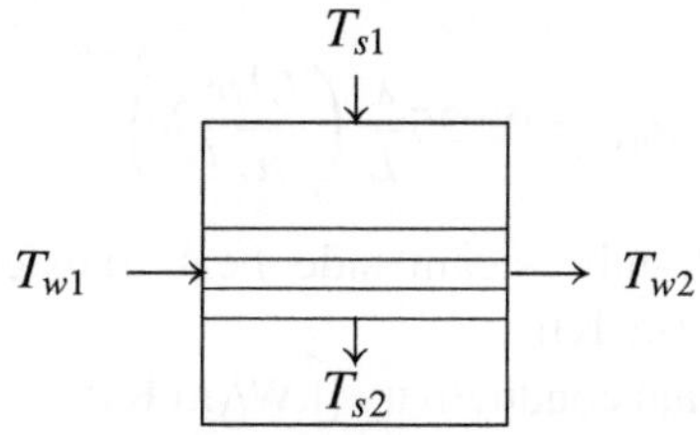

It can be assumed that the steam in the condenser is at a uniform temperature equal to the saturation temperature T_s' corresponding to the total pressure in the condenser

$$Ts_1 = Ts_2 = T_s'.$$

Then

$$\Delta T_2 - \Delta T_1 = Tw_2 - Tw_1$$

and

$$\Delta T_{LM} = \frac{Tw_2 - Tw_1}{\ln\left(\frac{T_s' - Tw_1}{T_s' - Tw_2}\right)}.$$

Since A is the total surface area of the tubes exposed to the steam,[3] the overall heat transfer coefficient U is a "per unit area" coefficient. It does not take account of tube bank effects or arrangements.

For film condensation on a horizontal tube and Reynolds number < 2,100, the condensing heat transfer coefficient is given from the Nusselt correlation [17]

$$\alpha_{hx} = 0.73\frac{\lambda}{d_o}\left(\frac{d_o^3\rho^2 g\,\mathbf{r}}{\lambda\,\mu_b\,\Delta T}^{0.25}\right)^{0.25} \tag{15.2}$$

or, with $\Gamma = W_F/2L$,

$$\alpha_{hx} = 0.76\frac{\lambda}{d_o}\left(\frac{d_o^3\rho^2 g}{\mu_b \Gamma}\right)^{1/3}. \tag{15.3}$$

The properties λ and ρ are quoted for the liquid state.

For film condensation on a vertical tube,

$$\alpha_{hx} = 0.943\frac{\lambda}{L}\left(\frac{L^3\rho^2 g\,\mathbf{r}}{\lambda\,\mu_b\,\Delta T}^{0.25}\right)^{0.25} \tag{15.4}$$

---

[3]This may be a single tube, a bank of tubes or the entire condenser. The individual cases are discussed a little later.

or,

$$\alpha_{hx} = 0.925\frac{\lambda}{L}\left(\frac{L^3\rho^2 g}{\mu_b \Gamma}\right)^{1/3}, \tag{15.5}$$

where

α_{hx}	single-tube steam side heat transfer coefficient [kW/(m^{2}K)]
λ	thermal conductivity [kW/(m K)]
d_o	tube outside diameter [m]
L	heated length [m]
ρ	density [kg/m^3]
μ_b	coefficient of viscosity of vapour [kg/(m s)]
g	acceleration due to gravity = 0.98 [m/s^2]
ΔT	temperature differential across the film
W_F	mass flow rate of condensate per tube (condensation loading) [kg/s]
$\mathbf{r}$	latent heat of condensation = $(h'' - h')$ [kJ/kg]

Grouping the terms which are functions of the condenser geometry or are otherwise constant over the range of pressures to be expected, the expression for the heat transfer coefficient may be written as

$$\alpha_{hx} = c_0\, \rho^{0.5} \mathbf{r}^{0.25} \Delta T^{-0.25}. \tag{15.6}$$

This is the theoretical heat transfer coefficient and is known to be conservative. In practice, the coefficient c_0 must be matched to actual performance data. For a multi-zone treatment, a different value of α_{hx} will apply to each zone to accommodate the local variations of condensate film thickness on the CW tubes. Perry [17] gives a nomogram for a number of single-component substances from which the condensing heat transfer coefficient for horizontal and vertical tubes may be read off, given condensing film temperature, temperature differential (vapour to cooling medium), tube diameter and tube count, for a single vertical bank column. For example, for a film temperature of 50 °C (122 °F) (saturation temperature of the steam), a tube diameter of 1.0 in. (25 mm), temperature differential of 27 °C (81 °F) and a tube count of 1 (single tube), the nomogram yields a heat transfer coefficient of 1,400 Btu/(h ft^2 F) (7,950 J/(s m^2 K)). For a tube count of 20, this reduces to around 850 Btu/(h ft^2 F) (4,826 J/(s m^2 K)).[4]

The Heat Exchange Institute is an industry grouping of heat exchanger manufacturers created to promote the benefits of standardisation within the heat exchanger industry. The institute publishes a set of Standards defining a unified basis for the types, design, construction, performance definition and test of a variety of heat exchangers used within the power generation and process industries. One of these

[4] To convert from Btu/(h ft^2 F) to J/(s m^2 K), multiply by 5.6780.

Standards deals with steam surface condensers [101] and contains a chart showing the variation of the overall heat transfer coefficient against cooling water velocity in a single tube, for a variety of tube metals and a cooling water inlet temperature of 70 °F (21 °C). The overall heat transfer coefficient U is calculated from the formula

$$U = c\sqrt{v}F_t,$$

where v is the cooling water velocity through the tube [ft/s] and c is a constant dependent on the tube diameter. F_t is a factor which allows for the tube material and accounts for the differing thermal conductivities of each metal. It is tabulated in the Standard for metals typically used in condenser manufacture. Other correction factors to be applied allow for surface fouling and for inlet temperatures other than 70 °F . Applying this formula and its various correcting factors to a condenser with Admiralty brass tubes of 7/8 in. OD x 18 BWG gives, for a water velocity through a single tube of 7 ft/s (2 m/s), $U = 543.7$ Btu/(h ft^2 F) or 3,087 J/(s m^2 K).

The condensing heat transfer coefficient for a bank of N_t tubes is related to the single-tube coefficient α_{hx}, assuming that the single tube is the top tube of a vertical column of N_t horizontal tubes and that condensate falling from the higher tubes does not accumulate on the lower tubes, by

$$\alpha_N = \alpha_{hx} N_t^{-1/4}.$$

For the earlier example, this would have given a heat transfer coefficient for the 50-tube arrangement of around 500 Btu/(h ft^2 F). However, this correlation is recognised as returning conservatively large figures, and Murty [103] has proposed a formula for the calculation of the overall heat transfer coefficient U_b for a tube bank of horizontal tubes as

$$\frac{1}{U_b} = \frac{A_o}{A_i}\left[\frac{\alpha_i - U_s}{\alpha_i U_s}\right] N_t^{1/12} + f_o + \left[\frac{1}{\alpha_i} + f_i\right]\frac{A_o}{A_i}, \tag{15.7}$$

where
- A_i is the tube inner surface area [m^2]
- A_o is the tube outer surface area [m^2]
- α_i is the tube inner heat transfer coefficient [J/(s m^2 K)]
- U_s is the overall heat transfer coefficient for the single tube [J/(s m^2 K)]
- f_O is the fouling factor for the shell side
- f_i is the fouling factor for the tube side

Applying this formula to the same example gives $U_b = 2{,}150$ J/(s m^2 K).

15.1.2 The Development of the Condenser Modelling Equations

In volumetrically large components, spatial effects contribute significantly to the dynamic behaviour of the equipment and to the steady-state conditions reached after transients have decayed. The following development uses a spatial control volume method to develop a generic approach to the modelling of these components.

The internal volume of the condenser may be considered as consisting of three major volumes:

1. A free steam space into which the steam flows from the turbine exhaust, condenser flash tank and turbine steam bypass system are introduced and to which the air extraction manifold is connected by extraction points located at opposite sides of the condenser
2. A tube sheet zone through which the tubes carrying the cooling water (CW) are arranged transverse to the downward flow of steam
3. A condensate hotwell into which condensate falls from the CW tubes and from which the condensate extraction pumps remove the condensate under control of the condensate hotwell level controller

These are treated as follows for modelling purposes:

1. The free steam space above the CW tubes is treated as a single space of volume V_0 in which steam and non-condensible gas (air) are present as a homogeneous mixture. Steam enters the space (a) as turbine exhaust and (b) as the vapour fraction leaving the condenser flash tank. Steam leaves this space as a downward flow to the CW tube zones. No condensation of the steam takes place in this space. It is used to define the local steam pressure on the basis of local mass balance and turbine exhaust conditions (flow and wetness). The steam flow distribution profile entering the CW tube zone is predefined.
2. The space occupied by the CW tubes is divided into two independent banks of four zones each. Physically, the condenser tubes are divided into multiple tube banks for reasons of simplified construction. Cooling water is supplied evenly to each bank from a plenum formed by the condenser end cover plate. CW from the two CW pumps is introduced into the inlet plenum, passed through the tubes and exhausted via an outlet plenum back to either an outlet canal or cooling tower.

   Steam is free to move through or across the tube banks and from one bank to the other. Each tube bank is divided into four zones, giving a steam space of eight zones. Steam enters the tube zone from the upper steam space and begins condensing on the CW tubes. The exhaust steam from the turbine is assumed saturated and is already wet (typically 85 % dry at full turbine load). Steam leaves each CW tube cell either as condensate, which falls through to the hotwell zone, or as uncondensed steam which passes to a neighbouring cell. Theoretically, all steam entering the tube zone is condensed. Therefore, no steam leaves the lower cells to enter the hotwell zone. This will clearly not be the case but is a reasonable assumption for the model and is used to define the lower boundary condition for cells 5–8.

3. The hotwell zone is a relatively small volume in which condensate falling from the CW tubes is collected. The hotwell pressure is used for the calculation of condensate flow to the condensate extraction pumps and may be taken as the mean pressure of cells 5–8 of the CW tube zone.

   The level of condensate in the hotwell is calculated simply from the mass balance of condensate entering and leaving the hotwell, using the hotwell geometry to convert from mass to level. The net inflow consists of the condensate leaving the tubes plus the water fraction leaving the condenser flash tank. The temperature of the condensate in the hotwell is derived from an energy balance of the incoming and outgoing flows. The water phase from the flash tank will be at saturation temperature. The condensate falling from the CW tubes will be sub-cooled to a temperature somewhere between saturation and tube temperature. In the absence of any better information, it is reasonable to assume that the temperature of the condensate produced in a particular cell is the mean of the saturation temperature in and the CW outlet temperature from that cell. Since the CW flow, heat transfer and outlet temperatures are calculated separately for each of the eight cells, the hotwell energy balance must take each of the eight cell condensate flows into account.

   Under some adverse operating conditions, the hotwell level can increase and cover the lower tubes of cells 5–8. The heat transfer area of these cells will be reduced proportionately.

Since the efficiency of the turbine is strongly affected by the condenser back pressure, the presence of air in the condenser has a marked adverse effect on turbine performance. These effects are discussed in more detail in a later section.

The arrangement of cells into which the condenser is divided is shown by Fig. 15.2.

The general mass balance for the j-th cell of volume V_j is written as

$$\frac{\mathrm{d}M_{s,j}}{\mathrm{d}t} = \sum(\dot{m})_j, \tag{15.8}$$

where $M_{s,j}$ is the mass of steam in the j-th cell and $\sum(\dot{m})_j$ is the net sum of all steam flows entering the cell space, either by mass transfer to the neighbouring cells or by loss through condensation. By convention, a negative flow is one leaving the cell. Note that condensate falling from a higher cell does not contribute to the cell mass balance. Writing

$$M_{s,j} = V_j/v_j,$$

where v_j is the specific volume of steam in the j-th cell and V_j is the cell's volume, the mass balance Equation 15.8 can be written as

$$\frac{\mathrm{d}v_j}{\mathrm{d}t} = -\frac{v_j^2}{V_j}\sum(\dot{m})_j. \tag{15.9}$$

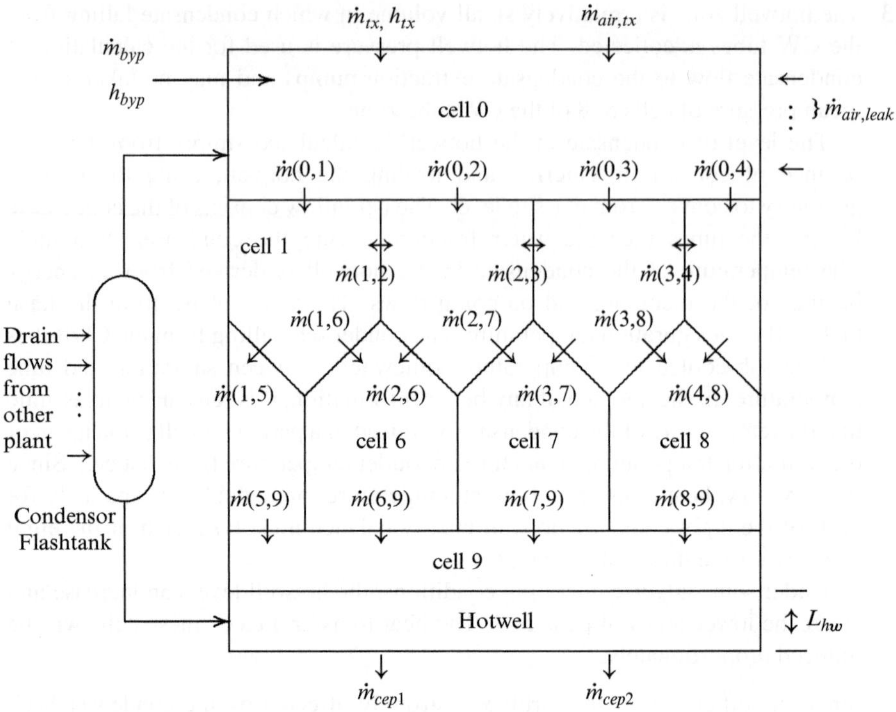

Fig. 15.2 Distribution of cells and flows in the condenser

The time derivative of specific volume may be replaced by the derivatives of pressure and enthalpy as follows:

$$\frac{\mathrm{d}v_j}{\mathrm{d}t} = \frac{\partial v}{\partial p}\frac{\mathrm{d}p_j}{\mathrm{d}t} + \frac{\partial v}{\partial h}\frac{\mathrm{d}h_j}{\mathrm{d}t}. \tag{15.10}$$

Assuming that steam acts under these conditions as a perfect gas,

$$\frac{\partial v}{\partial p} = -\frac{v''}{\gamma p}, \tag{15.11}$$

$$\frac{\partial v}{\partial h} = \frac{v'' - v'}{h'' - h'}, \tag{15.12}$$

where γ is the adiabatic index and is 1.3 for saturated steam. Rearranging the mass balance equation gives

$$\frac{\partial v}{\partial p}\frac{\mathrm{d}p_j}{\mathrm{d}t} + \frac{\partial v}{\partial h}\frac{\mathrm{d}h_j}{\mathrm{d}t} = -\frac{v_j^2}{V_j}\sum(\dot{m})_j \tag{15.13}$$

or

$$\frac{V_j}{v_j^2}\frac{\partial v}{\partial p}\frac{\mathrm{d}p_j}{\mathrm{d}t}+\frac{V_j}{v_j^2}\frac{\partial v}{\partial h}\frac{\mathrm{d}h_j}{\mathrm{d}t}=-\sum(\dot{m})_j. \tag{15.14}$$

The general energy balance for the jth cell may be written as

$$\frac{\mathrm{d}(M_j h_j)}{\mathrm{d}t}=\sum(\dot{m}h)_j+V_j\frac{\mathrm{d}p_j}{\mathrm{d}t}-\dot{Q}_j, \tag{15.15}$$

where $\sum(\dot{m}h)_j$ is the net sum of enthalpy flows entering the cell boundaries and $\dot{Q}_j$ is the rate of heat transfer out of the cell.

Noting that

$$\begin{aligned}\frac{\mathrm{d}(M_j h_j)}{\mathrm{d}t}&=M_j\frac{\mathrm{d}h_j}{\mathrm{d}t}+h_j\frac{\mathrm{d}M_j}{\mathrm{d}t}\\&=\frac{V_j}{v_j}\frac{\mathrm{d}h_j}{\mathrm{d}t}+h_j\sum(\dot{m})_j\end{aligned}$$

this equation may be rearranged as

$$-V_j\frac{\mathrm{d}p_j}{\mathrm{d}t}+\frac{V_j}{v_j}\frac{\mathrm{d}h_j}{\mathrm{d}t}=\sum(\dot{m}h)_j-h_j\sum(\dot{m})_j-\dot{Q}_j. \tag{15.16}$$

Writing

$$\begin{aligned}b_{11}&=\frac{V_j}{v_j^2}\frac{\partial v}{\partial p}_j & b_{12}&=\frac{V_j}{v_j^2}\frac{\partial v}{\partial h}_j\\ b_{21}&=-V_j & b_{22}&=\frac{V_j}{v_j}\end{aligned}$$

the mass and enthalpy equations for the j-th cell may be combined into a single matrix equation

$$\begin{bmatrix}b_{11}\ b_{12}\\ b_{21}\ b_{22}\end{bmatrix}\begin{bmatrix}\frac{\mathrm{d}p_j}{\mathrm{d}t}\\ \frac{\mathrm{d}h_j}{\mathrm{d}t}\end{bmatrix}=\begin{bmatrix}-\sum a_{ij}p_i+p_j\sum_i a_{ij}+\dot{m}_{c,j}\\ \sum(\dot{m}h)_j-h_j\sum(\dot{m})_j)-\dot{Q}_j\end{bmatrix}. \tag{15.17}$$

The net mass flow term may be expanded in terms of the pressures in the j-th and adjoining i-th cell(s) and the flow admittance a_{ij} between the cells. A linear representation is appropriate in view of the very small pressure differentials expected between adjoining cells.

$$\begin{aligned}\sum(\dot{m})_j&=\sum_i\dot{m}_{ij}-\dot{m}_{c,j}\\&=\sum_i a_{ij}(p_i-p_j)-\dot{m}_{c,j}\end{aligned}$$

$$= \sum_i a_{ij} p_i - p_j \sum_i a_{ij} - \dot{m}_{c,j} \tag{15.18}$$

$$= \sum \dot{m}_{in} - \sum \dot{m}_{out} - \dot{m}_{c,j}, \tag{15.19}$$

where

$\dot{m}_{c,j}$ is the local condensation flow in the cell
$\dot{m}_{in}$ is the total mass flow into the j-th cell (>0) $(a_{ij}(p_i - p_j) > 0)$
$\dot{m}_{out}$ is the total mass flow out of the j-th cell (<0) $(a_{ij}(p_i - p_j) < 0)$

The net enthalpy flow term $\sum(\dot{m}h)_j$ may be expanded in terms of the flows to/from the j-th cell and the enthalpies in the cells. A flow from the i-th to the j-th cell is denoted $\dot{m}_{i,j}$, while a flow from the jth to the ith cell is denoted $\dot{m}_{j,i}$.

$$\sum(\dot{m}h)_j = \sum_i \dot{m}_{i,j} h_i - h_j \sum \dot{m}_{out}) + \dot{m}_{c,j}(h'' - h'). \tag{15.20}$$

Equation 15.17 may now be written as

$$\mathbf{B}_j \begin{bmatrix} \frac{dp_j}{dt} \\ \frac{dh_j}{dt} \end{bmatrix} = \begin{bmatrix} \sum_i a_{ij} & 0 \\ 0 & -\sum_i \dot{m}_{i,j} + \dot{m}_{c,j} \end{bmatrix} \begin{bmatrix} p_j \\ h_j \end{bmatrix} + \begin{bmatrix} -\sum_i a_{ij} p_i \\ \sum \dot{m}_{i,j} h_i \end{bmatrix} + \begin{bmatrix} \dot{m}_{c,j} \\ \dot{m}_{c,j}(h'' - h') - \dot{Q}_j \end{bmatrix}. \tag{15.21}$$

The inverse of the matrix $\mathbf{B}_j$ is

$$\mathbf{B}_j^{-1} = D_j^{-1} \begin{bmatrix} v_j & -\frac{\partial v}{\partial h}_j \\ v_j^2 & \frac{\partial v}{\partial p}_j \end{bmatrix},$$

where

$$D_j = V_j \left(\frac{1}{v_j} \frac{\partial v}{\partial p}_j + \frac{\partial v}{\partial h}_j \right).$$

Using the inverse of $\mathbf{B}_j$, Eq. 15.21 may be written as

$$D_j \begin{bmatrix} \frac{dp_j}{dt} \\ \frac{dh_j}{dt} \end{bmatrix} = \begin{bmatrix} v_j \sum_i a_{ij} & \frac{\partial v}{\partial h}_j (\sum_i \dot{m}_{i,j} - \dot{m}_{c,j}) \\ v_j^2 \sum_i a_{ij} & -\frac{\partial v}{\partial p}_j (\sum_i \dot{m}_{i,j} - \dot{m}_{c,j}) \end{bmatrix} \begin{bmatrix} p_j \\ h_j \end{bmatrix}$$

$$+\begin{bmatrix} -v_j \sum_i a_{ij} p_i - \frac{\partial v}{\partial h}_j \sum_i \dot{m}_{i,j} h_i \\ -v_j^2 \sum_i a_{ij} p_i + \frac{\partial v}{\partial p}_j \sum_i \dot{m}_{i,j} h_i \end{bmatrix}$$

$$+\begin{bmatrix} v_j & -\frac{\partial v}{\partial h}_j \\ v_j^2 & \frac{\partial v}{\partial p}_j \end{bmatrix} \begin{bmatrix} \dot{m}_{c,j} \\ \dot{m}_{c,j}(h'' - h') - \dot{Q}_j \end{bmatrix}. \tag{15.22}$$

Noting that flow is exchanged only between adjoining cells, the index i is taken only over the cells adjoining cell j.

When the index j is taken over all cells $j \in (0,8)$ (except the hotwell for which boundary conditions only are defined), Eq. 15.22 defines a matrix equation of dimension 18 in the vectors **p** and **h**. Defining a composite variable $\mathbf{x} = \mathtt{col}(\mathbf{p},\mathbf{h})$, we have

$$\mathbf{D}\frac{d\mathbf{x}}{dt} = \mathbf{C}\mathbf{x} + \mathbf{Y}, \tag{15.23}$$

where **D** is an 18-element diagonal matrix $[D_0 D_1 \cdots D_8]$. **C** is an 18×18 matrix of coefficients, and **Y** is an 18-element vector incorporating the condensation flow and net heat transfer effects in the individual cells.

We may assume that the coefficients of **C**, **D** and **Y** are known, having been evaluated using the condenser conditions computed at the previous time step. Equation 15.23 may therefore be treated as a linear matrix differential equation whose solution develops the transient behaviour of the steam conditions in each of the condenser cells.

Boundary and subsidiary conditions are defined as follows:

Cell 0 The constituent equation is

$$D_0 \begin{bmatrix} \frac{dp_0}{dt} \\ \frac{dh_0}{dt} \end{bmatrix} = \begin{bmatrix} v_0 \sum_{i=1}^{4} a_{i0} & \frac{\partial v}{\partial h}(\dot{m}_{tx} + \dot{m}_{ft}) \\ v_0^2 \sum_{i=1}^{4} a_{i0} & -\frac{\partial v}{\partial p}(\dot{m}_{tx} + \dot{m}_{ft}) \end{bmatrix} \begin{bmatrix} p_0 \\ h_0 \end{bmatrix}$$

$$+\begin{bmatrix} -v_0 \sum_{i=1}^{4} a_{i0} p_i - \frac{\partial v}{\partial h} \sum_{i=1}^{4} \dot{m}_{i,0} h_i \\ -v_0^2 \sum_{i=1}^{4} a_{i0} p_i + \frac{\partial v}{\partial p} \sum_{i=1}^{4} \dot{m}_{i,0} h_i \end{bmatrix}$$

$$+\begin{bmatrix} v_0 & -\frac{\partial v}{\partial h} \\ v_0^2 & \frac{\partial v}{\partial p} \end{bmatrix} \begin{bmatrix} (-\dot{m}_{tx} + \dot{m}_{ft}) + \dot{m}_c \\ \dot{m}_{tx} h_{tx} + \dot{m}_{ft} h'' \end{bmatrix}.$$

Inlet conditions to the cell are defined by the turbine exhaust conditions $\dot{m}_{tx}$ and h_{tx} and the vapour flow from the flash tank $\dot{m}_{ft}$ entering the condenser with vapour saturation enthalpy h''.

Cell 9 This cell is treated in simplified fashion in view of the imponderables inherent in the physical description of the process. Steam enters this space from the tube bundles but does not condense to any notable degree either on the condenser casing or at the steam-condensate interface. Since the steam that flows into and out of this region are probably highly turbulent, with little well-defined direction, the cell steam pressure p_9 is taken as approximating the minimum of the pressures in the adjoining cells (5–8).

Similarly, the steam enthalpy in the cell will approximate the mean of the adjoining cells since it might be expected that steam will be exchanged between these and cell 9 in both directions. The steam in cell 9 can be taken to be a well-mixed blend of that coming from cells 5–8.

To ensure numerical stability, both p_9 and h_9 are made to follow their respective target values with a first-order lag. The lag time is typically in the range of 15–40 s.

Given that p_9 and h_9 are defined as boundary conditions, there is no constituent equation for cell 9. p_9 enters the equation for cells 5–8 via the calculation of the flows $\dot{m}_{i,9}$ for $i = 5, 8$. h_9 enters the energy balance equations for cells 5–8 as the source enthalpy for back flows from cell 9 to those cells.

Calculation of Flows In view of the small pressure differentials between cells. it will be permissible to use the linear form

$$\dot{m}_{ij} = a_{ij}\,[p_i - p_j] \tag{15.24}$$

in place of the exact form

$$\dot{m} = a_{ij}\sqrt{p_i - p_j}$$

for the calculation of mass flow from cell i to cell j.

15.1.3 Heat Transfer

Equation 15.1 is used to define the heat transferred from the condensing steam to the cooling medium along the cooling water (cw) path through the tubes enclosed within the j-th cell. It is assumed that all cells see the same inlet cooling water temperature T_{cwi}.

$$\dot{Q}_j = U_j\, A_j\, \Delta T_{LM,j}$$

and

$$\Delta T_{LM,j} = \frac{(T_{cwx} - T_{cwi})_j}{\ln\left(\frac{T_s' - T_{cwi}}{T_s' - T_{cwx}}\right)_j}.$$

The overall heat transfer coefficient U_j is calculated for each cell as discussed previously.

This heat is removed by the cooling water flow $\dot{m}_{cw}$ which is evenly distributed across the individual tubes of the bank. The grouping of the tubes corresponds to the cell divisions described in the preceding sections. The cooling water is warmed by the heat transfer and leaves its tube bank with temperature $T_{cwx,j}$. Then,

$$\dot{Q}_j = c_w \dot{m}_{cw,j} (T_{cwx,j} - T_{cwi}).$$

Equating the right-hand sides of the two equations in $\dot{Q}_j$ yields

$$T_{cwx,j} = T_{cwi} + (T'_{s,j} - T_{cwi}) \left[1 - \exp\left(\frac{U_j A_j}{\dot{m}_{cw,j} c_w}\right)\right]. \tag{15.25}$$

The cooling water temperature rise across the j-th tube bank is then

$$\Delta T_{cw,j} = (T'_{s,j} - T_{cwi}) \left[1 - \exp\left(\frac{U_j A_j}{\dot{m}_{cw,j} c_w}\right)\right] \tag{15.26}$$

and the heat removed from the j-th cell is

$$\dot{q}_{cw,j} = \dot{m}_{cw,j}\, c_w\, \Delta T_{cw,j}. \tag{15.27}$$

The total heat removed from the entire condenser is

$$\dot{Q}_{cw} = \sum_{j=1}^{8} \dot{q}_{cw,j}. \tag{15.28}$$

Both the heat transfer within and the cooling water outlet temperature from each cell are determined once the heat transfer coefficient in the cell is determined and the cooling water flow and its inlet temperature are known.

15.1.4 Heat Transfer in the Presence of Non-condensible Gases

The presence of non-condensible gases, in this case predominantly air, will reduce the heat transfer coefficient from its zero gas value. Two effects are the main contributors and from the simulation point of view can be treated independently.

Non-condensible gas molecules tend to accumulate in the coolest regions of the condenser, here the cooling tube surfaces. The vapour molecules must then diffuse through the gas layer. Since this imposes an extra element along the heat transfer path, it can be treated as an additional series term α_g within the calculation of the overall heat transfer coefficient U. Equation 15.7 may then be augmented to include this term.

$$\frac{1}{U_b} = \frac{A_o}{A_i}\left[\frac{\alpha_i - U_s}{\alpha_i U_s}\right] N_t^{1/12} + f_o + \frac{1}{\alpha_g} + \left[\frac{1}{\alpha_i} + f_i\right]\frac{A_o}{A_i}. \tag{15.29}$$

The analytical prediction of the numerical value of α_g for a given condenser under any operational conditions is extremely difficult. A condenser is a physically large component, conditions within the condenser vary in all three dimensions, vapour and gas distributions and their thermodynamic states are strongly dependent on flow velocities and directions, effectiveness of gas removal, and so on. Physically based and semiempirical methods do exist which can accurately predict these factors and are used for condenser design, but, as noted elsewhere [47], they are excessively complex for this form of simulation treatment. Several authors [44, 47, 72, 73, 98] quote Berman and Fuks [99] as the source of the following empirical expression for α_g:

$$\alpha_g = \frac{a D_p}{d_o} Re_g^{1/2}\left(\frac{p}{p - p_s}\right)^b p^{1/3}\left(\frac{\rho_g h_{lg}}{T_g}\right)^{2/3}\frac{1}{(T_g - T_c)^{1/3}}, \tag{15.30}$$

where,

for $Re_g > 350$, $b = 0.6$ and $a = 0.82$, and
for $Re_g < 350$, $b = 0.7$ and $a = 0.52$

and the individual symbols have the following meanings:

Re_g	Reynolds number of the gas = $v\rho_g L/\eta_g$
D_p	steam-in-air diffusivity coefficient
d_o	outer diameter of a single tube [m]
p	total pressure of the vapour/gas mixture = $p_s + p_g$ [Pa]
p_g	partial pressure of the gas [Pa]
p_s	partial pressure of the steam vapour [Pa]
η_g	absolute viscosity of the gas [Pa.s]
ρ_g	density of the gas [kg/m^3]
T_g	temperature of the gas [°C]
T_c	temperature of the steam-condensate interface [°C]
h_{lg}	latent heat of condensation of the vapour [kJ/kg]

The diffusivity term $D_p = D/R\mathcal{T}$, $\mathcal{T}$ being K, where D is given empirically from the equation (due to Fuller [100])

$$D = 0.00011756552\frac{\mathcal{T}^{1.75}}{p}.$$

The following figure displays the influence of partial gas pressure and flow velocity on the thermal resistance $1/\alpha_g$ as calculated by Eq. 15.30. Values are shown for a range of partial gas pressures of 1–50 Pa and a range of vapour/gas velocities from 0.5 to 4.5 m/s. The power plant condenser is assumed to be operating at a total pressure of 10 kPa abs (10,000 Pa) and contents at 47 °C.

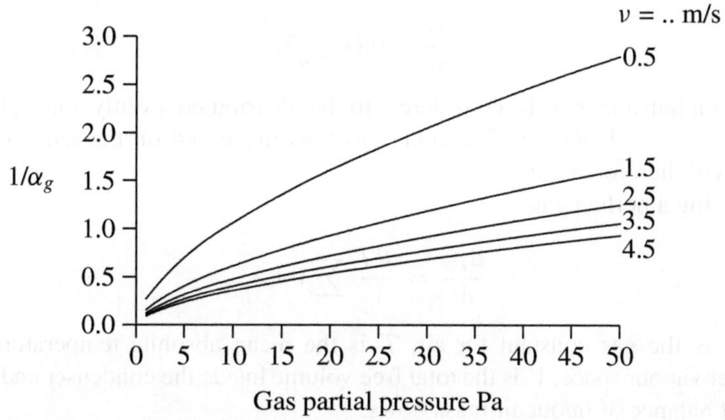

The improvement seen with increasing flow velocities is due to the improved removal of the gas layer from around the tubes. The rapid deterioration of heat transfer with increasing gas pressures is clearly evident with even a very small presence of gas introducing significant thermal resistance.

The second effect arises as follows. For a given total pressure, a gas pressure contribution reduces that of the steam with a consequential depression in local saturation temperature and a reduction in the temperature differential driving the transfer of heat from the steam to the coolant tube wall.

Calculation of Gas Pressure

From the ideal gas law

$$pV = MR\mathcal{T},$$

where V is the volume occupied by the gas [m^3], M is the mass of gas [kg] and R is the gas constant for the particular gas under consideration. For the air/steam mixture considered here, both V and $\mathcal{T}$ are common to both components. Hence,

$$\frac{p_a}{p_s} = \frac{m_a}{m_s}\frac{R_a}{R_s}.$$

The gas constant R is related to the universal gas constant **R** by

$$R = \frac{\mathbf{R}}{\mathcal{M}},$$

where $\mathcal{M}$ is the molar weight of the gas. Therefore, for air,

$$\frac{m_a}{m_s} = \frac{p_a}{p_s}\frac{\mathcal{M}_a}{\mathcal{M}_s}$$

$$= \frac{28.966}{18.016}\frac{p_a}{p_s}$$

giving

$$\frac{p_a}{p_s} = 0.662 \frac{m_a}{m_s}. \tag{15.31}$$

Non-condensible gas is considered to be distributed evenly throughout the condenser. The calculation of a single gas pressure based on the total condenser volume will therefore suffice.

Assuming a perfect gas

$$\frac{\mathrm{d}p_g}{\mathrm{d}t} = \frac{R\mathcal{T}}{V} \sum(\dot{m}_a),$$

where R is the gas constant for air, $\mathcal{T}$ is the mean absolute temperature of the condenser vapour space, V is the total free volume inside the condenser and $\sum(\dot{m}_a)$ is the net balance of in/out air mass flows.

For the present purpose, we will consider the presence of air only. The mass flow balance is the difference between total airflow into the vapour space and that removed by the vacuum maintaining units (air extractors). Under normal operating conditions, a constant flow of air enters the condenser through a variety of sources such as:

- Turbine glands not sealed
- L.P. heater leaks
- Vacuum breaking valve leaking
- Condenser body leak
- Doors on drains receivers open
- Deaerator venting
- Steam feed pump condenser venting

The flow rates associated with each of these items might be expected to lie in the range 0.1 to around 2 kg/s.

The air outflow from the condenser will depend on the number of extraction units in operation and their operating conditions. With a condenser in normal condition, one air extractor is usually sufficient to maintain an adequate condenser back pressure. Under moderate leak conditions, two extraction units may be needed. The leakage through the open vacuum breaker is massive and will result in total loss of vacuum to atmospheric pressure with a time constant of around 6 minutes.

15.1.5 Solution of the State Equations

Equation 15.22 may be solved in discrete time by replacing $\mathrm{d}/\mathrm{d}t$ by its simple discrete time equivalent and using a fully implicit formulation.

$$\mathbf{D}\frac{\mathbf{x}^{n+1} - \mathbf{x}^{n}}{\Delta t} = \mathbf{C}\,\mathbf{x}^{n+1} + \mathbf{Y},$$

where the coefficient matrices **D** and **C** and the forcing vector **Y** are evaluated using **x**, $\dot{Q}$ and $\dot{m}_c$ evaluated at the n-th time step. Then

$$[\mathbf{D} - \Delta t\ \mathbf{C}]\mathbf{x}^{n+1} = \mathbf{x}^n + \Delta t\ \mathbf{Y}^n.$$

This requires the solution of an 18×18 matrix at each time step. This may be reduced to the solution of two 9×9 matrix equations (not tridiagonal) by decoupling the pressure and enthalpy component equations of Eq. 15.22. Then

$$D_j \frac{\mathrm{d}p_j}{\mathrm{d}t} = v_j p_j \sum_i a_{ij} - v_j \sum_i a_{ij}\, p_i + v_j \dot{m}_{c,j} + \frac{\partial v}{\partial h}_j \left[\sum_i \dot{m}_{ij} - \dot{m}_{c,j}\right] h_j - \frac{\partial v}{\partial h}_j \left[\sum_i \dot{m}_{i,j} h_i + \dot{m}_{c,j}(h'' - h') - \dot{Q}_j\right] \quad (15.32)$$

and, given the relationships of Eq. 15.19,

$$D_j \frac{\mathrm{d}h_j}{\mathrm{d}t} = v_j^2 \sum_i \dot{m}_j - \frac{\partial v}{\partial p}_j \left[\sum_i \dot{m}_{ij} - \dot{m}_{c,j}\right] h_j + \frac{\partial v}{\partial p}_j \left[\sum_i \dot{m}_{i,j} h_i + \dot{m}_{c,j}(h'' - h') - \dot{Q}_j\right]. \quad (15.33)$$

Replacing thc time derivative by its discrete time equivalent and writing each equation in implicit form for the variable to be solved, these equations reduce to

$$\left[1 - \frac{\Delta t}{D_j} v_j \sum_i a_{ij}\right] p_j^{n+1} + \frac{\Delta t}{D_j} v_j \sum_i a_{ij}\, p_i^{n+1} = p_j^n + \frac{\Delta t}{D_j} v_j \dot{m}_{c,j} + \frac{\Delta t}{D_j} \frac{\partial v}{\partial h}_j \left[\sum_i \dot{m}_{ij} - \dot{m}_{c,j}\right] h_j - \frac{\Delta t}{D_j} \frac{\partial v}{\partial h}_j \left[\sum_i \dot{m}_{i,j} h_i + \dot{m}_{c,j}(h'' - h') - \dot{Q}_j\right] \quad (15.34)$$

and

$$\left[1 - \frac{\Delta t}{D_j} \frac{\partial v}{\partial p}_j \left[\sum_i \dot{m}_{ij} - \dot{m}_{c,j}\right]\right] h_j^{n+1} = h_j^n$$

$$+ \frac{\Delta t}{D_j} \left[v^2 \sum_i (\dot{m}_j) + \frac{\partial v}{\partial p}_j \left[\sum_i \dot{m}_{i,j} h_i + \dot{m}_{c,j} (h'' - h') - \dot{Q}_j \right]\right]. \quad (15.35)$$

The matrix of coefficients is formed by taking j over the range 0 to 8, using p_9 and h_9 as boundary values. The resulting coefficient matrices are symmetrically banded and can best be solved using LU decomposition.

15.1.6 Thermodynamic States and Condensation

So far, no assumptions have been made regarding the thermodynamic state of the steam in the condenser, and consequently two equations have been derived for the two defining state vectors **p** and **h**. The two Equations 15.34 and 15.35 are sufficient for the calculation of these two quantities for each cell given a mechanism for the calculation of condensation flow. However, that mechanism is not readily available, and a simplifying assumption is made at this point.

Examination of the pressure and energy Equations 15.32 and 15.33 shows that steady-state conditions can be achieved only if simultaneously,

$$\sum \dot{m}_{ij} = \dot{m}_{c,j}$$

and

$$\sum_i \dot{m}_{i,j} h_i + \dot{m}_{c,j} (h'' - h') - \dot{Q}_j = 0. \quad (15.36)$$

The first implies a net mass balance in each cell. The second implies that the condensation flow is always of such magnitude that the cell energy balance is satisfied. The cell enthalpy will therefore always be the saturation enthalpy corresponding to the cell pressure p_j and the condensation flow to produce this result is defined by the equation

$$\dot{m}_{c,j} = \frac{1}{(h'' - h')} [\dot{Q}_j - \sum_i \dot{m}_{i,j} h_i]. \quad (15.37)$$

With this value of condensation flow, the energy Equation 15.35 need not be used as the energy balance will always be satisfied.

The presence of the term $(\frac{\Delta t}{D_j} v_j \dot{m}_{c,j})$ in the pressure equation implies that the spatially distributed condensation creates a pressure gradient through the condenser.

15.2 Air-Cooled Condensers

An ACC consists of a number of cells, each associated with a single fan driving air through banks of finned tubes arranged either horizontally or inclined in an A- or V-shaped frame. Smaller units, which can include roof-mounted condensers, might use a V-frame arrangement to save space.

In large condensers the most common arrangement of fans and condensing tubes is the A-frame, depicted by the following figure. For the purposes of this discussion, we will focus on this arrangement though the principles apply equally to any.

The ACC depicted by Figs. 15.3 and 15.4 consists of 36 cells arranged as 6 rows, each containing 6 fan cells. Steam from the inlet header is distributed through overhead ducts to each fan cell and fills the condensing tubes. Heat removed by the fan airflow causes the steam to condense. The condensate runs down the inside of the tubes to be collected in the bottom headers and then to the common condensate collection tank (hotwell).

One side of one central section in each cell does not connect to the steam inlet header but is closed at its top end and connected to the air extraction manifold. Non-condensible gases accumulate in this bank and are removed from the top. The sections without the gas take-off are sometimes referred to as the primary sections, the gas take-off section as the secondary. This reflects the predominant direction of steam flow within each bank of tubes.

The modelling analysis of the ACC proceeds on the basis of a single fan cell, with the following assumptions:

- The total steam flow from the turbine exhaust and other sources is uniformly distributed, first to each parallel bank of cells and then individually to each cell and to each tube in each cell.
- The airflow is distributed uniformly across all tubes in a bank.

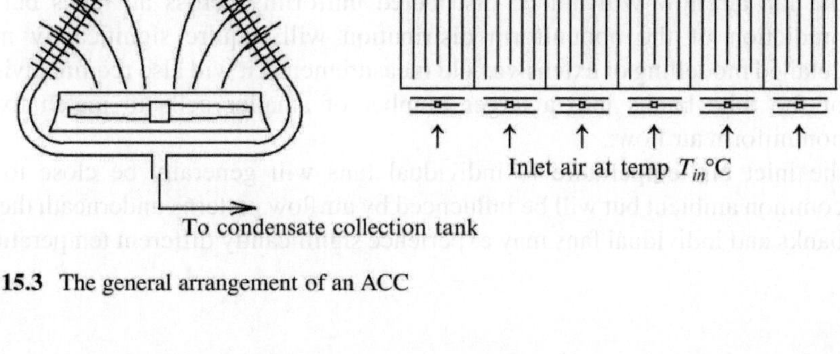

Fig. 15.3 The general arrangement of an ACC

Fig. 15.4 A large ACC during construction (courtesy of Millmerran Operating Company, Australia)

- Condensate does not accumulate on the inner walls of the tubes but is removed quickly to the collecting headers.
- Airflow interactions between adjacent fan cell can be ignored.
- All cells have the same inlet air temperature.
- Air and other non-condensible gases are distributed uniformly though all tubes such that it is valid to assume the same air/gas partial pressure in all tubes.
- Steam in the tubes is saturated.

These assumptions permit the treatment of a single cell, and the overall performance of the ACC can be obtained as a summation taken over all cells. However, not all of these assumptions are necessary, and differences between individual cells can be accommodated, if such effects are known in advance or if specific effects are to be investigated. For example,

(i) the fan air flow will not be distributed uniformly across all tubes but the prediction of the nonuniform distribution will require significantly more detailed modelling or extensive field measurements; it will also require division of the tube banks into a larger number of smaller cells to match to the nonuniform air flow;
(ii) the inlet air temperature to individual fans will generally be close to the common ambient but will be influenced by air flow patterns underneath the fan banks and individual fans may experience significantly different temperatures;

this is well illustrated by van Staden et al. [103] who have investigated the influence onwind strength and direction on the recirculation of hot air from fan discharge back to the fan inlet.

15.2.1 Heat Transfer

Consider a single cell consisting of a group of tubes associated with a single fan. Treating this as a cross-flow heat exchanger, we can relate inlet and outlet conditions on both air and steam sides via the *LMTD* heat transfer Equation 15.1.

$$\dot{Q} = U \cdot A \cdot \Delta T_{LM}. \tag{15.38}$$

With steam conditions defined by the saturation temperature T_s',

$$\Delta T_{LM} = \frac{T_{a2} - T_{a1}}{\ln\left(\frac{T_s' - T_{a1}}{T_s' - T_{a2}}\right)},$$

where T_{a1} is the inlet and T_{a2} the outlet airflow temperature.

For an ACC consisting of several cells of tube banks, the temperature of the airflow leaving the j-th bank may be written as

$$T_{ax,j} = T_{ai,j} + (T_{s,j} - T_{ai,j})\left[1 - \exp\left(\frac{U_j A_j}{\dot{m}_{air,j}\, cp_{air}}\right)\right], \tag{15.39}$$

where U_j is the heat transfer coefficient of the j-th cell and A_j is its heat transfer area exposed to the airflow. The air temperature differential across the j-th bank $\Delta T_{a,j}$ is

$$\Delta T_{a,j} = (T_{s,j} - T_{ai,j})\left[1 - \exp\left(\frac{U\, A}{\dot{m}_{air,j}\, cp_{air}}\right)\right].$$

The heat removed from the j-th cell is

$$\dot{q}_{hxa,j} = \dot{m}_{air,j}\, cp_{air}\, \Delta T_{a,j}$$

and the total heat removed from all cells is

$$\dot{Q}_{hxa} = \sum_{j}^{all\ cells} \dot{q}_{hxa,j}.$$

The air mass flow from the j-th fan is a function of fan speed. Fans are usually operated on or off, with a transient variation of flow during fan run-up and shutdown. Some fans may have two selectable speeds to provide a more precise regulation of condenser conditions.

The heat transfer coefficient U may be estimated from the physical dimensions and fin arrangements of the tube bank and airflow from the fan. Calculation of heat transfer coefficients for finned tubes is discussed in Sect. 8.2.8.

On the steam side of the cell, inside the several tubes comprising the tube bank, the heat removed by the airflow is provided by condensation of the steam. Assuming saturation conditions, this yields the condensation flow in a single cell as

$$\dot{m}_{cd} = \frac{\dot{q}_{hxa}}{h'' - h'}, \tag{15.40}$$

where h'' and h' are the saturation enthalpies of the steam and water phases, respectively, at the pressure within the tubes. The pressures within the several cells will adjust to ensure that the total condensation flow in all cells equals the total of all steam flows entering the condensing steam spaces. Some variation of pressures in individual cells will be observed, reflecting their differing heat flows. The steam temperature used in Eq. 15.39 is the saturation temperature corresponding to the local cell pressure.

15.2.2 Calculation of Cell Pressures

Figure 15.3 shows a typical connection of an ACC to the turbine exhaust. Unlike WCCs which are connected directly to the turbine exhaust flange, the ACC is located outside the turbine enclosure, some distance from the turbine exhaust. The turbine exhaust steam is ducted to the ACC through a large-diameter duct, up to some 50 m in length. At the ACC, steam is distributed to the individual parallel condensation zones and cells. The reduced steam flow past each zone is matched by a reduction in duct diameter.

At each cell, steam flows into the parallel tubes of each of the two banks associated with a fan. Condensate formed on the inside surface of the tubes flows down to collecting headers at the bottom of the tubes for removal to the condensate collection tank (CCT), located adjacent to the condensing cells. Drainage pumps remove to the CCT water which accumulates in the turbine exhaust duct. This is mainly dropout from the wet fraction of the turbine exhaust flow. Steam from other sources such as the condenser flash tank or feed pump turbine is introduced into the turbine exhaust duct. A small diameter pressure equalisation line connects the CCT to the turbine exhaust duct.

Figure 15.5 shows nine node pressures defined within the steam flow network. This number has been chosen to provide one node in each zone, one at both the common inlet and drainage headers and one in the CCT. An ACC with, say, nine zones would be represented by 12 nodes. Within a zone, all cells are assumed to see equal pressures. Condensation creates pressure differentials throughout the

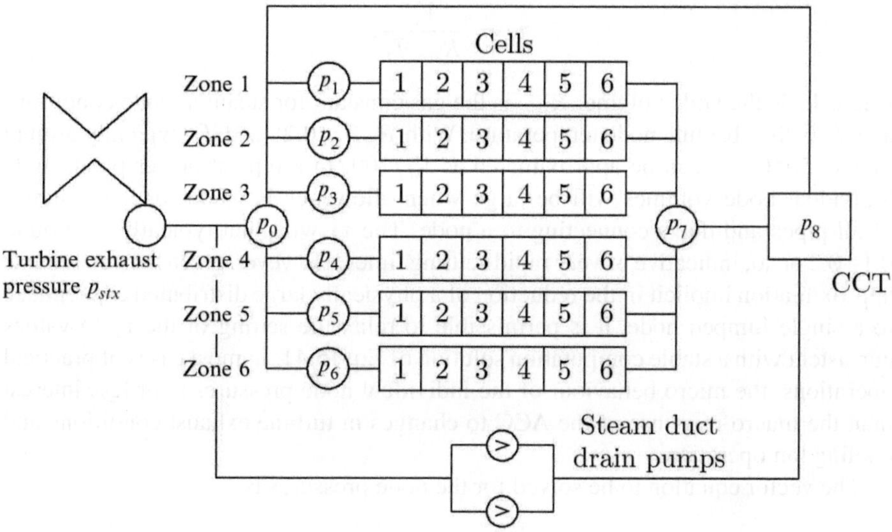

Fig. 15.5 ACC steam flow network

ACC which, though small, are sufficient to create substantial steam flows to each zone and cell. Condensate accumulating in the CCT is removed by the condensation extraction pumps.

As all pressure differentials are small and all pressures low—generally less than atmospheric, given the vacuum conditions within the condenser—we may assume flow linearly dependent on Δp rather than the more correct square root relationship.

The pressure at each node is determined from a simple mass balance. For the j-th node,

$$\tau_j \frac{\mathrm{d}p_j}{\mathrm{d}t} = \sum_i k_{ij}(p_i - p_j) - \sum_j \dot{m}_{cd}, \tag{15.41}$$

where p_i are the pressures in the connecting nodes and $\sum_j \dot{m}_{cd}$ is the total condensation flow rate in all cells of the j-th zone. The coefficients k_{ij} are flow coefficients selected to match node pressures and flows to the known steam turbine exit pressure and flow. Because the flow admittances throughout the ACC are extremely large, it is not necessary that the individual k coefficients be accurately calculated. A first approximation can be obtained by applying the conventional pressure loss calculation of Eq. 6.7, taking note of the large-diameter pipes and the low Reynolds numbers involved.

Expanded for all nodes, this equation generates for this ACC a 9×9 matrix equation to be solved for the individual node pressures. As these include the CCT pressure, no separate equation is required for the pressure in the tank. The single boundary condition is the turbine exhaust pressure p_{stx}. The coefficients τ_j are time constants with unit seconds per kPa, calculated ideally as

$$\tau_j = \frac{V_j}{R_{stm}\mathcal{T}_j},$$

where V_j is the node volume, R_{stm} is the gas constant for steam at node conditions and $\mathcal{T}_j$ is the absolute node temperature. With $R_{stm} = 0.28$ and $\mathcal{T}_j$, typically around 320 K (50 °C) τ_j can be approximated as $V_j/100$. In a typical power plant, ACC individual node volumes will be large when allowance is made for the volumes of all pipes and ducts connecting to a node. The τ_j will then typically be around 0.1–0.2 or so, indicative of very rapid settling times. However, given the very coarse approximation implicit in the reduction of a physically large distributed capacitance to a single lumped node, it is permissible to relax the setting of the τ_j to values consistent with a stable computation solution of Eq. 15.41. In most cases of practical operations, the micro behaviour of the individual node pressures is of less interest than the macro-response of the ACC to changes in turbine exhaust conditions and cooling fan operation.

The vector equation to be solved for the node pressures is

$$\vec{\tau}\frac{\mathrm{d}\mathbf{p}}{\mathrm{d}t} = \mathbf{K}\,\mathbf{p} + \dot{\mathbf{m}}_{cd},$$

where $\vec{\tau}$ indicates a vector of the individual values of τ_j. Individual condensation flows are given by the air-side thermal balance, Eq. 15.40. This equation may be solved by any of the methods described previously. The coefficient matrix $\mathbf{K}$ is symmetric and positive definite which allows the use of Cholesky decomposition if a time discretisation method is employed.

Chapter 16
Deaerators and Feedwater Heaters

Deaerators and feedwater heaters are important components of any power utility boiler condensate and feedwater supply system. Deaerators are found in all industrial plants using fired boilers for process steam production. Feedwater heaters of the type described here are usually found only in power generating plants using steam turbines as they rely on steam bled from the turbine as their heat source. While fundamentally different in purpose and design, these two components share the principle of steam condensation as the underlying method of heat transfer. In the deaerator, this is implemented as direct contact spray condensation while the feedwater heater utilises steam condensation on the outer surface of water-bearing tubes.

16.1 Deaerators

Water going to boilers operating at elevated pressures must be free of dissolved gases, particularly oxygen. The presence of dissolved oxygen in concentrations exceeding some eight parts per billion (ppb) in boiler feedwater can cause serious corrosion and lead to tube failures. The initial removal of dissolved gases from the boiler feedwater is the primary task of the deaerator. Final reduction of the oxygen content to allowable levels is the task of chemical water treatment.

In power plant applications the deaerator is a physically large piece of equipment consisting of two main items, each a tank (Fig. 16.1). The smaller deaeration head tank is mounted directly on top of the larger water collection tank, sometimes called the feedwater tank. This tank serves the important task of ensuring a short-term supply of feedwater to the boiler should the condensate feed from the condenser be lost, and is sized accordingly. The incoming condensate stream enters the upper tank where it is decomposed into a spray of droplets by one or more spray heads. The droplets fall under gravity onto a series of horizontal racks or perforated trays, being heated along the way by condensation of steam and releasing dissolved gases.

P. O'Kelly, *Computer Simulation of Thermal Plant Operations*,
DOI 10.1007/978-1-4614-4256-1_16,

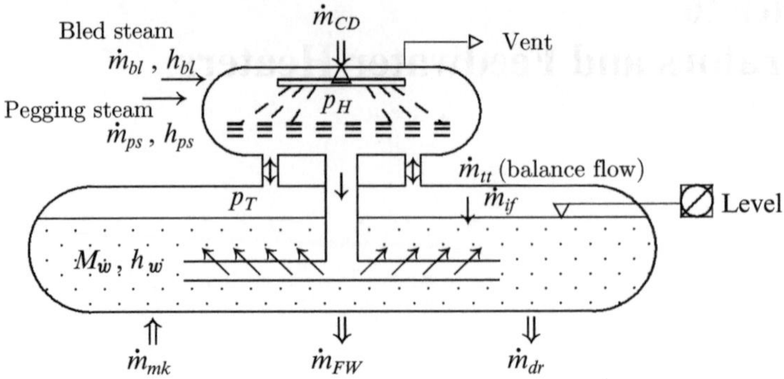

Fig. 16.1 Deaerator general arrangement

On the trays they are further mixed with steam, in either cross- or counter-flow depending on the particular design, where the mechanical separation of dissolved gases is completed. Having passed through the tray zone, the droplets, emerging as a collection of individual mini-streams, form a common stream which flows under gravity through interconnecting pipes to the water collection tank. This flow is distributed across the collection tank below the water level.

Dissolved gases are removed by the tank vents. A small amount of steam is also removed through the vents and is recovered in the vent condenser.

Water in the collection tank maintains a bulk temperature defined by the temperature of the incoming condensate. The water surface, being in contact with steam, holds a higher temperature. The interfacial differential temperature produces a steady condensation flow across the interface. The collection tank water can be brought to its initial operating temperature by either electrical heating or an auxiliary steam supply, where provided.

For a deaerator in a power generation unit, the heat transferred to the droplets in the head tank is supplied by steam bled from the turbine. In lower-pressure units this may be taken after the HP turbine exhaust (cold reheat) or in higher-pressure units after the IP turbine exhaust. During start-up, steam will not be available from this source and must be supplied from an auxiliary steam supply. This is sometimes referred to as pegging steam as it is used to "peg" the deaerator pressure or temperature at some reference value, usually around 110 kPa or 120 °C, until supply can be changed to the turbine. Independent of the source of supply, steam introduced into the deaerator head tank is mixed vigorously with the droplet cloud. It provides the energy to raise the incoming condensate to the head tank saturation temperature and the mass to replenish the steam lost by condensation and through the head tank vents.

The level of water in the collection tank is regulated by controlling the incoming condensate flow. Excursions of level beyond upper and lower limits initiate dump (> an upper limit) to the condenser hotwell or other storage (the drain flow $\dot{m}_{dr}$) or makeup (< a lower limit) from demineralised storage (makeup flow $\dot{m}_{mk}$).

Due to the very high rates of heat and mass transfer, most steam condensation and desorption of gases take place in the droplet zone and can be considered complete once the droplets have reached the trays. The hold-up of liquid on the trays will be ignored, and for analysis purposes, arrival of the droplet stream at the trays will be treated as equivalent to their arrival in the water collection tank.

In the steady state the balance flow between the two steam spaces will ensure that both tanks have almost equal pressures. The condensate flow to the collection tank is driven by the gravitational head difference between the bottom of the head tank and the collection tank water level. Under transient conditions the two tank pressures will diverge, in either direction, causing changes in the local saturation temperatures in both tanks. However these transient differences are of short duration.

16.1.1 Derivation of the Describing Equations

The derivation of the describing equations is based on the following assumptions:

1. Wet saturated steam occupies both the head tank free space and the water collection tank space above the water level.
2. All condensate collecting in the head tank flows without delay into the collection tank.
3. The two tanks exchange a steam flow $\dot{m}_{tt}$ proportional to the instantaneous pressure differential between them. This pressure differential will always be very small, allowing the use of a linear rather than the more correct square root relationship. Then,

$$\dot{m}_{tt} = k_{tt}(p_H - p_T). \tag{16.1}$$

4. The interface flow across the water/steam interface in the collection tank is assumed proportional to the difference in temperature between the water surface and steam temperatures.

$$\dot{m}_{if} = k_{if} A_{if} (T_w - T_s), \tag{16.2}$$

   where A_{if} is the water surface area in contact with the steam phase. k_{if} is a constant selected to match actual plant behaviour. Positive flow indicates evaporation into the steam space; negative flow indicates condensation to the water mass.
5. The mean drop size r_0 is a parameter of the spray nozzles and is known from design data. It will typically be around 2.5 mm. δt is the mean fall time and will generally lie in the range 1–2 s.
6. The condensate flow from the head tank includes the condensate flow entering the head tank plus the steam condensed on the droplets.

Head Tank

Steam Mass Balance—Pressure The rate of change of the mass of steam M_s in the head tank is given by a simple sum of in- and outflows.

$$\frac{\mathrm{d}M_s}{\mathrm{d}t} = \dot{m}_{bl} + \dot{m}_{ps} - \dot{m}_{iCD} - \dot{m}_{tt}. \tag{16.3}$$

We can also write

$$\frac{\mathrm{d}M_s}{\mathrm{d}t} = -\frac{V_{sH}}{v_s^2}\frac{\partial v_s}{\partial p}\frac{\mathrm{d}p_H}{\mathrm{d}t}.$$

Noting from Eq. 4.34 that

$$\frac{\partial v_s}{\partial p} = -\frac{v_s}{\gamma p}$$

this can be written as

$$\frac{\mathrm{d}M_s}{\mathrm{d}t} = \frac{V_{sH}}{\gamma\, v_s p_H}\frac{\mathrm{d}p_H}{\mathrm{d}t} = \frac{V_{sH}}{\gamma R_s \mathcal{T}_{sH}}\frac{\mathrm{d}p_H}{\mathrm{d}t},$$

where $\mathcal{T}_{sH}$ is the temperature of the steam in degrees Kelvin and R_s is the gas constant for saturated steam (0.4619 kJ/kg/K). The adiabatic index γ can be assumed 1.3 for saturated steam.

Defining $\tau_H = V_{sH}/(\gamma R_s \mathcal{T}_{sH})$, the head tank pressure equation may be written as

$$\tau_H \frac{\mathrm{d}p_H}{\mathrm{d}t} = \dot{m}_{bl} + \dot{m}_{ps} - \dot{m}_{iCD} - \dot{m}_{tt}. \tag{16.4}$$

After substitution for $\dot{m}_{tt}$ from Eq. 16.1, this becomes

$$\tau_H \frac{\mathrm{d}p_H}{\mathrm{d}t} = \dot{m}_{bl} + \dot{m}_{ps} - \dot{m}_{iCD} - k_{tt}\, p_H + k_{tt}\, p_T. \tag{16.5}$$

Under typical operating conditions, $p_H = 400\,\mathrm{kPa}$, $\mathcal{T}_H = 417\,\mathrm{K}$, $v_s = 0.4579\,\mathrm{m^3/kg}$ and τ_H is around 0.004 V_{sH} (kg/s) per (kPa/s). This indicates that head tank pressure will change rapidly in response to a mass flow imbalance. However, since pressure changes imply changes in saturation temperature and therefore condensation rates, these will be damped by the effects of condensation coupling.

Steam Energy Balance—Enthalpy The energy balance for the steam in the head tank is described by

$$\begin{aligned}\frac{\mathrm{d}}{\mathrm{d}t}(M_s h_s) &= M_s \frac{\mathrm{d}h_s}{\mathrm{d}t} + h_s \frac{\mathrm{d}M_s}{\mathrm{d}t}\\ &= \dot{m}_{bl} h_{bl} + \dot{m}_{ps} h_{ps} - \dot{m}_{iCD} h_s - \dot{m}_{tt} h_s + V_{sH}\frac{p_H}{dt} - \dot{q}_{amb},\end{aligned}$$

where $\dot{q}_{amb}$ is the rate of heat loss to ambient.

From Eq. 16.3, we have

$$h_s \frac{\mathrm{d}M_s}{\mathrm{d}t} = h_s(\dot{m}_{bl} + \dot{m}_{ps} - \dot{m}_{iCD} - \dot{m}_{tt})$$

from which follows

$$\begin{aligned} M_s \frac{\mathrm{d}h_s}{\mathrm{d}t} &= -h_s(\dot{m}_{bl} + \dot{m}_{ps} - \dot{m}_{iCD} - \dot{m}_{tt}) \\ &\quad + \dot{m}_{bl}h_{bl} + \dot{m}_{ps}h_{ps} - \dot{m}_{iCD}h_s - \dot{m}_{tt}h_s + V_{sH}\frac{\mathrm{d}p_H}{\mathrm{d}t} - \dot{q}_{amb} \\ &= h_s(\dot{m}_{bl} + \dot{m}_{ps} - \dot{m}_{iCD}) \\ &\quad + \dot{m}_{bl}h_{bl} + \dot{m}_{ps}h_{ps} + V_{sH}\frac{\mathrm{d}p_H}{\mathrm{d}t} - \dot{q}_{amb}. \end{aligned} \tag{16.6}$$

The compression energy term $V_{sH}\,\mathrm{d}p_H/\mathrm{d}t$ can be ignored as it is small in comparison with the other contributions.

The ambient heat loss makes no contribution to the short-term dynamic behaviour of the deaerator but is retained for its influence on longer-term cooling after shutdown.

Note that this calculation, unlike that used for the condensers and feedwater heaters, makes no assumption of saturation conditions within the head tank. The internal rate of condensation is determined from more fundamental considerations than simply the need to preserve saturation. The model is then able to handle large and rapid transients which may cause short-term departures from ideal saturation conditions.

Water Collection Tank

Steam Mass Balance—Pressure The rate of change of steam pressure in the collection tank is determined by the balance between the steam exchange between the two tanks and the water surface interface flow. The rate of change of the mass of steam M_{sT} in the collection tank is given from

$$\frac{\mathrm{d}M_{sT}}{\mathrm{d}t} = \dot{m}_{tt} + \dot{m}_{if}.$$

As done above for the head tank, introduce $\tau_{sT} = V_{sT}/(R_s T_{sT})$ and replace $\dot{m}_{tt}$ from Eq. 16.1. Then

$$\tau_{sT}\frac{\mathrm{d}p_T}{\mathrm{d}t} = -k_{tt}p_T + k_{tt}p_H + \dot{m}_{if}. \tag{16.7}$$

Unlike the head tank, the volume of steam in the collection tank is not constant and varies with water level.

Water Mass Balance The rate of change of water mass in the collection tank is given from a simple mass balance of flows in and out.

$$\frac{\mathrm{d}M_w}{\mathrm{d}t} = \dot{m}_{CD} + \dot{m}_{iCD} + \dot{m}_{mk} - \dot{m}_{FW} - \dot{m}_{dr} - \dot{m}_{if}. \quad (16.8)$$

Water Energy Balance—Enthalpy From the energy balance of the collection tank water,

$$\begin{aligned}\frac{\mathrm{d}}{\mathrm{d}t}(M_w h_w) &= M_w \frac{\mathrm{d}h_w}{\mathrm{d}t} + h_w \frac{M_w}{\mathrm{d}t} \\ &= \dot{m}_{CD} h_{CD} + \dot{m}_{iCD}(h_s - h') + \dot{m}_{mk} h_{mk} - (\dot{m}_{FW} - \dot{m}_{dr}) h_w.\end{aligned}$$

The equation for the rate of change of the enthalpy of the water then follows as

$$M_w \frac{\mathrm{d}h_w}{\mathrm{d}t} = \dot{m}_{CD}(h_{CD} - h_w) + \dot{m}_{iCD}(h_s - h' - h_w) + \dot{m}_{mk}(h_{mk} - h_w). \quad (16.9)$$

The following form is better suited to computation.

$$\left(\frac{M_w}{\Sigma(\dot{m})_w}\right)\frac{\mathrm{d}h_w}{\mathrm{d}t} + h_w = \Sigma(\dot{m}h)_w \qquad |\Sigma(\dot{m})_w| > 0, \quad (16.10)$$

where

$$\Sigma(\dot{m})_w = \dot{m}_{CD} + \dot{m}_{iCD} + \dot{m}_{mk} - \dot{m}_{if}$$

and

$$\Sigma(\dot{m}h)_w = \dot{m}_{CD} h_{CD} + \dot{m}_{iCD}(h_s - h') + \dot{m}_{mk} h_{mk} - \dot{m}_{if}(h'' - h').$$

It can normally be assumed that the enthalpy of the water in the collection tank is less than the saturation enthalpy h' for the tank pressure p_T. This implies the absence of any steam phase in the water. However, should p_T decrease below the pressure for which the water enthalpy h_w exceeds saturation enthalpy, a steam phase will develop immediately within the water. The resulting steam mass fraction x for the tank water is given by

$$x = \frac{h_w - h'}{h'' - h'}, \quad (16.11)$$

where the saturation properties h'' and h' correspond to the tank pressure p_T. The mean water specific volume v_{ws} becomes

$$v_{ws} = x v'' + (1 - x) v'$$

and the water volume will change from $M_w v_w$ to $M_w v_{ws}$. Since $v_{ws} > v_w$, the volume will increase and the water level will rise sharply. As this change in fluid specific volume will be experienced by the column of water in the downcomer, part of this water may convert to steam. The pressure at the feedwater pump inlet will decrease with the following consequences:

(a) The weight of the column will decrease and cause a reduction in the pump's NPSH margin.
(b) The two-phase friction drop will increase, causing a decrease in flow rate to the pump(s).

The aggregated effect of these changes could result in the trip of the feedpump(s) by cavitation protection.

16.1.2 Calculation of the Internal Rate of Condensation

The following development treats the deaerator as a direct contact spray condenser. The introduction of a fine cloud of steam droplets into a steam-filled space promotes intensive heat and mass transfer from the steam to the droplets. This is a result of the very large interfacial surface area exposed by the droplets, the high contact velocity between steam and droplets and the high degree of turbulence in the region of droplet formation.

Following Marto (Rohsenow et al. [44], Chap. 14), we may assume that the droplets are spherical and that the rate of condensation of steam onto the drop's surface is driven by the rate of heat diffusion into the drop. Interfacial heat transfer resistance is assumed negligible, allowing the assumption that the drop surface temperature reaches that of the steam instantaneously. The bulk temperature of the drop then approaches its surface temperature as steam transfers heat to the drop, cooling the steam and heating the drop. Both being water, the condensed steam forms a film on the surface of the drop whose radius increases as it accretes condensate. Butterworth and Hewitt [63] quote Brown [64] as the source of the equation which describes the rate at which the drop bulk temperature approaches its surface temperature.

$$\frac{T_{l1} - T_{l2}}{T_g - T_{l1}} = 1 - \frac{6}{\pi^2} \sum_{n=1}^{\infty} \frac{1}{n^2} \exp\left(-\frac{n^2 \pi^2 D_T\, t}{r_0^2}\right),$$

where T_{l2} is the temperature of the drop at time t, T_{l1} is the initial drop temperature and T_g is the temperature of the surrounding steam. D_T is the thermal diffusivity of the drop

$$D_T = \frac{\lambda}{c_{pl}\rho_l} \quad \text{m}^2\text{/s}$$

λ is the coefficient of thermal conductivity, c_{pl} is the specific heat capacity of the drop fluid and ρ_l is its density.

The thermal diffusivity is a function of temperature, as summarised by the following table. The second or higher-order dependencies of properties on pressure have been ignored. All properties are quoted for the saturated liquid state.

T deg C	λ W/m/K	C_{pl} kJ/kg/K	ρ kg/m^3	D_T m^{2}s$^{-1}\cdot 10^{-7}$
0	0.558	4.226	999.8	1.321
50	0.647	4.178	988.04	1.567
100	0.690	4.191	958.31	1.718
150	0.692	4.215	916.93	1.791
200	0.672	4.262	864.68	1.823
250	0.636	4.343	799.23	1.832
300	0.542	4.483	712.45	1.697
350	0.456	4.790	574.38	1.657

The tabulated values can be closely approximated as a simple function of temperature. With $\bar{T} = T/100$

$$D_T = (1.3388 + 0.46812\bar{T} - 0.10738\bar{T}^2) \cdot 10^{-7}. \tag{16.12}$$

The previous reference (Marto, in Rohsenow et al. [44], Chap. 14) quotes Jacobs and Cook [102] as the source of an equation describing the rate at which the drop radius increases with time as it falls through the condensing steam.

$$\frac{r_d(t)}{r_0} = \left\{1 + Ja\left[1 - \frac{6}{\pi^2}\sum_1^\infty \frac{1}{n^2}\exp(-n^2\pi^2\tau t)\right]\right\}^{\frac{1}{3}}, \tag{16.13}$$

where τ is the Fourier Number

$$\tau = \frac{\alpha_l}{r_0^2}. \tag{16.14}$$

Ja is the Jacobs Number given as

$$Ja = \frac{v'_{cond}c_{pl}(T_s - T_i)}{v'_{drop}(h'' - h')}, \tag{16.15}$$

where
- v'_{cond} is the specific volume of the condensate film
- v'_{drop} is the specific volume of the droplet
- c_{pl} is the specific heat capacity of the condensate film
- T_s is the steam temperature
- T_i is the temperature of the spray flow

For most purposes, it may be assumed that $v'_{cond} = v'_{drop}$.

Substitution of τ from Eq. 16.14 in Eq. 16.13 gives the mean radius of the drops reaching the trays. Computation experience shows that only three or four terms of the summation need be retained as the exponent decreases rapidly with increasing n. With $r_d(\delta t)$ the drop radius after δt seconds, the mass of a single droplet after its fall is $(4/3)\pi\, r_d(\delta t)^3 \rho$ and the change in mass over the fall time δt is

$$\delta M_{CD} = \frac{4}{3}\pi \left(r_d(\delta t)^3 - r_0^3 \right) \rho'.$$

Given the assumed mean drop radius r_0, the volume of the drop is $V_{dr} = (4/3)\pi r_0^3$ and its mass is $M_{dr} = V_{dr}/v'$. The number of drops formed per second can be estimated as $N_{dr} = \dot{m}_{CD}/M_{dr}$.

The condensation flow rate per drop is then $M_{iCD}/\delta t$ and the total condensation rate for all droplets is

$$\dot{m}_{iCD} = N_{dr} \frac{\delta M_{iCD}}{\delta t}. \tag{16.16}$$

16.1.3 Organisation of the Equations for Solution

The four state variables which describe the dynamics of the deaerator are p_H, p_T, h_s and h_w. Of these, h_w varies slowly compared to the other three and can be calculated independently using Eq. 16.10.

Define the state vector $\mathbf{\Psi} = \mathtt{col}\{p_H, p_T, h_s\}$ and the vector of forcing functions $\mathbf{F}$. Then,

$$\frac{\mathrm{d}}{\mathrm{d}t}\mathbf{\Psi} = \mathbf{A}\,\mathbf{F}, \tag{16.17}$$

where

$$A(1,1) = -k_{tt}/\tau_H \qquad A(1,2) = k_{tt}/\tau_H \qquad A(1,3) = 0$$

$$A(2,1) = k_{tt}/\tau_H \qquad A(2,2) = -k_{tt}/\tau_H \qquad A(2,3) = 0$$

$$A(3,1) = -v_s k_{tt}/\tau_H \quad A(3,2) = v_s k_{tt}/\tau_H \quad A(3,3) = -\Sigma(\dot{m})_s/M_s$$

and

$$F(1) = (\Sigma(\dot{m})_s - \dot{m}_{iCD})/\tau_H$$

$$F(2) = \dot{m}_{if}/\tau_T$$

$$F(3) = \Sigma(\dot{m}h)_s + \tfrac{v_s}{\tau_H}(\Sigma(\dot{m}_s - \dot{m}_{iCD})/M_s$$

Given the input and boundary conditions defined by the initial state of the deaerator and the interfacing variables from the external systems, this equation can be solved using any suitable method.

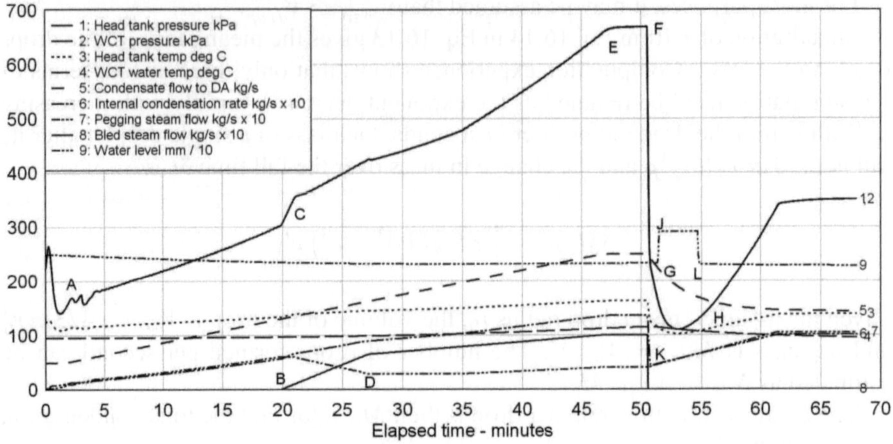

Fig. 16.2 Deaerator trends: plant start and turbine trip

16.1.4 Model Performance

The performance of the deaerator model is demonstrated by the simulation of two practically significant phases of plant operation. The evolution of the plant variables is shown by Fig. 16.2.

The procedure follows one aspect of a simplified start-up of a steam turbo-generating unit. It starts (point A on the figure) with the deaerator head tank temperature initially set to 100 °C and then regulated to around 112 °C by automatic control of the pegging steam. The variations in pressure reflect the manual shutdown of support steam as the automatic control of pegging steam took over. The condensate system is in recirculation operation with a condensate flow of 50 kg/s and an incoming temperature of around 85 °C. The water level is automatically controlled around 2,300 mm.

Start-up of the unit is simulated by ramping the condensate flow linearly to a full-load flow of 250 kg/s. The set-point temperature for the deaerator head tank temperature is ramped at the same rate to a temperature giving a head tank pressure of 670 kPa at full load. At a condensate flow of 130 kg/s, the bled steam line to the turbine is opened (point B), and the pegging steam valve closes automatically. The transient excess steam flow causes a sharp rise in pressure (point C) which is corrected as the pegging steam valve shuts (point D). The start continues to full load (point E).

The trend curves show the internal condensation $\dot{m}_{iCD}$ and bled steam flows increasing in line with the condensate flow. The head tank and WCT pressures increase with the increasing bled steam pressure in the turbine. The pegging steam valve, which in practice is held slightly open to keep the valve and piping warm and had been closed to its minimum position, opens further at higher loads to maintain set-point temperature.

At point F, the turbine is tripped. The bled steam valve shuts abruptly, removing the support steam flow to the deaerator. The incoming condensate flow decreases sharply, and the condensate temperature, now deprived of the LP feedwater heaters, decreases rapidly towards the condenser hotwell temperature (around 55 °C). Given the lag in opening of the pegging steam valve, the deaerator pressure drops rapidly as the steam is quenched by the rapidly cooling condensate (point G). The WCT water temperature starts decreasing towards the temperature of the incoming condensate (point H).

The WCT pressure reduces below the saturation pressure of the water which rapidly develops a steam fraction, causing a sharp rise in water level (point J). As the pegging steam valve opens (point K), it re-establishes support steam to the deaerator head tank, and normal pressure and temperature conditions are restored. The steam phase in the WCT water collapses when the pressure again exceeds the local saturation pressure and the water level returns to normal (point L).

If this were to happen in a real plant, it can be confidently expected that the feedwater pumps would be tripped by "low NPSH" protection.

16.2 Feedwater Heaters

In its Technical Data Sheet #127, the Heat Exchange Institute describes a feedwater heater as

> a heat exchanger designed to preheat boiler feedwater by means of condensing steam extracted ("bled") from a steam turbine.

The purpose of the feedwater heater is to improve the overall efficiency of the generating unit. A power generating unit usually has several feedwater heaters in the condensate and feedwater systems. They are classified as high, intermediate or low pressure, depending on the normal operating pressure. Heaters in the condensate system—between the condenser hotwell and the deaerator—have a low operating pressure and are termed *low-pressure* heaters. Heaters located after the feedwater pumps have higher operating pressures and may be termed *high-* or *intermediate*-pressure heaters. Plants provided with a feedwater booster pump may have one or more feedwater heaters between the booster and main pumps. These will be connected to the turbine intermediate pressure cylinder and are the intermediate pressure heaters. Those after the main pump will be connected to the high-pressure turbine or its exhaust and are high-pressure heaters.

Most feedwater heaters follow the same construction principles. A large number of U-shaped tubes are individually welded into a thick baseplate—the tube sheet. Together with the tube, sheet they form a single unit which can be inserted into and withdrawn from a cylindrical shell. At the feedwater inlet, the cavity facing the open tube ends is divided by a curved plate into an inlet and an outlet plenum. Feedwater enters the inlet plenum and is distributed into the tubes. At the outlet of the tubes, the water is guided through the outlet plenum to the discharge nozzle or stub.

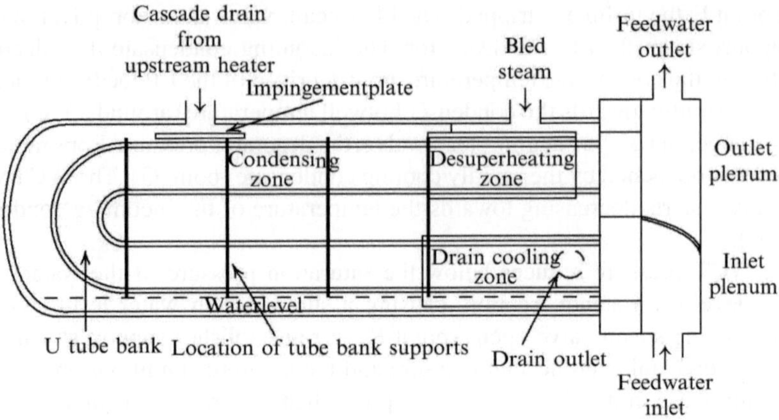

Fig. 16.3 Feedwater heater horizontal arrangement

The inlet and outlet plenums and tube sheet form the closure at the water end. The shell is closed at the other end by a rounded dome. All tubes are manufactured from the same material (for HP heaters usually a carbon-molybdenum high-temperature steel) and have the same internal diameter and wall thickness. The number of tubes in the bank is ultimately limited by the minimum allowable bend radius. Within the bank, the individual tubes are supported by a series of spacers and brackets and are restrained against movement and vibration. The tubes are protected from damage by the high-speed incoming steam flow by impingement plates against which the steam flow is initially directed.

Two mounting arrangements are usual. Figure 16.3 illustrates a horizontally mounted heater. This is the most common form for all pressure types but some high-pressure heaters may be mounted with vertical tubing in order to optimise floor space utilisation. In the horizontal heater, condensate forms a film on the horizontal tubes and drops through the tube bank to the bottom of the shell. The lower tubes will be affected by condensate dropping from higher tubes, but given the turbulence of steam flows within the shell and between the tubes, this is less of a concern than in, say, a large condenser. With vertical tubes, the condensate forms a film on the individual tubes which flows down the tubes to collect on the tube sheet. A second but rarely found vertical type uses a reverse flow of condensate, from bottom to top. In this arrangement, the upflowing condensate will be in contact with some tubes over the complete flow path.

16.2.1 Heat Transfer Zones

A feedwater heater can be manufactured with up to three distinct functional zones. The bulk of heat transfer and condensation takes place in the *condensing zone* in which the tubes are in contact with steam at or near its saturation temperature. In condensing onto the tubes, the steam gives up its latent heat of condensation which is transferred through the tube wall to the feedwater flowing inside. Some heaters have a *drain-cooling* or sub-cooling zone in which the incoming water is initially heated by condensate on its way out of the heater. The difference between the temperatures of the incoming water and the departing drain condensate is the drain cooler approach (DCA) temperature. Heaters provided with superheated bled steam often include a *desuperheating* zone in which the steam superheat is transferred to the outgoing water flow before condensation commences. Despite the apparent benefit of the larger temperature differential, heat transfer from superheated steam is less effective than from saturated steam due in part to the poorer heat transfer coefficient and to the lower quantity of heat available for transfer. The final feedwater temperature leaving a heater with a desuperheating zone may be higher than the saturated steam temperature in the shell, giving rise to a negative terminal temperature difference (TTD).

The following figure depicts the idealised profiles of temperatures of steam, drain condensate and water flows through a feedwater heater equipped with three zones.

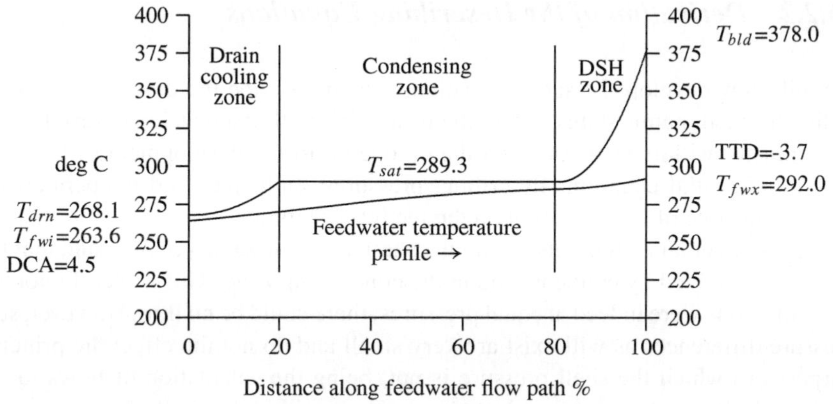

In order to maximise heat recovery through a chain of feedwater heaters, the condensate from the higher-pressure heaters is cascaded down through the chain. The flow is driven by the pressure differential between consecutive heaters and is regulated by a control valve which maintains a defined water level in its reference heater. Being at a temperature and pressure higher than the receiving heater, the cascading condensate will flash to steam and is therefore introduced into the condensing zone of the receiving heater. By condensing in the lower-pressure heater the condensate gives up its latent heat of condensation which is again captured by the feedwater. Only at the lowest pressure heater is the residual latent heat lost from the system through the drain.

With the bled steam valve(s) open, the operating pressure in the heater shell is defined by the pressure at the turbine bleed point. The bled steam flow will match the condensation rate in the heater. This is a self-regulating process. An increase in condensation rate will cause an initial decrease in shell pressure which, for a given bleed point pressure in the turbine, will result in an increase in bled steam flow. The lower shell pressure corresponds to a lower saturation temperature and a reduced heat transfer. The condensation rate will decrease and the pressure will again rise, reducing the bled steam flow to match the available condensation flow and the pressure/flow balances will be re-established. The distribution of heat load over the zones is illustrated by the following table. The data is taken from design data for a low- and a high-pressure heater installed in a large generating unit.

	LP heater		HP heater	
Zone	kJ/s	% of total	kJ/s	% of total
Desuperheating	–	–	5.3	11.9
Condensing	29.1	95.7	36.85	82.9
Drain cooling	1.31	4.33	2.3	5.2

16.2.2 Derivation of the Describing Equations

We will now develop the set of equations which describe the dynamic behaviour of the thermodynamic states within the heater. We will first develop descriptions of heat transfer within each zone and then move on to the development of the time-based differential equations describing pressures, enthalpies and temperatures in terms of the rate of condensation on the feedwater tubing.

It is assumed that all points within the shell are at the same pressure and that this pressure is defined by condensation in the condensing zone. This is clearly not true as, if all points were indeed at equal pressures, there could be no flow. However, such pressure differences as will exist are very small and do not the effect the principle purposes to which the shell pressure is put, being the calculation of flows to and from the heater and of steam and water properties within the shell.

The following diagram illustrates the principle of its connection within the feedwater circuit.

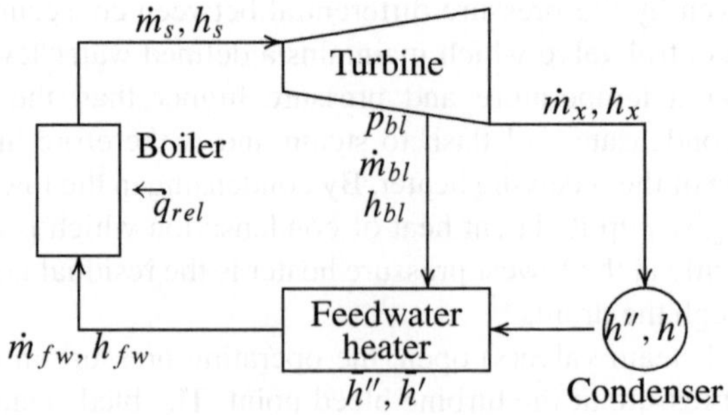

The diagram shows a steam turbine supplied from a boiler with steam flow $\dot{m}_s$ [kg/s] and enthalpy h_s [kJ/kg]. The turbine exhausts a flow $\dot{m}_x$ with exhaust enthalpy h_x to the condenser. The single bled steam point on the turbine supplies $\dot{m}_{bl}$ of steam with enthalpy h_{bl} to a feedwater heater. $\dot{m}_{fw}$ of feedwater arrives at the boiler inlet with enthalpy h_{fw}. The rate of heat release in the furnace $\dot{q}_{rel}$ is sufficient to evaporate the incoming feedwater and raise the steam enthalpy to h_s.

16.2.3 Heat Transfer in the Various Zones

The Desuperheating Zone Bled steam from the turbine enters the heater in the desuperheating zone. The arrangement of variables is shown by the following diagram.

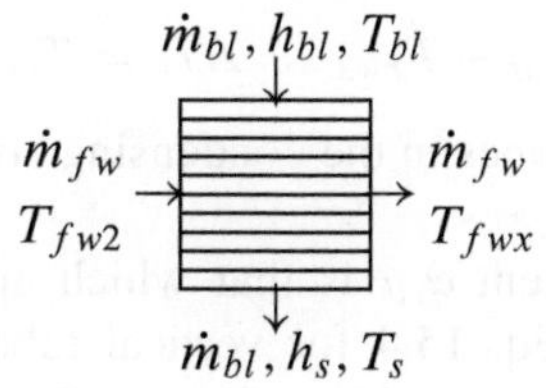

Feedwater enters the zone at temperature T_{fw2}, the temperature with which it leaves the condensation zone. Heat transferred is calculated from

$$\dot{q}_{ds} = A_{ds}\alpha_{ds}\,\mathtt{LMTD}_{ds}, \tag{16.18}$$

where

$$\mathtt{LMTD}_{ds} = \frac{\Delta T_1 - \Delta T_2}{\ln\left(\frac{\Delta T_1}{\Delta T_2}\right)}$$

and

$$\Delta T_1 = T_{bl} - T_{fwx} \qquad \Delta T_2 = T_s - T_{fw2}$$

T_s is the steam temperature leaving the zone after desuperheating. It is usually a few degrees above saturation temperature as, by design, not all the superheat is removed.

In the absence of condensation the heat transfer coefficient α_{ds} is the standard Hausen coefficient, Eq. 8.9. Should the bled steam have no degree of superheat, the case during start-up and normal for LP heaters, the desuperheating zone is treated as a condensing zone.

Condensing Zone Feedwater enters the zone from the drain-cooling zone with temperature T_{fw1} and leaves with T_{fw2}. Steam enters the zone with enthalpy close to or equal to saturation and begins to condense.

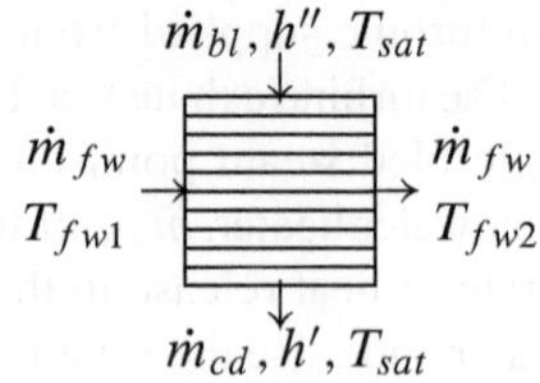

Heat transferred is calculated from

$$\dot{q}_{cd} = A_{cd}\alpha_{cd}\,\mathtt{LMTD}_{cd}, \tag{16.19}$$

where

$$\mathtt{LMTD}_{cd} = \frac{\Delta T_3 - \Delta T_4}{\ln\left(\frac{\Delta T_3}{\Delta T_4}\right)}$$

and

$$\Delta T_3 = T_{sat} - T_{fw2} \qquad \Delta T_4 = T_{sat} - T_{fw1},$$

where it is assumed that all tubes in the condensing zone are in contact with steam at saturation temperature.

The heat transfer coefficient α_{cd} is that which applies to film condensation, Eq. 15.2 for horizontal or Eq. 15.4 for vertical tubes. Since the functional dependence on the local process variables of each of these correlations is the same, Eq. 15.6 can be used, with the coefficient c_0 selected to match the known heater performance data.

Drain-Cooling Zone On the assumption that the tubes are completely covered by condensate, no condensation will occur in the zone. Should a low water level expose some tubes to steam, the steam will be at saturation temperature but, moving at a low velocity, will transfer little heat compared to those tubes in contact with water. Should the heater contain little or no condensate, there will be little or no heat transfer.[1]

Feedwater enters the zone from the external system with temperature T_{fwi} and leaves at T_{fw1}. Condensate enters from the condensate pool on the bottom of the heater shell (for a horizontal heater) or on top of the tube sheet (for a vertical heater) with saturation temperature and a flow rate equal to the drain flow from the heater.

Heat transferred is calculated from

$$\dot{q}_{dc} = A_{dc}\alpha_{dc}\,\mathtt{LMTD}_{dc}, \tag{16.20}$$

---

[1]The water level in a feedwater heater of either orientation is restricted to very low values for operational reasons of turbine protection which will isolate the bled steam and cascade inlet flows should a high level develop. This is not the case for the spray condenser (Fig. 11.4) which uses the partial covering of tubes by condensate to self-regulate the water level.

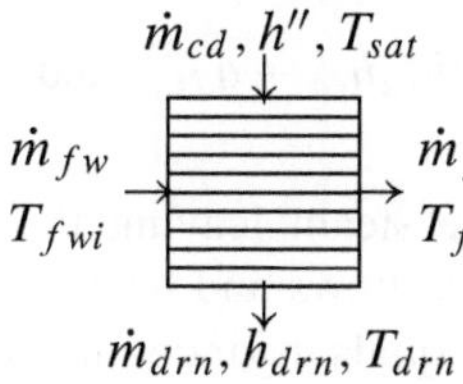

where

$$\mathtt{LMTD}_{dc} = \frac{\Delta T_5 - \Delta T_6}{\ln\left(\frac{\Delta T_5}{\Delta T_6}\right)}$$

and

$$\Delta T_5 = T_{sat} - T_{fw1} \qquad \Delta T_5 = T_{sat} - T_{fwi},$$

where it is assumed that all tubes in the zone are in contact with water at drain temperature.

The heat transfer coefficient α_{dc} is the usual Colburn-type coefficient for water/tube transfer, Eq. 8.5.

16.2.4 Heat Transfer Dynamics

In each of the zones, heat transfer is calculated as if it responds instantaneously to changes in temperature or flow. This is a consequence of the assumed homogeneity of the zone contents but is not the real behaviour. The rate of change of bulk properties will follow the spread of heat flux through the zone. In the absence of a more detailed multidimensional approach, we can assume an empirical first-order lag with which the calculated heat transfer follows its steady-state target value. Lag times lie typically in the range of several seconds.

16.2.5 Mass and Energy Balances

The Desuperheating Zone

Mass Balance This will always be satisfied as it is assumed that the bled steam will pass through without condensation, that is, no loss of mass.

Energy Balance Omitting the compression energy term from the general energy balance allows us to write

$$M_{ds}\frac{\mathrm{d}h_{ds}}{\mathrm{d}t} = \dot{m}_{bl}h_{bl} + \dot{m}_{ca}h_{ca} - \dot{q}_{ds} \quad \text{and} \quad M_{ds} = \frac{V_{ds}}{v_s} \tag{16.21}$$

from which the enthalpy of the steam leaving the desuperheating zone can be calculated directly, given the inlet flows and heat transfer. Alternatively, assuming that the heat transferred exactly equals the incoming superheat, steam will leave the desuperheating zone with vapour saturation enthalpy, giving

$$\dot{q}_{ds} = \dot{m}_{bl}(h_{bl} - h'') \tag{16.22}$$

The Condensing Zone

Mass Balance

$$\begin{aligned}
\frac{\mathrm{d}M_{cd}}{\mathrm{d}t} &= \frac{\mathrm{d}}{\mathrm{d}t}\left(\frac{V_{cd}}{v_s}\right) = V_{cd}\frac{\mathrm{d}}{\mathrm{d}t}\left(\frac{1}{v_s}\right) \quad \text{(assuming constant steam volume)} \\
&= -\left(\frac{V_{cd}}{v_s^2}\right)\frac{\mathrm{d}v_s}{\mathrm{d}t} \\
&= -\frac{V_{cd}}{v_s^2}\left(\frac{\partial v}{\partial p}\frac{\mathrm{d}p}{\mathrm{d}t} + \frac{\partial v}{\partial h}\frac{\mathrm{d}h}{\mathrm{d}t}\right) \\
&= \dot{m}_{bl} + \dot{m}_{ca} - \dot{m}_{cd} = \Sigma(\dot{m})_{cd}\,.
\end{aligned}$$

Energy Balance Treating the condensing zone as a closed system, the energy balance is defined by the flows of energy into and out of the zone. Energy enters the zone as bled and cascade steam flows. The condensation process removes energy from the steam space at rate equal to the condensation flow times the vapour saturation enthalpy. The latent heat of condensation is released back to the steam space and is removed by heat transfer to the tubes. The energy contained in the condensate liquid is accumulated in the condensate drain mass.

$$\begin{aligned}
\frac{\mathrm{d}}{\mathrm{d}t}(M_{cd}h_{cd}) &= M_{cd}\frac{\mathrm{d}h_{cd}}{\mathrm{d}t} + h_{cd}\frac{\mathrm{d}M_{cd}}{\mathrm{d}t} \\
&= \dot{m}_{bl}h_{bl} + \dot{m}_{ca}h_{ca} + V_{cd}\frac{\mathrm{d}p}{\mathrm{d}t} + \dot{m}_{cd}(h'' - h') - \dot{q}_{cd} \\
&= \Sigma(\dot{m}h)_{cd} + \dot{m}_{cd}(h'' - h') - \dot{q}_{cd}\,,
\end{aligned}$$

where $\Sigma(\dot{m}h)_{cd} = \dot{m}_{bl}h_{bl} + \dot{m}_{ca}h_{ca}$. Setting

$$\dot{m}_{cd} = \frac{\dot{q}_{cd}}{(h'' - h')} \quad \text{and} \quad M_{cd} = \frac{V_{cd}}{v_s} \tag{16.23}$$

the preceding equation becomes

$$\left(\frac{V_{cd}}{v_s}\right)\frac{\mathrm{d}h_{cd}}{\mathrm{d}t} - V_{cd}\frac{\mathrm{d}p}{\mathrm{d}t} = -h_{cd}\,\Sigma(\dot{m})_{cd} + \Sigma(\dot{m}h)_{cd}. \tag{16.24}$$

From the mass balance, we have

$$-\left(\frac{V_{cd}}{v_s^2}\right)\frac{\partial v}{\partial p}\frac{\mathrm{d}p}{\mathrm{d}t} - \left(\frac{V_{cd}}{v_s^2}\right)\frac{\partial v}{\partial h}\frac{\mathrm{d}h_{cd}}{\mathrm{d}t} = \Sigma(\dot{m})_{cd}. \tag{16.25}$$

Replace $\partial v/\partial p$ by $-v_s/\gamma p$. Then the above equation may be written as

$$\left(\frac{V_{cd}}{\gamma RT_s}\right)\frac{\mathrm{d}p}{\mathrm{d}t} - \left(\frac{V_{cd}}{v_s^2}\right)\frac{\partial v}{\partial h}\frac{\mathrm{d}h_{cd}}{\mathrm{d}t} = \Sigma(\dot{m})_{cd} \tag{16.26}$$

using the gas equation $pv = RT$.

Equations 16.24 and either 16.25 or 16.26 may be solved simultaneously for the condensing zone steam pressure and enthalpy, given the incoming flows and enthalpies and the zone heat transfer.

A simplification can be made, with little effect on the observed results, by ignoring the compression energy term $V\,\mathrm{d}p/\mathrm{d}t$ in Eq. 16.24. We can then write

$$\left(\frac{V_{cd}}{v_s}\right)\frac{\mathrm{d}h_{cd}}{\mathrm{d}t} = \dot{m}_{bl}(h'' - h_{cd}) + \dot{m}_{ca}(h_{ca} - h_{cd}), \tag{16.27}$$

which can be solved directly for h_{cd}. This formulation shows that the condensing steam zone enthalpy will approach a final value somewhere between the saturation enthalpy of vapour and the enthalpy of the incoming cascade flow.

We can substitute $\left(\frac{V_{cd}}{v_s}\right)\frac{\mathrm{d}h_{cd}}{\mathrm{d}t}$ from Eq. 16.27 into Eq. 16.26 to obtain an equation for $\mathrm{d}p/\mathrm{d}t$.

$$\left(V_{cd}\frac{\partial v}{\partial p}\right)\frac{\mathrm{d}p}{\mathrm{d}t} = \frac{\partial h}{\partial p}\left[\dot{m}_{bl}(h'' - h_{cd}) + \dot{m}_{ca}(h_{ca} - h_{cd})\right] + \Sigma(\dot{m})_{cd}, \tag{16.28}$$

which can be solved directly for the pressure in the shell.

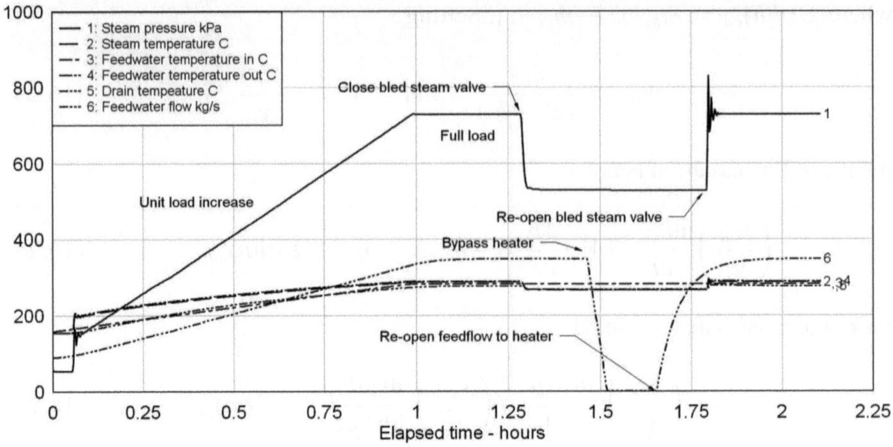

Fig. 16.4 Feedwater heater model: pressure and temperatures

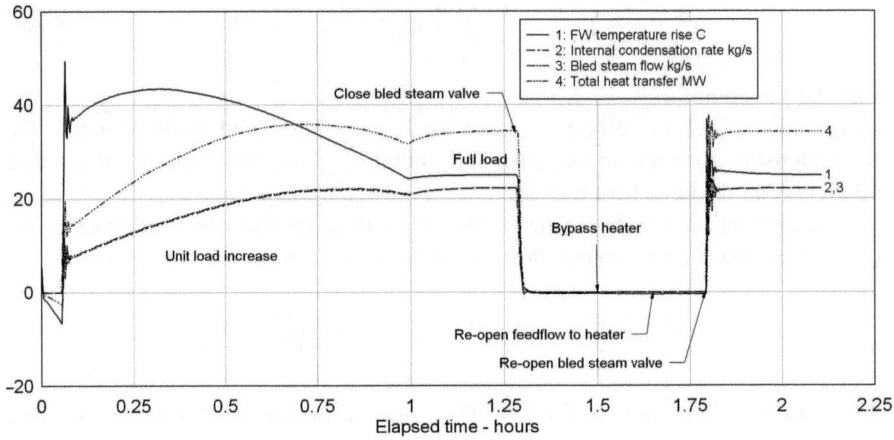

Fig. 16.5 Feedwater heater model: flows and heat transfer

The Drain-Cooling Zone

Mass Balance

$$\frac{\mathrm{d}M_{dr}}{\mathrm{d}t} = \dot{m}_{cd} - \dot{m}_{dr}. \tag{16.29}$$

Energy Balance

$$M_{dr}\frac{\mathrm{d}h_{dr}}{\mathrm{d}t} = \dot{m}_{cd}(h' - h_{dr}) - \dot{q}_{dc}. \tag{16.30}$$

16.2.6 Model Performance

Figures 16.4 and 16.5 show the performance of the feedwater heater model during a typical start of a power station generating unit. The procedure commences with the unit on around 90 kg/s steam and feedwater flow and ready for the first HP heater to be placed into service. Placing a feedwater heater into service normally follows a careful procedure of establishing water flow through the heater, setting up drains level controllers and *slowly* opening the bled steam valve. This minimises thermal shocks and avoids undue perturbations to the drains level controllers which can propagate to the downstream heaters. The simulation exercise is not inhibited by these practicalities, and the bled steam valve was opened abruptly to impose maximum strain on the numerical solution procedure, a vector formulation of two-step midpoint integration.

With the feedwater flow established and the bled steam valve open, the inputs to the heater—incoming feedwater temperature and flow, bled steam enthalpy and pressure—were calculated by drivers which mimic the expected behaviour of a real unit start-up. Loading took 60 min to full load, and during the loading phase, it can be seen that the bled steam and internal condensation flows remained close to equal. Conditions were allowed to settle for 10 min after which the bled steam valve was abruptly shut to simulate trip of the heater. Bled steam and with it condensation flows reduced rapidly to zero, as did the temperature differential across the heater (outlet temperature equalised with the input). Some 10 min later, the heater was bypassed and isolated (no feedwater flow through the heater). A further 10 min later, feedflow was re-established in preparation for the heater's return to service. This was completed after another 10 min by (again abruptly) opening the bled steam valve. During this procedure the unit load was maintained and no change was made to the unit's feedwater flow.

Chapter 17
Simulation of Complex Networks

The concept of a *network* has been introduced in Chap. 2 as consisting of

- Branches, a branch being a contiguous series of components through which the working fluid flows,
- Nodes, being points of connection of branches with each other,
- Linking stubs, being points of connection to external interfacing components or atmosphere.

Each network is characterised by its working medium, which will usually, but not necessarily, be the same in each branch of the network, and by the treatment of the working fluid as incompressible or compressible. It is usual to treat fluids such as water, fuel and lubricating oils and slurries as incompressible for computation purposes. Low-pressure gas networks which work at pressures close to atmospheric can be treated as incompressible if they contain fans. If they contain compressors they are probably not operating close to a uniform pressure and should be treated as compressible. Steam and higher pressure gas networks should be treated as compressible, as should any branch involving phase change.

The issue of compressibility is not confined to questions of operating pressure range. Some low pressure gas systems are built using very large-diameter pipes extending over significant distances. Pressure changes in these pipes require the in- and outflows of considerable masses of gas. Examples of such systems are the blast furnace and coke oven gas networks around steelworks. Because of their large volumes and filling requirements, such networks should be treated as compressible even though their operating pressure ranges lie within the otherwise allowable incompressible range.

While it is almost always valid to treat the furnace air supply network, including forced draft, primary and seal air fans, hot and cold air ducts, dampers and air heater ducting, as incompressible, since treatment of the system dynamics adds little to simulation realism or accuracy, the same cannot be said of the flue gas system. With its large ducts and close dynamic interaction with the furnace, the flue gas system has a noticeable influence on longer-term furnace pressure variations.

P. O'Kelly, *Computer Simulation of Thermal Plant Operations*,
DOI 10.1007/978-1-4614-4256-1_17,

However, since acoustic wave effects associated with gas compressibility have been excluded from this treatment, the "organ-pipe" effects sometimes observed as a severe oscillation of furnace pressure following a rapid disturbance in the furnace or flue gas system cannot be reproduced by the modelling equations derived in this chapter.

The division of networks according to compressibility applies only to the method of calculation of the pressures and flows throughout the network. No distinction in made is the calculation of energy and temperature distributions although clearly, compression energy need not be considered for incompressible flows.

17.1 Incompressible Flow Networks

The assumption of incompressibility in each branch implies equal mass flow at every point between any given pair of nodes. Initially, branch flow inertia will be neglected ($d\dot{m}/dt = 0$) allowing the instantaneous flow to be determined from the branch lumped admittance and the differential pressure across the branch. The mass flow network can then be developed on the basis of continuity of mass at each node. Assume the network consists of n_{br} branches connecting internal nodes to each other, n_{bc} branches connecting internal nodes to external interfaces, and n_{nd} nodes.

Without assuming incompressibility of the fluid in the node[1], continuity at the jth node requires that

$$\sum_{i}^{n_{br}} \dot{m}_{ij} + \sum_{s}^{n_{bs}} \dot{m}_{sj} + \dot{m}_{ext} = \frac{dM_j}{dt} \tag{17.1}$$

where,

- $\dot{m}_{ij}$ is the mass flow from node i to node j
- $\dot{m}_{sj}$ is the mass flow from an external source s to node j
- $\dot{m}_{ext}$ is a fixed flow (independent of node pressure) entering the node from an external source (−ve flow = outflow)
- M_j is the mass of fluid in the jth node

Neglecting flow momentum, the mass flow through the ij-th branch connecting nodes i and j is given from

$$\dot{m}_{ij}^2 = \mathcal{A}_{ij}(p_i - p_j + g_{ij}) \tag{17.2}$$

[1]Had we assumed incompressibility of the fluid in the node, we would have been forced to set $d\dot{m}_j/dt = 0$ for each node. This would lead to an algebraic system for the solution of flows. This system can become singular for some perfectly normal equipment configurations and is therefore to be avoided. Alternatively, we could allow the volume of the node to vary, the sometimes called "rubber-hose" effect, and also to be avoided.

Similarly, the mass flow through the sj-th branch connecting the external source s and node j is given from,

$$\dot{m}_{sj}^2 = \mathcal{A}_{sj}(p_s - p_j + g_{sj}) \tag{17.3}$$

where,

$\mathcal{A}_{ij}$ is the branch admittance from node i to node j
$\mathcal{A}_{sj}$ is the branch admittance from source s to node j
g_{ij} is the combination of geodetic pressure height difference between nodes i and j plus any developed heads in the branch
g_{sj} is the same for the branch connecting source s to node j

With $M = \rho V$, the right-hand side of Eq. 17.1 can be expanded as,

$$\frac{\mathrm{d}M_j}{\mathrm{d}t} = V_j \frac{\partial \rho_j}{\partial p_j}\frac{\mathrm{d}p_j}{\mathrm{d}t} + V_j \frac{\partial \rho_j}{\partial T_j}\frac{\mathrm{d}T_j}{\mathrm{d}t}$$

Replacing $\mathrm{d}p_j/\mathrm{d}t$ by its discrete-time equivalent allows us to write

$$\frac{\mathrm{d}M_j}{\mathrm{d}t} = C_j \left(\frac{p_j^{n+1} - p_j^n}{\Delta t} \right) + \dot{m}_{th,j} \tag{17.4}$$

where $C_j = V_j \frac{\partial \rho_j}{\partial p_j}$ and is directly related to fluid compressibility ($\partial\rho/\partial p > 0$). The flow term $\dot{m}_{th,j}$ arises from the thermal expansion of the fluid and couples the mass and heat flow networks at the nodes.

Equation 17.2 can be linearised by noting that

$$\sqrt{\Delta p} = \frac{\Delta p}{\sqrt{|\Delta p|}} \quad \text{where } |\Delta p| > 0$$

Then,

$$\dot{m}_{ij} = A_{ij}(p_i - p_j + g_{ij}) \tag{17.5}$$

where

$$A_{ij} = \left[\frac{\mathcal{A}_{ij}}{(p_i - p_j + g_{ij})^n} \right]^{\frac{1}{2}} \tag{17.6}$$

The index n implies use of these quantities calculated at the previous time step.

Similarly for the source branches,

$$\dot{m}_{sj} = A_{sj}(p_s - p_j + g_{sj}) \tag{17.7}$$

where

$$A_{sj} = \left[\frac{\mathcal{A}_{sj}}{(p_s - p_j + g_{sj})^n} \right]^{\frac{1}{2}} \tag{17.8}$$

Combining Eqs. 17.1, 17.2, 17.5 and 17.7 and collecting like terms gives

$$-\frac{\Delta t}{C_j}\sum_i A_{ij}\,p_i^{n+1} + \left(1 + \frac{\Delta t}{C_j}\sum_i A_{ij} + \frac{\Delta t}{C_j}\sum_s A_{sj}\right)p_j^{n+1} = p_j^n$$
$$+\frac{\Delta t}{C_j}\left(\sum_s A_{sj}\,p_s^n + \sum_i A_{ij}\,g_{ij}^n + \sum_s A_{sj}\,g_{sj}^n\right) - \frac{\Delta t}{C_j}\left(\dot m_{th,j}^n + \dot m_{ext}\right) \tag{17.9}$$

Equation 17.9 represents a set of n_{nd} equations in the node pressures p_j, given the external source pressures p_s and branch pressure heads g_{ij} and g_{sj} for each branch connecting to the jth node. The equation is written concisely in matrix form as

$$[\mathbf{I} + \mathbf{A_{dm}}]\,\mathbf{P^{n+1}} = \mathbf{P}^n + \frac{\Delta t}{C_j}[\mathbf{A_{bc}}\,\mathbf{P}_s + \mathbf{A_{dh}}\,\mathbf{G} - \mathbf{M_{th}} + \mathbf{M_{ext}}]$$
$$= \mathbf{P}^n + \frac{\Delta t}{C_j}\mathbf{S} \tag{17.10}$$

where each element on the RHS is evaluated at time n and

$\mathbf{P}$	is a vector of node pressures at times $(n + 1)$ and n
$\mathbf{P_s}$	is a vector of (external) source pressures
$\mathbf{G}$	is a vector of net pressure heads in each branch
$\mathbf{M_{th}}$	is a vector of thermal expansion flows
$\mathbf{M_{ext}}$	is a vector of fixed external flows
$\mathbf{A_{dm}}$	is a matrix of admittances of branches connected to each internal node
$\mathbf{A_{bc}}$	is a matrix of admittances of branches connected to each external source
$\mathbf{A_{dh}}$	is a matrix of admittances of branches connecting to each node
$\mathbf{I}$	is the unit matrix

The matrix $\mathbf{A_{dm}}$ is the network admittance matrix. The vector $\mathbf{S}$ is the network source vector.

Equation 17.10 is linear, and the admittance coefficient matrix is usually symmetric and positive-definite. Since all admittances have been linearised, solution of this equation is iterative, the iterations continuing until the node pressures have converged to within an acceptable tolerance. In practice very few iterations are required unless the system has been subjected to a large and sudden displacement from its previous state, and even then, convergence is rapid but transiently may exhibit noticeable dynamic errors. The individual branch flows are computed from Eq. 17.5 using the resulting node pressures.

Because the main diagonal elements are always ≥ 1 in magnitude (1 + some positive admittance which can be zero) the coefficient matrix $\mathbf{A_{dm}}$ will never be singular even if a specific node becomes totally isolated, that is, if the admittance of each branch joining the node is zero. For example, Eq. 17.10 will yield a stable solution for p_0 even if all three valves of Fig. 17.1 are closed.

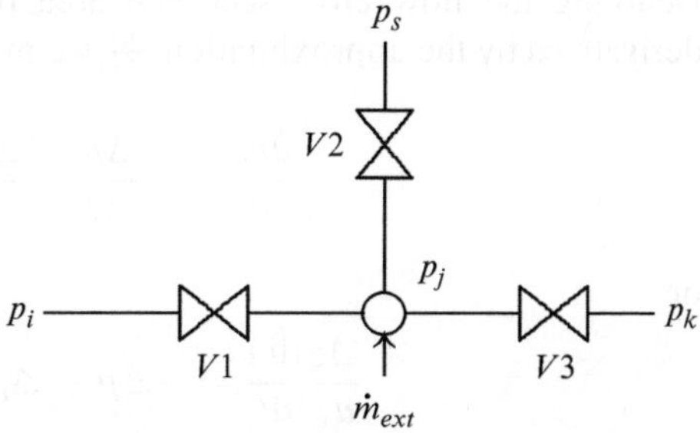

Fig. 17.1 Flow connections at a node

17.1.1 Inclusion of Flow Momentum

Because flow momentum has been neglected, in response to dynamically changing conditions, Eq. 17.10 generates a series of steady-state solutions to the network pressures and flows. In most cases, this is sufficient. However, as noted earlier, this can give rise to dynamic errors when things change rapidly. To avoid this, it is necessary to include flow inertia in the derivation of the node pressure and branch flow equations.

From Chap. 5 we have the original momentum conservation Eq. 5.11

$$\frac{\partial \phi}{\partial t} = -\left(1 + \phi^2 \frac{\partial v}{\partial p}\right)\frac{\partial p}{\partial z} - \phi^2 \frac{\partial v}{\partial h}\frac{\partial h}{\partial z} - 2\frac{\phi}{v}\frac{\partial v}{\partial t} - \left.\frac{\partial p}{\partial z}\right|_{loss} - \rho g \sin \beta$$

Restricting attention to the case $\phi^2 \frac{\partial v}{\partial p} \ll 1$ and neglecting the thermal expansion effects introduced by the term $\phi^2 \frac{\partial v}{\partial h}\frac{\partial h}{\partial z}$, this equation reduces to

$$\frac{\partial \phi}{\partial t} = -\frac{\partial p}{\partial z} - 2\frac{\phi}{v}\frac{\partial v}{\partial t} - \left.\frac{\partial p}{\partial z}\right|_{loss} - \rho g \sin \beta$$

or

$$\frac{\partial \phi}{\partial t} = -\frac{\partial p}{\partial z} - 2\frac{\phi}{v}\frac{\partial v}{\partial p}\frac{\partial p}{\partial t} - \left.\frac{\partial p}{\partial z}\right|_{loss} - \rho g \sin \beta$$

For incompressible flow $\frac{\partial v}{\partial p} = 0$. For compressible flow $\frac{\partial v}{\partial p} \approx -\frac{v}{p}$ and we have

$$\frac{\partial \phi}{\partial t} = -\frac{\partial p}{\partial z} + 2\frac{\phi}{p}\frac{\partial p}{\partial t} - \left.\frac{\partial p}{\partial z}\right|_{loss} - \rho g \sin \beta$$

We can assume for the present purpose that $\frac{\partial p}{\partial z} \gg 2\frac{\phi}{p}\frac{\partial p}{\partial t}$ which allows us to write

$$\frac{\partial \phi}{\partial t} = -\frac{\partial p}{\partial z} - \left.\frac{\partial p}{\partial z}\right|_{loss} - \rho g \sin \beta$$

Denoting the flow cross-sectional area by a_x, and replacing the spatial partial derivatives by the approximation $\frac{\Delta p}{\Delta z}$ we may write

$$\frac{1}{a_x}\frac{\mathrm{d}\dot{m}}{\mathrm{d}t} = -\frac{\Delta p}{\Delta z} - \left.\frac{\Delta p}{\Delta z}\right|_{loss} - \rho\, g\, \sin\beta$$

or

$$\frac{\Delta z}{a_x}\frac{\mathrm{d}\dot{m}}{\mathrm{d}t} = -\Delta p - \Delta p|_{loss} - \rho\, g\, \Delta z \sin\beta$$

Then, for a flow $\dot{m}_{ij}$ between two points i and j having pressures p_i and p_j

$$\begin{aligned} I_M \frac{\mathrm{d}\dot{m}_{ij}}{\mathrm{d}t} &= p_i - p_j + g_{ij} - \Delta p|_{loss} \\ &= \Delta p_{ij} - \Delta p|_{loss} \end{aligned} \tag{17.11}$$

where $g_{ij} = \rho\, g\, \Delta z \sin\beta$.

The coefficient $I_M = \Delta z / a_x$ is related to the parameter known as *inertance*[2]. Inertance is defined by the pressure gradient required to produce a given *volumetric* flow rate,

$$\Delta p = \rho \frac{\Delta z}{a_x}\frac{\mathrm{d}\dot{Q}}{\mathrm{d}t} = I \frac{\mathrm{d}\dot{Q}}{\mathrm{d}t}$$

with the inertance I given from

$$I = \rho \frac{\Delta z}{a_x}$$

It follows that

$$I_M = \frac{\Delta z}{a_x} = \frac{I}{\rho}$$

and is independent of the flow medium.

$\Delta p|_{loss}$ is the flow-dependent friction pressure loss across the branch, given by the standard equation

$$\mathcal{F}_{dp} = \Delta p|_{loss} = \frac{1}{\mathcal{A}_{ij}}\dot{m}_{ij}^2 \tag{17.12}$$

In Eq. 17.11 replace the continuous derivative by its discrete time equivalent to give

$$I_M \left(\frac{\dot{m}_{ij}^{n+1} - \dot{m}_{ij}^{n}}{dt} \right) = \Delta p_{ij}^{n+1} - \mathcal{F}_{dp} \tag{17.13}$$

[2]The term *inertance* is also found in the fields of acoustics and respiratory physiology.

Linearisation of the Friction Loss Term In an earlier section, linearisation of the friction pressure drop was handled by using the substitution

$$\sqrt{\Delta p^{n+1}} = \frac{\Delta p^{n+1}}{\sqrt{\Delta p^{n}}}$$

Here, we will use linearisation of $\mathcal{F}_{dp}$ by operating on $\dot{m}^2$ and consider two alternatives.

Version 1: $(\dot{m}^{n+1})^2 \approx \dot{m}^{n+1}\dot{m}^{n}$ Substituted into Eq. 17.13, this leads to

$$\dot{m}^{n+1} = \left(\frac{I_M \mathcal{A}_{ij}}{I_M \mathcal{A}_{ij} + \Delta t\, \dot{m}^n_{ij}}\right) \dot{m}^n_{ij} + \left(\frac{\Delta t\, \mathcal{A}_{ij}}{I_M \mathcal{A}_{ij} + \Delta t\, \dot{m}^n_{ij}}\right) \Delta p^{n+1}_{ij} \tag{17.14}$$

Version 2: Taylor Series Expansion Approximate $\mathcal{F}_{dp}$ by the first two terms of a Taylor series expansion about the known $\dot{m}^n_{ij}$.

$$\mathcal{F}^{n+1}_{dp} \approx \mathcal{F}_{dp}(\dot{m}^n_{ij}) + \left(\dot{m}^{n+1}_{ij} - \dot{m}^n_{ij}\right) \frac{\partial(\mathcal{F}_{dp})}{\partial \dot{m}_{ij}}.$$

With substitution of the derivative

$$\frac{\partial(\mathcal{F}_{dp})}{\partial \dot{m}_{ij}} = \frac{2}{\mathcal{A}_{ij}} \dot{m}_{ij}$$

this reduces to

$$\mathcal{F}^{n+1}_{dp} \approx \left(\frac{2}{\mathcal{A}_{ij}} \dot{m}^n_{ij}\right) \dot{m}^{n+1}_{ij} - \frac{1}{\mathcal{A}_{ij}} (\dot{m}^n_{ij})^2$$

Again, substituted into Eq. 17.13 this leads to

$$\dot{m}^{n+1}_{ij} = \left(\frac{I_M\, \mathcal{A}_{ij} + \Delta t\, \dot{m}^n_{ij}}{I_M\, \mathcal{A}_{ij} + 2\Delta t\, \dot{m}^n_{ij}}\right) \dot{m}^n_{ij} + \left(\frac{\Delta t\, \mathcal{A}_{ij}}{I_M\, \mathcal{A}_{ij} + 2\Delta t\, \dot{m}^n_{ij}}\right) \Delta p^{n+1}_{ij} \tag{17.15}$$

By setting $\dot{m}^{n+1}_{ij} = \dot{m}^n_{ij}$, it can easily be shown that Eqs. 17.14 and 17.15 reduce to Eq. 17.2 in the steady state.

The bracketed coefficient terms in Eqs . 17.14 and 17.15 can be handled more simply by denominating them by ϑ and θ, respectively. Then, for Eq. 17.15,

$$\theta_{ij} = \frac{\Delta t\, \mathcal{A}_{ij}}{I_M\, \mathcal{A}_{ij} + 2\Delta t\, \dot{m}^n_{ij}} \tag{17.16}$$

and

$$\vartheta_{ij} = \frac{I_M\, \mathcal{A}_{ij} + \Delta t\, \dot{m}^n_{ij}}{I_M\, \mathcal{A}_{ij} + 2\Delta t\, \dot{m}^n_{ij}} \tag{17.17}$$

and an equivalent pair for Eq. 17.14.

For either form of linearisation, we can write

$$\dot{m}_{ij}^{n+1} = \vartheta_{ij}\,\dot{m}_{ij}^{n} + \theta_{ij}\,\Delta p_{ij}^{n+1} \tag{17.18}$$

Returning to Eq. 17.4, we have

$$p_j^{n+1} = p_j^n + \frac{\Delta t}{C_j}\sum_i \dot{m}_{ij} - \frac{\Delta t}{C_j}(\dot{m}_{th} - \dot{m}_{ex})$$

Summing $\dot{m}_{ij}$ from Eq. 17.18 over all branches connecting to the jth node, we can write

$$\sum_i \dot{m}_{ij}^{n+1} = \sum_i \vartheta_{ij}\,\dot{m}_{ij}^{n} + \sum_i \theta_{ij}\,p_i^{n+1} - \sum_i \theta_{ij}\,p_j^{n+1} + \sum_i \theta_{ij}\,g_{ij}^{n}$$

Combining this with Eq. 17.4, and separating terms dealing with internal branches from those dealing with branches to external sources to match the style of Eq. 17.10, we arrive at

$$\begin{aligned}
&\frac{\Delta t}{C_j}\sum_i \theta_{ij}\,p_i^{n+1} + \left(1 + \frac{\Delta t}{C_j}\sum_i \theta_{ij} + +\frac{\Delta t}{C_j}\sum_s \theta_{sj}\right) p_j^{n+1}\\
&\quad = p_j^n + \frac{\Delta t}{C_j}\left(\sum_i \vartheta_{ij}\,\dot{m}_{ij}^{n} + \sum_s \theta_{sj}\,p_s^n + \sum_i \theta_{ij}\,g_{ij}^{n} + \sum_s \theta_{sj}\,g_{sj}^{n}\right)\\
&\quad -\frac{\Delta t}{C_j}\left(\dot{m}_{th,j}^{n} + \dot{m}_{ex}\right)
\end{aligned} \tag{17.19}$$

Taken over all nodes, Eq. 17.19 builds the equivalent of Eq. 17.10 but now includes the effects of flow inertia. Given the new node pressures and branch pressure differentials the new branch flows are computed from Eq. 17.18. All the coefficients of Eq. 17.19 can be computed using flows and admittances known from the previous time step.

Comparing with Eq. 17.10 we see that the two equations are identical in form. The introduction of flow momentum has added inertial terms to the calculation of the equation's coefficients. The other differences stem from the differing methods of admittance linearisation.

I_M appears as a factor multiplying $\mathcal{A}_{ij}$. Calling this product IA we can observe the following.

Equation 17.15 can be used to calculate a stable sequence for $\dot{m}^{n+1}$ if the coefficient of $\dot{m}^n$ is <1. This will always be the case though for $IA \gg 1$ this coefficient will approach 1 closely.

Conversely if $IA \ll 1$ then $\vartheta \approx \frac{1}{2}$ and $\theta \approx \frac{1}{2}\frac{\mathcal{A}_{ij}}{\dot{m}_{ij}^n}$. Then

$$\dot{m}_{ij}^{n+1} \approx 0.5\dot{m}_{ij}^n + 0.5\left(\frac{\mathcal{A}_{ij}}{\dot{m}_{ij}^n}\right)\Delta p_{ij}^{n+1} \tag{17.20}$$

For the first linearisation method, these results are valid for $IA \gg 1$. For $IA \ll 1$ however the effect of flow inertia disappears, and flow follows Δp with no or negligible delay.

For a pipe or duct section, admittance is proportional to a_x^2

$$\mathcal{A} = \frac{2\,d_h a_x^2}{\xi L v}$$

which, with $I_M = L/a_x$, gives for the product IA

$$IA = \frac{2\,d_h a_x}{\xi\, v}$$

17.1.2 Numerical Solution

In most cases, the topology of the physical system leads to the coefficient matrix of Eq. 17.10 being symmetric and positive definite. In this case the preferred solution method will be Cholesky decomposition. However, this situation cannot be guaranteed in every case and the best generic solution method, which will work virtually every time, is LU decomposition, preferably with pivoting.

17.2 Compressible Flow Networks

This section describes the derivation of a set of discrete-space equations for the computation of the time-varying pressures, flows, enthalpies and pressures within a complex network of branches carrying a compressible working medium. The starting point for these derivations is the set of conservation equations whose formulation in terms of one-dimensional discrete-space ordinate has been outlined in Sect. 5.3. The derivation applies generically to any arrangement of equipment within a branch. It follows from a consideration of the generic set of three cells shown by the following figure for a tube bank located within a gas duct transferring heat to a secondary fluid.

side-flow	$\dot{m}_{sf}$ ↕	$\dot{m}_{sf}$ ↕	$\dot{m}_{sf}$ ↕
	p^n_{j-1}	p^n_j	p^n_{j+1}
gas duct	T^n_{j-1}	T^n_j	T^n_{j+1}
	$\dot{m}_{j-1} \leftrightarrow$	$\dot{m}_j \leftrightarrow$	$\dot{m}_{j+1} \leftrightarrow$
heat transfer	⇕	⇕	⇕
gas− >outer metal	$\dot{q}^n_{hxmo,j-1}$	$\dot{q}^n_{hxmo,j}$	$\dot{q}^n_{hxmo,j+1}$
	⇕	⇕	⇕
	Tmo^n_{j-1}	Tmo^n_j	Tmo^n_{j+1}
	⇕	⇕	⇕
conduction through	$\dot{q}^n_{mm,j-1}$	$\dot{q}^n_{mm,j}$	$\dot{q}^n_{mm,j+1}$
tube wall	⇕	⇕	⇕
	Tmi^n_{j-1}	Tmi^n_j	Tmi^n_{j+1}
heat transfer	⇕	⇕	⇕
− > secondary fluid	$\dot{q}^n_{hxmi,j-1}$	$\dot{q}^n_{hxmi,j}$	$\dot{q}^n_{hxmi,j+1}$
	⇕	⇕	⇕
secondary fluid	$T^n_{ref\ j-1}$	$T^n_{ref\ j}$	$T^n_{ref\ j+1}$

The symbols have the following meanings.

p	pressure in the jth spatial cell [kPa]
T	temperature in the jth spatial cell [°C]
h	specific enthalpy of the working fluid in the jth spatial cell [kJ/kg]
v	specific volume of the working fluid in the jth spatial cell [m^3/kg]
$\dot{m}_j$	mass flow between the $(j-1)$th and jth cells [kg/s]
$\dot{m}_{sf}$	side-flow mass in to or out of the cell, [kg/s]
h_{sf}	specific enthalpy of the side-flow into the jth cell [kJ/kg]
Tmi	temperature of wall inner zone metal in the jth spatial cell [°C]
Tmo	temperature of wall outer zone metal in the jth spatial cell [°C]
$\dot{q}_{hxmi}$	heat transfer between the primary working fluid and the wall in the jth spatial cell [kW]
$\dot{q}_{hxmo}$	heat transfer between the wall and the secondary working fluid in the jth spatial cell [kW]
$\dot{q}_{mm}$	heat transfer through the wall from the outer to the inner zones in the jth spatial cell [kW]
$Q = Mh$	total heat in the working fluid in the jth spatial cell [J]
M	total mass of working fluid in the jth cell [kg]
V	total volume of working fluid in the jth cell [m^3]
A_{hxmi}	heat transfer area on the inner surface of the jth cell wall [m^2]
A_{hxmo}	heat transfer area on the outer surface of the jth cell wall [m^2]
$A_{hxinsul}$	heat transfer area—lagging to ambient—on the outer surface of the jth cell lagging [m^2]
a_{xsect}	mass flow cross-section area per tube in jth cell [m^2]

T_{ref} is the reference temperature seen by $\dot{q}_{hxmi,j}$, either local ambient temperature for a pipe, pump, damper or ducting or secondary fluid temperature for a heat exchanger.

Two types of cell variables—cell centre and cell boundary variables—are defined [7]. Cell variables defined at cell centres are denoted by full indices. Cell variables defined at the boundaries are denoted by half-indices. Cell centre variables are determined from mass and energy balance equations which treat flows which cross the boundaries. Cell edge or boundary variables link the adjacent cells and are defined in terms of the adjacent cell centre variables.

17.2.1 The Conservation Equations: Flow Momentum Neglected

To avoid clutter, variables associated with the jth cell will in general not be indexed. Indexing of the jth cell variable will be used selectively to avoid ambiguities.

Mass Conservation From Eq. 5.36, Sect. 5.3.1, we have

$$\begin{aligned}\frac{\mathrm{d}M}{\mathrm{d}t} &= \sum(\dot{m})_j \\ &= \frac{\mathrm{d}}{\mathrm{d}t}\left(\frac{V}{v}\right) \\ &= -\frac{V}{v^2}\frac{\mathrm{d}v}{\mathrm{d}t}\end{aligned}$$

where

$$\sum(\dot{m})_j = \dot{m}_{j-1} - \dot{m}_j + \dot{m}_{sf}$$

Given that

$$v = f(p, h)$$

we may write

$$\frac{\mathrm{d}v}{\mathrm{d}t} = \frac{\partial v}{\partial p}\frac{\mathrm{d}p}{\mathrm{d}t} + \frac{\partial v}{\partial h}\frac{\mathrm{d}h}{\mathrm{d}t}$$

and

$$\frac{\partial v}{\partial p}\frac{dp}{dt} + \frac{\partial v}{\partial h}\frac{dh}{dt} = -\frac{v^2}{V}\sum(\dot{m})_j. \tag{17.21}$$

Equation 17.21 may be rearranged into

$$\frac{V}{v^2}\frac{\partial v}{\partial p}\frac{dp}{dt} + \frac{V}{v^2}\frac{\partial v}{\partial h}\frac{dh}{dt} = \dot{m}_{j-1} - \dot{m}_j + \dot{m}_{sf}. \tag{17.22}$$

Energy Conservation The energy balance for the jth cell may be written

$$\frac{d(Mh)}{dt} = \dot{m}_{j-1}h_{j-\frac{1}{2}} - \dot{m}_j h_{j+\frac{1}{2}} + \dot{m}_{sf}h_{sf} + V\frac{dp}{dt} - \dot{q}. \tag{17.23}$$

where the net heat flow $\dot{q}$ out of the cell is composed as $\dot{q} = \dot{q}_{hxm} + \dot{q}_{amb} - \dot{q}_{gen}$. $\dot{q}_{hxm}$ is the heat exchanged across the boundaries of the cell (+ve is conventionally out of the cell). $\dot{q}_{gen}$ is the heat generated within a cell by friction losses, chemical reactions, etc. $\dot{q}_{amb}$ is the heat transfer between the fluid and its surroundings. $h_{j-\frac{1}{2}}$ and $h_{j+\frac{1}{2}}$ are the enthalpies of the flows crossing the left and right hand boundaries of the jth cell respectively. Note that a positive $\dot{m}_{sf}$ is directed into the cell.

Expanding the left-hand side of Eq. 17.23 leads to

$$M\frac{dh}{dt} + h\frac{dM}{dt} = \dot{m}_{j-1}h_{j-\frac{1}{2}} - \dot{m}_j h_{j+\frac{1}{2}} + \dot{m}_{sf}h_{sf} + V\frac{dp}{dt} - \dot{q}$$

which, after substituting for $\frac{dM}{dt} = \sum(\dot{m})_j$, reduces to

$$M\frac{dh}{dt} = \dot{m}_{j-1}(h_{j-\frac{1}{2}} - h_j) - \dot{m}_j(h_{j+\frac{1}{2}} - h_j) + \dot{m}_{sf}(h_{sf} - h_j) + V\frac{dp}{dt} - \dot{q}. \tag{17.24}$$

The cell boundary enthalpies $h_{j-\frac{1}{2}}$ and $h_{j+\frac{1}{2}}$ may be expressed in terms of cell-centre enthalpies in the adjacent cells by defining,

$$h_{j-\frac{1}{2}} = \frac{(1+\varphi_{j-1})}{2}h_{j-1} + \frac{(1-\varphi_{j-1})}{2}h_j$$

$$h_{j+\frac{1}{2}} = \frac{(1+\varphi_j)}{2}h_j + \frac{(1-\varphi_j)}{2}h_{j+1},$$

where φ can take values in the range $(-1, +1)$ and reflects the direction of flow. The general form is,

$$\varphi_j = a \text{ if } \dot{m}_j > 0; \quad \varphi_j = -a \text{ if } \dot{m}_j < 0.$$

Full donor cell differences are obtained if $a = 1$. Selection of $a = 0$ results in the usual centre differencing.

The preceding equations may be rearranged as

$$\begin{aligned} h_{j-\frac{1}{2}} - h_j &= \tfrac{(1+\varphi_{j-1})}{2}(h_{j-1} - h_j) \\ h_j - h_{j+\frac{1}{2}} &= \tfrac{(1-\varphi_j)}{2}(h_j - h_{j+1}). \end{aligned} \tag{17.25}$$

With these substitutions and M replaced by V/v, Eq. 17.24 becomes

$$\begin{aligned} -V\frac{\mathrm{d}p}{\mathrm{d}t} + \frac{V}{v}\frac{\mathrm{d}h}{\mathrm{d}t} &= \frac{(1+\varphi_{j-1})}{2}\dot{m}_{j-1}(h_{j-1} - h_j) \\ &\quad + \frac{(1-\varphi_j)}{2}\dot{m}_j(h_j - h_{j+1}) - \dot{q} + \dot{m}_{sf}(h_{sf} - h_j). \end{aligned} \tag{17.26}$$

The Coupled Mass and Energy Equations

The mass and energy conservation equations for the jth cell may be grouped into a single matrix equation. Writing

$$\begin{aligned} c_1 &= \frac{(1+\varphi_{j-1})}{2}(h_{j-1} - h_j), \\ c_2 &= \frac{(1-\varphi_j)}{2}(h_j - h_{j+1}), \\ c_3 &= h_{sf} - h_j. \end{aligned}$$

the composite equation can be written

$$\begin{bmatrix} \frac{V}{v^2}\frac{\partial v}{\partial p} & \frac{V}{v^2}\frac{\partial v}{\partial h} \\ -V & \frac{V}{v}. \end{bmatrix} \begin{bmatrix} \frac{\mathrm{d}p}{\mathrm{d}t} \\ \frac{\mathrm{d}h}{\mathrm{d}t} \end{bmatrix} = \begin{bmatrix} -1 & 1 \\ c_1 & c_2 \end{bmatrix} \begin{bmatrix} \dot{m}_{j-1} \\ \dot{m}_j \end{bmatrix} + \begin{bmatrix} -1 & 0 \\ c_3 & -1 \end{bmatrix} \begin{bmatrix} \dot{m}_{sf} \\ \dot{q} \end{bmatrix} \tag{17.27}$$

Inverting the left-hand side coefficient matrix and rearranging give

$$\begin{bmatrix} \frac{\mathrm{d}p}{\mathrm{d}t} \\ \frac{\mathrm{d}h}{\mathrm{d}t} \end{bmatrix} = \frac{1}{D}\begin{bmatrix} v & -\frac{\partial v}{\partial h} \\ v^2 & \frac{\partial v}{\partial p} \end{bmatrix} \left[\text{right-hand side of Eq. 17.27} \right] \tag{17.28}$$

where,

$$D = V\left[\frac{1}{v}\frac{\partial v}{\partial p} + \frac{\partial v}{\partial h}\right]. \tag{17.29}$$

Since $\left|\frac{1}{v}\frac{\partial v}{\partial p}\right| > \frac{\partial v}{\partial h}$ and $\frac{\partial v}{\partial p}$ is always negative, so too is the determinant D.

The analytical solution of Eq. 5.19 of Sect. 5.2 revealed the high rate of decay of the pressure/flow transients compared to the rate of thermal effects. Subject to certain limitations which will be discussed later, we will initially ignore flow inertial effects and use the simplest form of the momentum equation:

$$\dot{m}_{j-1}^2 = \Omega_{j-1}\,(p_{j-1} - p_j + g_{j-1}),$$

where Ω_{j-1} is the flow admittance of the cell. Using the linearisation

$$\sqrt{\Delta p} = \frac{\Delta p}{\sqrt{\Delta p}}$$

and introducing the linearised cell admittance $\mathcal{A}$ we have

$$\dot{m}_{j-1} = \mathcal{A}_{j-1}\,(p_{j-1} - p_j + g_{j-1}),$$

where, for $|p_{j-1} - p_j + g_{j-1}| > 0$,

$$\mathcal{A}_{j-1} = \sqrt{\frac{\Omega_{j-1}}{|p_{j-1} - p_j + g_{j-1}|}}.$$

As used previously, g_j is the sum of developed and geodetic heads for the cell.

Any cell may be connected to a side-flow, either as an in- or outflow. The side flow $\dot{m}_{sf}$ to or from the jth cell is given from

$$\dot{m}_{sf} = \mathcal{A}_{sf}\,(p - p_{ref}),$$

where p_{ref} is the pressure at the other end of the side-flow path and $\mathcal{A}_{sf}$ is the admittance expressed in terms of an equivalent flow cross section,

$$\mathcal{A}_{sf} = \sqrt{\frac{\zeta\, A_{x,sf}^2}{|p - p_{ref}|}} \quad \text{for } |p - p_{ref}| > 0. \tag{17.30}$$

The expansion and reorganisation of Eq. 17.28 is facilitated by the following definitions:

$$a_1 = v + \frac{\partial v}{\partial h}\frac{(1 + \varphi_{j-1})}{2}(h_{j-1} - h_j),$$

$$a_2 = v - \frac{\partial v}{\partial h}\frac{(1 - \varphi_{j-1})}{2}(h_j - h_{j+1}),$$

$$a_3 = v\frac{\partial v}{\partial h}(h_{sf} - h_j). \tag{17.31}$$

and,

$$b_1 = \frac{\partial v}{\partial p}\frac{(1+\varphi_{j-1})}{2}\dot{m}_{j-1},$$

$$b_2 = \frac{\partial v}{\partial p}\left(-\frac{(1+\varphi_{j-1})}{2}\dot{m}_{j-1} + \frac{(1-\varphi_j)}{2}\dot{m}_j - \dot{m}_{sf}\right),$$

$$b_3 = -\frac{\partial v}{\partial p}\frac{(1-\varphi_j)}{2}\dot{m}_j. \tag{17.32}$$

from which it is apparent that,

$$b_2 = -b_1 - b_3 - \dot{m}_{sf}\frac{\partial v}{\partial p}.$$

The Decoupled Mass and Energy Equations

We can utilise the widely differing response rates of the underlying dynamics associated with pressure and enthalpy (or temperature) variations within the system to treat the mass and energy systems via separate but coupled pressure and enthalpy equations. Equation 17.28 may be expanded and separated into its two constituent equations as follows.

The Pressure Equation

$$\begin{aligned} D_j\frac{\mathrm{d}p_j}{\mathrm{d}t} &= -\,a_{1,j}\,\mathcal{A}_{j-1}\;p_{j-1} + [-a_{1,j}\,\mathcal{A}_{j-1} + a_{2,j}\,\mathcal{A}_j - a_{3,j}\,\mathcal{A}_{sf}]\,p_j \\ &\quad -a_{2,j}\,\mathcal{A}_j\;p_{j+1} - a_{3,j}\,\mathcal{A}_{sf}\;p_{j,ref} + \frac{\partial v}{\partial h}\dot{q}_j \\ &\quad +a_{1,(j-1)}\mathcal{A}_{j-1}\mathrm{d}p_{j-1} - a_{1,j}\,\mathcal{A}_j\mathrm{d}p_j. \end{aligned} \tag{17.33}$$

The Enthalpy Equation

$$D_j\frac{\mathrm{d}h_j}{\mathrm{d}t} = -v_j^2 S(\dot{m}_j) + b_{1,j}h_{j-1} + b_{2,j}\,h_j + b_{3,j}\,h_{j+1} + \frac{\partial v}{\partial p}(\dot{m}_{sf}\,h_{sf} - \dot{q}_j), \tag{17.34}$$

where $S(\dot{m}_j) = \dot{m}_{j-1} - \dot{m}_j$

17.2.2 Solution of the Branch Equations

The first step is to reorganise Eqs. 17.33 and 17.34 into discrete-time form by replacing the continuous derivatives by their first-order discrete-time approximations. The original equations are then rearranged into implicit form.

The following coefficients are defined.

$$B_{j,j-1} = -\frac{\Delta t}{D} a_1 \mathcal{A}_{j-1}$$

$$B_{j,j+1} = \frac{\Delta t}{D} a_2 \mathcal{A}_j$$

$$B_{j,j} = 1 - B_{j,j-1} - B_{j,j+1}$$

$$C_{j,j-1} = -\frac{\Delta t}{D} b_1$$

$$C_{j,j+1} = -\frac{\Delta t}{D} b_3$$

$$C_{j,j} = 1 + \frac{\Delta t}{D} b_2$$

$$Y_j = \frac{\Delta t}{D}\left[v\, S(\dot{m}) + a_3 \dot{m}_{sf} - \frac{\partial v}{\partial h}\left((\dot{m}h)_{in} - \dot{m}_{cd}\, h'' - \dot{q}\right)\right]$$

$$Z_j = \frac{\Delta t}{D}\left[v^2 S(\dot{m}_j) + \frac{\partial v}{\partial p}\left((\dot{m}h)_{in} - \dot{m}_{cd}\, h'' - \dot{q}\right)\right]$$

The condensation term $\dot{m}_{cd}\, h''$ applies only to a condensing working medium such as steam, for which phase changes must be considered.

The pressure and enthalpy equations may then be written compactly as,

$$B_{j,j-1}\, p_{j-1}^{n+1} + B_{j,j}\, p_j^{n+1} + B_{j,j+1}\, p_{j+1}^{n+1} = p_j^n + Y_j. \tag{17.35}$$

$$C_{j,j-1}\, h_{j-1}^{n+1} + C_{j,j}\, h_j^{n+1} + C_{j,j+1}\, h_{j+1}^{n+1} = h_j^n + Z_j. \tag{17.36}$$

Equations 17.35 and 17.36 form a coupled set of mass and energy balance equations. The coefficients of each depend on the solution of the other through the coefficients a_1, a_2, b_1, etc., and the equation pair must be solved iteratively.

As j is taken over all n_{cell} contiguous cells in a branch, the pressure and enthalpy equations form two tridiagonal matrix equations. Boundary conditions are required at each end ($j = 1$ and $j = n_{cell}$). These are provided by the appropriate state variables of the interconnecting components.

When written out for all j for, say, a 7-cell section, Eqs. 17.35 and 17.36 form a tridiagonal matrix (the enthalpy equation is shown as the example).

$$\begin{bmatrix} C_{1,1} & C_{1,2} & & & & & \\ C_{2,1} & C_{2,2} & C_{2,3} & & & & \\ & C_{3,2} & C_{3,3} & C_{3,4} & & & \\ & & C_{4,3} & C_{4,4} & C_{4,5} & & \\ & & & C_{5,4} & C_{5,5} & C_{5,6} & \\ & & & & C_{6,5} & C_{6,6} & C_{6,7} \\ & & & & & C_{7,6} & C_{7,7} \end{bmatrix} \begin{bmatrix} h_1 \\ h_2 \\ h_3 \\ h_4 \\ h_5 \\ h_6 \\ h_7 \end{bmatrix}^{n+1}$$

$$= \begin{bmatrix} h_1 \\ h_2 \\ h_3 \\ h_4 \\ h_5 \\ h_6 \\ h_7 \end{bmatrix}^{n} + \begin{bmatrix} C_{1,0}h_0 \\ 0 \\ 0 \\ 0 \\ 0 \\ 0 \\ C_{7,8}h_8 \end{bmatrix}^{n} + \begin{bmatrix} Z_1 \\ Z_2 \\ Z_3 \\ Z_4 \\ Z_5 \\ Z_6 \\ Z_7 \end{bmatrix}^{n}. \quad (17.37)$$

where h_0 and h_8 are boundary conditions and are the enthalpies either in interfacing components or in internal nodes of the network.

Solution of this equation yields the map of working fluid enthalpies for each of the branch cells.

A similar matrix equation is constructed for pressures, with the C coefficients replaced by B coefficients, the Z vector replaced by the Y vector and with pressure boundary conditions defined in place of enthalpy boundary conditions.

Inclusion of Thermal Diffusion

The form of the advection equation makes no provision for the transfer of thermal energy between cells in the absence of flow. Once flow ceases in a cell or branch, enthalpy changes depend only on heat exchange across the boundaries with ambient or interfacing systems. These exchanges are very small and the attendant changes in enthalpy and temperature are correspondingly small and slow. Plant operations frequently result in the isolation of branches through which the working medium had previously been flowing. The medium within these isolated branches will cool, both by this slow ambient heat loss and by diffusion of energy throughout the stationary working medium. The ambient heat loss is included in the foregoing treatment but the diffusion influence is not. For this purpose, we must add a diffusion term to the original energy equation. Equation 17.23 then becomes

$$\frac{\mathrm{d}(Mh)}{\mathrm{d}t} = \dot{m}_{j-1}h_{j-\frac{1}{2}} - \dot{m}_j h_{j+\frac{1}{2}} + \dot{m}_{sf}h_{sf} + V\frac{\mathrm{d}p}{\mathrm{d}t} - \dot{q} + c_{dif}\frac{\partial^2 h}{\partial z^2}. \quad (17.38)$$

Using the spatial discretisation

$$\frac{\partial^2 h_j}{\partial z^2} = \frac{h_{j-1} - 2h_j + h_{j+1}}{\Delta z^2}$$

diffusion may be included by the modification of the C-coefficients as follows.

$$C_{j,j-1} = -\frac{\Delta t}{D}(b_1 - c_{dif}/\Delta z^2),$$

$$C_{j,j+1} = -\frac{\Delta t}{D}(b_3 - c_{dif}/\Delta z^2),$$

$$C_{j,j} = 1 + \frac{\Delta t}{D}(b_2 - 2c_{dif}/\Delta z^2).$$

where Δz is the length of the cell. The diffusion coefficient c_{dif} is given from Eq. 10.1

$$c_{dif} = \frac{\lambda}{c_p \rho}.$$

λ is the coefficient of thermal conductivity of the material, ρ is its density and c_p the specific heat capacity. For water, c_{dif} is typically around 1.5e-7. For saturated steam at 10 bar, c_{dif} = 2.76e-6 and at 160 bar, 1.45e-7. For superheated steam it lies in the range of 1e-6 to 5e-7. Since the observable effects of diffusion are slow and noticeable only during cooldown, a phase marked by very low or zero mass flows as temperatures tend towards those of their surroundings, the choice of a value for c_{dif} around 1.0e-7 is not critical.

Treatment of Boundary Conditions

Each branch has a left-and a right-hand end. Conventionally, the flow into the left-hand end and leaving the right-hand end cell is positive. A branch connects at each end to either an internal network node or an external device. Denoting by the index -1 the pressure or flow enthalpy at the left-hand end and by $+1$ that at the right-hand interface, we can write for the left-hand end

$$p_0^{n+1}\left(1 + \frac{\Delta t}{D_0}\mathcal{A}_0\right) - \frac{\Delta t}{D_0}\mathcal{A}_0 p_1^{n+1} = p_0^n + \frac{\Delta t}{D_0}\dot{m}_{-1}$$

$$= p_0^n + \frac{\Delta t}{D_0}\mathcal{A}_{-1}(p_{-1} - p_0)$$

and for the right hand end (Nth cell)

$$p_N^{n+1}\left(1+\frac{\Delta t}{D_N}\mathcal{A}_N\right)-\frac{\Delta t}{D_N}\mathcal{A}_N p_{N-1}^{n+1} = p_N^n+\frac{\Delta t}{D_N}\dot{m}_{+1}$$
$$= p_N^n+\frac{\Delta t}{D_N}\mathcal{A}_{+1}(p_N-p_{+1}).$$

17.2.3 Network of Multiple Branches

Equation 17.37 defines the map of steam enthalpies for a set of n_{cell} cells in a single branch. A network consists of n_{br} branches. The coefficient matrix of the composite equation for the entire network is composed of a set of coefficient submatrices, each of the form of Eq. 17.37. The network enthalpy state vector is composed of a set of subvectors, one for each branch and of dimension equal to the number of cells in the respective branch. The composite network equation is then of the form

$$\begin{bmatrix} C_1 & & & & \\ & C_2 & & & \\ & & \ddots & & \\ & & & C_{ibr} & \\ & & & & \ddots \\ & & & & & C_{nbr} \end{bmatrix}\begin{bmatrix}\vec{h}_1\\ \vec{h}_2\\ \vdots\\ \vec{h}_{ibr}\\ \vdots\\ \vec{h}_{nbr}\end{bmatrix}^{n+1} = \begin{bmatrix}\vec{h}_1\\ \vec{h}_2\\ \vdots\\ \vec{h}_{ibr}\\ \vdots\\ \vec{h}_{nbr}\end{bmatrix}^{n} + \begin{bmatrix}h\vec{b}c_1\\ h\vec{b}c_2\\ \vdots\\ h\vec{b}c_{ibr}\\ \vdots\\ h\vec{b}c_{nbr}\end{bmatrix}^{n} + \begin{bmatrix}\vec{Z}_1\\ \vec{Z}_2\\ \vdots\\ \vec{Z}_{ibr}\\ \vdots\\ \vec{Z}_{nbr}\end{bmatrix}^{n}. \quad (17.39)$$

The presence of the $\vec{}$ above a symbol denotes a complete subvector for a single branch. $h\vec{b}c_{ibr}$ denotes a vector of boundary conditions for the (ibr)th branch. This vector will have non-zero entries in only the first and last elements.

At the points (nodes) at which branches join, boundary conditions are defined by the node enthalpies and pressures. Boundary enthalpies are required only for points of inflow to the branch and are selected for each branch on the basis of branch entry flow direction.

17.2.4 Calculation of Node Pressures

Node pressures may be included either explicitly as boundary conditions to the network branches or implicitly as elements of the network pressure vector.

The node pressures $\bar{p}_k$ are adjusted to ensure that the net mass flow at the kth node is balanced at each time step. The net flow balance will be exact for incompressible flow. For compressible nodes transient imbalances of inlet and outlet flows will reflect the mass changes within the node volumes. Denoting by M the mass of working medium in the kth node, conservation of mass requires

$$\frac{\mathrm{d}M}{\mathrm{d}t} = \sum_i \mathcal{A}_i(p_i - \bar{p}_k) + \Sigma(\dot{m}_{ext}), \tag{17.40}$$

where $\mathcal{A}_i$ is the (linearised) admittance of the flow cell linking the ith cell and kth node and p_i is the pressure in the ith cell adjacent to the kth node. This will be either the first cell if the left-hand branch end connects to the node or the last cell if the right-hand end connects. $\Sigma(\dot{m}_{ext})$ is the sum of flows entering or leaving the node to external branches or systems. With V_k the volume of the kth node, $M = V_k/v_k$, giving

$$\begin{aligned}\frac{\mathrm{d}M}{\mathrm{d}t} &= \frac{\mathrm{d}}{\mathrm{d}t}\left(\frac{V_k}{v_k}\right)\\ &= -\left(\frac{V_k}{v_k^2}\right)\frac{\mathrm{d}v_k}{\mathrm{d}t}\\ &= -\left(\frac{V_k}{v_k^2}\right)\frac{\partial v}{\partial p}\frac{\mathrm{d}\bar{p}_k}{\mathrm{d}t}\\ &= \left(\frac{V_k}{\gamma\,\bar{p}_k v_k}\right)\frac{\mathrm{d}\bar{p}_k}{\mathrm{d}t}.\end{aligned}$$

External Treatment of Node Pressures

With this substitution for $\mathrm{d}M/\mathrm{d}t$ and replacement of $\mathrm{d}\bar{p}_k/\mathrm{d}t$ by its discrete time equivalent, Eq. 17.40 can be expressed in implicit form, giving for $\bar{p}_k^{n+1}$

$$\bar{p}_k^{n+1} = \frac{\left[\bar{p}_k^n + \frac{\Delta t}{\tau_k}\left(\sum_i \mathcal{A}_i\, p_i^n + \Sigma(\dot{m}_{ext})\right)\right]}{\left[1 + \frac{\Delta t}{\tau_k}\sum_i \mathcal{A}_i\right]}, \tag{17.41}$$

where

$$\tau_k = \frac{V_k}{\gamma\,\bar{p}_k\, v_k} = \frac{V_k}{\gamma R T_k}. \tag{17.42}$$

Internal Treatment of Node Pressures

A composite network pressure vector **P** is formed by augmenting the cell pressure vector **p** by the vector of node pressures $\bar{\mathbf{p}}$

$$\mathbf{P} = [\mathbf{p}\ \bar{\mathbf{p}}]^T.$$

For a boundary cell, Eq. 17.35 expresses the jth cell pressure in terms of its adjacent upstream cell $(j-1)$ and its connecting boundary pressure $\bar{p}_k$.

$$B_{j,j-1}\, p_{j-1}^{n+1} + B_{j,j}\, p_j^{n+1} + B_{j,j+1}\, \bar{p}_k^{n+1} = p_j^n + Y_j$$

or, expanding the B coefficients,

$$-\frac{\Delta t}{D_j} a_1 \mathcal{A}_{j-1} p_{j-1}^{n+1} + \left(1 + \frac{\Delta t}{D_j} a_1 \mathcal{A}_{j-1} + \frac{\Delta t}{D_j} a_2 \mathcal{A}_j\right) p_j^{n+1} + \frac{\Delta t}{D_j} a_2 \mathcal{A}_j\, \bar{p}_k^{n+1}$$
$$= p_j^n + Y_j. \qquad (17.43)$$

With some minor simplifications

$$D = V\left[\frac{1}{v}\frac{\partial v}{\partial p} + \frac{\partial v}{\partial h}\right] \approx \frac{V}{v}\frac{\partial v}{\partial p} \approx -\frac{V}{v}\frac{v}{p} = -\frac{V}{p}.$$

Therefore, given $p\,v = R\mathcal{T}$,

$$-\frac{\Delta t}{D_j} a_1 = -\Delta t \frac{R\mathcal{T}_j}{V_j}.$$

Similarly

$$\frac{\Delta t}{D_j} a_2 = -\Delta t \frac{R\mathcal{T}_j}{V_j}$$

and the cell equation becomes

$$-B_{j,j-1} p_{j-1}^{n+1} + (1 + B_{j,j-1} + B_{j,k}) p_j^{n+1} - B_{j,k} \bar{p}_k^{n+1} = p_j^n + Y_j^n, \qquad (17.44)$$

where

$$B_{j,j-1} = \Delta t \frac{R\mathcal{T}_j}{V_j} \mathcal{A}_{j-1}.$$

$$B_{j,k} = \Delta t \frac{R\mathcal{T}_j}{V_j} \mathcal{A}_j.$$

For the kth node, the mass balance Eq. 17.40 can be rearranged into

$$\left(1+\frac{\Delta t}{\tau_k}\sum_i \mathcal{A}_k\right)\bar{p}_k-\frac{\Delta t}{\tau_k}\sum \mathcal{A}_i\, p_i=\bar{p}_k^n+\frac{\Delta t}{\tau_k}\Sigma(\dot{m}_{ext}). \tag{17.45}$$

The augmented network equation can then be written

$$\begin{bmatrix}\mathbf{B}_{11} & \mathbf{B}_{12}\\ \mathbf{B}_{21} & \mathbf{B}_{22}\end{bmatrix}\begin{bmatrix}\mathbf{p}\\ \bar{\mathbf{p}}\end{bmatrix}^{n+1}=\begin{bmatrix}\mathbf{p}\\ \bar{\mathbf{p}}\end{bmatrix}^{n}+\begin{bmatrix}\mathbf{Y}\\ 0\end{bmatrix}+\begin{bmatrix}\mathbf{p}_{bc}\\ \frac{\Delta t}{\tau_k}\sum(\dot{m})\end{bmatrix}, \tag{17.46}$$

where the elements of the submatrices $\mathbf{B}_{11}$, $\mathbf{B}_{12}$, $\mathbf{B}_{21}$ and $\mathbf{B}_{22}$ are given from Eqs. 17.44 and 17.45.

Equation 17.46 is nearly tridiagonal. $\mathbf{B}_{11}$ is purely tridiagonal and $\mathbf{B}_{22}$ is purely diagonal. $\mathbf{B}_{12}$ and $\mathbf{B}_{21}$ contain the terms linking the branch end cells to their interfacing nodes and, in general, are sparse. The solution method described in Sect. 3.7.5 can be applied here.

It turns out that, not only is it not necessary to solve the augmented equation, it is advantageous to un-bundle the single augmented equation into two separate equations, one for the branch cell pressures $\mathbf{p}$ and one for the node pressures $\bar{\mathbf{p}}$, and solve them sequentially. The procedure then becomes

Solve:

$$\mathbf{B}_{11}\,\mathbf{p}^{n+1}=\mathbf{p}^n+\mathbf{Y}+\mathbf{p}_{bc} \tag{17.47}$$

for $\mathbf{p}^{n+1}$. Then

Calculate $\bar{\mathbf{p}}^{n+1}$ from:

$$\bar{\mathbf{p}}^{n+1}=\mathbf{B}_{22}^{-1}\left[\bar{\mathbf{p}}^n-\mathbf{B}_{21}\mathbf{p}^{n+1}+\frac{\Delta t}{\tau_k}\sum(\dot{m})\right]. \tag{17.48}$$

The advantage of the unbundled approach over the augmented is computation speed—the un-bundled approach is much faster—and numerical stability. Both Eqs. 17.47 and 17.48 are absolutely stable and allow the use of actual physical dimensions in the calculation of the equation coefficients. As a result they converge rapidly and stably. The off-diagonal elements of Eq. 17.46 introduce stability issues and their magnitude must be limited to less than the main diagonal elements to preserve diagonal dominance.

17.2.5 Calculation of Node Enthalpies and Temperatures

For incompressible flow, enthalpies at a mixing point depend only on the incoming stream enthalpies and on the time taken for the streams to mix homogeneously. These mixing times will be extremely short, particularly for high flow rates and small mixing volumes. At very low flow rates, mixing times could be longer, but enthalpy changes are unlikely to be large.

For the jth network node,

$$\tau_j \frac{\mathrm{d}h_j}{\mathrm{d}t} = \sum_i \dot{m}_i h_i - h_j \sum_i \dot{m}_i$$
$$= \sum_i \dot{m}_i (h_i - h_j),$$

where τ is the assumed mixing time constant [seconds], $\dot{m}_i$ is the mass flow from the ith branch into the jth node, h_i is the enthalpy of the flow leaving the jth branch and h_j is the medium enthalpy at the jth node.

Replacing the derivative by its discrete-time equivalent, we may write

$$h_j^{n+1} = h_j^n + \frac{\Delta t}{\tau_j} \sum_i \dot{m}_i (h_i - h_j)$$

which is written in fully implicit form as

$$h_j^{n+1} = \frac{[h_j^n + \frac{\Delta t}{\tau_j} \sum_i \dot{m}_i h_i^n]}{[1 + \frac{\Delta t}{\tau_j} \sum_i \dot{m}_i]}. \tag{17.49}$$

For a flow leaving a branch in the positive sense, h_i is the medium enthalpy in the last branch cell. For a flow leaving a branch in the negative sense, h_i is the medium enthalpy in the first branch cell.

17.2.6 Wall Temperature Calculation

The working medium, incompressible or compressible, flowing through the network is contained by the walls of a variety of components including pipes, valves, ducts, dampers, heat exchangers and so on that make up the physical network. These exchange heat with the medium—in the case of heat exchangers, with media on both sides of the wall—and with their surroundings, frequently through an insulating layer. The heat flows thus exchanged are represented in the preceding equations as $\dot{q}_{hxm}$, calculated for each cell from the local medium-wall temperature differential, assuming normal convective heat transfer. This means that the wall metal temperatures must be calculated simultaneously with the working fluid temperatures for each cell in the network.

The calculation method and arrangement of wall temperatures described in Sect. 10.3 is suitable for this purpose and can be linked directly with the network cell structure described above. The result is an integrated procedure that yields, within the scope of a single calculation, a complete map of the pressures, flows, enthalpies, temperatures and derived properties for the working medium, together with two wall metal temperatures, for each cell and node of the complete network.

17.2.7 Application to a Typical Steam Flow Network

The methods of the preceding sections are illustrated by an example based on a real steam flow and heat exchange network. The following schematic depicts the parallel steam paths from a drum or water separators to the HP turbine of a power-generating unit. All branches are treated as compressible.

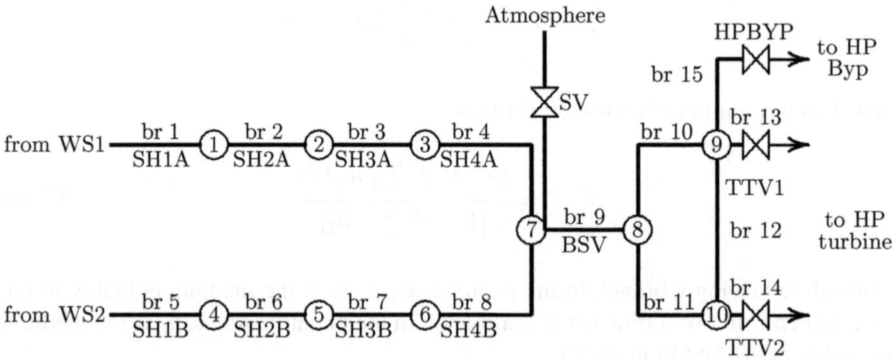

Branches 1–8 contain superheaters connected by pipes and headers. Superheaters are treated as tubular heat exchangers with a tube side (steam) and a duct or gas side. Each superheater is represented by a series of five contiguous cells. Connecting pipes contain between one and three cells, depending on length, are 250–300 mm NB, with a wall thickness of 7 mm and between 5 and 35 m long. Nodes 1–6 represent spray desuperheaters. Node 7 is the boiler outlet header which connects to node 8 via a short branch which includes the boiler stop valve (BSV). A safety valve (SV) to atmosphere is connected to node 7. Branches 10 and 11 are the long steam pipes (55 m, 9 cells) which terminate at the HP turbine stop/throttle valves. The two valve chests are connected by a short crossover pipe (branch 12). Steam flows to the HP turbine from nodes 9 and 10 through each of two throttle valves TTV1 and TTV2 are adjusted by modulating the valve positions to follow a preset loading program. In this exercise the turbine itself is not simulated. For the purpose of the demonstration, boundary conditions are defined by preset ramps of pressures as might be followed during a typical start of a generating unit but without the long holding periods normally used for turbine warming.

The associated (simplified) gas flow network is shown by the following figure.

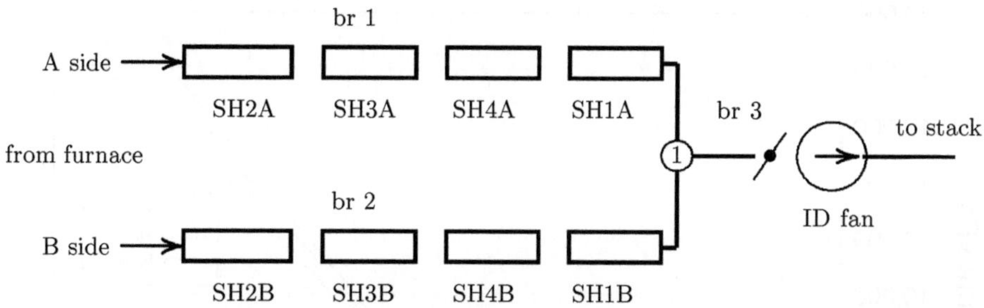

Obvious simplifications include omission of reheaters, economiser and airheater(s) from the gas flow paths and use of a single induced draft fan. One or more reheater tube banks might be interleaved with the superheater banks ahead of the primary superheater. It is to be noted that, following normal plant design, the primary superheaters SH1A/B are placed first in the steam flow path (lowest steam temperatures) and last in the gas flow path (lowest gas temperatures) and are connected in a counter-flow arrangement.

Network Behaviour: Flow Momentum Ignored

The calculation covers the following scope of operations which might form part of a unit start. Variations of pressures, flows and temperatures are shown. Following a typical start procedure, boiler steam pressure was initially increased to around 3,100 kPa (30 bar) using only igniters, giving a furnace exit gas temperature of around 450°C. With the boiler stop valve open, steam flows into the main steam system remained small, being only steam used to fill and pressurise the piping and superheater tubing and to make up steam lost as condensate through the boiler and steam line drains and vents. Steam pressure at the turbine stop valves was held at 3,100 kPa (30 bar) (A)[3] by modulating the position of the HP bypass valve, connected to node 9. During this time, the turbine might be accelerated to some intermediate speed, typically 1,000 rpm (for a 50 Hz system), and held there for initial warming while steam flows through the system increased to equal that passing through the bypass plus any losses through drain valves and vents. After a wait period of some 15 minutes, the pressure was increased to around 6,100 kPa (60 bar) by increasing the HP bypass valve pressure setpoint. The bypass valve closed (B) and steam flows reduced to their previous small values. Once the stop valve pressure reached its target the bypass valve again opened (C) to regulate pressure. Along the way, the first coal pulverisers would have been started, producing a rapid rise of furnace exit temperature to around, say, 650°C. The turbine would now be accelerated to synchronous speed. After synchronising to the grid (D) and initial loading (E), pressure was raised uniformly to the normal operating pressure of

[3]These letters refer to the those used on the trend graphs to identify points in the procedure.

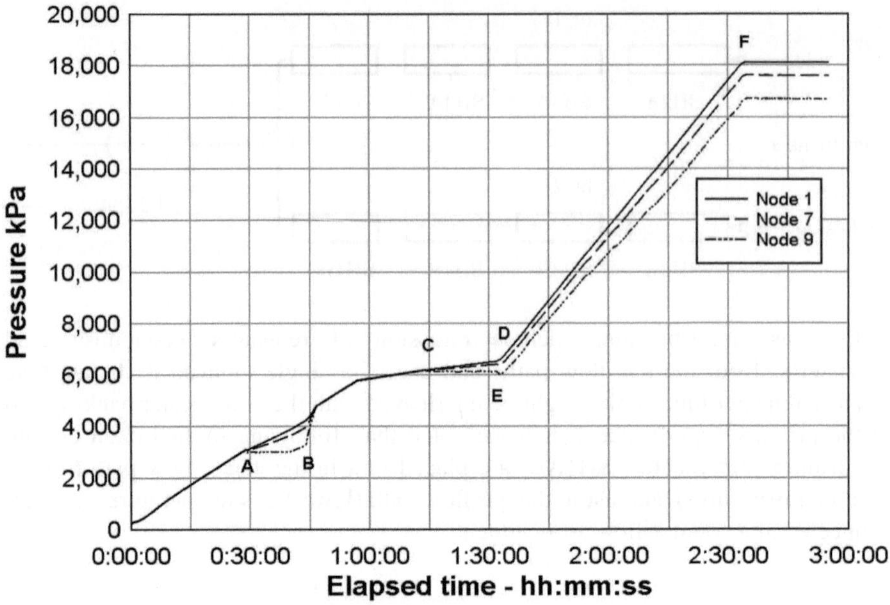

Fig. 17.2 Network pressure trends

18,700 kPa (185 bar) (F) at 2 bar/min while loading the turbine to 100 %. Additional pulverisers will have been started to match the increasing firing with increasing unit load. Each pulveriser start brought a rapid increase in furnace exit temperature which continued to increase with increased firing up to around 60 % load, after which it remains relatively constant at (here) 1,050°C. Full-load steam flow to the turbine was 330 kg/s and flue gas flow 300 kg/s (Figs. 17.2 and 17.3).

The increasing spread among the individual pressure trends indicates the increasing pressure drops through the system with increasing steam flows.

Figure 17.4 shows trends of individual steam temperatures. In the generation of these results, a very simplified method was used to match firing rate, and with it furnace exit temperature, to steam demand. As a consequence, the gas temperature entering the superheater duct changes in a somewhat unrealistic manner. The trends shown by Fig. 17.4 are not to be taken as representative of real plant behaviour. They do however allow realistic interpretation, given the other parameters of the simulated start. For example, during periods of low steam flow through the tubes, the superheater tube metal and steam temperatures rise quickly, possibly into alarm ranges, before reducing quickly once the steam flow increases as the bypass valve opens. The increase in steam temperatures during loading is typical (though excessive) and reflects the higher heat input into the steam as both firing rate and in particular the radiant contribution to the secondary and tertiary superheaters increase more rapidly than the steam flow through them. In practice, it would be expected that the steam temperature control desuperheating sprays (not included here) would

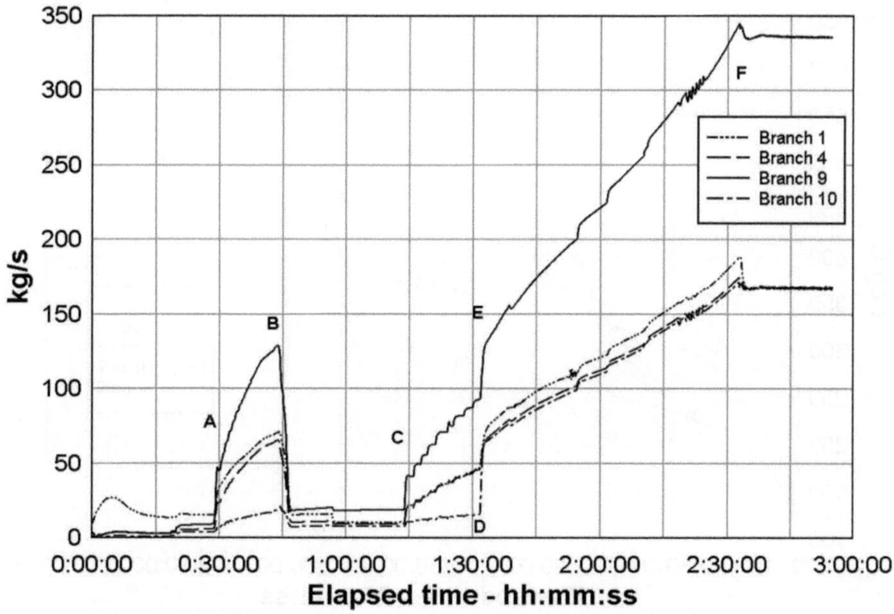

Fig. 17.3 Network flow trends

intervene to maintain set point values. As the steam flow increased beyond some 80 % of full load, the steam temperatures reduced to their normal operating values.

An allowance has been made in these calculations for the radiant heat contributions to the secondary and tertiary superheaters. Without this contribution, convective heat transfer alone would yield a lower boiler outlet temperature for this arrangement of superheaters and gas temperatures, as shown by the following simple calculation.

On the gas side,

Furnace exit temperature is set to 1,050°C
Flue gas temperature to the economiser is 415°C
Flue gas flow is 300kg/s
Flue gas mean specific heat is around 1.25 kJ/kg/C

Convective heat transfer gas side is

$$300 \times 1.25 \times (1{,}050 - 415) = 238{,}080\,\text{kJ/s}.$$

On the steam side,

Boiler steam pressure is 18,700 kPa
Boiler drum steam temperature = saturation temperature is then 359°C
Boiler drum saturation specific enthalpy is 2,493.8 kJ/kg
Total steam flow is 330 kg/s

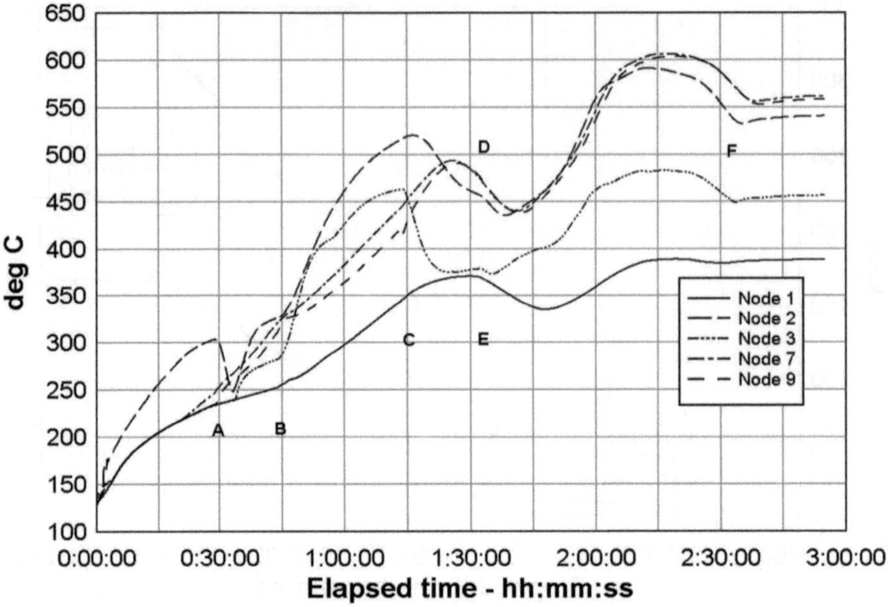

Fig. 17.4 Network temperature trends

The increase in steam specific enthalpy from drum to boiler outlet, assuming all the convective heat from the flue gas is transferred to the steam, will be 238,080 kJ/s / 330 kg/s or 721 kJ/kg. The specific enthalpy of steam at boiler outlet is then (2,493.8 + 721) or 3,215.3 kJ/kg. At the boiler outlet pressure of 16,700 kPa, this indicates a temperature of 477 °C, considerably less than the more typical 540 °C expected. The difference is made up by the radiant component which can be estimated as follows.

For the same pressure and 540°C, the steam enthalpy will be 3,382.1 kJ/kg or an increase of 166.8 kJ/kg. For the given steam flow of 330 kg/s, the radiant component to all radiant superheaters must add 55,044 kJ/s, say 55 MW. As we saw in Chap. 11, on the basis of some assumed distribution of radiant heat, around 35 MW might be transmitted to the secondary superheater in a boiler of similar dimensions and rating to the one in discussion here. It should be noted that the radiant component is transferred to the secondary and tertiary superheaters, which are the first elements in the flue gas path. The resulting higher tube temperatures will reduce the convective heat transfer to these superheaters and raise the gas temperatures seen further down the gas path, increasing the convective components of the later superheaters and reducing overall the requisite radiant components. These figures are therefore indicative only but they are consistent with those of the earlier chapter.

Since this is a discussion of the simulation method rather than of boiler design and operation, the numerical procedure itself is of primary interest. A good indicator

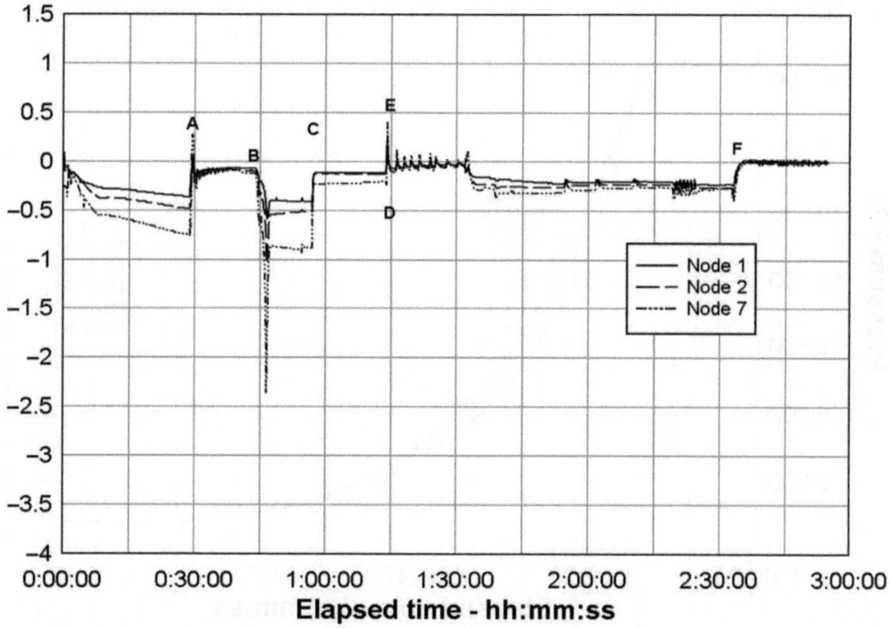

Fig. 17.5 Network node mass balance trends

of numerical performance is provided by the individual node mass balances. The formulation of the mass balance equation is intended to yield a steady-state solution showing zero mass balance for each node. Transient deviations from zero indicate adjustments to the node mass. Figure 17.5 shows the trends of the instantaneous node mass balances for selected nodes during these simulated operations. As can be seen, the node balances remain close to zero during transients and are practically zero during steady-state. Short-lived excursions away from zero accompany rapid changes in branch flows, such as at point B when the HP bypass valve closed.

The same operations were simulated using the augmented Eq. 17.46 with the same network parameters and node and branch tuning parameters. Visually, at the scale offered by these diagrams, no differences are observable. The node mass balance deviations produced by the fully integrated method were around half those produced by the unbundled method except in response to the rapid pressure loss (see below) for which both were almost identical. The network tuning parameters were chosen to give convergence of the iterative near-tridiagonal matrix solution within 30 iterations, though usually less than 10 sufficed to satisfy the convergence criteria. The unbundled method allowed up to five iterations but for most situations needed only one or two.

The numerical method was tested in an extreme case by the introduction of an unscheduled fault of unspecified nature that caused an extremely rapid loss of pressure to around 1,000 kPa and trip of the turbine. The salient features of this

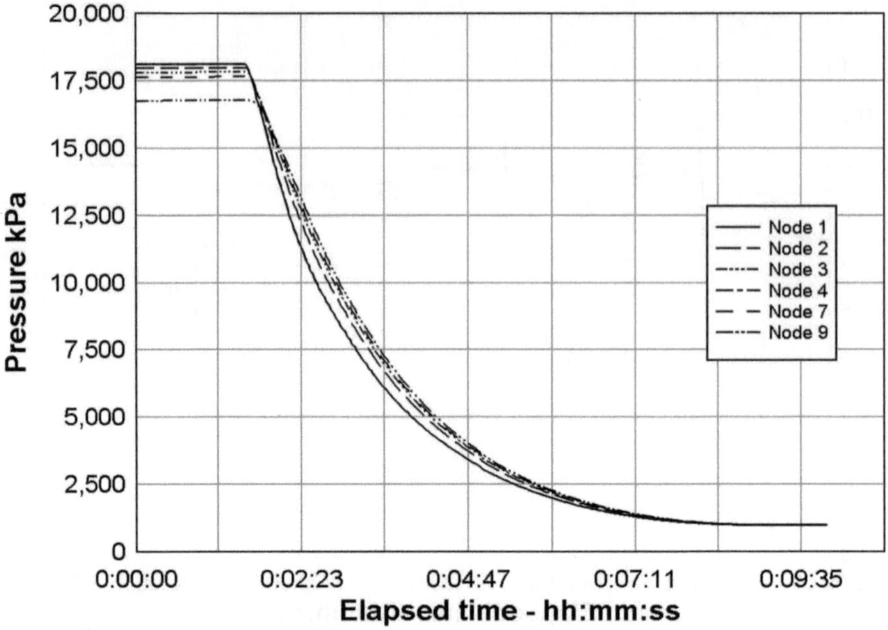

Fig. 17.6 Node pressure trends during an extreme pressure reduction

event are shown by Figs. 17.6 and 17.7. Turbine steam flows reduced to zero within a few seconds as the turbine valves shut. A pressure loss of this magnitude and speed might be expected only as a result of an extreme malfunction event, such as a steam pipe fracture, whose accurate reproduction is beyond this method. The event does however illustrate the dynamic response and numerical stability of the method, in even this extreme case.

Transiently, the node mass balances increased well beyond the values achieved when handling more normal rates of change, peaking at around 80 kg/s for node 3, though for a very short duration. The error at node 9, directly influenced by the rapid reduction of flow to the turbine, persisted for much longer.

The same extreme event was rerun using double the iteration count, and the results are shown by Figs. 17.8 and 17.9. The benefits of the higher iteration count are immediately apparent from the reduction in peak mass balance errors by around 30 % and the halving of their duration. However, the node mass balance deviations in this extreme case exceed those which might be expected given the physical node volumes and indicate the presence of significant dynamic errors. In the general case, increasing the iteration count will not be sufficient as the real-time constraint will limit the number of iterations to less than that required for convergence. Indeed, in the light of later comments, it must be expected that the solution will never converge, regardless of the number of iterations.

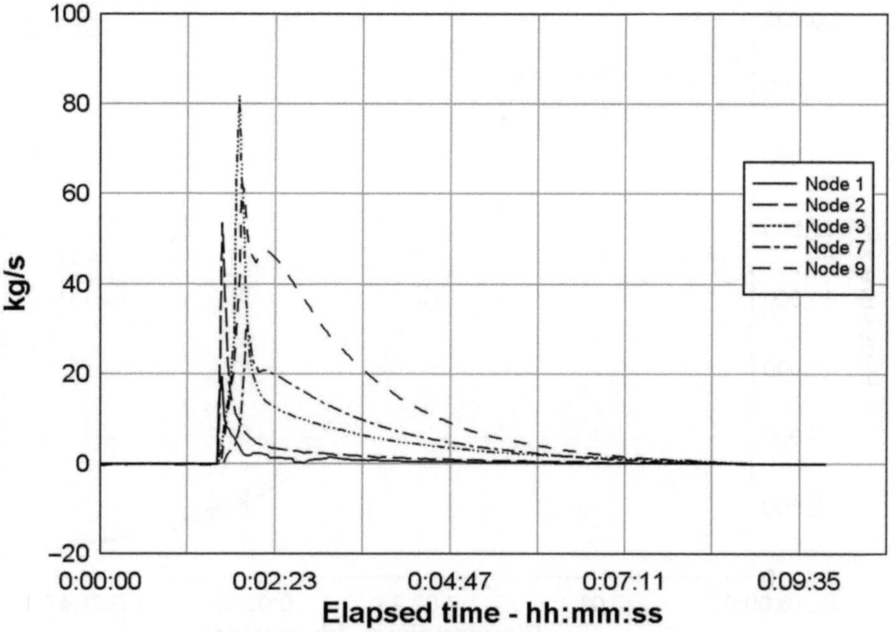

Fig. 17.7 Node mass balances during an extreme pressure reduction

Finally, a severe but operationally "normal" event is the opening of a boiler safety valve. Figures 17.10 and 17.11 show the responses of pressures and mass balances of the nodes closest to the SV attachment point to abrupt opening and subsequent closure of the boiler safety valve SV. Around 230 kg/s of steam flowed through the fully opened valve, being some 70 % of the boiler full-load capacity. The valve remained open for about 2 min before re-closing. Although by operational standards a major disturbance, this event is readily handled by the network computation. It should be noted that the illustrated behaviour is not that expected from the opening of a safety valve. For this simplified example, the boiler pressure is held constant, and the network solution shows the response to a sudden increase in steam flow which can be supplied by the boiler without loss of pressure. The reduction in pressure at the node connected to the valve is equal to the increased frictional pressure loss corresponding to the increased steam flow.

The inability of the method as described to yield node mass balances consistent with physical node volumes in extreme cases may be attributed to the neglect of flow momentum. Extremely rapid large pressure changes induce large and sudden forces on the working fluid masses which do not respond instantaneously, as assumed if momentum is ignored. Should dynamically accurate simulation of these extreme cases be the objective, the approach described above must be extended to include treatment of flow momentum. However, the illustrated numerical results suggest that, for all normal and even major but not extreme plant upsets, neglect of momentum will yield satisfactory results.

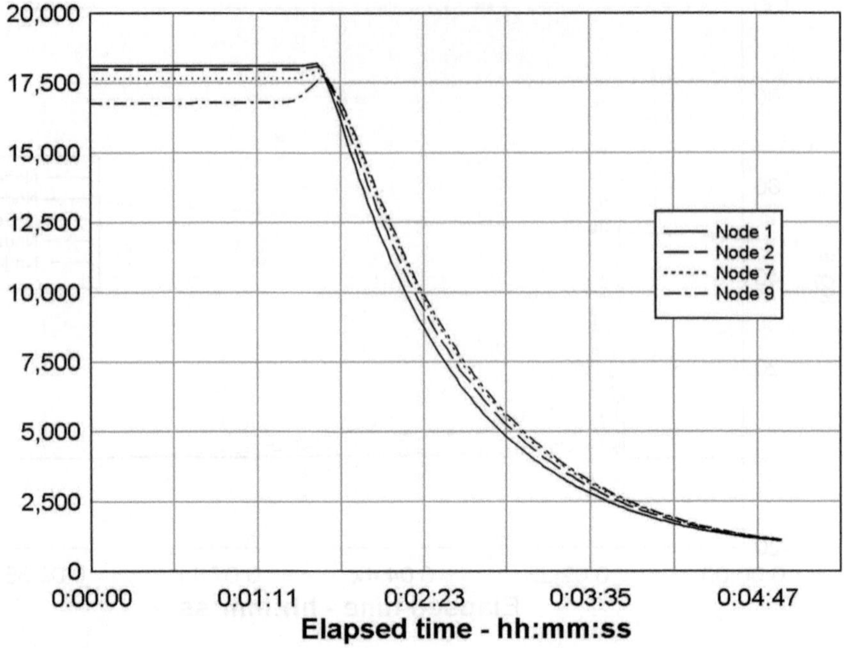

Fig. 17.8 Node pressures with increased iteration count

17.2.8 Inclusion of Flow Momentum

Equation 17.18 was derived for incompressible flow in a network branch. For compressible flow in a network cell, the same form applies but with the branch parameters ($\dot{m}_{ij}$, $\mathcal{A}_{ij}$ and I_M) replaced by their cell equivalents. Then

$$\dot{m}_j^{n+1} = \vartheta_j\, \dot{m}_j^n + \theta_j\, \Delta p_j^{n+1} \tag{17.50}$$

with

$$\theta_j = \frac{\Delta t\, \mathcal{A}_j}{I_M\, \mathcal{A}_j + 2\Delta t\, \dot{m}_j^n} \tag{17.51}$$

and

$$\vartheta_j = \frac{I_M\, \mathcal{A}_j + \Delta t\, \dot{m}_j^n}{I_M\, \mathcal{A}_j + 2\Delta t\, \dot{m}_j^n} \tag{17.52}$$

and an equivalent pair for the linearisation $\dot{m}^2 \approx \dot{m}^{n+1} \dot{m}^n$.

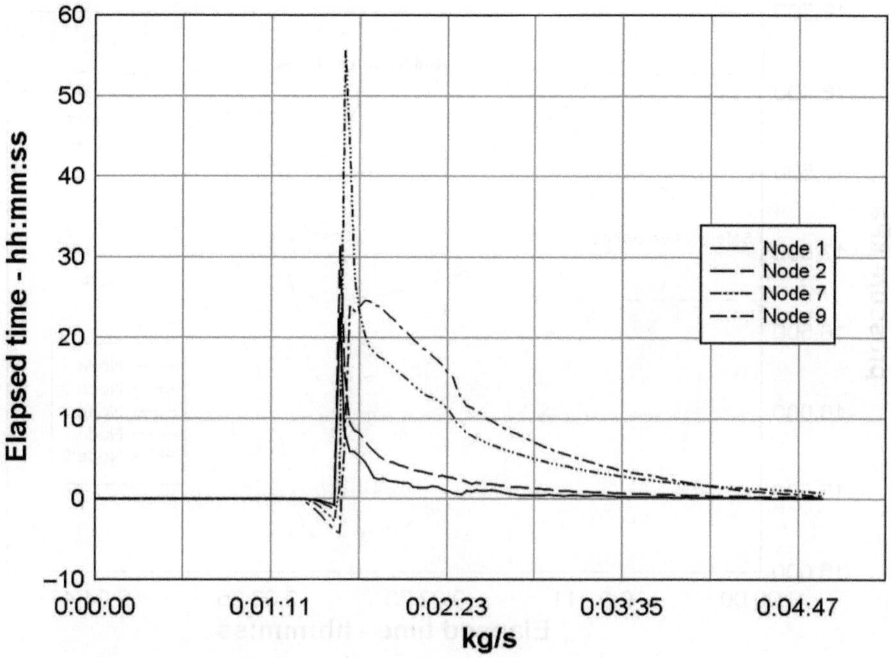

Fig. 17.9 Node mass balances with increased iteration count

Considering only the pressure equation from Eq. 17.28, and using Eq. 17.50 for $\dot{m}_{j-1}$ and $\dot{m}_j$, we arrive at the following equation for the pressure in the jth cell.

$$-a_1\frac{\Delta t}{|D|}\theta_{j-1}p_{j-1}^{n+1}+\left(1+a_1\frac{\Delta t}{|D|}(\theta_{j-1}+\theta_j)\right)p_j^{n+1}-a_2\frac{\Delta t}{|D|}\theta_j p_{j+1}^{n+1}$$
$$=p_j^n+a_1\frac{\Delta t}{|D|}\left(\vartheta_{j-1}\dot{m}_{j-1}^n-\vartheta_j\dot{m}_j^n\right). \qquad (17.53)$$

Expanded for all cells in a branch this equation generates a tridiagonal matrix equation to be solved for all the cells in the branch, now including flow momentum. Boundary conditions are applied in the usual way.

The equation for the k-th node pressure including flow momentum is,

$$\left(1+\frac{\Delta}{\tau_k}\sum_j\theta_j\right)\bar{p}_k^{n+1}=\bar{p}_k^n+\frac{\Delta}{\tau_k}\left(\sum_j\vartheta_j\dot{m}_j^n+\sum_j\theta_j p_j^{n+1}+\sum\dot{m}_{ext}\right) \qquad (17.54)$$

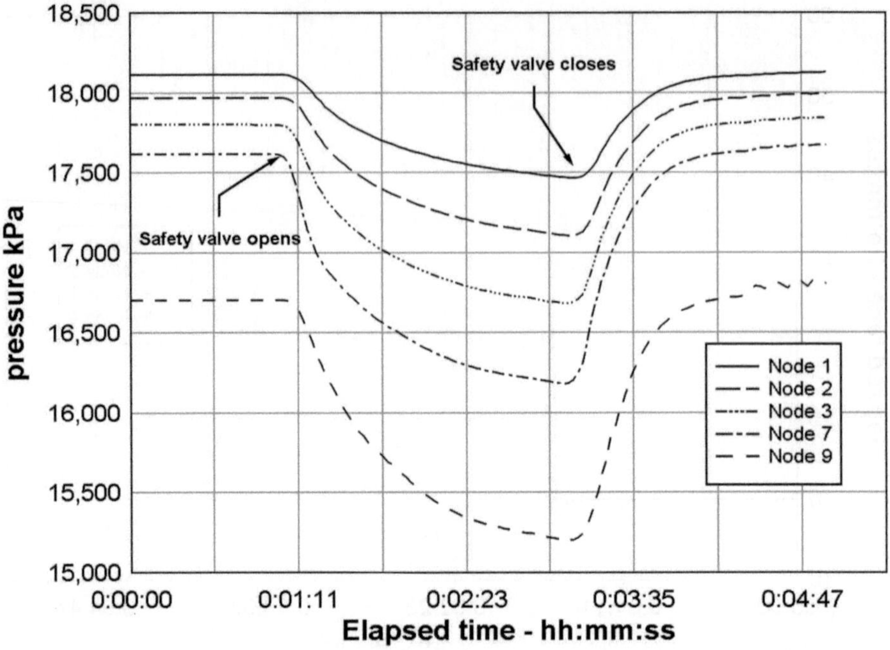

Fig. 17.10 Node pressure trends—operation of the SV

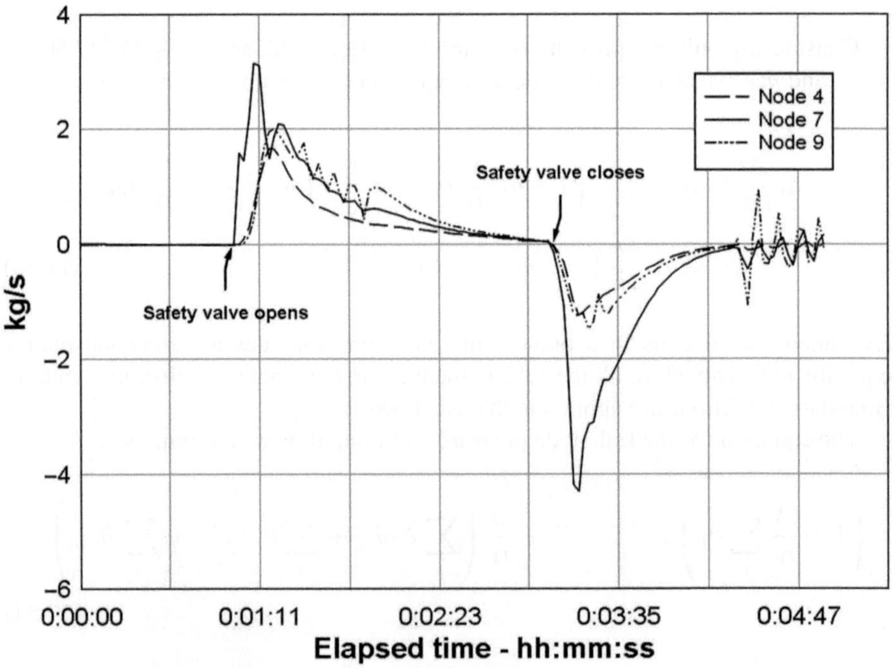

Fig. 17.11 Node mass balance trends—operation of the SV

Network Behaviour: Flow Momentum Included

The same network was subjected to the same start-up procedure, this time with flow momentum included. The results are shown by Figs. 17.12 and 17.13.

The variations of network pressures with momentum included are visually indistinguishable from those without momentum included[4]. The variations of node mass balances show differences of detail, but magnitudes remain small, within the same limits. These results allow the conclusion that, for these rates of change, flow momentum is not a significant factor.

The same however cannot be said of more rapid events. Figures 17.13 and 17.14 show the same results as Figs. 17.6 and 17.7, but now with momentum included. Most striking is the reduction of the magnitude of the node mass balances from around 80 kg/s with a tail >2 kg/s persisting for several minutes, to a sharp peak of less than 2.5 kg/s and a tail of <0.5 kg/s. The duration of the transient, being the time to achieve steady state, increased from around 7 to around 15 min as the effect of momentum caused things to change more slowly. These results allow the conclusion that large rapid transient events cannot be accurately reproduced or predicted if flow momentum is ignored.

It is interesting to note that inclusion of flow momentum adds very little to the computation load. The dimensions and population of the describing coefficient matrices are the same with or without momentum, there being only the minor addition of computation of the ϑ and θ terms (Fig. 17.15).

[4]The slightly closer spacing of the individual traces in Fig. 17.12 is the result of the use of somewhat larger pipes in this case.

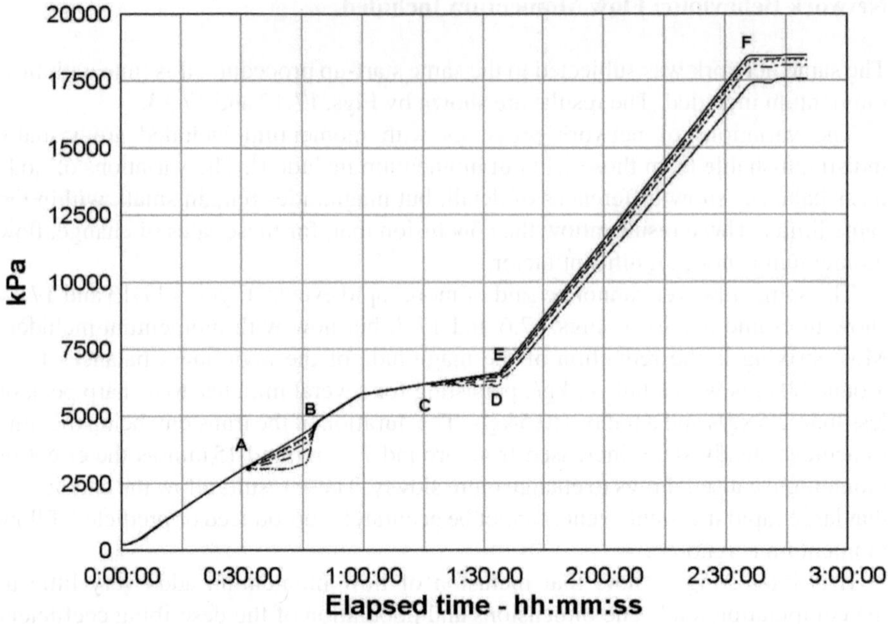

Fig. 17.12 Network pressures during start, including flow momentum

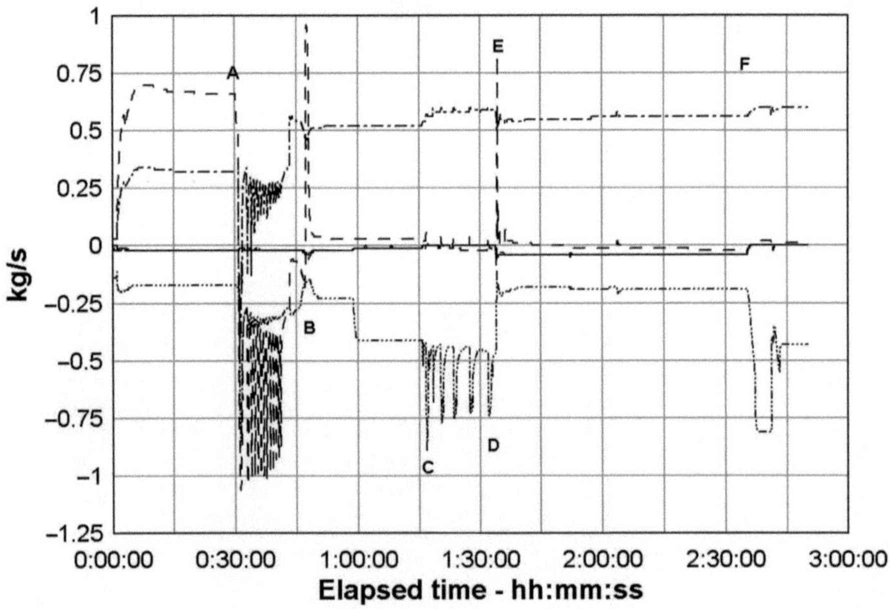

Fig. 17.13 Network node balances during start, including flow momentum

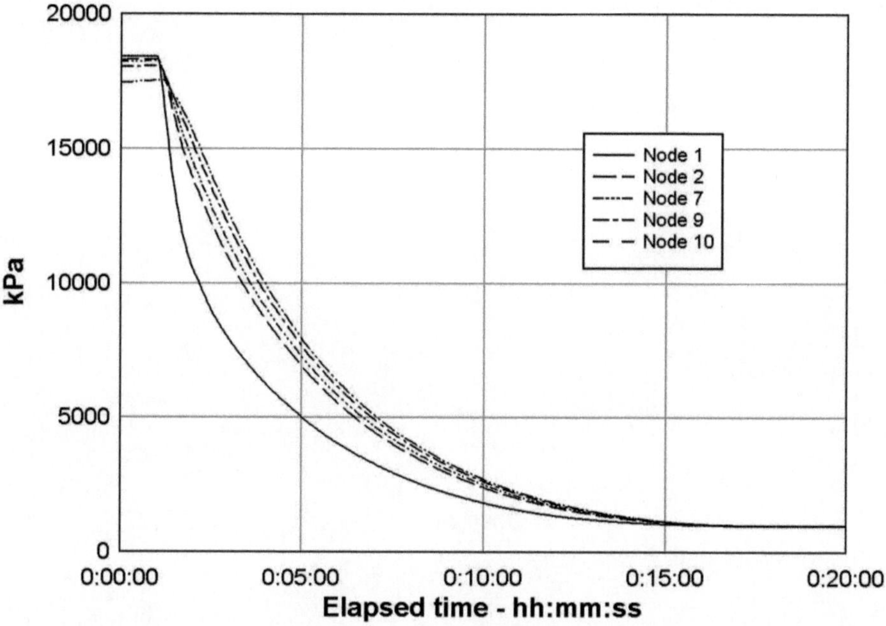

Fig. 17.14 Network pressures during rapid depressurisation

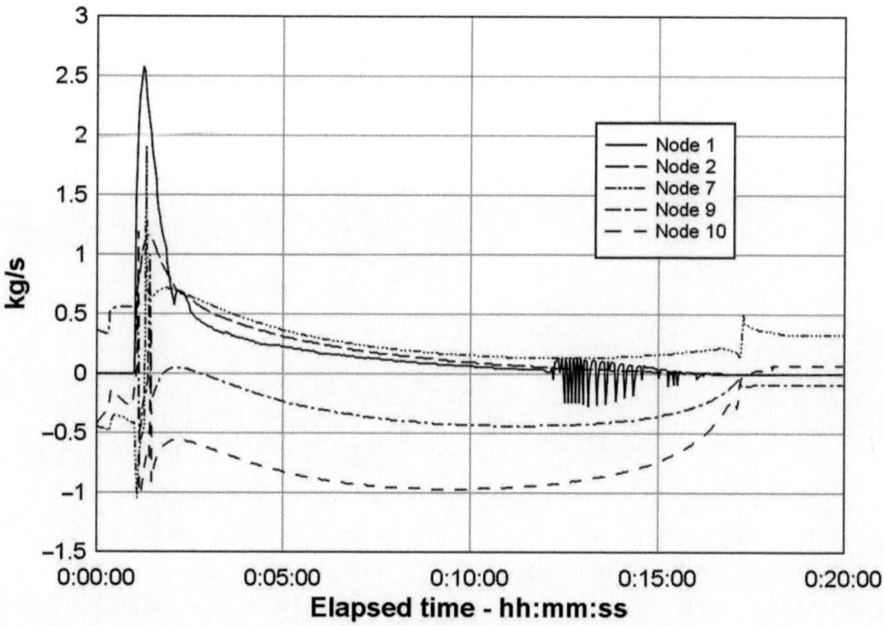

Fig. 17.15 Node balances during rapid depressurisation

Appendix A
Solution of the Distributed Energy Equation

The distributed energy equation can be written in terms of the variable u where u can be any of the advected quantities T, h or x.

$$\frac{\partial u}{\partial t} + \nu \frac{\partial u}{\partial z} + k\, u = \theta(z). \tag{A.1}$$

θ includes the temperature of the heat source, most frequently the wall of the tube enclosing the flow. For the purposes of this analysis, we will consider θ as a linear ramp function of z along the length of the flow path

$$\theta = \theta_0 + \dot{\theta} z,$$

where $\dot{\theta} = \mathrm{d}\theta/\mathrm{d}z$ is the linear spatial gradient of θ.

Taking Laplace transforms of each term of Eq. A.1 yields, where $\bar{u}(z, p)$ is the Laplace transform of $u(z, t)$ and p is the frequency domain variable,

$$p\bar{u} - u(z, 0) + \nu \frac{\mathrm{d}\bar{u}}{\mathrm{d}z} + k\bar{u} = \frac{\theta_0}{p} + \frac{\dot{\theta}}{p} z$$

or

$$\nu \frac{\mathrm{d}\bar{u}}{\mathrm{d}z} + (p + k)\bar{u} = \frac{\theta_0}{p} + \frac{\dot{\theta}}{p} z + u(z, 0),$$

where $u(z, 0)$ is the initial profile of $u(z, t)$ defined at $t = 0$ and is a known but not necessarily an analytic function of z. This can be rearranged into

$$\left(\frac{\nu}{p + k}\right) \frac{\mathrm{d}\bar{u}}{\mathrm{d}z} + \bar{u} = \frac{\theta_0}{p(p + k)} + \frac{\dot{\theta}\, z}{p(p + k)} + \frac{u(z, 0)}{(p + k)}$$

or

$$\tau_z \frac{\mathrm{d}\bar{u}}{\mathrm{d}z} + \bar{u} = \frac{\theta_0}{p(p + k)} + \frac{\dot{\theta}\, z}{p(p + k)} + \frac{u(z, 0)}{(p + k)}, \tag{A.2}$$

P. O'Kelly, *Computer Simulation of Thermal Plant Operations*,
DOI 10.1007/978-1-4614-4256-1,

where

$$\tau_z = \frac{\nu}{p+k}.$$

Equation A.2 is a first-order linear ordinary differential equation whose solution is,

$$\begin{aligned}\bar{u}(z,p) &= c_0 \mathrm{e}^{-z/\tau_z} + \mathrm{e}^{-z/\tau_z}\frac{1}{\tau_z}\int \mathrm{e}^{z'/\tau_z}\left[\frac{\theta_0}{p(p+k)} + \frac{\dot{\theta}\, z'}{p(p+k)} + \frac{u(z,0)}{(p+k)}\right]\mathrm{d}z' \\ &= c_0 \mathrm{e}^{-z/\tau_z} + \mathrm{e}^{-z/\tau_z}\frac{1}{\tau_z}\left[\frac{\theta_0}{p(p+k)} + \frac{u(z,0)}{(p+k)}\right]\int \mathrm{e}^{z'/\tau_z}\mathrm{d}z' \\ &\quad + \frac{\dot{\theta}}{p(p+k)}\mathrm{e}^{-z/\tau_z}\frac{1}{\tau_z}\int \mathrm{e}^{z'/\tau_z} z' \mathrm{d}z.\end{aligned}$$

Noting that

$$\int \mathrm{e}^{z'/\tau_z} z' \mathrm{d}z' = \tau_z\, \mathrm{e}^{z/\tau_z}(z - \tau_z)$$

this last equation may be written

$$\bar{u}(z,p) = c_0 \mathrm{e}^{-z/\tau_z} + \frac{\theta_0 + \dot{\theta} z}{p(p+k)} + \frac{u(z,0)}{(p+k)} - \frac{\dot{\theta}\nu}{p(p+k)^2} \tag{A.3}$$

with c_0 a constant to be chosen to satisfy one of the boundary conditions. For this, we use the boundary condition on $u = u(0,0)$ at the inlet $z = 0$. Then

$$\bar{u}(0,p) = \frac{u(0,0)}{p}$$

and we can write

$$\bar{u}(0,p) = c_0 + \frac{\theta_0}{p(p+k)} + \frac{u(0,0)}{(p+k)} - \frac{\dot{\theta}\nu}{p(p+k)^2} = \frac{u(0,0)}{p}$$

from which c_0 follows:

$$\begin{aligned} c_0 &= u(0,0)\left[\frac{1}{p} - \frac{1}{(p+k)}\right] - \frac{\theta_0}{p(p+k)} + \frac{\dot{\theta}\nu}{p(p+k)^2} \\ &= u(0,0)\frac{k}{p(p+k)} - \frac{\theta_0}{p(p+k)} + \frac{\dot{\theta}\nu}{p(p+k)^2} \\ &= \frac{1}{p(p+k)}[k\, u(0,0) - \theta_0] + \frac{\dot{\theta}\nu}{p(p+k)^2}. \end{aligned}$$

Therefore

$$\bar{u}(z,p) = \frac{e^{-z/\tau_z}}{p(p+k)}[k\,u(0,0) - \theta_0] + \frac{(\theta_0 + \dot{\theta} z)}{p(p+k)} + \frac{u(z,0)}{(p+k)} + \frac{e^{-z/\tau_z}\dot{\theta}\nu}{p(p+k)^2} - \frac{\dot{\theta}\nu}{p(p+k)^2}. \quad \text{(A.4)}$$

Since $\tau_z = \nu/(p+k)$ this may be written

$$\bar{u}(z,p) = e^{-k\,(z/\nu)}\frac{e^{-p\,(z/\nu)}}{p(p+k)}[k\,u(0,0) - \theta_0] + e^{-k\,(z/\nu)}\frac{e^{-p\,(z/\nu)}\dot{\theta}\nu}{p(p+k)^2} + \frac{(\theta_0 + \dot{\theta} z)}{p(p+k)} + \frac{u(z,0)}{(p+k)} - \frac{\dot{\theta}\nu}{p(p+k)^2}.$$

The expression for $u(z,t)$ in the time domain follows from the inverse transform.

$$\begin{aligned} u(z,t) &= u(z,0)e^{-kt} + \frac{1}{k}\left(\theta_0 + \dot{\theta} z\right)\left(1 - e^{-kt}\right) \\ &\quad - \frac{1}{k^2}\dot{\theta}\nu\left(1 - e^{-kt} - k\,t\,e^{-kt}\right) \\ &\quad + \frac{1}{k}e^{-k\,z/\nu}\left(1 - e^{-kt}\right)[k\,u(0,0) - \theta_0]\,H(t - z/\nu) \\ &\quad + \frac{1}{k^2}e^{-k\,z/\nu}\left(1 - e^{-kt} - k\,t\,e^{-kt}\right)\dot{\theta}\nu\,H(t - z/\nu), \end{aligned}$$

where $H(t - z/\nu)$ is the Heaviside operator.

Starting from some defined time $t = 0$, this equation can be used to calculate the evolution of $u(z,t)$ for increasing t, in both the spatial ordinate and time, given the initial distribution $u(z,0)$.

For a constant driving input, $\dot{\theta} = 0$, and $u(z,t)$ reduces to

$$\begin{aligned} u(z,t) &= u(z,0)e^{-kt} + \frac{\theta_0}{k}\left(1 - e^{-kt}\right) \\ &\quad + \frac{1}{k}e^{-k\,z/\nu}\left(1 - e^{-kt}\right)[k\,u(0,0) - \theta_0]\,H(t - z/\nu). \end{aligned} \quad \text{(A.5)}$$

Appendix B
A High-Pressure Thermocouple Pocket

Simulators frequently have had difficulty in reproducing actual plant behaviour of steam temperature measurements under all operating conditions. A frequent criticism of these measurements has been that they respond "too fast," that is, they do not display the lags characteristic of temperature measurements derived from thermocouples[1] located in thick-walled pockets. Attempts to correct this by imposing delay functions of up to third order have been partially successful but cannot reproduce all of the steam-flow dependent characteristics of the measurements and do not simulate the true process. The model described here goes somewhat further down the path of doing this.

The errors shown by thermocouple pockets are influenced both by the *wall deficit*, being the difference between the temperature on the inside pipe wall and the flowing fluid temperature, and by the temperature differential between the thermocouple inside the tube and the working medium outside. The wall deficit is influenced by the heat loss from the pipe wall into which the pocket is inserted which, for a well-insulated pipe, will be minimal. As long as the pocket tip extends into the bulk temperature zone of the fluid stream, measurement errors at high-flow velocities will be minimal but will increase at lower flow velocities.

B.1 Description of the Thermocouple Pocket

The pocket consists of a tapered cylindrical steel block inserted through the wall of the steam pipe and protruding some several block diameters into the steam flow. The probe is either welded or screwed into the pipe wall, making good thermal and sealing contact with the wall. The upper section of the probe extends beyond the pipe wall and is attached to some form of electrical termination to signal

[1] Also known as "thermowells."

P. O'Kelly, *Computer Simulation of Thermal Plant Operations*,
DOI 10.1007/978-1-4614-4256-1,

cabling. The probe is drilled out along its long axis with a thin hole into which the thermocouple junction is inserted. The intention is that the thermocouple makes good thermal contact with the tip of the probe and records the tip metal temperature. This temperature is close to and dynamically follows the temperature of the steam.

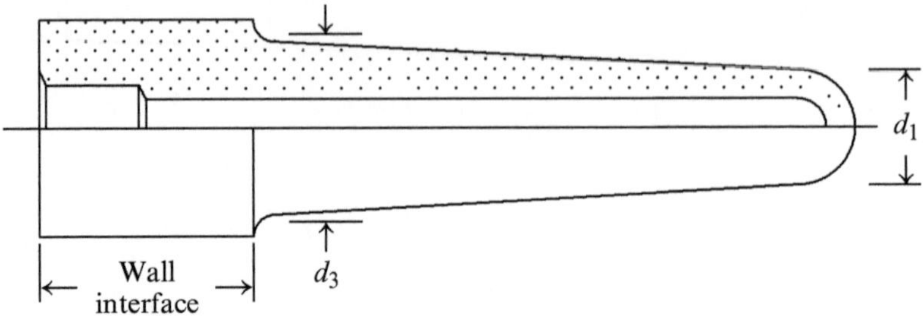

The probe is generally made of a material similar to that of the wall into which it is inserted. In high-temperature, high-pressure applications, this is almost always austenitic chrome–nickel or chrome–nickel–molybdenum stainless steel. Some designs insert a section of lower conductivity steel (higher chrome content) between the tip and the pipe wall in order to improve the probe's measuring accuracy.

The probe protrudes into and across the steam flowing axially along the pipe. Under normal flow conditions, the probe tip is located in a region of high steam velocity and heat transfer is high. Under low mass flow conditions, this will not be the case and the probe accuracy will deteriorate. The probe will experience high heat transfer from the steam along that part of its length in contact with the steam. Heat will flow by conduction along the probe to the pipe wall. The section of the probe in direct contact with the pipe wall will be at the temperature of the wall.

The pipe wall temperature will differ from that of the adjacent steam and from the probe tip temperature because:

- The tip will be in a region of higher steam temperature as the steam temperature towards the centre of the flow will be higher than at the pipe wall.
- The pipe wall loses heat to ambient.

B.1.1 Derivation of the Model Equations

The probe is treated as a series arrangement of three cells along the probe's axis, as shown by Fig. B.1.

A heat balance on the jth cell gives

$$c_m M_j \frac{dT_j}{dt} = \alpha_j A_j (T_s - T_j) + a_{j-1,j} \frac{\lambda_{j-1,j}}{l_{j-1,j}} (T_{j-1} - T_j) - a_{j,j+1} \frac{\lambda_{j,j+1}}{l_{j,j+1}} (T_j - T_{j+1}),$$

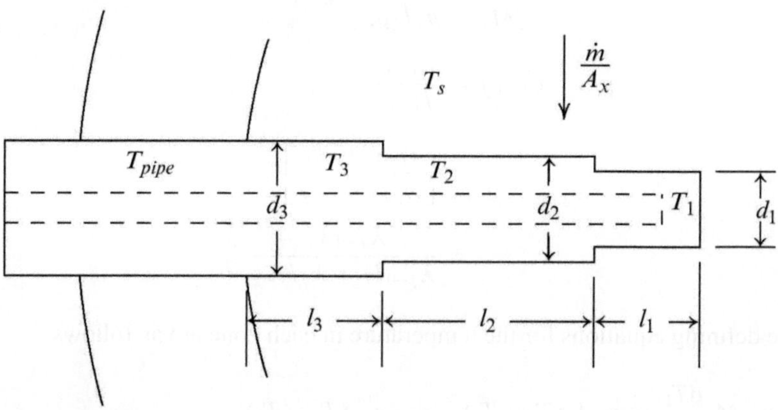

Fig. B.1 3-cell representation of a thermocouple pocket

where

- c_m is the specific heat of the probe material (assumed the same for all cells)
- M_j is the mass of the probe material in the cell
- T_j is the mean temperature of the probe material
- α_j is the coefficient of heat transfer from the steam to the probe
- A_j is the area of the probe surface exposed to heat transfer from the steam
- T_s is the local steam temperature, assumed radially uniform across the stream
- $a_{j-1,j}$ is the probe cross-section area used for heat conduction calculation from the $(j-1)$th to the jth cell
- $\lambda_{j-1,j}$ the probe thermal conductivity used for heat conduction calculation from the $(j-1)$th to the jth cell, noting that each cell may have a different conductivity

With

- l_j the length [m] of the jth cell
- d_j the diameter of the probe in the jth cell (allowing taper)
- ρ the density of the material in each cell, assumed equal in all cells

we can write

$$A_j = \pi d_j l_j,$$

$$a_j = \pi d_j^2,$$

$$a_{j-1,j} = 0.5(a_{j-1} + a_j),$$

$$
\begin{aligned}
M_j &= a_j l_j \rho, \\
c_{j-1,j} &= \frac{\lambda_{j-1,j}}{l_{j-1,j}} \\
&= \left[\frac{l_{j-1}}{\lambda_{j-1}} + \frac{l_j}{\lambda_j}\right]^{-1} \\
&= \frac{\lambda_{j-1}\lambda_j}{\lambda_{j-1} l_j + \lambda_j l_{j-1}}.
\end{aligned}
$$

The defining equations for the temperature in each zone are as follows:

$$
\begin{aligned}
c_m\, M_1 \frac{\mathrm{d}T_1}{\mathrm{d}t} &= \alpha_1 A_1 (T_s - T_1) - a_{1,2}\, c_{1,2} (T_1 - T_2), \\
c_m\, M_2 \frac{\mathrm{d}T_2}{\mathrm{d}t} &= \alpha_2 A_2 (T_s - T_2) + a_{1,2}\, c_{1,2} (T_1 - T_2) - a_{2,3}\, c_{2,3} (T_2 - T_3), \\
c_m\, M_3 \frac{\mathrm{d}T_3}{\mathrm{d}t} &= \alpha_3 A_3 (T_s - T_3) + a_{2,3}\, c_{2,3} (T_2 - T_3) - a_{3,4}\, c_{3,4} (T_3 - T_4). \quad \text{(B.1)}
\end{aligned}
$$

The pipe temperature T_p is not influenced significantly by heat flow from the probe but is influenced by the local steam temperature and does influence the probe temperatures. It is defined by heat flows independent of the probe, such as longitudinal conduction along the pipe and convection from the local steam. It can therefore be taken as defining a boundary condition for the probe. For the purpose of demonstration of the probe calculation, we will include a dynamic calculation of the pipe temperature via a local pipe wall heat balance and adjust its parameters to reproduce actual plant observations. We then can write

$$
c_p\, M_1 \frac{\mathrm{d}T_p}{\mathrm{d}t} = \alpha_p A_p (T_s - T_p) - a_{3,4}\, c_{3,4} (T_3 - T_p) + \alpha_{loss} A_{po} (T_p - T_{amb})
$$

and adjust α_p to suit. α_{loss} and A_{po} are the ambient loss coefficient and outer pipe surface area, respectively.

Heat Transfer from Steam to the Probe

Steam flows across the probe normal to the probe axis. Convective heat transfer is given by Eq. 8.12 of Sect. 8.2.6.

$$
\dot{q}_{hx} = \alpha_{hx}\, A_{hx}\, (T_s - T_m)
$$

for $m = 1, 2$ or 3. The heat transfer coefficient is given by Eq. 8.15.

The specific mass flow ϕ is given from

$$\phi = \frac{\dot{m}_s}{A_x} \quad \text{with} \quad A_x = \frac{\pi}{4} d_p^2$$

and d_p is the pipe inner diameter.

B.2 Calculated Thermocouple Pocket Behaviour

Computational results are obtained by solving Eq. B.1 for the following conditions and pocket dimensions and material.

Steam temperature (T_s)	495.0 °C
Tip metal temperature (T_1)	494.996 °C
Centre metal temperature (T_2)	494.989 °C
Inner metal temperature (T_3)	494.4 °C
Pipe metal temperature (T_4)	492.95 °C
Ambient temperature (T_{amb})	40 °C
Diameter—tip section ($d1$)	0.025 m
Diameter—centre section ($d2$)	0.03 m
Diameter—inner section ($d3$)	0.038 m
Length—tip section ($l1$)	0.015 m
Length—centre section ($l2$)	0.12 m
Length—inner section ($l3$)	0.015 m
Thermal conductivity—tip section (λ_1)	0.042 kW/m K
Thermal conductivity—centre section (λ_2)	0.015 kW/m K
Thermal conductivity—inner section (λ_3)	0.035 kW/m K
Density - all sections	8000 kg/m^3

Figure B.2 shows calculated trend curves for the response of the pocket with the parameters listed above to a step change in steam temperature T_s from 358 °C to 368 °C and steam flow through the pipe of 200 kg/s. The response of the indicated steam temperature exhibits a time constant of around 60 s, with negligible steady-state measurement error.

Also shown by this figure are trends of measured indicated steam and pipe wall temperatures, measured by a similar thermocouple pocket and steam flow, in a plant of similar rating [105].

Figure B.3 shows the same calculated information as Fig. B.2 but at the much reduced steam flow of 20 kg/s. The time constant of the indicated steam temperature has increased to around 300 s and the steady state error to some 0.25°C . This error is still quite small, but the increased time delay is significant.

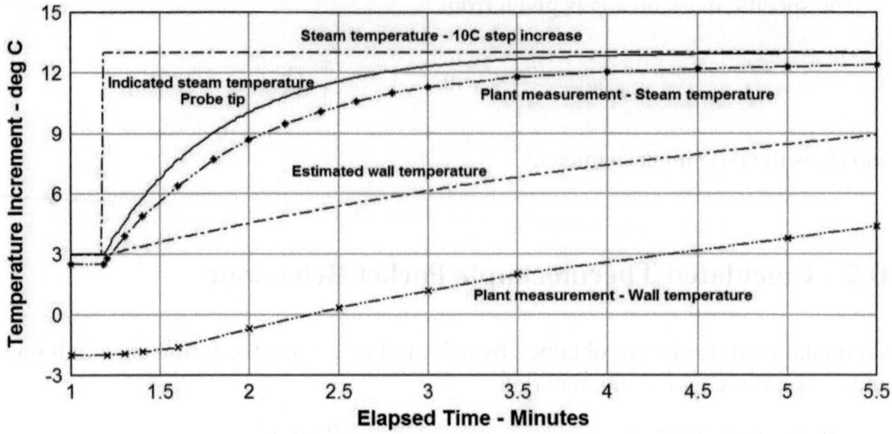

Fig. B.2 Steam temperature step response—200 kg/s steam flow

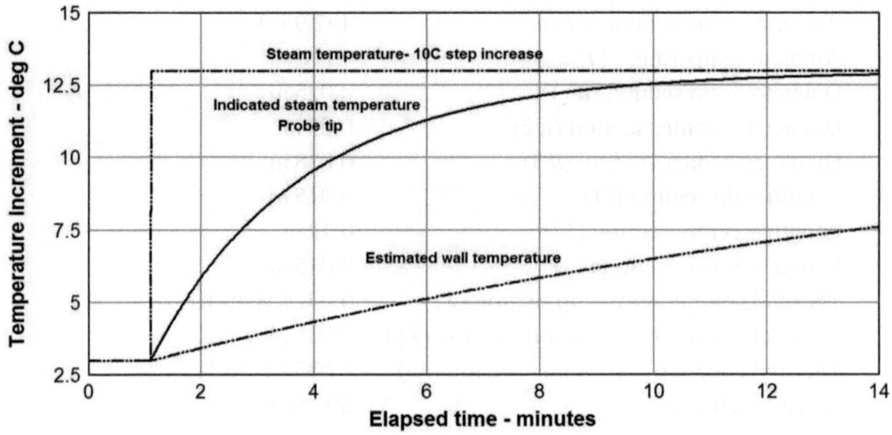

Fig. B.3 Steam temperature step response-20 kg/s steam flow

References

1. W.H. Press et al *Numerical Recipes in C*, 2nd Edition, Cambridge University Press, 1992
2. G.H. Golub, C.F. van Loan *Matrix Computations*, 3rd Edition John Hopkins Univ. Press, 1996
3. L. Fox and D.F. Mayers *Computing Methods for Scientists and Engineers*, Monographs on Numerical Analysis, Oxford University Press, Reprint, 1977
4. A. Quarteroni, R. Sacco, F. Saleri *Numerical Mathematics*, 2nd Edition, Springer Texts in Applied Mathematics No. 37, 2007
5. G. Bader, P. Deuflhard *Numerische Mathematik* 41, pp 373–398, 1983
6. P.J. Roache *Computational Fluid Dynamics*, Revised Edition Academic Press, 1982
7. A. Prosperetti, G. Tryggvason (Edit.) *Computational Methods for Multiphase Systems*, Cambridge University Press, Cambridge, UK, 2009
8. N. Gershfeld *The Nature of Mathematical Modeling*, Cambridge Univ. Press, 1999
9. Steinmueller *Steam Generation*, Vulkan Verlag, Essen, 2nd English Edition, 1994
10. Wagner, Kurse *Properties of water and steam: the industrial standard IAPWS-IF97 for the thermodynamic properties and supplementary equations for other properties; tables based on these equations.*, Springer Berlin, 1998
11. K. Scheffler, N, Rosner, J. Straub, U. Grigull *Der Neue Internationale Standard der dynamischen Viskosität von Wasser und Dampf*, Brennstoff-Wärme-Kraft 30 Nr. 2, February 1978
12. J.P. Holman *Heat Transfer*, McGraw Hill New York, 1991
13. P.J. Stuttaford, P.A. Rubini *Assessment of a Radiative Heat Transfer Model for a New Gas Turbine Combustor Preliminary Design Tool*, 35th Aerospace Sciences Meeting AIAA 97-0294, 1997
14. S.P. Fuss, A. Hamins *An Estimate of the Correction to be Applied to radiant Flame Measurements due to Attenuation by Atmospheric* CO_2 *and* H_2O, Fire Safety Journal 37 pgs 181–190, 2002
15. P. Basu, C. Kefa, L. Jestin *Boilers and Burners Design and Theory*, Mechanical Engineering Series, Springer Verlag, 2000
16. P.K. Swamee, A.K. Jain *Explicit Equation for Pipe Flow Problems*, Journal of Hydraulic Division, ASCE, Vol 102, No. 5, May:657–664, 1976
17. *Perry's Chemical Engineers' Handbook*, 6th Edition McGraw-Hill
18. *Fan Engineering*, Buffalo Forge Company, Buffalo, NY, USA, 1961
19. B. Eck *Ventilatoren*, 5th Edition, Springer Verlag Berlin
20. B. Eck *Fans*, Pergamon Press London
21. A.J. Osiadacz *Simulation and Analysis of Gas Networks*, Gulf Publishing, 1987
22. Sir Frank Whittle *Gas Turbine Aero-Thermodynamics*, Pergamon Press, Oxford, UK, 1981
23. R.K. Turton *Principles of Turbomachinery*, Chapman and Hall, 2nd Edition, 1985

P. O'Kelly, *Computer Simulation of Thermal Plant Operations*,

DOI 10.1007/978-1-4614-4256-1,

24. C.E. Brennen *Hydrodynamics of Pumps*, Oxford University Press, OUP/Concepts Edi, White River Jn, VT, USA, 1994
25. M.P. Boyce *Gas Turbine Engineering Handbook* Butterworth-Heinemann 2nd Edition, 2002
26. L.G. Tetu *Improving Centrifugal Compressor Performance by Optimizing Diffuser Surge Control (Variable Diffuser Geometry) and Flow Control (Inlet Guide Vane) Device Settings* International Compressor Engineering Conference 2004, Paper 1719, Purdue University
27. F. Willems *Modeling and Control of Compressor Flow Instabilities* Report No. WFW 96.151, Eindhoven University of Technology, Faculty of Mechanical Engineering, 1997
28. E.M. Greitzer *Surge and Rotating Stall in Axial Flow Compressors, Part I: Theoretical Compression System Model*, Journal of Engineering for Power, 98:190–198, 1976
29. F.K. Moore, E.M. Greitzer *A Theory of Post-Stall Transients in Axial Compressor Systems: Part II-Application*, Journal of Engineering for Gas Turbines for Power, 108:231–239, 1986
30. J.T. Gravdahl, O. Egeland *Control of the Three State Moore-Greitzer Compressor Model using a Close-Coupled Valve*, Proc. 1997 European Control Conference, July 1997
31. J.T. Gravdahl, O. Egeland *A Moore-Greitzer Axial Compressor Model with Spool Dynamics*, Proc 36th IEEE Conference on Decision and Control, 1997
32. A. Hafaifa, A. Daoudi, M. Guemana *SCADA for Surge Control*, Control, pgs 69–71, March 2011
33. C. Hayashi *Nonlinear Oscillations in Physical Systems*, McGraw-Hill, 1964
34. G.G. Mejeoumov *Improved Cement Quality and Grinding Efficiency by Means of Closed Mill Circuit Modeling* Ph.D Thesis, Texas A&M University, 2007
35. R.C. Juvinall, K.M. Marshek *Fundamentals of Machine Component Design*, John Wiley and Sons, Inc., New York, 1991
36. A.M. Trzynadlowski *Control of Induction Motors*, Academic Press, San Diego, CA, USA, 2001
37. T. Perrotin, D. Clodic *Fin Efficiency Calculation in Enhanced Fin-and-Tube Heat Exchangers in Dry Conditions*, International Congress of Refrigeration 2003, Washington D.C. USA
38. K.T. Hong, R.L. Webb *Calculation of Fin Efficiency for Wet and Dry Fins*, HVAC&R Research, Vol 2, No. 1:pp27–41, 1996
39. S.A. Habbits, T.J. Sheer, H.N. Jawurek, M. Lander, W. Schmitz *Simulation and Measurement of the Thermal Performance of Rotary Regenerative Boiler Air Heaters*, Proc. of the 11th International Heat Transfer Conference, Vol. 6, Aug 1998, Korea
40. F. Bowman *Introduction to Bessel Functions*, Dover Publications, 2nd Edit. New York, 1958
41. I.N. Sneddon *Fourier Transforms*, McGraw-Hill Series in Pure and Applied mathematics, 1951
42. H.S. Carslaw, J.C. Jaeger *Conduction of Heat in Solids*, Oxford University Press, 2nd Edit. 1959 Clarendon Press, NY, USA, 1947
43. W.M. Rohsenow, J.P. Hartnett, E.A. Ganic *Handbook of Heat Transfer Fundamentals*, McGraw-Hill Book Company, London, 1981
44. W.M. Rohsenow, J.P. Hartnett, Y.I. Cho *Handbook of Heat Transfer*, 3rd. Edition McGraw-Hill Book Company, New York, 1998
45. L.M. Jiji *Heat Conduction*, Jaico Publishing Mumbai, 2003
46. M.N.Özisik *Heat Conduction*, 2nd Edit. Wiley, 1993
47. S. Kakac (Ed.) *Boilers, Evaporators and Condensers*, Wiley Interscience, 1991
48. H.C. Hottel, A.F. Sarofim *Radiative Transfer*, McGraw-Hill, 1967
49. K. Kuehlert, U. Renz *A Comprehensive Radiation Model for Numerical Simulation of Pulverised Coal Flames*, Proc. 11th International Heat Transfer Conference (Seoul Korea) Vol 7, 1988
50. S. Zaichik *Application of a Diffusion-Inertia Model for 3-Dimensional Numerical Simulation of Solid Fuel Combustion in Furnace Chambers*, Proc. 11th International Heat Transfer Conference (Seoul Korea) Vol 7, 1988
51. I.T. Shvets et al *Heat Engineering*, Mir Publishers Moscow - English Translation, 1975
52. R.B.Stull *Meteorology for Scientists and Engineers*, 2nd Edition, Brooke/Cole-Thomson Learning, 1999

53. I.W. Smith *The Combustion of Coal Chars: A Review*, 19th Symposium (International) on Combustion, The Combustion Institute, pp 1045–1065, 1982
54. A. McKenzie, I.W. Smith, G.A.D. Szpindler, J. Institute of Fuel, 47, 75, 1974
55. K.J. Åstrom, R.D. Bell *Simple Drum-Boiler Models*, IFAC Symposium Power Systems, Modelling and Control Applications, Brussels, Sept 1988
56. R.D. Bell, K.J. Åstrom *A Non-Linear Model for Steam Generation Process*, IFAC 12th World Congress, Sydney Australia, 1993
57. K.J. Åstrom, R.D. Bell *A Fourth Order Non-Linear Model for Drum-Boiler Dynamics*, 13th Triennial World Conference, San Francisco USA, 1996
58. *Steam - Its Generation and Use* The Babcock & Wilcox Company, 1978
59. S.S. Bogdanovic, S.M. Jovanovic *Simulation and Modelling of Once-through Benson and Sulzer Steam Generators*, IFAC Symposium Power Systems, Modelling and Control Applications, Brussels, Sept. 1988
60. W.J. Peet, T.K. Leung *Development and Application of a Dynamic Simulation Model for a Drum Type Boiler with Turbine Bypass*, International Power Conference, Singapore, March 1995
61. W. Traupel *Thermische Turbomachinen*, Vols I and II, 3rd Edition, Springer Verlag, Berlin
62. A.S. Leyzerovich *Steam Turbines for Modern Fossil-Fuel Power Plants*, Fairmont Press, 2008
63. D. Butterworth and G.F. Hewitt *Two Phase Flow and Heat Transfer*, Harwell Series, Oxford Uni Press, 1977
64. G. Brown *Heat transmission by condensation of steam on a spray of water drops*, Inst. Mech. Engrs. Proc. Discussion on heat transfer, pp 49–52, 1951
65. J.R.S. Thom *Prediction of pressure drop during forced circulation boiling of water*, Int. J. of Heat and Mass Transfer 7 (1964) pgs. 709–824
66. R.W. Lockhart, R.C. Martinelli *Proposed Correlation of Data for Isothermal Two-Phase, Two-Component Flow in Pipes*, Chemical Engineering Progress Symposium 45(1), pp. 39–48
67. R.C. Martinelli, D.B. Nelson *Prediction of Pressure Drop during Forced-Circulation Boiling of Water*, Transactions of the ASME 70(6) 45(1), pp. 695–702
68. G.B. Wallis *One-Dimensional Two-Phase Flow*, McGraw-Hill Book Company, New York, 1996
69. D. Chisholm *A Theoretical Basis for the Lockhart-Martinelli Correlation for Two-Phase Flow*, Int. J. Heat and Mass transfer, 10(12), pp. 1767–1778, 1973
70. D. Chisholm *The Influence of Mass Velocity on Friction Pressure Gradients during Steam-Water Flow*, Thermodynamic and Fluid Mechanics Group Convection, Inst. of Mech. Engrs Proceedings, Bristol Vol. 182 Pt. 3H, pp. 336–341, 1968
71. D. Chisholm *Pressure Gradients during the Flow of Evaporating Two-Phase Mixtures in Smooth Tubes and Channels*, Int. J. Heat and Mass Transfer 16(2), pp. 347–358
72. S.M. Ghiaasiaan *Two-phase Flow, Boiling and Condensation in Conventional and Miniature Systems*, Cambridge University Press, 2008
73. A. Grzebielec, A. Rusowicz *Thermal Resistance of Steam Condensation in Horizontal Tube Bundles*, Journal of Power Technologies 91(1) pp. 41–48, 2011
74. J.G. Collier *Convective Boiling and Condensation, 2nd. Edition*, McGraw-Hill Book Company, New York, 1996
75. C.J. Baroczy *A Systematic Correlation for Two-Phase Pressure Drop*, Chemical Engineering Progress Symposium 62(44), pp. 232–239, 1966
76. S.G. Bankhoff *A Variable Density Single Fluid Model Two-Phase Flow with Particular Reference to Steam-Water*, J. Heat Transfer 11(Series 13) 265–272 1960 Meeting, Ispra Italy, 1979
77. L. Friedel *Improved Friction Pressure Drop Correlations for Horizontal and Vertical Two Phase Pipe Flow*, Paper E2, European Two Phase Flow Group Meeting, Ispra Italy ,1979

78. A. Cicchitti, C. Lombardi, M. Silvestri, G. Soldaini, R. Zavatorelli *Two-phase Cooling Experiments - Pressure Drop, Heat Transfer and Burn-out Measurements*, Energia Nucleare 7(6) 407–425, 1960
79. K.E. Gungor and R.H.S. Winterton *A general correlation for flow boiling in tubes and annuli*, International Journal of Heat and Mass Transfer, 29(3), 351–358, 1986
80. K.E. Gungor, R.H.S. Winterton *Simplified general correlation for saturated flow boiling and comparisons of correlations with data*, Chemical Engineering Research and Design. 65 148–156, 1987
81. D. Steiner, J. Taborek *Flow Boiling Heat Transfer of Single Components in Vertical Tubes*, Heat Transfer Eng., 13, pp. 4368, 1992
82. *VDI Wärme-Atlas*, VDI-Verlag Düsseldorf Germany, 1993
83. A.J. Ghajar *Non-Boiling Heat Transfer in Gas-Liquid Flow in Pipes - A Tutorial*, J. of the Braz. Soc. of Mech. Sci. and Eng.27, 1, pp. 46–73, Jan-March 2005
84. J.C. Chen *A correlation for boiling heat transfer to saturated fluids in convective flow*, ASME Paper 63-HT-34, Boston, 1963
85. H. Muller-Steinhagen, K. Heck *A Simple Pressure Drop Correlation for Two-Phase Flow in Pipes*, Chem. Engr. Process, 20, pp.297–308, 1986
86. G. Riemenschneider *Analyse der Anlagendynamik eines Steinkohlbefeurerten Gross-dampferzeugers mit vorgeschalteter Gasturbine*, Energieerzeugung, Reihe 6 Nr. 228, 1989
87. TRAC-PD2 *An Advanced Best-Estimate Computer Program for PWR LOCA Analysis*, NUREG/CR-2054, Los Alamos Scientific Laboratory
88. S.M. Zivi *Estimation of steady-state steam void fraction by means of the principle of minimum entropy production*, J. Heat Transfer 86, 247–52, 1964
89. A.A. Armand, G.G. Treschev *Investigation of the Resistance During the Movement of Steam Water Mixtures in a Heated Boiler Pipe at High Pressures*, AERE-Lib/Trans. 816, Jan. 1959
90. M.A. Woldesemayat, A.J. Ghajar *Comparison of void fraction correlations for different flow patterns in horizontal and upward inclined pipes*, Int. Journal of Multiphase Flow, 33 pp. 347–370 Elsevier, 2007
91. R. Doležal *Two-phase pressure loss in heated boiler tubes*, VGB Kraftswerktechnik 52 (1972) Nr 1, pgs 11–15
92. R. Doležal *Vorgänge beim Anfahren eines Dampferzeugers*, Vulkan Verlag, 1977
93. B.P. Vitalis, P.J. Hunt *Constant and Sliding Pressure Options for New Supercritical Plants*, Technical Publication, Riley Power, Power-Gen International, Las Vegas, 2005
94. M. Palkes, E.S. Sadlon, A. Salem *State-of-the-Art Large Capacity Sliding Pressure Supercritical Steam Generators*, ABB SPERI Power Generation Conference, 1994
95. H.G. Kwatny, J.W. Bauerle *Simulation analysis of the stability of coal-fired furnaces at low loads*, Second IFAC Workshop on Modelling and Control of Electric Power Plants, Drexel University, Philadelphia Sept 16–18, 1986
96. F. Brandt *Brennstoffe und Verbrennungsrechung*, FDBR Fachbuchreihe Band 1, Vulkan Verlag, 1981
97. S.M. Cho *Furnace Combustion and Heat Transfer in Large Utility Boilers*, Vol 1 Proc 11th International Heat Transfer Conference, Korea, 1998
98. I. Nedelkovski, I. Vilos, T. Geramitcioski *Finite Element Solution of Navier-Stokes Equations for Steam Flow and Heat Transfer*, World Academy of Science and Engineering and Technology 5, 2005
99. L.D. Berman, S.N. Fuks *Mass Transfer in Condensers with Horizontal Tubes when Steam Contains Air*, Teploenergetika, 5, pp. 66–74, 1958
100. E.N. Fuller, P.D. Schettler, J.C. Giddings *New Method for Prediction of Binary Gas Phase Diffusion Coefficients*, Industrial and Engineering Chemistry, 58(5), 19, 1966
101. Heat Exchange Institute *Standards for Steam Surface Condensers*, 9th Edition, 2001
102. H.R. Jacobs, D.S. Cook *Direct Condensation on Non-circulating Drop*, Proc. 6th Int. Heat Transfer Conf., Toronto, 2, pp.389–393, 1978

103. K.N. Murty *Surface Condensers 1. Find the most compact surface condenser*, Chemical Engineering, January 18, 1988
104. J. van Standen, L. Pretorius, M.P. Meyer *Simulation of Heat Exchange in Large Air Cooled Condensers*, Proc 11th International Heat Transfer Conference Vol 6, Korea, 1998
105. P. Johnman, K. Hitze, E. Fyvie, H. Morris, D. Gosden, B. Taber *Eraring Power Station*, Technical Paper, Advanced Process Control, The Warren Centre, Sydney University, Australia, October 1987

Index

P. O'Kelly, *Computer Simulation of Thermal Plant Operations*,
DOI 10.1007/978-1-4614-4256-1,

MIX
Papier aus verantwortungsvollen Quellen
Paper from responsible sources
FSC® C105338

If you have any concerns about our products,
you can contact us on
ProductSafety@springernature.com

In case Publisher is established outside the EU,
the EU authorized representative is:
Springer Nature Customer Service Center GmbH
Europaplatz 3, 69115 Heidelberg, Germany

Printed by Libri Plureos GmbH
in Hamburg, Germany